REFERENCE

THE NEW BOOK OF
POPULAR SCIENCE

THE NEW BOOK OF
POPULAR
SCIENCE

VOLUME 3 DISCARD

Physical Sciences
General Biology

Library of Congress Cataloging in Publication Data

Main entry under title:
The New book of popular science.
     p.   cm.
    Includes bibliographical references.
    Summary: Discusses the major sciences and their
applications in today's world.
    Contents: v.   1.   Astronomy & space science,
mathematics.
    ISBN 0-7172-1219-X (6 v. set)
    1.   Science—Popular works.   2.   Technology—Popular
works.   3.   Natural history—Popular works.   I. Grolier
Incorporated.
Q162.N437 1994   93-42488
500—dc20   CIP

PRINTED
IN
U.S.A.

Volume 3

Contents

PHYSICAL SCIENCES

George Roos, DPI
U.S.I.S.

Answers to questions about matter and energy are sought through many kinds of experiments. Sometimes the results can lead to great innovations. One such innovation was the development of the transistor after many years of experiments in physics and chemistry. Above: manufacturing a transistor.

THE WORLD OF MATTER AND ENERGY

People, animals, trees, bacteria, rocks, the oceans, the air, and the stars and planets are all different forms of what scientists call *matter*. To put it as simply as possible, matter is that which occupies space. We have no difficulty in recognizing certain forms of matter. We can see and touch solids, like wood, iron, and marble, and liquids, like water, gasoline, and milk. We know that muscular effort is required to move these materials and that they can be weighed. It is not so easy to identify gases as matter, because many gases are invisible. Yet if you blow against the palm of your hand, you can feel the onrush of the gases contained in your lungs. You will find, too, that a container weighs more when it is full of air—a mixture of gases—than when the air has been removed by means of a vacuum pump.

It is clear enough, then, that matter occupies space and has weight. But what, exactly, is matter? We know a vast number of different materials—different kinds of matter. Concrete, iron, milk, illuminating gas, cosmetics, butter, blood, a cat's fur, a man's teeth—have these anything in common?

MATTER

The problem seems to become more complicated than ever when we consider that many apparently pure substances are not so pure as we had imagined. From a distance a slab of concrete appears to be a uniform material of white or grayish color. Upon close examination, however, we find that it is made up of several kinds of matter. There are particles of gravel or crushed stone imbedded in cement. A panful of sea water looks like a simple, uniform substance. Yet when the water evaporates, various salts are left behind, including sodium chloride, or common table salt. Suppose we now pass an electric current through molten sodium chloride. We break

it down into the metal called sodium and the gas called chlorine.

ELEMENTS

This breaking down of apparently simple substances into other substances and then breaking down these others into still others might seem to lead nowhere. But if we continue the process long enough, we shall find that all matter consists of about one hundred pure substances which cannot be decomposed into anything simpler by ordinary methods of analysis. We call such substances *elements*. Elements are made up of identical atoms. Atoms are, in turn, made up of smaller particles. Atoms cannot be separated by ordinary chemical means, however.

Matter is anything that occupies space—a boy, a ball, a building. Matter has energy, or the capacity to do work. Here a boy does work; he applies a force to a ball, causing it to move across the schoolyard.

Lou Jones

Here are some familiar elements:

Aluminum	Cobalt	Nitrogen
Antimony	Copper	Oxygen
Arsenic	Fluorine	Phosphorus
Barium	Gold	Platinum
Beryllium	Helium	Potassium
Bismuth	Hydrogen	Radium
Boron	Iodine	Silicon
Bromine	Iridium	Silver
Cadmium	Iron	Sodium
Calcium	Lead	Sulfur
Carbon	Lithium	Tin
Cerium	Magnesium	Tungsten
Cesium	Manganese	Uranium
Chlorine	Mercury	Zinc
Chromium	Nickel	

Under ordinary conditions many of these elements are solids—copper, gold, iron, and lead, for example. Others are liquids—bromine, mercury. Still others are gases—oxygen, nitrogen. The elements of which matter consists are distributed throughout the universe.

Eight elements make up almost 99 per cent of the earth's crust, as the following table shows:

Element	Per Cent
Oxygen	46.6
Silicon	27.7
Aluminum	8.1
Iron	5.0
Calcium	3.6
Sodium	2.9
Potassium	2.6
Magnesium	2.1
All others	1.4

The oceans are composed of water (made up of the elements hydrogen and oxygen) and various salts. Air is almost entirely a mixture of oxygen and nitrogen, with a few other elements in small quantities.

Astronomers have succeeded in identifying in the sun and the other stars a great number of elements that we know on the earth. They have not found any that are not now known to exist on the earth. A number of meteorites have crashed through our atmosphere and have landed on the earth. More than fifty of the elements known on earth have been identified in these visitors from outer space.

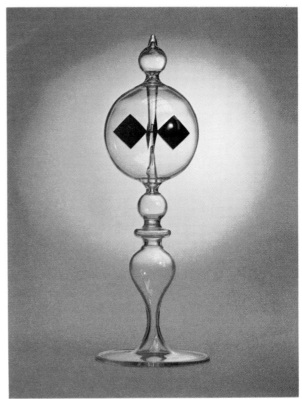

Michel Cambazard

Scientists use special tools to study matter and energy. This device, called a radiometer, is used to measure the intensities of various forms of radiant energy.

COMPOUNDS

Some of the materials that are familiar to us are elements in pure form—gold, silver, lead, mercury, aluminum, among others. Other substances are combinations of two or more elements. When elements combine in such a way that an entirely new substance is formed, we call this new substance a *compound*. Thus, the gases oxygen and hydrogen combine to form water, a liquid compound. The solid potassium and the gases nitrogen and oxygen are found combined in saltpeter (potassium nitrate), a solid compound.

In compounds the elements are always combined in a definite proportion. Pure water always has the same proportion of oxygen and hydrogen. Saltpeter always has the same proportion of potassium, nitrogen, and oxygen. Ammonia always has the same proportion of nitrogen and hydrogen. In all these cases the elements have combined to form a compound according to what we call the *law of definite proportions*.

A compound has properties all its own. Thus pure water, a liquid compound, which cannot catch fire, is made up of hydrogen, an element that burns readily, and oxygen, an element that supports burning. In its uncombined state sodium is a soft, silvery white metal. Chlorine is a greenish poisonous gas. In combination they form salt, an indispensable article of our diet. The properties of plaster of Paris have little in common with those of the elements—calcium, sulfur, oxygen, and hydrogen—of which plaster of Paris is composed.

The total number of compounds is very great. Those of carbon are particularly numerous. Already something like a million compounds of carbon are known and scientists are continuously preparing new ones in their laboratories. On the other hand, the elements helium, neon, argon, krypton, xenon, and radon rarely form stable compounds. They are called *noble gases,* or *inert gases,* because they generally do not combine with other elements. As for the rest of the elements, they combine with some elements and not with others.

Many materials are made up entirely of a single element or a single compound. Each of the substances that we call zinc, gold, silver, iron, and copper is composed of a single element—zinc, gold, silver, iron, and copper respectively. Salt is a compound, made up of sodium and chlorine. Water is a compound, made up of hydrogen and oxygen. Marsh gas, or methane, is a compound made up of carbon and hydrogen.

MIXTURES

A great many materials are not made up of a single pure substance but are a *mixture* of pure substances. Syrup, for example, is a mixture of sugar (a compound of carbon, hydrogen, and oxygen) and water (a compound of hydrogen and oxygen).

Just how does a mixture differ from a compound? For one thing, a compound is formed as the result of a *chemical change* or a chemical reaction. A mixture results from an entirely different and far less drastic sort of change—a *physical change*.

To illustrate what we mean by a chemi-

Sugar dissolved in coffee is an example of a solution. After stirring the sugar in the hot coffee, individual particles of sugar cannot be detected.

cal change, let us suppose that we put two gases, hydrogen and oxygen, in a container in the proper proportions. If we apply a spark to the mixture, there will be a violent reaction, producing heat, light, and sound. (Incidentally we urge our readers not to try the experiment, as it is a dangerous one.) Now we shall have in the container a compound—plain, ordinary water. This compound has come into being as the result of a chemical reaction involving the two elements hydrogen and oxygen.

A chemical reaction also takes place when a compound is broken down, freeing its elements. Such a change occurs when common salt is decomposed into sodium and chlorine or when water is decomposed into hydrogen and oxygen. We tell more about chemical changes, or reactions, in the article, "Chemical Reactions."

Unlike a chemical change, a physical change does not create a new substance. It merely produces certain alterations in the old substances. We bring about a physical change, for example, when we compress the air in an automobile tire, or when we melt sugar, or when we turn on electric current and cause the tungsten filament in an electric bulb to become white-hot. The air,

Inflating a tire with air is a physical change.

Freshly mixed concrete is an example of a heterogeneous mixture.

the sugar, and the tungsten have acquired certain new properties. The air is denser, the sugar particles have been broken down into smaller units, and the tungsten is hotter. But in each of these cases no new substance has been formed. We still have air, sugar, and tungsten.

When a mixture appears to be made up of a single substance, even when viewed under the microscope, we call it a *solution,* or a *homogeneous* (same kind) *mixture.* Thin syrup is a good example of a solution. We obtain it by stirring sugar in warm water. In the course of time we can no longer detect the particles of sugar.

Many solutions are liquids, but not all, by any means. Certain alloys of metal are just as truly solutions as are syrup and salt water. Monel metal, used for kitchen equipment, is such an alloy. This mixture contains 66 per cent nickel, 31.5 per cent copper, 1.35 per cent iron, and smaller amounts of manganese, silicon, and carbon.

Certain mixtures, when examined closely under the microscope or even, in certain cases, by the naked eye, are very obviously not uniform throughout but contain different particles. Such mixtures are called *heterogeneous* (different kind). Concrete is a good example of a heterogeneous mixture. So is a nut-chocolate candy bar.

ENERGY

We now have at least a partial answer to the question, "Of what does matter consist?" We know that it is made up of about one hundred elements and that these, singly or joined together in the form of compounds or mixtures, are the materials out of which all things on earth or in the heavens are made.

But after all, it is not enough to know of what things are made. It is important, also, to know what they do. Let us suppose you wish to buy an automobile. It is interesting to know that the chassis of the car is made of steel, that its cylinder block is of cast iron, that its pistons are of aluminum, that it has copper wiring and platinum contact points, that its steering wheel is of plastic, that its tires are made up of natural and synthetic rubber. You may be suitably impressed by the sleek appearance of the car and you may be reassured by the sturdiness of its parts. But after all, you are going to buy this combination of elements and compounds and mixtures primarily because its motor can make the wheels turn and because, as the wheels turn, the car bears you and the members of your family with great rapidity from place to place.

A scientist would say of the automobile when it is in motion that it has a certain capacity for doing *work* — that it can exert a force through a certain distance. A turbine has this capacity for doing work when it is spinning rapidly; a hammer, as it descends toward a nail; an avalanche, as it hurtles down a sheer mountain slope. The elephant has a capacity for doing work when the momentum of its charge carries it crashing through the underbrush of its jungle home; the grasshopper as it alights upon a blade of grass; that microscopic blob of matter, the ameba, as it thrusts its way through the waters in which it makes its home.

Scientists have a special name for the capacity for doing work. They call it *ener-*

Large transformers like this one at a power plant are used to change raw energy into usable power such as electricity.

gy. The worlds of living and non-living things alike are brimful of energy. Each and every living and non-living thing has the capacity for performing work or can be made to have the capacity.

KINETIC AND POTENTIAL ENERGY

If a body is in motion and capable of performing work at once, we say that it possesses *kinetic energy.* Kinetic comes from the Greek word *kinein,* meaning "to move." An automobile moving along a highway, a baseball thrown by a pitcher, and an ocean wave traveling towards the shore are examples of bodies possessing kinetic energy. In all these cases there is physical motion.

Other bodies possess *potential energy.* That is, they have stored up within them the capacity for doing work at some future time. A good example of such a body is a pile driver, a machine in which a heavy weight is lifted up to a considerable height and then allowed to fall upon a pile or post that is to be driven into the earth. When the weight is being lifted, work is being done on it and potential energy is being stored in it. When the weight is at the top of the machine it is not capable of doing work immediately, so that it still possesses only potential energy. It becomes capable of doing immediate work only when it starts moving downward toward the post. Its potential energy has now been transformed into kinetic energy.

FORMS OF ENERGY

The energy that fills the universe assumes a great variety of forms. Among the most striking of these is *radiant energy* — energy in the form of radiation. The sun is an important source of this kind of energy. The earth receives but a small fraction of the outpouring of radiation from the sun. Yet this tiny share suffices to maintain life upon the earth, for the *light* that is radiated from the sun is used by plants to manufacture their own food. Plants are dependent upon this operation for their very existence; so are animals, for they eat plants or else they eat the animals that have eaten these plants.

Heat is another form of radiant energy. The heat that is radiated from the sun is ultimately responsible for the rain that gives needed moisture to plant life. The sun's heat causes water in the oceans, lakes, and rivers to evaporate. As the water vapor rises in the air, it is often condensed in the form of clouds, made up of water droplets. Sometimes these water droplets combine to form still larger drops and then they fall as rain, or snow, or sleet, or the other forms of precipitation.

The changes that occur when substances undergo chemical changes cause the release of energy in the form of heat, light, and the sudden expansion of gas. This is known as *chemical energy*. *Electrical energy* is released when the particles of matter called electrons move along a conductor or freely through space. There is energy stored up, too, in the magnetic forces that surround the earth. The vibrations known as *sound waves* possess energy. We become aware of them when they strike our eardrums.

There are tremendous stores of energy in the core of an atom. Atoms are the basic particles of matter, the building blocks of elements. When the core of the atom is split or when the cores of the atoms merge, vast quantities of energy are released. According to the theories of modern science, *atomic,* or *nuclear, energy* accounts for the radiant energy of the sun and the other stars. Man has succeeded in producing atomic energy here upon the earth by splitting the cores of certain atoms.

FORCE

When energy is released, a variety of forces enter into play. A good working definition of *force* is that it is anything that produces motion in a body, or that changes the motion of a body, or that alters its shape. You are exerting force when you step into a car and push down its springs, or when you strike a tennis ball with your racket, or when you stretch a rubber band.

Forces act on certain bodies through direct contact. When you push on a table to move it across the floor, your hand touches the table as it exerts force upon it. Forces exerted by direct contact also come into play when a batter strikes a baseball, or when a violinist draws his bow across the strings of his instrument, or when an automobile driver shifts gears.

Some forces act upon a given object from a distance. These may be called *action-at-a-distance* forces. When a plane is in the air, the force of gravitation, acting through a considerable distance, is constantly pulling at it. We say in such a case that the plane is within the gravitational *field* of the earth. When a body is within an electrical or a magnetic field, it is also subject to action-at-a-distance forces.

CONVERSION OF MATTER AND ENERGY

There was a time when matter and energy were considered to be two entirely different aspects of being. Scientists held, too, that the quantities of matter and energy in the world have always remained the same from the beginning of time. They believed in the law of the conservation of matter, which stated that matter might be changed into different forms of matter but

A new development in technology: Doctors use liquid crystals to see the areas of body heat in a patient's hand.

Bernard Gefroy

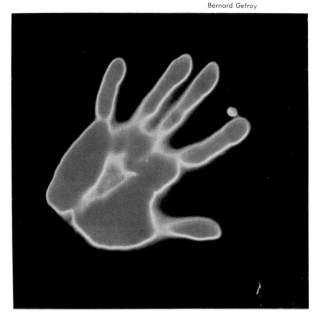

Scientific Research Staffs Ford Motor Company

A laser beam becomes concentrated and therefore stronger, as it is directed through a glass lens. The air around it becomes ionized.

never destroyed. They believed in a similar law to the effect that energy could neither be created nor destroyed but only transformed into other forms of energy.

Today scientists no longer believe that there is an impassable barrier between matter and energy. They have come to think of these as two separate phases of the same phenomenon. Furthermore they have modified the old laws of the conservation of matter and energy. They now believe that at the level of subatomic particles of matter, matter can be transformed into energy and energy into matter.

Matter is changed into energy when larger atoms are split into smaller ones (atomic fission) or when smaller ones combine to form larger atoms (atomic fusion). The great mathematician and physicist Albert Einstein calculated the amount of energy that appears when matter disappears by the following equation:

$$E = mc^2$$

in which E stands for the gain of energy, m the loss in mass, and c the velocity of light, roughly 300,000 kilometers per second.

The equation works equally when energy disappears or when it appears. The mass, or quantity of matter, of a moving particle increases in proportion to its velocity. This means that increased energy, in the form of increased velocity, is converted into matter. The difference in mass involved is insignificant until the velocity of the moving body approaches that of light.

Scientists, then, have had to revise their ideas about the conservation of matter and energy. But they still maintain that neither is destroyed when one is transformed into the other. There is still as much matter, *or its equivalent in energy,* as there ever was. Furthermore, there is still as much energy, *or its equivalent in matter,* as there ever was. To this extent, at least, the old laws of conservation still hold.

STUDYING MATTER AND ENERGY

Every science deals with some aspect or other of matter and energy or both. Astronomy, for example, studies the kinds of matter in the stars, planets, satellites, asteroids, comets, meteorites, and nebulae. It also considers the energy of the heavenly bodies as they rush through space. Biology is concerned with the composition of the matter that makes up the world of living things; it is no less interested in the energy that is released when the front legs of a praying mantis fasten on a grasshopper than when sap rises in a tree.

There are two sciences, however, that are especially concerned with the basic problems of matter and energy. These two are chemistry, the "matter science," and physics, "the energy science."

CHEMISTRY

Chemistry studies the composition and the properties of substances and the transformations that they undergo. Chemists must learn to recognize and identify with absolute accuracy all the pure substances with which they work. It is not enough for them to say, for example, that a transparent solid on the laboratory table is colorless, too hard to scratch with the fingernail, and yet not too hard to scratch with a diamond. That description would apply equally well to glass or quartz or any one of a number of plastics. The chemist would have to identify the substance by means of a series of searching tests. For example, he would subject it to intense heat. If it burned, he would know definitely that it could not be quartz or glass, for, although both of these substances would soften and melt in a hot flame, they would not burn. The chemist would then apply other tests to identify the substance.

Chemists must also become familiar with the properties of pure substances. They must find the answers to such questions as, "What is the melting point of this substance? Its boiling point? Is it flammable? Does it conduct electricity? Is it soluble in water? Is it subject to corrosion? Is it poisonous?" Armed with this knowledge, they are able to suggest numberless applications for different substances. Thus, a hard metal that does not corrode readily may make a good tip for fountain-pen points. A material that is a poor conductor of heat may be suitable for insulating pipes. A nonpoisonous, nonflammable liquid that dissolves grease but not textiles may be useful as a drycleaner's solvent.

The chemist aims not only to describe the properties of matter but to know the reasons for such properties. The chemist wants to know why one metal is hard and another soft, why salt dissolves in water but not in gasoline, and why sulfur does not conduct electricity as well as metals do. If it is known why a substance has a certain valuable property, it may be possible to manufacture a new material that will possess the desired property to a high degree.

Finally, the chemist is interested in the transformation of matter and in the principles that govern such transformations. Why does wood burn, but not asbestos? Why do iron and steel corrode more easily than gold or platinum? How may one prevent corrosion? By finding out the laws governing these transformations, the chemist can bring about more effectively certain changes, as well as prevent others.

PHYSICS

The "energy science" of physics deals particularly with the forces released by energy. The subdivision of physics called *mechanics* studies the forces acting on bodies at rest or in motion—the forces exerted on bridges and dams, on bullets and baseballs, on speedboats and airplanes. Another subdivision of physics deals with heat phenomena: the study of how heat moves from one place to another and of what happens when we add heat to one substance or remove it from another. The physicist is also interested in sound—the vibrations that move from one place to another through the medium of a substance like air or water.

The physicist studies radiations of various kinds: visible light, X rays, ultraviolet rays, infrared rays, and radio waves. He deals with electricity at rest and in motion and its relation to magnetism. Another branch of physics is *electronics,* which is concerned with the control of free electrons—electrified particles with a negative charge. One of the most important branches of physics nowadays is *nuclear physics,* which involves the study of the nucleus, or core, of the atom. In all of these fields the physicist is primarily interested in the discovery of basic principles, which are then used by the engineer in making machines and gadgets of all kinds.

NO HARD-AND-FAST SEPARATION

There is today no hard-and-fast line between chemistry and physics. Certain chemical changes interest the physicist, and certain physical changes interest the chemist. Hence it is not always clear whether a given scientific investigation is primarily a chemical or a physical one. Both physicists and chemists have concerned themselves with the transformation of gases into liquids at low temperatures. Both have attempted to explain why some substances are more readily liquefied than others. As a matter of fact, a separate science, *physical chemistry,* was established in the latter part of the nineteenth century to deal with areas in which physics and chemistry overlap. When the chemist and the physicist attack the same problem, each considers it from his own point of view. For example, in studying electricity, the physicist deals with the forces and the mechanisms involved. The chemist examines the changes undergone by matter when an electric current is passed through a solution.

GASES, LIQUIDS, AND SOLIDS

We have already learned that animals, plants, rocks, water, the air, the planets, and our sun are simply different aspects of matter. We shall now turn our attention to the small units of matter called *molecules*.

The name molecule comes from a Latin word meaning "a little mass," or "a little quantity." A molecule of a substance is not merely a little quantity. It is the smallest possible quantity of that substance. To illustrate the point, suppose we deposit a single drop of water on a glass slide. Suppose now that we divide this drop and then subdivide it into smaller and smaller units. Each droplet would still be water and not some new form of matter. Eventually, we would have a water particle that could no longer be divided and still be water. This smallest particle would be a molecule.

The molecule is not the tiniest unit of matter. Molecules can be broken up into smaller units called *atoms*. Each water molecule, for instance, is composed of two atoms of hydrogen and a single atom of oxygen. But if we broke up all the molecules in a glass of water into their atoms, we would no longer have a glassful of water. Instead we would only have two gases, hydrogen and oxygen, which are as different from liquid water as they can be.

Molecules are the building blocks of gases and liquids and many solids as well. Gases, liquids, and solids make up the three states of matter. According to some scientists, plasmas, which are gases composed of charged particles, make up a fourth state of matter. The gas escaping from the burner of a gas oven, the liquid in a gasoline tank, and the ice of a skating rink are all composed of molecules. This may seem rather strange in view of the obvious differences between gases, liquids, and solids. But the fact is that there is no hard-and-fast dividing line between these three states of matter. The same substance may be a solid, or a liquid, or a gas, depending upon the temperature. Water, for example, is most familiar to us in the liquid form. When it freezes, it becomes a solid—ice. When it boils, it becomes a gas—steam. What is true of water is true of other substances.

Strictly speaking, then, it is not accurate to call this or that substance a solid, or a liquid, or a gas, because most substances can be found in one or another of the states of matter under certain conditions of temperature or pressure. However, most bodies are solids, or liquids, or gases under ordinary conditions, and we refer to them as such. Thus we call iron a solid, even though it will melt and become a liquid at the temperature of 1,535° Celsius. We call carbon dioxide a gas even though it can be transformed into the solid that we call dry ice.

KINETIC THEORY OF MATTER

Molecules are certainly quite unusual building blocks, for they are constantly in motion, except at the theoretical temperature that we call absolute zero, the equivalent of −273.16° Celsius. The laws governing this motion have been set forth in what is known as the *kinetic theory,* or *kinetic-molecular theory,* of matter. It offers a perfectly natural explanation for many phenomena that were once thought to be very mysterious.

According to the kinetic theory, the molecules of a gas in a container are continually bumping against one another and also against the walls of the container. Think, if you will, of a half-dozen billiard balls, kept constantly in motion on a billiard table. Each ball will move in a straight line until it hits another ball or the side of the table.

We are all familiar with the three states of matter in which water can exist. Far left: steam—water in the gaseous state—condenses into water droplets as it leaves the kettle. Left: liquid water. Right: ice—water in the solid state.

After such a collision, it will bounce off at an angle, moving in a new direction and often with a different speed. The gas molecules follow much the same sort of chaotic pattern, except that they move in the three dimensions of space and are not confined to the two dimensions of a table surface.

We can see the effect of this constant movement of molecules by examining smoke through a microscope. We first admit the smoke into a closed container which has transparent windows. If we now illuminate the container with a bright light and examine it under the microscope, we shall be able to see the individual particles of smoke dancing about continually in an erratic way. The smoke particles move about in this manner because they are being constantly bombarded by the molecules of the gases that make up the air in the container. The perpetual dance of particles is called *Brownian movement*, or motion, from Dr. Robert Brown (1773–1858), a 19th-century Scottish scientist, who first showed how general this kind of movement is.

The rate at which the molecules move depends upon the degree of hotness of a given substance. As a matter of fact, heat is simply the kinetic energy, or energy of motion, of the molecules of a substance. If there is little heat in a given quantity of gas, the motion of the molecules is slow. If there is more heat, the molecules will dart to and fro more rapidly. They will strike the walls of the container more often and with greater force.

TEMPERATURE SCALES

The degree of hotness of a substance is called its *temperature*. We measure it by means of the device called the *thermometer*. In one thermometer system—the Celsius scale—the freezing point of water is marked 0° and the boiling point is marked 100°. On the Fahrenheit scale, the freezing point of water is marked 32° and the boiling point 212°.

The rate of movement of molecules gradually diminishes as the temperature drops. When we reach approximately the −273.16° Celsius mark, the motion of the molecules stops and there is no longer any

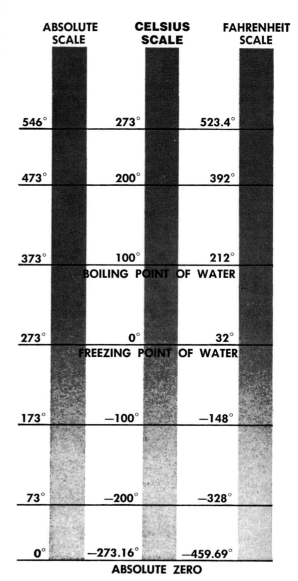

How the absolute (Kelvin), Celsius, and Fahrenheit temperature scales compare.

heat to measure. This point is known as *absolute zero*, or 0° absolute, or 0° K, because it marks the absolute end of the temperature scale. The absolute scale is more commonly called the Kelvin scale, after Lord Kelvin, the famous 19th-century English scientist who first proposed it. We have not quite succeeded in producing this temperature in our laboratories, but we have come to within 20-millionths of a degree of it.

The temperature 0° K is used as the basis of the absolute scale. In this we begin with 0° K and we reckon in degrees of centigrade. For example, 273° K would correspond to 0° Celsius, 280° K would correspond to 7° Celsius, and so on. To change temperature in degrees Celsius to degrees absolute we simply add 273.

PRESSURE

The steady bombardment of gas molecules striking the walls of a container produces constant pressure. The air in the room where you are sitting presses upon every bit of floor, wall, ceiling, furniture, and so on with a force of about one kilogram per square centimeter. The floor and walls do not collapse because there is air all around them and they are pushed equally in various directions. The degree of pressure—the rate of bombardment of the molecules—increases as a gas becomes hotter and decreases as it becomes colder.

Throughout all changes of temperature and pressure, the number of molecules of gas within a tightly sealed container will remain the same. At the temperature of 100° Celsius the molecules will bombard the walls of the container more vigorously than at the temperature of 0° Celsius. In either case, however, there will be the same number of molecules and they will fill the container. If we reduced the size of the container by half, the molecules would still find space for themselves within the reduced volume. But at the same temperature, they would press more vigorously upon their cramped quarters. If, on the other hand, we were to double the size of the container, the molecules would spread throughout the added space. At the same temperature, they would not exert as much pressure as before.

We know, then, that the molecules of a gas exercise continuous pressure. The pressure depends upon temperature and volume. The volume of a gas container can be increased or decreased without affecting the number of the molecules in the container.

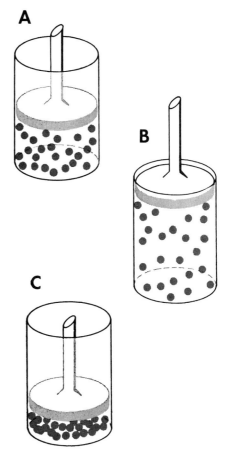

Boyle's law, one of the ideal gas laws, in operation. There is a given number of molecules of gas in the cylinder under the piston (A). When the piston is raised (B) so that the volume of space in which the molecules are enclosed is doubled, the pressure is half as great. If the piston that is shown in A is pushed down, as in C, so that the volume of the space containing the molecules is halved, the gas pressure is doubled. We assume that the temperature remains the same throughout.

IDEAL GAS LAWS

To express the relationship among the pressure, temperature, and volume of gases, scientists have established three simple laws. These hold to a certain extent for all gases, but they would apply with absolute accuracy only to theoretical, or perfect, or "ideal" gases, which do not really exist in nature. We therefore call the three laws in question the ideal gas laws.

The first one is called *Boyle's law,* because it was announced in 1660 by the English scientist Robert Boyle. According to this law, if the temperature remains the same throughout, the pressure upon a confined body of gas is inversely proportional to its volume. That means that if we reduce the volume by half, the pressure is doubled. If we reduce the volume to one-tenth the former volume, there will be ten times as much pressure.

The second ideal gas law was developed independently by two early nineteenth-century French scientists—Jacques-Alexandre-César Charles and Joseph-Louis Gay-Lussac. Hence it is called either *Charles' law* or *Gay-Lussac's law.* It states that if the pressure of a gas remains unchanged, the volume is directly proportional to the absolute temperature. To maintain a pressure at a certain value, the volume of a gas would have to be doubled if the absolute temperature were doubled.

According to the third ideal-gas law, if the volume of a gas remains unchanged, the pressure is directly proportional to the absolute temperature. Suppose that a given amount of gas in a container exercises a pressure of one kilogram per square centimeter at 273° K. At 546° K, the pressure would be twice as great—that is, two kilograms.

MANY DEVIATIONS

These, then, are the three ideal gas laws. As we have noted, they would apply exactly only to theoretical gases which are not found in nature. The gases that actually exist deviate from the laws to a greater or a lesser extent. The nineteenth-century Dutch chemist Johannes Diderik van der Waals made a careful study of these deviations. He summed them up in a mathematical formula that is generally called *van der Waals' equation.* In this formula, he took into account two factors that are not considered in the ideal-gas laws: (1) the attraction between molecules, and (2) the fact that molecules occupy a part of the total space in a given container.

The attraction between molecules is due in part to gravitation, the force that

holds the planets in their orbits around the sun and that causes meteorites to fall to the earth. But this force is far too weak in the case of molecules to account for all the actual deviations from ideal behavior. Some more powerful force must also be at work. Now, as you will read in another article, the molecules are made up of atoms, and the atoms, in turn, contain tiny, electrically charged particles. When two molecules are close together, the electrical attraction between them is particularly strong. At the same time, the force of gravitation also operates. This combination of forces is known as the *van der Waals' forces.* They are most powerful when the molecules are closest to each other.

The fact that gas molecules must occupy space is obvious. We can even calculate their size. Now the ideal laws assume that the space taken up by the gases under varying conditions of pressure or temperature is free or unoccupied space. The truth is that the molecules of gas themselves actually take up part of the supposedly unoccupied space allowed for in the ideal laws.

Van der Waals' equation, therefore, modifies the original ideal gas laws, but only to a certain extent. The laws still suffice to give us a good idea of how molecules of gases react to changes in temperature, volume, and pressure.

MOLECULES OF LIQUIDS

As we have seen, even in gases there is a certain amount of attraction between molecules because of gravitation or electrical forces. As the temperature of a gas is lowered, the molecules become increasingly sluggish. They do not wander so far apart and the forces of attraction between them become stronger. When the temperature is low enough, the molecules of the gas will cluster together and the gas will become a liquid. If the temperature is lowered still more, the movements of the molecules will become still slower and the forces of attraction will become still stronger. The liquid will then become a solid.

A substance occupies much less space when it is a liquid than when it is a gas. As we have just seen, the molecules of a liquid

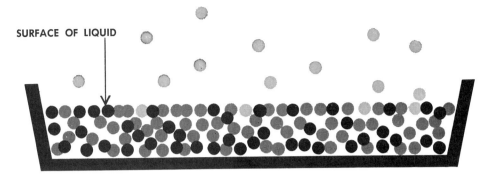

SURFACE OF LIQUID

How a liquid evaporates. The molecules of the liquid do not all have the same kinetic energy (energy of motion). Some of the molecules with the highest kinetic energy will break away from the attraction of the slower molecules at the surface. They will then escape from the liquid and will constitute a gas.

are much closer together than those of a gas. That is why it is so difficult to compress liquids. Enormous pressures bring about only slight decreases in volume.

Another important difference between liquids and gases is that liquids show a definite horizontal surface. The molecules in the body of a liquid strongly attract those on the surface and thus keep them under constant tension. This is known as *surface tension*. It causes the liquid to shrink to as small an area as possible. It makes the liquid behave as if it had a thin skin. Thus bits of paper or even small steel needles can be made to float on water. When you squeeze the bulb of a medicine dropper containing water, you will notice that each falling drop tends to take the form of a sphere. Surface tension is the cause. The inner molecules of the drop attract the outer molecules so strongly that they are drawn into the smallest possible space.

LIQUIDS CAN BECOME GASES

You all know that when damp clothes are hung out on a line in the open air, the clothes will gradually become dry. You know, too, that if you leave a bowl of water in a well-heated room, the level of water in the bowl will gradually be lowered and in time the water will disappear entirely. In both cases, *evaporation* has taken place. Evaporation really means "vapormaking,"

or "gasmaking." When a liquid evaporates, its molecules escape from the liquid and make their way into the open air. In other words, the moisture of the wet clothes and the liquid in the bowl of water have turned into the invisible and odorless gas called *water vapor*.

The kinetic theory explains how evaporation takes place. We know that all molecules have a certain amount of heat, or kinetic energy, or energy of motion (except at absolute zero, of course). The molecules in a liquid do not all have the same amount of kinetic energy at a given moment and therefore they are not all traveling at the same speed. At a given moment, then, certain molecules at the surface of a liquid will be comparatively slow-moving, while others will be darting about more rapidly. Some of the molecules with the highest kinetic energy will break away from the attraction of the slower molecules and will escape into the open space above the surface. This is the process of evaporation.

Naturally, the escape of the molecules having the highest kinetic energy will lower the average kinetic energy of the molecules that remain in the liquid. Therefore, if you put a thermometer in the water after evaporation has been going on for some time, it will register a lower temperature than before. To put it into other words, the liquid has now become somewhat cooler.

As the temperature of a liquid is raised, the rate of evaporation is increased. The reason is that the average kinetic energy of the molecule becomes higher, and the rate at which molecules break away from each other increases.

Different liquids show different rates of evaporation, because the forces of attraction between molecules are not the same in all liquids. The molecules of ether, for example, have less attraction for each other than do those of water, and therefore there will not be so much pull upon a molecule about to escape from ether. Hence the speed necessary for an ether molecule to break through the surface will be lower than the corresponding speed for a water molecule. This means that at the same temperature a greater proportion of ether molecules will have enough speed to break through the surface, and the rate of evaporation will be greater.

GASES CAN BECOME LIQUIDS

We have shown that by increasing the temperature of a liquid we can cause it to evaporate — that is, to turn into a gas. We can reverse the process just as effectively. If we lower the temperature of a gas, we cause the molecules of the gas to have less kinetic energy. They are slowed down, and at the slower speed the force of attraction between them holds them closer together. As the cooling continues, the gas changes to a liquid. This effect is called *condensation* — a most appropriate term, since liquids are much denser than gases.

Nature offers many examples of condensation. When warm water vapor in the air rises to cooler levels, it condenses and forms the cluster of tiny water droplets that we call a cloud. Dew is another product of condensation. So is the moisture that forms on windowpanes when the temperature of a room is raised and the glass of the panes is still cold because of its contact with the outer air.

We see condensation at work in the process known as *distillation*. First we heat a liquid or solid in one part of a retort. The vapor that forms then condenses in another part of the retort and is collected in the form of a liquid. Since different substances have different rates of condensation, we can separate the various components of a liquid or solid by collecting each in turn as it condenses.

BOILING POINT OF LIQUIDS

When the temperature of a liquid is raised enough, the liquid boils. This means that bubbles of vapor, containing millions of molecules, form below the surface. In order for such bubbles to be produced, the pressure of the vapor inside them must be equal to the pressure of the air upon the surface of the liquid. If the air pressure is greater, the bubble will collapse. The boiling point of a liquid, then, is the temperature at which the vapor pressure of the liquid is equal to the atmospheric pressure upon the liquid at the surface.

The pressure of the atmosphere at sea level is about one kilogram per square centimeter. We call this pressure one atmosphere. The boiling point of a liquid, therefore, is the temperature at which its vapor pressure equals one atmosphere. The boiling point of water is 100° Celsius; of ethyl alcohol, 78.5° Celsius; of mercury, 356.9° Celsius.

If the pressure on the surface of the water is reduced, the boiling point will be lower. Now the higher up we go from sea level, the lower the atmospheric pressure becomes, since there is less air to press upon us. Hence, while water boils at sea level at a temperature of 100° Celsius, its boiling point on a mountain three kilometers high is only 90° Celsius. Since the temperature of water does not rise as a result of boiling, it would take you longer to cook potatoes or carrots in boiling water on the mountain than it would at sea level.

CRITICAL TEMPERATURE

Suppose we put a quantity of water in a steel tank which is so strong that it will not burst, no matter how much heat is applied to it. If we keep raising the temperature of the water more and more, the molecules of vapor will become more numerous and will exert increasing pressure upon the surface of the liquid. The vapor molecules will be

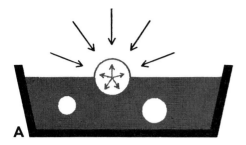 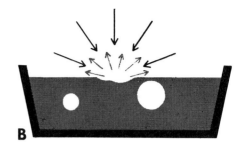

The liquid shown in the above drawing has been subjected to heat, and as the temperature has steadily risen, bubbles have begun to form. Each of the bubbles contains a great many molecules which are all in the gaseous state (A). A bubble at the surface of the liquid can remain as a bubble only if the pressure of the atmosphere upon it is as great as the pressure of the vapor inside the bubble. If the two pressures in question are unequal, the bubble will burst (B).

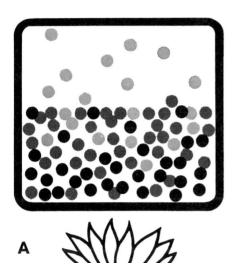 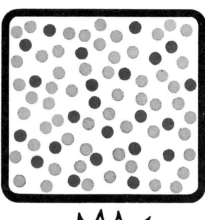

Heat is applied to the liquid in a closed container (A). Molecules of the liquid jump from the surface and become vapor. As the temperature rises more and more, the vapor becomes denser, while the density of the liquid decreases, meaning that the molecules are farther apart. Finally, at the so-called critical temperature, the average distance between molecules in the vapor and the liquid is the same; the boundary between the vapor and liquid has disappeared (B).

getting closer together since there will be so ..any more of them within about the same space as before. On the other hand, since the density of a liquid decreases with increased temperature, the molecules in the liquid will be getting farther apart. Finally, at 374° Celsius, the average distance between molecules in both the liquid and vapor states becomes the same. Therefore the boundary line between liquid and gas — in other words, the surface of the liquid — will disappear. The temperature at which this occurs is called the *critical temperature*. As we have seen, the critical temperature of water is 374° Celsius. The critical temperature of ammonia is 132.4° Celsius; of carbon dioxide, 31.1° Celsius; of oxygen, −118.8° Celsius; of nitrogen, −147.1° Celsius. Above the critical temperature, a liquid becomes a gas.

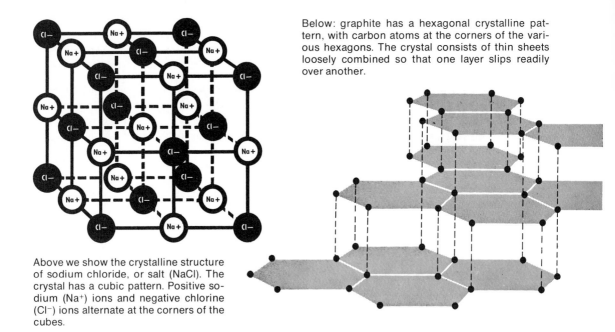

Below: graphite has a hexagonal crystalline pattern, with carbon atoms at the corners of the various hexagons. The crystal consists of thin sheets loosely combined so that one layer slips readily over another.

Above we show the crystalline structure of sodium chloride, or salt (NaCl). The crystal has a cubic pattern. Positive sodium (Na$^+$) ions and negative chlorine (Cl$^-$) ions alternate at the corners of the cubes.

CRYSTALLINE SOLIDS

Most solids are made up of *crystals* — bodies that have a definite number of planes, or flat surfaces, arranged in a particular way. The shape of the crystal will depend upon the substance. If you look at a grain of salt through a magnifying glass, against a dark background, you will see that it is a crystal in the shape of a perfect little cube. Other crystals show complicated patterns with a great number of planes. Solids with crystals as their basic units are called *true solids*.

True solids are not fluid, like gases or liquids. They are rigid, or comparatively rigid. Their components—molecules, or atoms, or ions (molecules or atoms that have acquired an electrical charge)—are joined together to form a framework that has been given the name of crystal lattice. Long ago, scientists deduced that this framework existed, but they could not prove experimentally that there was any such thing. However, X-ray analysis has proved the existence of the lattice beyond a doubt.

The crystals of many solids are made up of molecules that no longer wander about but that now have fixed posts in the crystal lattice. This holds true for iodine, camphor, menthol, and naphthalene, among other substances. The molecules of such solids have kinetic energy. That is, they move. But their motion does not represent a ceaseless change of place. Rather, it takes the form of vibration about a fixed position.

In the case of certain solids, the units that make up the lattice structure are not molecules. The crystals of metals are made up of atoms. Those of salts are made up of ions. The kinetic theory also applies to the atoms or ions of such solids, for they too vibrate within the crystal.

AMORPHOUS SOLIDS

When some liquids cool, they do not solidify into true solids with a fixed crystalline pattern. Instead, their internal structure is a haphazard affair. It is quite shapeless, compared with the beautifully geometric patterns of true solids. For this reason, we call such solids "shapeless," or, to use the technical term, "amorphous."

Glass is a good example of an amorphous solid. Liquid glass is a complex mixture of various compounds. It is viscous—

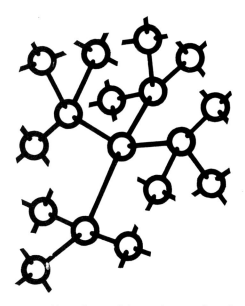

Like graphite, diamond is made up of carbon atoms. Each of these atoms is linked to four others, forming a three-dimensional pattern that is repeated throughout. This produces a sturdy framework.

that is, it flows and pours slowly, much like thick molasses. Even after the liquid hardens, it remains viscous. If you would apply even a small force to a piece of glass over a period of months, or years, the glass would flow noticeably at the end of that time. Because of the viscous nature of glass, the molecules do not have a chance to arrange themselves in a true crystalline pattern. Glass, however, may slowly become crystalline in structure over a period of many years. This phenomenon is called *devitrification* or "de-glassing."

MELTING POINT OF SOLIDS

If we raise the temperature of a crystalline solid, the vibration of its molecules, atoms, or ions becomes energetic enough to destroy the fixed crystal pattern throughout the whole structure of the material. The molecules, atoms, or ions, freed from their rigidly set places in the crystal lattice, move around freely. What was formerly a solid is now a liquid.

Most solids change into the liquid state when their temperature rises beyond what is called the *melting point*. As we shall see, some solids change directly into the gas-

eous state. At the melting point, the solid form of the substance and its liquid form are in equilibrium. The rate of melting of the solid is equal to the rate of freezing of the liquid. Thus if an ice-water mixture is held at the melting point, the weights of the ice and water will remain unchanged. Melting points for solids vary widely. That of ice is 0° Celsius; of lead, 327.4° Celsius; of silver, 960.5° Celsius; of iron, 1,535°. The melting point of any solid is the freezing point of its liquid form.

SUBLIMATION

If we scatter moth balls freely about a garment, a strong moth-ball odor will soon arise. The reason is that molecules have escaped from the surface of the original substance — a solid — and have become gas molecules. We give the name of *sublimation* to the change of state from solid to gas and from gas to solid. We say of substances that undergo such a change that they sublime. In general, crystals whose units are molecules, instead of atoms or ions, sublime most readily.

Solids vary a great deal in the rate at which they sublime. Salts and metals do so very slowly. Ordinary ice sublimes appreciably. A cake of ice will evaporate even if the temperature is below freezing. Dry ice — solid carbon dioxide — has a very high rate of sublimation. That is one reason why it is a more efficient cooling agent than ordinary ice. As it sublimes, it continually envelops the material that is to be cooled with a blanket of carbon-dioxide gas. This serves as effective insulation.

SPEED AND NUMBER OF MOLECULES

The kinetic theory gives us a clear idea of how molecules are affected by changes in temperature, volume, and pressure and how they change from one state of matter to another. There are other things that we know about molecules. We can calculate their speed, their number, their size, and their relative weights with a considerable degree of accuracy.

Of course, the speed of a molecule will depend upon the temperature. It will also be affected by collisions with other molecules. At any particular instant, therefore, the molecules in a gas will move at a wide range of speeds. The average is high—something like 1,600 kilometers an hour for an oxygen molecule at 0° Celsius. Heavy molecules generally move more slowly than light ones at the same temperature.

It is quite easy for the imagination to grasp a figure like 1,600 kilometers per hour, since airplanes have already reached speeds several times greater than that rate. It is much harder to have a clear understanding of the very great number of molecules and their very small size. For example, in every cubic centimeter of air about us there are some 30,000,000,000,000,000,000 molecules.

The great Italian chemist Count Amedeo Avogadro (1776–1856) calculated that in 18 grams of water there are 602,000,000,000,000,000,000,000 molecules. This huge figure is known as *Avogadro's number*. It has proved to be invaluable in many calculations. It may be abbreviated to read 6.02×10^{23}. Why did he choose a weight of 18 grams? This is the *gram molecular weight* of water—that is, the atomic weight of one molecule of water expressed in grams. One gram molecular weight of any substance will contain this same number of molecules—6.02×10^{23}.

We can calculate the volume of a single water molecule by using Avagadro's number. We assume that the molecules are so closely packed in liquid water that the amount of empty space is negligible in comparison to the volume of the molecules themselves. Hence we may divide the total volume of water in 18 grams by the total number of molecules (6.02×10^{23}) in order to find the volume of one molecule—about 0.000,000,000,000,000,000,000,03 cubic centimeters, or 3×10^{-23} cubic centimeters. The actual volume of a molecule will be rather less than the figure given since there is a certain amount of free space between molecules in liquid water.

Of course, this calculation is in terms of the molecules of a liquid. We have already pointed out that liquid water can change into the gas called water vapor and into the solid called ice. The individual water molecule remains unchanged. It will have the same volume in the gaseous and solid states that it had when it was in the liquid state.

Avogadro has been immortalized by the famous number that bears his name. He is no less famous for his hypothesis about the relative number of molecules in gases. Avogadro's hypothesis is that "equal volumes of gases at the same temperature and pressure contain equal numbers of molecules." This means that at a given temperature and pressure, if 6.02×10^{23} molecules of water vapor occupy 50 liters, then 50 liters of any other gas, such as oxygen, will contain 6.02×10^{23} molecules.

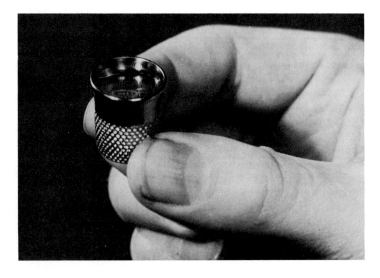

A thimbleful of water has approximately 602,000,000,000,-000,000,000,000 molecules.

PROPERTIES OF MATTER

by Eric Rodgers

A whiff of perfume, soap bubbles blown from a child's pipe, the floating of a needle on the surface of water—these phenomena are possible because of powerful forces acting upon the particles called *molecules*. Each molecule of a given substance is the smallest unit that retains all the properties of that substance. Molecules are the building blocks of liquids and gases and many solids as well. They are in constant motion, even in solids, except at the theoretical temperature called absolute zero.

VISCOSITY OF LIQUIDS

Liquids such as alcohol or gasoline flow very freely when poured from the mouth of a jug. Others, such as honey or castor oil, tend to creep out when poured. The difference is due to the friction of the molecules as they pass over each other. This internal friction is called *viscosity*. Honey and castor oil have a high viscosity. On the other hand, alcohol and gasoline have a low viscosity.

Suppose a board floats on the surface of a liquid in a shallow container, as shown in Figure 1. We could drag the board over the surface by applying a horizontal force F to it. Experiment shows that the greater the force F, the greater the velocity of the board. While the board is in motion, a thin layer of liquid clings to it and moves at the same speed. Another thin layer of liquid adheres to the bottom of the container and does not move at all. Layers of liquid in between move with velocities that increase from bottom to top, as indicated by the arrows in the diagram. The cohesive force between the layers causes the friction that we call viscosity.

Each portion of liquid flowing in a shallow container is continually changing shape. If it is cubical in shape at one instant (Figure 2, *a*), it will have a different shape (Figure 2, *b*) a short time later as it moves along. We might define *viscosity* as the resistance that a liquid offers to having its shape changed.

The viscosity of a liquid may be measured in several ways. One method uses two concentric cylinders. The space between them is filled with the liquid that is being tested. One cylinder is held in a fixed position while the other is made to rotate at a constant speed. The torque, or turning force, required to keep the moving cylinder rotating at a constant speed is measured. This torque varies with the viscosity of the liquid. Another method involves measuring the rate of flow of a liquid that escapes from a small opening.

The viscosity of liquids decreases as the temperature rises and increases as the temperature falls. The lubricating oil in an automobile engine on a very cold morning may be so viscous that it does not flow freely enough to form a protective film when the engine is first cranked. This is why one should not race the engine before it warms

It is important to know exactly how molecules will behave under different conditions. This metal part did not have the right properties to do its job.

The Franklin Institute Research Laboratories

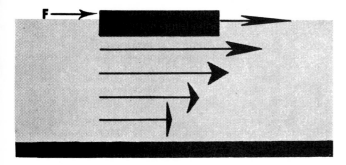

Fig. 1. When a board is moved along the surface of a liquid in a shallow container, the different water layers between the board and the bottom of the container also move, with different velocities. The cohesive force between layers brings about friction (resistance to flow).

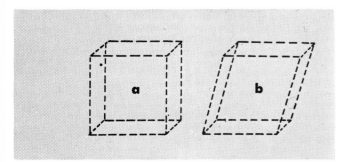

Fig. 2. Each portion of the liquid that flows in a shallow container changes its shape continually. Let us suppose, for example, that it is cubical at one instant (a). It will acquire a different shape (b) as it moves along.

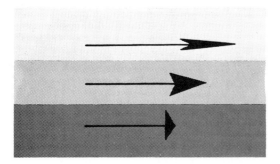

Fig. 3. We may think of the air above a surface as divided into a number of thin layers, each moving more rapidly than the one below it. The sliding of one layer over another produces friction between the different layers.

up on a cold morning. The viscosity of lubricating oils is indicated by numbers, a system devised by the Society of Automotive Engineers (SAE). The viscosity rating of a lubricating oil is important in selecting the proper oil for each use. For example, the viscosity of lubricating oil SAE 30 is approximately double that of SAE 20 when the two are at the same temperature.

VISCOSITY OF GASES

Gases also have viscosity, but it is much less than in liquids. As air moves over a surface, a thin layer of air in contact with the surface does not move with the rest of the air. We may think of the air immediately above the surface as being divided into thin layers, each moving faster than the one below it. This sliding gives rise to viscous friction.

There is an important difference between the viscosity of liquids and that of gases. In liquids, the cohesive forces between layers cause viscosity. Cohesive forces could hardly account for the viscosity of gases since gas molecules are so much farther apart on the average than liquid molecules. We must, therefore, seek another explanation.

In Figure 3 we see that the upper layer of air is moving forward faster than the lower one. In addition to the forward motion, individual molecules in each layer are moving with a random motion. This results in a free exchange of molecules between the two layers. The exchange produces the same result as the cohesive forces in liquids. An increase in temperature in the gas would increase the random motion, and it would therefore increase the exchange rate and the viscosity.

SURFACE TENSION IN LIQUIDS

Any molecule within a liquid is attracted by its neighboring molecules. These forces operate at very short ranges, so that only molecules in the immediate vicinity exert any appreciable force on a given molecule. A molecule that is not near the surface, but in the body of the liquid, is completely surrounded by other molecules. It is equally attracted in all directions by its

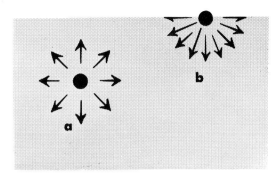

Fig. 4. *a* and *b* represent molecules within a liquid. *a* is equally attracted in all directions by its neighbor molecules. *b*, at the surface, has no molecules above it; therefore there is a net downward attraction.

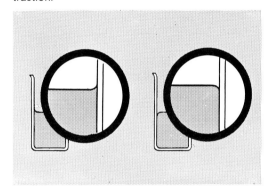

Figs. 5-6. In Figure 5, left, we see how the surface of the water in a glass container curves upward wherever it comes in contact with the glass. Figure 6, right, shows the surface of the mercury in a glass container curving downward where it touches the glass.

neighbors and the resultant pull in any one direction is zero (Figure 4, *a*). However, a molecule at the surface (Figure 4, *b*) does not have liquid molecules above it and consequently experiences a net attraction downward. Since all molecules on the surface experience this attractive force downward, the liquid acts as though it were covered with a tightly stretched elastic membrane, tending to shorten itself to the greatest possible extent. A given quantity of liquid, therefore, tends to assume the shape that has the smallest surface area. In this respect the surface behaves as though it were under tension. Thus, we speak of the *surface tension* of the liquid.

Surface tension accounts for many interesting phenomena. Small droplets of liquid tend to be spherical because for a given volume of liquid the spherical shape requires the smallest surface area. The pull on the surface of a liquid produces the effect of a thin skin covering it. Small insects are often observed to walk or run on the surface of water. A needle can be made to float on the surface, even though the needle may be nearly ten times as dense as water.

COHESION AND ADHESION

The molecules of a substance are held together by an attractive force that is known as *cohesion*. The name *adhesion* is given to the attraction between molecules of different substances. The adhesive force between a liquid and the air at its surface is small. But if a liquid surface is in contact with a solid one, the adhesive force may be quite large. In many cases this force is greater than the cohesive force of the liquid. The surface of water in a glass container curves upward where it comes in contact with the glass, as shown in Figure 5. The adhesive force between glass and water is stronger than the cohesive force of the liquid, and it pulls the water upward around the edge.

On the other hand, if we place some mercury in a glass container, we find that the cohesive force within the mercury is much greater than the adhesive force between mercury and glass. The result is that the cohesive force pulls the mercury down around the rim (Figure 6).

If we put a glass rod in water and then pull it out, we note that the glass is wet because water adheres to it. If we insert the rod in a basin containing mercury, no mercury will adhere to the rod when we pull it out. Where the cohesive force in a liquid is greater than the adhesive force between the liquid and a solid, as in the case of mercury and glass, the liquid will not wet the solid. Where the adhesive force is greater, as in the case of water and glass, the solid will become wet.

Suppose we put the lower end of an open capillary tube — one with a very small bore — in a liquid. If the liquid wets the tube, it will rise in it (Figure 7); if it does not

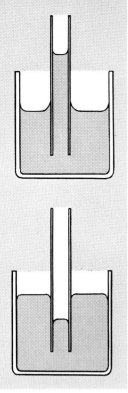

Fig. 7. If a small-bore tube becomes wet when put in a liquid, the liquid will rise in the tube.

Fig. 8. If a small-bore tube does not become wet when it is put in a liquid, this liquid will sink in the tube.

wet the tube, it will sink (Figure 8). The extent to which a liquid will rise or sink will depend on the size of the tube. 'The smaller the diameter of the tube, the higher the liquid will rise or sink in it. The name *capillarity* is given to the action of the liquid in a capillary tube.

The phenomenon of capillarity plays an extremely important part in the flow of water in a living plant. It also causes water to rise toward the surface through the capillary openings between grains of soil. Farmers in dry areas try to keep the soil on the surface loose rather than closely packed so as to increase the size of the capillary openings and reduce the flow of moisture to the surface, where it evaporates and is lost to the farmer.

If the end of a towel is placed in a pan of water, the water soon creeps up by capillary action to wet the entire towel. Capillary action is at work, too, when the kerosene in a lamp climbs up the wick.

Surface tension is affected by temperature. It is reduced as the temperature rises. Impurities in solution also bring about a reduction of surface tension. So do modern synthetic detergents. As a result they greatly increase the wetting ability of water. Many materials that are not wet at all by pure water become soaked when a small

quantity of detergent is dissolved in the liquid. A duck can swim easily because its feathers are not wet by water. If it tries to swim in water to which detergent has been added, its feathers will become soaked and it will drown if not rescued. We can show the effects of synthetic detergents by putting drops of pure water and drops containing a detergent on clean paraffin. The pure water will not wet the paraffin and the drops will not change their shape appreciably (Figure 9, *a*). The drops containing the detergent, however, will wet the paraffin and the water will spread out (Figure 9, *b*).

Antiseptics used for cuts and other punctures of human flesh should have a low surface tension, as well as good germ-killing ability. If the surface tension were high, not all the damaged surface would be wet by the liquid. Alcohol and most other antiseptics have a low surface tension.

DIFFUSION

If a bottle of ammonia is opened in one corner of a room, it can soon be smelled all over the room. The spaces between air molecules are large compared to the space occupied by the molecules themselves. The ammonia gas molecules will pass through the space between some air molecules and will collide with others. In time the ammonia molecules will be spread throughout the entire volume. This mixing of gases is called *diffusion*. Liquids also diffuse. If a lump of sugar is placed at the bottom of a vessel of water, the sugar will dissolve after a time. It will slowly diffuse throughout the water so that a sample of water taken from any part of the container will taste sweet.

If a small amount of gas is released in a

Fig. 9. If a drop of pure water is placed on some clean paraffin, it will not wet the paraffin and the drop will not change appreciably in shape *(a)*. If a drop of water containing detergent is put on the paraffin, the paraffin will become wet. As a result the water will spread over the paraffin and its shape will change radically *(b)*.

a   b

container from which the air has been removed, it diffuses more rapidly than when the air is at atmospheric pressure. This is mainly because the gas molecules collide less frequently at the reduced pressure.

The diffusion rate of a gas depends on its molecular weight: the lighter the gas, the more rapid the diffusion. This also applies to diffusion through a porous wall. Hydrogen and other light gases usually can diffuse readily through thin pottery. If two gases are placed in a closed vessel made of material that allows the gases to diffuse through its walls, the lighter gas will diffuse through walls more rapidly than the heavier gas.

Diffusion is usually a two-way process. When two gases or liquids mix by diffusion, each moves from the place where it is strongly concentrated to the place where the concentration is not so great. If two liquids are separated by an impermeable membrane, each liquid will start immediately diffusing into the other as soon as the membrane is removed. The process will continue until each liquid is spread uniformly throughout the entire region.

Certain substances are porous to one material, but not to another. For example, water will diffuse through a thin slice of potato, but sugar molecules will not. Close an open tube at the bottom part with a thin slice of potato. Put some strong sugar solution in the tube. If the lower end of the tube is placed in water, the water will diffuse through the membrane into the tube because of the strong water concentration below. The sugar molecules, however, cannot pass through the membrane to diffuse into the water below. If the tube has the shape shown in Figure 10, enough water will pass through the membrane to cause the liquid level in the tube to rise to a point considerably above the surface of the water in the beaker. A membrane of this sort is called a *semipermeable membrane*. The selective diffusion through such a membrane is known as *osmosis*.

ADHESIVE FORCES IN GASES

There is an adhesive force between gases and solids. If air is pumped out of a glass tube until the pressure is low and the tube is then sealed, the pressure gradually diminishes as time passes. This is because the molecules have become attached to the walls of the tube—a process called *adsorption*. The air can be driven from the walls with heat: a long baking is necessary if it is desired to drive out practically all the air. Certain tubes used in radio and other electronic devices are designed to operate at a high vacuum. While air is being pumped from these tubes, they are usually baked at a temperature about as high as the tube will stand. The purpose is to drive out the adsorbed gases. If they were not disposed of in the manufacturing process, small quantities would be expelled later by the heat from the tube's filament. This might greatly shorten the life of the tube, or interfere with its proper operation.

Gases are absorbed somewhat by liquids with which they are in contact. Some gases are particularly affected in this way by certain liquids. Water vapor, for example, is readily absorbed in sulfuric acid.

AN ANALYSIS OF SOLID SUBSTANCES

Solids can be divided into two classes, on the basis of their internal structure. Most solids are composed of aggregates of small crystals and are called *crystalline substances*. In crystals the atoms are arranged in an orderly manner in units that are repeated throughout the crystal. The

Fig. 10. An open tube is closed at the bottom with a thin slice of potato; strong sugar solution is put in it. The lower end of the tube is put into water. The water will diffuse into the tube; the sugar molecules will not pass out of it. Hence the liquid level in the tube will rise.

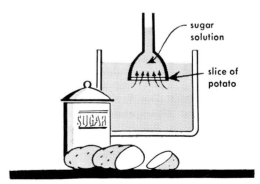

simplest crystals have a cubical pattern. The atoms lie at the corners of the cubes, which are repeated throughout the crystal, as shown in Figure 11. Rock salt crystallizes in just this way. If small grains of salt are viewed through a low-powered microscope, the cubical nature of the small crystals is apparent. There are a number of other crystal types. Figure 12 shows the hexagonal crystal pattern, which we find in ordinary graphite and many other substances. Graphite is pure carbon, as is diamond. They differ because diamond shows an entirely different crystal pattern (Figure 13).

As one examines a piece of calcite or rock salt, its crystalline structure is evident. That of iron or aluminum, however, is not so clear. The reason is that these two metals are composed of many very small crystals that are usually oriented almost at random. The presence of the minute crystals is revealed by X-ray examination of the materials.

Substances that are not crystalline in nature are called *amorphous*. Glass and pitch are examples. They act more like highly viscous liquids than like solids. Amorphous substances do not have definite melting temperatures. As they are heated, they gradually soften until they flow sluggishly. Crystalline substances, however, melt at definite temperatures. Ice, for example, remains rigid while heat is being applied until it changes into water at 0° Celsius. Below the melting point, the atoms within a crystal vibrate about their fixed positions, but remain at the same average distances apart. The higher the temperature, the greater is the intensity of the vibrations. At a definite temperature they become so energetic that the atoms break away from their positions. The crystalline solid flows like a liquid as it melts.

PHYSICAL PROPERTIES OF SOLIDS

Most of the physical properties of solids, such as hardness, tensile strength, brittleness, and elasticity, depend upon the type and arrangement of the tiny crystals of which they are composed. The forces acting between atoms in a crystal are such that the atoms resist having their average dis-

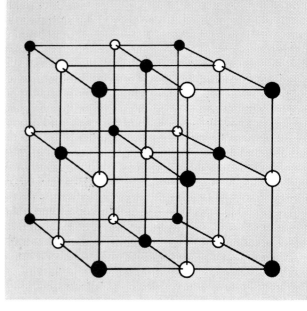

Fig. 11. A crystal of rock salt (sodium chloride) has a cubical pattern; the atoms lie at the corners of the cubes.

tance from each other changed. They resist any change in the shape of the solid of which they form a part. The result is that large forces are required to change the shape or volume of a solid.

All solids yield somewhat to forces applied to them. Some solids will jump back to their original shape or size when the force is removed. Others will not. Those that return more or less to the original shape are called *elastic*. A perfectly elastic substance would regain its original form upon the removal of the force. No substance is perfectly elastic, but many are nearly so, provided they are not distorted too much. Beyond a certain degree of distortion, called the *elastic limit,* there is not even an approximate return to the former state.

A solid may be distorted in several ways. Its length or its volume may be changed. It may be twisted, bent, or sheared. The name *strain* is given to a deformation of this kind. The force acting upon the solid is called a *stress*. Perhaps the simplest type of distortion is that of a change of length. We refer to this condition as *tensile strain*. We define it as change in length divided by original length. If a thrust is applied to each end of a bar, it will tend to become shorter. The ratio of change in

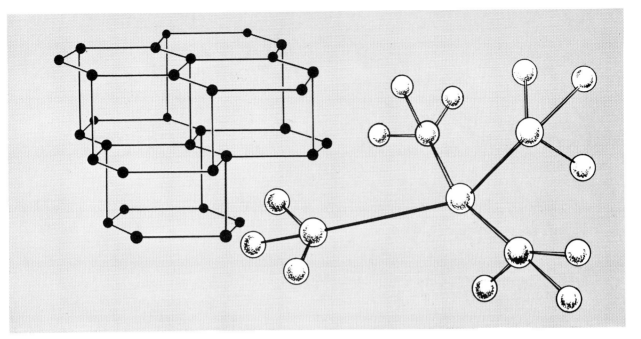

Fig. 12. The crystal of graphite (a form of carbon) has a hexagonal pattern.

Fig. 13. Diamond is another form of carbon. In its crystal lattice each carbon atom is joined to four neighbors.

length to the original length is the *compressive strain*. Certain compressive forces can bring about a change in volume. The ratio of the change of volume to the original volume is the *volume strain*.

Figure 14 illustrates *shearing strain*. A block that was originally a cube is subjected to forces that change its shape as indicated in the diagram. We define the strain as the displacement *x* divided by the height *y*. We can bring about shearing strain if we push horizontally on the top of a thick book that is lying on a table. The pages will slide slightly in a horizontal direction.

A law known as *Hooke's law* states that so long as the elastic limit is not passed for an elastic substance, strain is proportional to stress. The ratio of tensile stress to tensile strain is called *Young's modulus*. It depends only upon the nature of the material of which a piece of wire, say, is made, and not on the length or diameter of the material. The name *bulk modulus* is given to the ratio of volume stress to volume strain. The ratio of shearing stress to shearing strain is known as the *shear modulus*, or, sometimes, the *coefficient of rigidity*.

When we speak of the *limiting stress* of a substance, we usually have in mind the stress that must not be exceeded if we wish

the body to behave like an elastic substance. For example, in the case of tensile stress, there is a limiting stretching force which may not be safely exceeded. If the stress goes beyond the limiting stress, the wire will not even approximately regain its original length when the stretching force is removed. It is in danger of breaking.

A knowledge of the limiting stress of materials is extremely important when these materials are used in structures where they have to support heavy loads. In planning bridges and steel supports for skyscrapers, engineers consider each steel beam that is used. They determine whether there is a chance that it will be subjected to a stress exceeding its limiting stress. Usually a rather large safety factor is required by the engineer. He might, for example, demand that the cross section of a beam should be such that no conceivable loading would produce a stress as much as one-fourth the limiting stress. We would then say that he used a safety factor of four in his computation.

Often it is necessary to keep the safety factors small to save weight. The best example is in airplane construction. Since each kilogram added to the metal structure of an airplane reduces the possible pay-

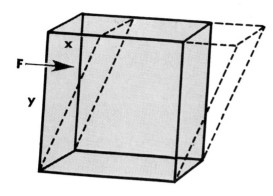

Fig. 14. Shearing strain. A block originally a cube is subjected to a force *(F)* that changes the block's shape.

load, it is important to keep the metal parts as small as possible and still allow for the necessary safety factors. Since these are computed closely in airplane construction, it is very important that none of the supporting parts should have any internal flaws. As a check for flaws, the parts that have to support big loads are often examined with X rays before their installation in the plane.

The tensile strength of any metal depends to a large extent on the size and arrangement of the tiny crystals of which it is composed. These can be controlled by proper heat treatment and by special processes such as rolling or drawing the metal out into sheets or rods at the proper temperature.

A number of other properties of solids, such as *ductility, malleability, hardness,* and *brittleness,* depend on the type and arrangement of the elementary crystals. Ductility is the property that determines to what extent a solid can be drawn out into fine threads. Tungsten is an extremely ductile solid material. Quartz is another very ductile material. The property of malleability refers to the extent to which a material can be made into thin sheets. Gold is one of the most malleable of all solids. Resistance to scratching of a surface is called hardness. Diamond, corundum, and topaz are all very hard. A scale called *Mohs' Scale* is used to indicate the degree of hardness of minerals. Brittleness refers to the tendency of a substance to crumble when subjected to a blow. A typical brittle substance is the familiar rock salt.

The spatial arrangement of molecules in a substance can often be determined with a transmission electron microscope.

The Franklin Institute Research Laboratories

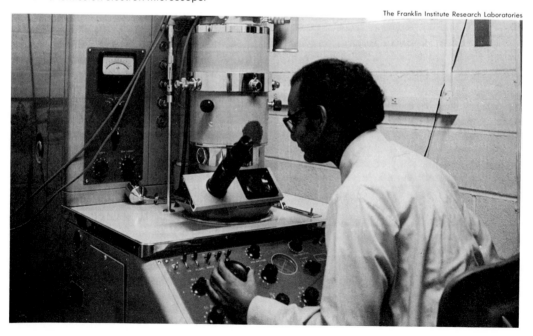

THE PERIODIC TABLE

by Anthony Standen

In a recent science-fiction tale, there was an account of certain space travelers who came upon the remains of an advanced civilization on a distant planet, where all life had been extinguished. The travelers found a number of books and other writings in a language and script totally unknown on earth. One of these documents contained the periodic table, in which chemical elements are arranged in periodic, or regularly recurring, patterns. Our voyagers were able to reconstruct the language by using only the periodic table as a basis for their task. This is fiction, of course, but it is plausible fiction. The table is so basic in both chemistry and physics, and it has such a characteristic shape, that it would be recognizable in any conceivable language.

It was Dmitri Ivanovich Mendeleev, a great 19th-century Russian chemist, who drew up the first satisfactory periodic table. He discovered certain natural patterns that revealed a deep underlying regularity in the variation of properties among the elements.

There are a little more than a hundred chemical elements, of which only about seventy were known in Mendeleev's time. Their properties vary greatly. The most obvious difference between them is that some twenty are classified as nonmetals, most of the rest as metals, and a few (sometimes called metalloids) as intermediate between metals and nonmetals. The nonmetals are in general lighter in weight than the metals.

It would be very convenient if the lighter elements were all nonmetals and if there were a gradual increase in metallic properties, as we proceeded to the higher elements, but that is not the case. How then are the chemical elements to be classified? This puzzle has been a fascinating challenge to scientists from Mendeleev's day to our own. Its gradual unraveling has been of great importance to modern science.

We give the periodic table on page 33 of this article and also, in what might seem to be a quite different version, on page 35. Actually, this second table presents the same information as the first one, with only a slight change in emphasis. Each element, in both tables, is represented by its chemical symbol. Hydrogen, for example, is given as "H"; helium, as "He"; lithium, as "Li." The small number above each symbol represents the atomic number of the element. The number following the symbol indicates atomic weight. The table on page 33 is referred to as the long, or extended, form. The one on page 35 represents the short form. The periodic table has also been set forth in a somewhat different arrangement.

REPEATING PATTERNS

The tables on pages 33 and 35 seem to present, at first sight, a bewildering pattern of symbols, numbers, and group names. It will be easier to understand the pattern if we consider only a small part of the table at the outset. We shall omit, for the time being, the first two elements—hydrogen and helium—and concentrate on the next sixteen. The chemical elements from lithium, Li, through argon, Ar, are listed below.

These sixteen elements begin with lithium, Li, which is a reactive, light metal. The next three elements, beryllium, Be, boron, B, and carbon, C, are all solids, and the

Tass, from Sovfoto

Dmitri Ivanovich Mendeleev

metallic properties diminish as we go along. Carbon is a nonmetal. We go on through the nonmetals nitrogen, N, and oxygen, O, to fluorine, F, which is a very reactive, gaseous nonmetal, and we finally come to neon, Ne, which is one of the inert gases.

We are now struck by a most significant fact: the eight elements of the second group (going from left to right) repeat the same pattern as the previous eight. Sodium, Na, is extremely similar to lithium, Li, directly above it. We note also that there is considerable resemblance between the members of each of the other vertical pairs—beryllium, Be, and magnesium, Mg; boron, B, and aluminum, Al; carbon, C, and

Valence +1	Valence +2	Valence +3	Valence ±4	Valence −3	Valence −2	Valence −1	Valence 0
3	4	5	6	7	8	9	10
Li 7	Be 9	B 11	C 12	N 14	O 16	F 19	Ne 20
11	12	13	14	15	16	17	18
Na 23	Mg 24	Al 27	Si 28	P 31	S 32	Cl 35	Ar 40
↑	↖	↗	↑	↖	↗	↑	↑
Light, reactive metals	Solids		Solid nonmetals	Nonmetals		Gaseous reactive nonmetals	Gaseous nonreactive nonmetals

silicon, Si; and so on. Toward the end of the two horizontal groups, the resemblance between the members of each pair is very close. Chlorine, Cl, is a reactive gas, strikingly similar to fluorine, F, which comes above it. Argon, Ar, is an inert gas, just like neon, Ne.

The numbers at the top of each of the columns in the diagram give the valence, or chemical binding power, of each pair of elements below them. The steady, progressive change in the valence, from +1 through +2 up to carbon which has a valence of either +4 or −4 (written as ±4), then through −3, −2 and −1 to zero, shows that the periodic table is not based on a few accidental resemblances between elements, but is intimately connected with their chemical behavior as well as their physical properties.

Given this arrangement of the elements from lithium, Li, with atomic number 3, to argon, Ar, with atomic number 18, any chemist could say which element ought to come next. It should be potassium, K, because it resembles closely both lithium, Li, and sodium, Na. And sure enough, taking the elements in order of atomic number, potassium does come next. It has atomic number 19. Any chemist can see, too, that under fluorine, F, and chlorine, Cl, should come bromine, Br, for this element bears a great resemblance to fluorine and chlorine. The next element should be krypton, Kr, an inert gas entirely similar to neon, N, and argon, Ar. But here an unexpected complication comes in.

NOT A SIMPLE PATTERN

A glance at the periodic table as presented on page 33 will show that there seem to be far too many elements after potassium before we reach bromine and krypton. Each of the first two rows, or periods as they are called—Li to Ne and Na to Ar—has eight elements, from reactive metal to inert gas. But the next row, instead of eight elements, has eighteen from the reactive metal potassium, K, to the inert gas krypton, Kr.

This does not mean that the periodic table has "broken down." It means that the principle of atomic structure that underlies the table is somewhat more complicated than one might suppose. For the regularity is continued, in a different way. Beneath the long row of eighteen elements, from K to Kr, is another row, of exactly eighteen elements again, from rubidium, Rb, to xenon, Xe, and this second row of eighteen elements matches the first. The members of each vertical pair of elements—K-Rb, Ca-Sr and so on through the row—show considerable similarities. The elements at the beginning of these two long periods resemble the elements at the beginning of the two short periods. Similarly, the long periods, like the short periods, end with a reactive nonmetal and an inert gas. It is rather more difficult to point to similarities between the "in-between" elements of the long periods and the elements of the two short periods. However, it should be noted that as we go from one element to the next in each long period, there is a steady gradation in properties.

It will be noted that group VIII is located in the middle of each long period. It differs from the others in that it contains three-element groupings. In the first long period, these elements are iron, Fe, cobalt, Co, and nickel, Ni, all with very similar properties. Immediately below these elements, we find the similar metals ruthenium, Ru, rhodium, Rh, and palladium, Pd. The third long period also has such a grouping of similar metals—osmium, Os, iridium, Ir, and platinum, Pt.

The long periods of eighteen elements each, as shown in the table on page 33, make for a rather cumbersome arrangement. The table on page 35, representing the short form of the periodic table, "doubles up" the long periods and hence is somewhat more convenient. The elements from K to Mn are written under the first seven groups in the short periods. You will note that "group" in the periodic table always refers to a vertical row; "period," to a horizontal one. At the right of these seven elements are the first three elements of group VIII—iron, Fe, cobalt, Co, and nickel, Ni. Then we start again with copper, Cu, under group I, zinc, Zn, under group II and so on up to bromine, Br, under group VII, following which we have the inert gas krypton, Kr, in group O. In this arrangement each group, beginning with the first long period, is divided into subgroups *a* and *b*.

In the short form of the periodic table, group I begins with lithium, Li, and sodium, Na, in the short periods. Then, going down the column in subgroup *a*, we find potassium, K, rubidium, Rb, and cesium, Cs, which bear a strong resemblance to lithium and sodium. In subgroup *b* of group I, we have copper, Cu, silver, Ag, and gold, Au. These are rather unlike lithium and sodium, for instead of being reactive metals they are conspicuously unreactive. Their only marked point of resemblance to lithium and sodium is that, like the latter elements, they usually have valence +1. In group I, therefore, the elements of the two short periods, Li and Na, are more closely akin to those of subgroup *a* than to those of subgroup *b*.

The same is true in the case of group VII. Reading down the column in this group, we have fluorine, F, chlorine, Cl, and, in subgroup VII*a*, bromine, Br, and iodine, I, which are all strikingly similar. In subgroup VII*b*, we have manganese, Mn, technetium, Tc, and rhenium, Re. These are metals and, generally speaking, unlike the nonmetals of VII*a*; yet they show certain resemblances in chemical behavior.

In groups I and II, then, the first two elements, reading downward, resemble the elements in subgroup *a* rather than those in subgroup *b*. At the other end of the periodic table, too, in groups VI and VII, the short-period elements resemble those in subgroup *a* rather than those in *b*. As we approach the middle of the table, it becomes harder to say which subgroup is more closely akin to the elements of a given short period. This is particularly true in the case of group IV.

SHORT SHORT PERIOD

What about the first two elements of the periodic table, hydrogen, H, and helium, He, which we bypassed before, when we began our discussion of the two short periods? We find that hydrogen and helium make up what might be called a "short-short" period. Hydrogen usually has a va-

lence of +1, as do the elements in group I. Otherwise, however, it does not resemble these elements, for it is not a metal but a gas. As it sometimes has a valence of −1, it can be said to have a certain kinship with the gaseous nonmetals of group VII, which also have a valence of −1. In the tables on pages 33 and 35, we have linked hydrogen to both group I and group VII, indicating that it has similarities with both groups.

Like the other periods—short and long—the "short-short" period ends with an inert gas. This is helium, He.

We now have, for the successive periods of the table, two elements ("short-short" period), eight elements (first short period), eight (second short period), eighteen (first long period), eighteen (second long period),—and then we are confronted with still another complication.

OTHER COMPLICATIONS ARISE

The third long period, as written out in the long form of the table on page 33, shows an asterisk after the *lanthanides*. Following the corresponding asterisk at the bottom of the table, there are the symbols of a whole row of fifteen elements, from lanthanum, La, to lutetium, Lu. These are the lanthanides, or rare-earth elements. The name "rare earths" is sometimes applied to the fourteen elements that follow lanthanum in the periodic table. After lutetium, Lu, the third long period goes on just like the first two long periods, and finally ends with the inert gas radon, Rn. Altogether, there are thirty-two elements in the period.

After radon, another long period—the fourth—begins. A double asterisk is affixed to the actinides. The corresponding double asterisk at the bottom of the table is followed by a number of elements called actinides or elements of the *actinium* series. They are all radioactive.

How can we explain the existence of the lanthanides and actinides in terms of atomic structure? We have already pointed out that the order in which the elements are arranged in the periodic table is that of their atomic numbers. The atomic number of an element is the number of positive charges in

PERIODIC TABLE- LONG FORM

"SHORT-SHORT" PERIOD

FIRST SHORT PERIOD

SECOND SHORT PERIOD

	Group I	Group II	Group III	Group IV	Group V	Group VI	Group VII	Group 0
								2 He 4
First short period	3 Li 7	4 Be 9	5 B 11	6 C 12	7 N 14	8 O 16	9 F 19	10 Ne 20
Second short period	11 Na 23	12 Mg 24	13 Al 27	14 Si 28	15 P 31	16 S 32	17 Cl 35	18 Ar 40

1
H 1

	Group I a	Group II a	Group IIIb	Group IVb	Group Vb	Group VIb	Group VIIb	Group VIII			Group I b	Group II b	Group IIIa	Group IVa	Group Va	Group VIa	Group VIIa	Group 0
FIRST LONG PERIOD	19 K 39	20 Ca 40	21 Sc 45	22 Ti 48	23 V 51	24 Cr 52	25 Mn 55	26 Fe 56	27 Co 59	28 Ni 59	29 Cu 64	30 Zn 65	31 Ga 70	32 Ge 73	33 As 75	34 Se 79	35 Br 80	36 Kr 84
SECOND LONG PERIOD	37 Rb 85	38 Sr 88	39 Y 89	40 Zr 91	41 Nb 93	42 Mo 96	43 Tc 99	44 Ru 101	45 Rh 103	46 Pd 106	47 Ag 108	48 Cd 112	49 In 115	50 Sn 119	51 Sb 122	52 Te 128	53 I 127	54 Xe 131
THIRD LONG PERIOD	55 Cs 133	56 Ba 137	57-71* Lanthanides	72 Hf 178	73 Ta 181	74 W 184	75 Re 186	76 Os 190	77 Ir 192	78 Pt 195	79 Au 197	80 Hg 201	81 Tl 204	82 Pb 207	83 Bi 209	84 Po 210	85 At 210	86 Rn 222
FOURTH LONG PERIOD	87 Fr 223	88 Ra 226	89-103** Actinides	104*** Transac- tinides														

*LANTHANIDES OR RARE-EARTHS

57 La 139	58 Ce 140	59 Pr 141	60 Nd 144	61 Pm 145	62 Sm 150	63 Eu 152	64 Gd 157	65 Tb 159	66 Dy 163	67 Ho 165	68 Er 167	69 Tm 169	70 Yb 173	71 Lu 175

**ACTINIDES

89 Ac 227	90 Th 232	91 Pa 231	92 U 238	93 Np 237	94 Pu 244	95 Am 243	96 Cm 247	97 Bk 247	98 Cf 251	99 Es 254	100 Fm 257	101 Md 256	102 No 254	103 Lr 260

***TRANSACTINIDES

104 Rf 257	105 Ha 260	(106)												

the nucleus, which is equal to the number of the electrons that surround the nucleus. In general, the higher the atomic number, the higher the atomic weight. There are a few exceptions, however. For example, argon, Ar, with atomic number 18 and atomic weight 40, is followed by potassium, K, with atomic weight 39. Cobalt, Co, with atomic number 27 and atomic weight 58.9, is followed by nickel, Ni, with atomic weight 58.7.

As we go from the lightest element, hydrogen, and increase the atomic number, the added electrons arrange themselves in shells. The first shell can hold only two electrons and is completed with the second element, helium. A completed shell always gives an inert gas. The second shell can hold eight electrons, and as this shell is successively filled, we get the eight elements in the first short period. The electrons in the outer shell are the ones that contribute to the chemical properties. The successive electrons in the first short period account for the regular change in valence, from $+1$ for lithium, up to ± 4 for carbon, and then down to -1 for fluorine.

As more and more electrons are added to the third shell, we get the eight elements of the second short period, ending with the inert gas argon, Ar. We now enter upon the first long period. More electrons are added to the third shell and only a few to the fourth shell. The inert gas krypton, Kr, marks the end of the first long period. In the second long period, from rubidium, Rb, through xenon, Xe, more electrons are added to the fourth shell and only a few to the fifth shell. The fourth shell takes on still more electrons in the third long period, and this accounts for the insertion of the fourteen elements (from cerium, Ce, through lutetium, Lu) that follow lanthanum, forming the lanthanide series.

The elements of the actinide series are also the result of adding electrons in an inner shell. After the actinides come the "transactinide" elements, beginning with 104. This is marked with a triple asterisk in the tables and is then shown at the bottom as the start of an entire new series of possible elements after 104.

Most known elements coming after uranium (atomic number 92) are man-made and unstable, subject to decay into other elements. However, it is possible that some heavy elements far beyond uranium may be stable and may possibly occur naturally.

TABLE IS INDISPENSABLE

Such then is the periodic table of the chemical elements. It fills an urgent need, for without some such classification it would be next to impossible to have a clear understanding of the more than one hundred elements that are now known. The chemist, in particular, finds the periodic table indispensable. Among other things, since it is a natural classification, based on definite properties of the elements, it can be used by the chemist to point the way to discoveries that can be confirmed in the laboratory.

A striking example of how the chemist can use the periodic table is afforded by the discovery of the Freons, widely used and completely nontoxic refrigerating fluids. The distinguished U.S. research chemist Thomas Midgley, Jr., noted that the older refrigerants, carbon dioxide, CO_2, ammonia, NH_3, sulfur dioxide, SO_2, and methyl chloride, CH_3Cl, all contained elements that were clustered in the upper part of the periodic table and more or less to the right. In search of a new refrigerant, he decided to experiment with fluorine, F, which is at the extreme right-hand corner of the table (not counting group O, which is composed of inert gases). Fluorine itself is extremely poisonous, but many of its compounds are not as dangerous.

Midgley succeeded in synthesizing a number of nontoxic compounds, all containing fluorine and with boiling points in just the right range to make excellent refrigerants. These compounds were given the name of Freons. The most widely used is CCl_2F_2, which has the chemical name dichlorodifluoromethane. It represents the methane molecule, CH_4, with two chlorine atoms replacing two of the hydrogen atoms of methane and two fluorine atoms replacing the other hydrogens.

PERIODIC TABLE-SHORT FORM

"SHORT-SHORT" PERIOD

	Group I	Group II	Group III	Group IV	Group V	Group VI	Group VII	Group VIII	Group 0
									2 He 4
FIRST SHORT PERIOD	3 Li 7	4 Be 9	5 B 11	6 C 12	7 N 14	8 O 16	9 F 19		10 Ne 20
SECOND SHORT PERIOD	11 Na 23	12 Mg 24	13 Al 27	14 Si 28	15 P 31	16 S 32	17 Cl 35		18 Ar 40
	Subgroup a / b	Subgroup a / b	Subgroup a / b	Subgroup a / b	Subgroup a / b	Subgroup a / b	Subgroup a / b		
FIRST LONG PERIOD	19 K 39 / 29 Cu 64	20 Ca 40 / 30 Zn 65	21 Sc 45 / 31 Ga 70	22 Ti 48 / 32 Ge 73	23 V 51 / 33 As 75	24 Cr 52 / 34 Se 79	25 Mn 55 / 35 Br 80	26 Fe 56 · 27 Co 59 · 28 Ni 59	36 Kr 84
SECOND LONG PERIOD	37 Rb 85 / 47 Ag 108	38 Sr 88 / 48 Cd 112	39 Y 89 / 49 In 115	40 Zr 91 / 50 Sn 119	41 Nb 93 / 51 Sb 122	42 Mo 96 / 52 Te 128	43 Tc 99 / 53 I 127	44 Ru 101 · 45 Rh 103 · 46 Pd 106	54 Xe 131
THIRD LONG PERIOD	55 Cs 133 / 79 Au 197	56 Ba 137 / 80 Hg 201	57-71* Lanthanides / 81 Tl 204	72 Hf 178 / 82 Pb 207	73 Ta 181 / 83 Bi 209	74 W 184 / 84 Po 209	75 Re 186 / 85 At 210	76 Os 190 · 77 Ir 192 · 78 Pt 195	86 Rn 222
FOURTH LONG PERIOD	87 Fr 223	88 Ra 226	89-103** Actinides	104-*** Transactinides					

1 H 1

*LANTHANIDES OR RARE-EARTHS

57 La 139	58 Ce 140	59 Pr 141	60 Nd 144	61 Pm 145	62 Sm 150	63 Eu 152	64 Gd 157	65 Tb 159	66 Dy 163	67 Ho 165	68 Er 167	69 Tm 169	70 Yb 173	71 Lu 175

**ACTINIDES

89 Ac 227	90 Th 232	91 Pa 231	92 U 238	93 Np 237	94 Pu 244	95 Am 243	96 Cm 247	97 Bk 247	98 Cf 251	99 Es 254	100 Fm 257	101 Md 256	102 No 254	103 Lr 260

***TRANSACTINIDES

104 Rf 257	105 Ha 261	(106)

THE CHEMICAL ELEMENTS

This article is a brief *Who's Who* of the chemical elements. Elsewhere you can read about how the elements fit into the periodic table and how they are related chemically. Here you are provided with a listing of their basic physical characteristics and their importance.

Each entry gives the element's name and the letter symbol that scientists use to represent it. Next comes the atomic number of the element—the number of protons that an atom of the element contains. The atomic mass, or weight, of the element is the number of protons and neutrons in one of its atoms. This number is a relative weight, useful to scientists. It is based on carbon-12, the most common carbon isotope. Its weight is taken as 12. All other atomic masses are relative to this number.

The element is next identified as a metal, nonmetal, or semi-metal. Where appropriate, melting and boiling points are given. The year when the element was identified, and by whom, follows this information. Thereafter are listed the abundance and sources of the element and— again, where appropriate—the uses of the element and its compounds and their biological significance.

ACTINIUM (Ac): Atomic number 89; atomic mass 227. Radioactive metal. Silver-white. m.p. 1,050°C; b.p. (est.) 3,200°C. Discovered in 1899 by A. Debierne.
Sources: Small amounts in uranium. Neutron bombardment of radium.
Uses: Actinium-225 is used as a radioactive tracer.
ALUMINUM (Al): Atomic number 13; atomic mass 26.98. Metal. Silver-gray. m.p. 659.7°C; b.p. 2,450°C. Believed isolated in 1825 by H. C. Oersted. Definitely recognized by F. Wöhler in 1827.
Sources: Not found free. Chief ore is bauxite.
Uses: Electrical wiring and equipment; structure of planes, autos, and other vehicles; and in various household products. Compounds used in medicine, water purification, and photography, and as ingredients of paints, dyes, abrasives, and synthetic gems.
AMERICIUM (Am): Atomic number 95; atomic mass 243. Radioactive metal. Silver-white. m.p. 100°C; Discovered in 1944 by G. Seaborg and associates.

Sources: Neutron bombardment of plutonium.
Uses: Radiography and research.
ANTIMONY (Sb): Atomic number 51; atomic mass 121.75. Metal. Silver-white. m.p. 630.5°C; b.p. 1,635°C. Discovered in 1604 by B. Valentine.
Sources: Sometimes found free, but chiefly as sulfide.
Uses: In alloys to increase strength and hardness. In chlorides, sulfides, and oxides for industry.
ARGON (Ar): Atomic number 18; atomic mass 39.95. Nonmetal; gas. f.p. −189.2°C; b.p. −185.7°C. Discovered in 1894 by W. Ramsay and J. Rayleigh. Makes up 0.94% of earth's atmosphere.
Sources: Liquid air; radioactive decay of potassium-40.
Uses: Specialized light bulbs; shield in arc welding; refining of certain metals.
ARSENIC (As): Atomic number 33; atomic mass 74.9216. Semi-metallic. Dark gray. m.p. 814°C. Discovered in 1250 by Albertus Magnus.
Sources: Realgar and orpiment; arsenates.
Uses: Insecticides, fireworks, and solar batteries; bronzing.
ASTATINE (At): Atomic number 85; atomic mass 211. Radioactive nonmetal. Synthesized in 1940 by E. Segré and associates. Approximately 27 grams occur in earth's crust.
Sources: Alpha bombardment of bismuth.
Biological significance: Concentrates in thyroid gland.
BARIUM (Ba): Atomic number 56; atomic mass 137.34. Metal. Silver-white. m.p. 850°C; b.p. 1,140°C. Discovered in 1808 by H. Davy.
Sources: Common in earth's crust; small amounts in seawater. Not found free.
Uses: Medicine, glassmaking, bleaching; ingredient in fireworks, fillers, insecticides, luminous paints.
Biological significance: Minute amounts in organisms.
BERKELIUM (Bk): Atomic number 97; atomic mass 243. Radioactive metal. Discovered in 1950 by S. Thompson and associates.
Sources: Irradiation of americium-241.
BERYLLIUM (Be): Atomic number 4; atomic mass 9.0122. Metal. Gray. m.p. 1,278°C; b.p. 2,970°C. Discovered in 1828 by F. Wöhler and A. Bussy.
Sources: Rare in earth's crust. Chief minerals are beryl, chrysoberyl, and phenakite.
Uses: Neutron absorption in nuclear reactors; structural material in space technology; X-ray technology.
BISMUTH (Bi): Atomic number 83; atomic mass 208.980. Metal. m.p. 271.3°C; b.p. 1,560°C. Discovered in 1753 by C. Geoffroy.
Sources: Found free. Common ore is bismuthinite. By-product of lead smelting.
Uses: Low-melting alloy for fuses and automatic sprinkler systems. In medicine for X-ray treatments, wound dressings, treatment of syphilis.
BORON (B): Atomic number 5; atomic mass 10.811. Nonmetal. m.p. 2,300°C; b.p. 2,550°C. Discovered

in 1808 by H. Davy and by L. Gay-Lussac and L. Thenard.

Sources: Not found free. Principle ores are kernite, borax, colemanite, ulexite.

Uses: Nuclear reactors, solar cells. Some compounds used in medicine.

BROMINE (Br): Atomic number 35; atomic mass 79.909. Nonmetal; liquid. Deep red. f.p. −7.2°C; b.p. 58.78°C. Discovered in 1826 by A. Balard.

Sources: Brines, salt deposits, and seawater.

Uses: Photography; medicine; antiknock gasoline.

CADMIUM (Cd): Atomic number 48; atomic mass 112.40. Metal. Bluish-white. m.p. 320.9°C; b.p. 767°C. Discovered in 1817 by F. Stromeyer.

Sources: Greenockite; zinc, copper, and lead ores.

Uses: Neutron absorption in atomic reactors; electroplating; medicine; in anticorrosion coatings, bearings; low-melting alloys; solders; dyes and pigments; fluorescent screens; batteries and cells; vapor lamps; and semiconductors.

Biological significance: Minute traces in organisms.

CALCIUM (Ca): Atomic number 20; atomic mass 40.08. Metal. m.p. 842°C; b.p. 1,480°C. Discovered in 1808 by H. Davy and independently by J. Berzelius and M. Pontin.

Sources: Not found free. Found in limestone, phosphate, gypsum, and fluorspar.

Uses: With its compounds, used in metallurgy; fixing nitrogen from air; bleaching; deodorizing; manufacture of lime, mortar, concrete, cement, bricks, glass, paint, and so on; production of acetylene gas; in fertilizers, drugs and dentifrices.

Biological significance: Important constituent of living organisms, found especially in shells, bones, and teeth. Forms about 2% of human body.

CALIFORNIUM (Cf): Atomic number 98; atomic mass 245. Radioactive metal. Discovered in 1950 by S. Thompson and associates.

Sources: Alpha bombardment of curium-242.

CARBON (C): Atomic number 6; atomic mass 12.01115. Nonmetal. Charcoal form, black; graphite form, gray; diamond form, colorless if pure. m.p. 3,500°C; b.p. 4,200°C. Known since prehistoric times.

Sources: May occur free as graphite or diamond. Often combined with other elements in carbonates, petroleum, coal, asphalt, natural gas, and so on. May be produced artificially.

Uses: Elementary carbon used in pencils, cutting and grinding equipment, jewelry, electrodes, radiation shields, building materials, printing inks, fuels; purifiers and filters, catalysts, coke; in rubber manufacture. Radioactive carbon-14 used in dating ancient remains. Carbon compounds have vast number of uses — in medicine; and in foods, beverages, cleaners, fuels, lubricants, among others.

Biological significance: Basic constituent of living organisms. Forms 18% of human body.

CERIUM (Ce): Atomic number 58; atomic mass 140.12. Metal. Steel-gray. m.p. 795°C; b.p. 3,257°C. Discovered in 1803 by J. Berzelius and W. d'Hisinger.

Sources: Minerals orthite, cerite, samarskite.

Uses: As a catalyst, and in alloys for jet engines. Radioisotope used to fuel thermoelectric generators.

Chlorine is prepared industrially by electrolysis of some of its compounds and is used in the manufacture of many products.

Oxide used in incandescent gas mantles; sulfate in chemical analysis.

CESIUM (Cs): Atomic number 55; atomic mass 132.905. Metal. Silver-white. f.p. 28.5°C; b.p. 705°C. Discovered in 1860 by R. Bunsen and G. Kirchhoff.

Sources: Rare. Not found free. Chief minerals are lepidolite, pollucite, carnallite, and certain mineral waters.

Uses: Chiefly as oxygen remover in vacuum tubes. Also in advanced electrical generators and photoelectric cells and as catalyst. Possible fuel in proposed plasma-jet engines. Compounds used in medicine and brewing, and as ingredients in mineral waters.

CHLORINE (Cl): Atomic number 17; atomic mass 35.453. Nonmetal. Greenish-yellow gas. f.p. −103°C; b.p. −34.6°C. Discovered by K. Scheele in 1774.

Sources: Rarely found free. Forms many solid mineral compounds. Metal chlorides often derived by evaporation of seawater or from salt deposits or brines. Derived by electrolysis or oxidation of compounds.

Uses: Chlorine and compounds are used in bleaching, disinfection, and rubber, chemical, and paper manufacture; and in antiseptics, pesticides, herbicides, drugs, foods, solvents, cleaners, refrigerants, explosives, fireworks, matches, paints, plastics, textiles.

Biological significance: Common in living organisms. Forms 0.15% of the human body. Highly toxic.

CHROMIUM (Cr): Atomic number 24; atomic mass 51.996. Metal. Gray. m.p. 1,890°C; b.p. 2,200°C. Discovered in 1797 by N. Vauquelin.

Sources: Chiefly chrome iron ore.

Uses: In hardening steel and manufacture of stainless steel and plating; as a catalyst.

COBALT (Co): Atomic number 27; atomic mass 58.9332. Metal. Gray with reddish tinge. m.p. 1,495°C; b.p. 3,100°C. Discovered in 1735 by G. Brandt.

Sources: Certain ores, usually sulfide or arsenide.

Uses: Machine tools, surgical instruments, dies, and

Copper sulfate solution is sprayed on these grapes in a French vineyard. The chemical helps keep parasites from destroying the crops.

so on. Cobalt salts used to produce permanent blue colors in porcelain, glass, and enamel. Radioactive forms in medical diagnostics and cancer therapy.

Biological significance: Important trace element in soil and in animal nutrition.

COPPER (Cu): Atomic number 29; atomic mass 63.54. Metal. Reddish. m.p. 1,083°C; b.p. 2,595°C. Known since prehistoric times.

Sources: Scarce; rarely native. Easily obtained from numerous minerals and compounds that are locally abundant.

Uses: Forms numerous alloys, the most important being brasses, bronzes, and gold and silver alloys. Used in electroplating; and in piping and plumbing, electrical wires and fittings, electronics, coins, jewelry, kitchenware, and chemical and industrial equipment. Copper compounds are used in analytical chemistry and water purification; and as ingredients in insecticides, drugs, paints, pigments.

Biological significance: Traces in living organisms; important in blood of certain mollusks and crustaceans.

CURIUM (Cm): Atomic number 96; atomic mass 242. Radioactive metal. Silver-white. m.p. 1,340°C. Discovered in 1944 by G. Seaborg and associates.

Sources: Alpha bombardment of plutonium.

Uses: Thermoelectric power generation for instrument operation in remote locations.

DYSPROSIUM (Dy): Atomic number 66; atomic mass 162.500. Metal. m.p. 1,407°C; b.p. 2,330°C. Discovered in 1886 by F. Lecoq de Boisbaudran.

Sources: Not yet isolated in free state. Found in minerals with granite or pegmatite veins.

EINSTEINIUM (Es): Atomic number 99; atomic mass 254. Radioactive metal. Discovered by A. Ghiorso and associates in 1952 in dust and debris of H-bomb explosion at Eniwetok.

Sources: Nuclear fusion reactions; by irradiation of plutonium.

ERBIUM (Er): Atomic number 68; atomic mass 167.36. Metal. m.p. 1,522°C; b.p. 2,500°C (approx.). Discovered in 1843 by C. Mosander.

Sources: Minerals normally found in granite or in pegmatite veins.

Uses: Nuclear control; in special alloys and room-temperature lasers.

EUROPIUM (Eu): Atomic number 63; atomic mass 151.96. Metal. m.p. 826°C; b.p. 1,490°C (approx.). Discovered in 1896 by E. Demarçay.

Sources: Obtained from samarium and gadolinium oxides.

Uses: As neutron absorber in nuclear reactors; in manufacture of lasers and color television.

FERMIUM (Fm): Atomic number 100; atomic mass 254. Radioactive metal. Discovered by A. Ghiorso and associates in 1952 in debris of H-bomb explosion at Eniwetok. Synthesized in 1953.

Sources: Nuclear fusion reactors; neutron irradiation of plutonium.

FLUORINE (F): Atomic number 9; atomic mass 18.9984. Nonmetal; gas. f.p. −219°C; b.p. −188°C. discovered in 1771 by K. Scheele.

Sources: Chiefly in fluorspar and cryolite.

Uses: As oxidizer in rocket fuels. Compounds used in etching glass and in air-conditioning systems.

Biological significance: Can help prevent tooth decay when added to drinking water.

FRANCIUM (Fr): Atomic number 87; atomic mass 223 (most stable isotope). Radioactive metal. Discovered by M. Perey in 1936.

Sources: Decay product of actinium.

GADOLINIUM (Gd): Atomic number 64; atomic mass 157.25. Metal. m.p. 1,312°C; b.p. 3,000°C. Discovered in 1880 by J. C. de Marignac.

Sources: Mineral gadolinite.

GALLIUM (Ga): Atomic number 31; atomic mass 69.72. Metal. Steel-gray. m.p. 29.78°C; b.p. 2,403°C. Discovered in 1875 by F. Lecoq de Boisbaudran.

Sources: Many zinc blendes and bauxites.

Uses: In high-temperature thermometers and semiconductors.

GERMANIUM (Ge): Atomic number 32; atomic mass 72.59. Semimetallic. Gray-white. m.p. 958.5°C; b.p. 2,830°C. Discovered in 1886 by C. Winkler.

Sources: Generally rare in earth's crust and in the sea. Not found free. Occurs in various minerals, in ores of other metals, in ash of certain coals, and in flue dusts of refineries.

Uses: In transistors and other electronic elements, low-melting and other alloys, infrared detectors, spectroscopes, lenses, phosphors for lamps; as catalyst. Also used in antibiotics and plant stimulants.

GOLD (Au): Atomic number 79; atomic mass 196.967. Metal, m.p. 1,063°C; b.p. 2,600°C. Most ductile and malleable of all metals. Good conductor of heat and electricity. Chemically inactive; resists corrosion. Known since prehistoric times.

Sources: Generally rare. Often occurs free in stream deposits, quartz veins, or with pyrites. May be present in ores of silver, copper, lead, nickel, and tellurium.

Uses: Formerly used in coinage; still a standard of international exchange. Used today in jewelry; in art, dentistry, surgery; and in electronic devices, and coatings. Compounds useful in photography. Radioactive form used to treat tumors.

HAFNIUM (Hf): Atomic number 72; atomic mass 178.49. Metal. m.p. 2,150°C; b.p. 5,400°C. Discovered in 1923 by D. Coster and G. von Hevesy.
Sources: All zirconium minerals.
Uses: Control rods in water-cooled nuclear reactors; lightbulb filaments; electrodes.

HELIUM (He): Atomic number 2; atomic mass 4.0026. Nonmetal; gas. f.p. −272.2°C; (25 atm); b.p. −268.9°C. Discovered in 1868 by P. Janssen when solar spectrogram was taken during a solar eclipse.
Sources: Abundant in universe. Widely distributed on earth, though usually in very small amounts. Obtained from natural gas and many radioactive minerals.
Uses: Inflating balloons; chromatography; treatment of certain respiratory diseases. In mixture with oxygen as artificial atmosphere for divers and astronauts. In liquid form as superconductor.

HOLMIUM (Ho): Atomic number 67; atomic mass 164.93. Metal. m.p. 1,470°C; b.p. 2,720°C. Discovered in 1878 by J. Soret and M. Delafontaine while studying spectral absorption bands.
Sources: Chiefly monazite. Also found in gadolinite.

HYDROGEN (H): Atomic number 1; atomic mass 1.00797. Can exist as metal or nonmetal. f.p. −259.14°C; b.p. −252.8°C. Discovered in 1766 by H. Cavendish. Three isotopes known: normal, deuterium (heavy hydrogen), and tritium (superheavy hydrogen). Most abundant element in universe, present in various celestial bodies and in space. Forms only 0.001% of earth's atmosphere, concentrated in outermost fringes.

The inert gases have been put to many uses. Here, welding in a helium atmosphere.

Sources: Common in earth's crust. Obtained by numerous methods, especially electrolysis of water or solutions of hydrogen-bearing compounds; and by decomposition of water and other hydrogen-containing compounds.
Uses: Elementary hydrogen is used in nuclear research, welding, hydrogenation of fats and oils, production of hydrocarbons and other organic compounds, production of ammonia, reduction of ores, metallurgy, chemical industry; and in rocket fuels. Compounds have numerous uses.
Biological significance: Forms over 10% of human body. Abundant in all living matter. Chemically active, producing vast array of compounds: water, acids, bases, alcohols, hydrocarbons, carbohydrates, and many other organic compounds and food substances.

INDIUM (In): Atomic number 49; atomic mass 114.82. Metal. m.p. 156.4°C; b.p. 1,450°C. Discovered in 1863 by F. Reich and T. Richter.
Sources: Chiefly in zinc blendes, sometimes with pyrites and siderite.
Uses: Electroplating and manufacture of bearings, and as a radiation detector. Alloys used for jewelry and in dentistry.

IODINE (I): Atomic number 53; atomic mass 126.9044. Nonmetal. Solid form, gray-black; gas, violet. m.p. 113.5°C; b.p. 184.35°C. Discovered in 1811 by B. Courtois.
Sources: From brines in salt wells and old sea deposits; from its iodides by treatment with sulfuric acid and an oxidizing agent.
Uses: In medicine, to prevent goiter and on external wounds. Compounds used in the chemical industry, sometimes in photography.
Biological significance: Concentrates in the thyroid gland of vertebrates.

IRIDIUM (Ir): Atomic number 77; atomic mass 192.2. Metal. White. m.p. 2,454°C; b.p. 4,500°C. Discovered in 1803 by S. Tennant.
Sources: Found free.
Uses: In high-temperature equipment. Alloyed with platinum for standard weights and measures; with osmium for pen tips and compass bearings and as a catalytic agent.

IRON (Fe): Atomic number 26; atomic mass 55.847. Metal. Silver-white. m.p. 1,535°C; b.p. 3,000°C (approx.). Known since prehistoric times.
Sources: In limited quantities in basalts and meteorites. Seldom free. Widely distributed in compounds. Oxides and taconite are usual ores. Obtained from ores by smelting. *Pure* iron is difficult to manufacture.
Uses: Ordinary iron and steels are fabricated into many products, such as construction elements, rails, auto and truck parts, ships, various wares, tools, and computer memory cores. Iron compounds are used in dyes, pigments, and medicines. Pure iron is used only in research and a few specialized applications.
Biological significance: Present in living tissues, especially vertebrate blood. Complex iron-containing organometallic compound—hemoglobin—carries oxygen to the cells.

KRYPTON (Kr): Atomic number 36; atomic mass

Iron and several other metals are used in the manufacture of steel. Metals are smelted and cast into usable forms in foundries such as this one.

83.80. Nonmetal; gas. f.p. −157.1°C; b.p. 152.9°C. Discovered in 1898 by W. Ramsay and M. Travers. One part krypton to every million parts of air.

Sources: Found with xenon in certain mineral springs; obtained from the atmosphere.

Uses: In incandescent bulbs, fluorescent tubes, lasers; and in high-speed photography.

LANTHANUM (La): Atomic number 57; atomic mass 138.91. Metal. m.p. 900°C; b.p. 3,454°C. Discovered in 1839 by C. Mosander.

Sources: Found in ores cerite, orthite, and monazite.

Uses: Metal alloys; rocket propellants.

LAWRENCIUM (Lr): Atomic number 103; atomic mass 260. Radioactive metal. Discovered in 1961 by A. Ghiorso and associates.

Sources: Bombardment of californium with boron nuclei.

Uses: Alpha emitter.

LEAD (Pb): Atomic number 82; atomic mass 207.19. Metal. Gray. m.p. 327.4°C; b.p. 1,620°C. Known since ancient times.

Sources: Scarce element; familiarity due to local concentration and easy workability. Chief ore is the sulfide galena. Also extracted from compounds.

Uses: In alloys such as brass, bronze, pewter, solder, fusible (low-melting) alloys, type metals, and bearing metals; as coating for less resistant metals; in piping, foil, roofing, sheathing, chemical equipment, radiation shielding. End product in radioactive series; thus used to date rocks and minerals. Compounds find wide application in glass and rubber manufacture, textile industry; in paints and pigments, batteries, munitions, gasoline additives, medicines, pesticides.

Biological significance: Highly toxic; affects digestive tract, muscular movement, and central nervous system.

LITHIUM (Li): Atomic number 3; atomic mass 6.941. Metal. White. m.p. 179°C; b.p. 1,317°C. Discovered in 1817 by A. Arfvedson.

Sources: Very abundant in universe, fairly abundant in earth's crust; scarcer in seawater. Not found free. Widespread in numerous minerals, mineral springs, and brines.

Uses: Metal and its alloys and compounds used in infrared and ultraviolet transmission, diamond synthesis; in ceramics, structural materials, purifiers, deoxidizers, dehumidifiers, catalysts, achromatic lenses, lubricants, electrical equipment; in medicine, including treatment of severe depression; source of tritium.

LUTETIUM ((Lu): Atomic number 71; atomic mass 174.97. Metal. m.p. 1,652°C; b.p. 3,327°C. Discovered in 1907 by G. Urbain and independently by C. von Welsbach.

Sources: Found in minerals gadolinite, euxenite, xenotime, and others.

Uses: Nuclear technology.

MAGNESIUM (Mg): Atomic number 12; atomic mass 24.305. Metal. White. m.p. 651°C; b.p. 1,107°C. Discovered in 1808 by H. Davy.

Sources: Third most abundant element by weight on earth. Chiefly concentrated below crust, much in seawater and in natural brines. Not found free. Forms numerous minerals and rocks, especially limestone, dolomite, magnesite, and various silicates. Commercially obtained by electrolysis of seawater and natural brines; also obtained from various solid minerals.

Uses: In alloy itself with other metals yields structural materials considerably lighter than aluminum and rivaling steel in strength, which are used in architecture; in aviation and automotive parts, bearings, die castings, kitchenware, and various other goods. Magnesium and its compounds are also used in metallurgy, organic chemistry, textile industry, and deoxidation and desulfurizing; in flares and lighting, munitions, refractory brick, ceramics, medicines, batteries, printing plates.

Biological significance: Important part of living tissues, especially chlorphyll in green plants.

MANGANESE (Mn): Atomic number 25; atomic mass 54.9380. Metal. Gray-white. m.p. 1,245,C; b.p. 2,097,C. Discovered in 1774 by D. Gahn.
Sources: Widely distributed in earth's crust. Principal source is the mineral pyrolusite.
Uses: Alloys with iron, copper, and nickel are important in industry. Compounds useful as antiseptics and in quantitative analysis. Pyrolusite gives color to glass.

MENDELEVIUM (Md): Atomic number 101; atomic mass 258 (most stable isotope). Radioactive metal. Chemical properties similar to those of thorium. Discovered in 1955 by A. Ghiorso and associates.
Sources: Bombardment of einsteinium with alpha particles.

MERCURY (Hg): Atomic number 80; atomic mass 200.59. Metal. Silver. f.p. $-38.87°C.$; b.p. 356.58°C. Known since ancient times.
Sources: Generally rare in earth's crust and in sea. Not found free. Chief ore is the sulfide cinnabar.
Uses: Mercury and its alloys (amalgams) and compounds have vast number of uses: in laboratory research; dentistry; illumination; leather and paper manufacture; treatment of gold and silver ores and of steel; textile chemistry; heating processes; and in electrical equipment, electronic devices, drugs, ammunition and detonators, paints, various measuring instruments, vacuum pumps, liquid sealing, catalysts, pigments, disinfectants, preservatives, pesticides and fungicides, bearing alloys, mirror coatings, and mercury turbines, motors, boilers, and atomic piles.
Biological significance: Element and most compounds are highly toxic, affecting the central nervous system.

MOLYBDENUM (Mo): Atomic number 42; atomic mass 95.94. Metal. Gray or black. m.p. 2,620°C; b.p. 5,560°C. Discovered in 1778 by K. Scheele.
Source: Not found free. Obtained from molybdenite and wulfenite.
Uses: Certain grades of tool steel and also in boiler plates and rifle barrels to increase toughness and tensile strength. Also used in radio grids, filaments, screens; and in dyeing glass, fur, and hair.

NEODYMIUM (Nd): Atomic number 60; atomic mass 144.24. Metal. Yellow. m.p. 1,029°C; b.p. 3,030°C. Discovered in 1885 by C. von Welsbach.
Sources: Cerite.
Uses: Alloys; coloring agent for glass used in lasers and astronomical lenses.

NEON (Ne): Atomic number 10; atomic mass 20.179. Nonmetal; gas. f.p. $-248.67°C$; b.p. $-245.9°C$. Discovered in 1898 by W. Ramsay and M. Travers. In atmosphere in proportion of 18 parts per million.
Sources: Obtained by liquefaction of air with separation by fractional distillation.
Uses: Electric signs and beacons (glows red-orange in a vacuum tube); lasers; in cryogenic research.

NEPTUNIUM (Np): Atomic number 93; atomic mass 237.0482. Radioactive metal. Silver-white. m.p. 640°C. Discovered in 1940 by E. McMillan and P. Abelson.
Sources: Found naturally in uranium ores; decay product of uranium-239. By-product in production of plutonium-239.
Uses: Neutron-detecting instruments.

NICKEL (Ni): Atomic number 28; atomic mass 58.71. Metal. White. m.p. 1,455°C; b.p. 2,900°C. Discovered in 1751 by A. Cronstedt.
Sources: Not scarce; inner core of earth may be nickel-iron alloy. Not usually found free. Chief minerals are chalcopyrite, pentlandite, and pyrrotite.
Uses: In many alloys, to which it imparts hardness, corrosion-resistance, electrical conductivity, and other properties. Used in chemical industry, manufacturing, and in plating, coinage, jewelry, kitchenware, batteries, electric and electronic equipment, wiring.

NIOBIUM (Nb): Atomic number 41; atomic mass 92.906. Metal. m.p. 2,468°C; b.p. 3,700°C. Discovered in 1801 by C. Hatchett.
Sources: Rare; found associated with iron in the mineral columbite.
Uses: Nuclear reactors and cryogenic research; in alloys in steel-making.

NITROGEN (N): Atomic number 7; atomic mass 14.0067. Nonmetal; gas. f.p. $-209.86°C$; b.p. $-195.8°$ C. Discovered in 1772 by D. Rutherford and by J. Priestley.
Sources: Abundant; forms 78% of atmosphere and also common in earth's crust, where it occurs in variety of minerals. Derived commercially by fractional distillation of liquid air and other methods.
Uses: Elementary nitrogen is used in production of ammonia; in steel hardening; supplying inert atmosphere for use in lamps, tubes, and various industrial processes; liquid form used as a cryogenic agent. Most familiar compounds are ammonia, saltpeter, niter, and nitric acid. These and other compounds used in explosives, fertilizers, cleaners, disinfectants, plastics, anesthetics, medicines, solvents.
Biological significance: Important constituent of organic matter, such as proteins and certain body wastes. Special kinds of soil and plant-root bacteria convert atmospheric nitrogen into nitrates and nitrites for plant use. Makes up 3% of human body.

NOBELIUM (No): Atomic number 102; atomic mass

A very thin section of tempered steel under the microscope. Many metals are analyzed in this way.

J. Cure

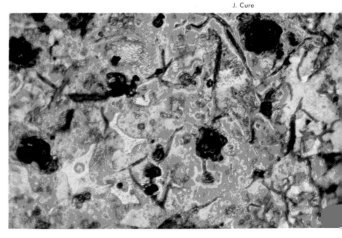

254 (most stable isotope). Radioactive metal. Discovered in 1958 by A. Ghiorso and associates.
Source: Bombarding curium with carbon-13 ions.

OSMIUM (Os): Atomic number 76; atomic mass 190.2. Metal. White m.p. 3,000°C; b.p. 5,500°C. Discovered in 1803 by S. Tennant.
Sources: Found in small amounts in platinum ores, but obtained chiefly from natural alloy osmiridium.
Uses: Alloy used for pen tips and machine bearings.

OXYGEN (O): Atomic number 8; atomic mass 15.9994. Nonmetal. Gas, colorless;liquid and solid, blue. f.p. −218°C; b.p. −183°C. Discovered in 1772 by K. Scheel. Makes up over 20% of atmosphere.
Sources: Abundant on the earth; forms 50% of earth's crust and 90% by weight of water. Also occurs in many solid minerals. Obtained by numerous methods.
Uses: For air supplies; for oxidizing in metallurgy and chemical industry; in disinfection; welding; gas manufacture. Liquid form (LOX) used in rocket fuels and explosives. Compounds numerous and useful.
Biological significance: Essential for higher forms of life. Makes up, on average, 20% of animal tissues and 40% of plant tissues. Constitutes 65% of human body, Supports animal breathing, but is actually poisonous in excess.

PALLADIUM (Pd): Atomic number 46; atomic mass 106.4. Metal. White. m.p. 1,554°C; b.p. 2,927°C. Discovered in 1803 by W. Wollaston.
Sources: Always found in platinum ores, often in nickel ores, sometimes in osmiridium and combined with gold.
Uses: In antimagnetic watches; parts of delicate balances; surgical instruments; as catalyst.

PHOSPHORUS (P): Atomic number 15; atomic mass 30.9738. Nonmetal. Yellow-white. m.p. 44.1°C; b.p. 280°C. Discovered in 1669 by H. Brand.

Neon and the other noble gases are components of our atmosphere. When separated from liquid air, they are put in tubes. When an electric current passes through the gas, it begins to glow.

R. Goorts — Air liquide

Sources: Not found free. Abundant in the earth's crust; found in seawater, combined with other elements in wide variety of minerals.
Use: Phosphorus and its compounds used in fertilizers; matches; munitions; fireworks; pesticides; metallic alloys; soaps; glass; chinaware; electroluminescent screens; water softeners; gasoline additives, and semiconductors; in medicine; and as radioactive tracers in industry and biology.
Biological significance: Essential to life. In animals, found in brain and nerve tissue, important in bone formation and muscle action; vital role in plant development.

PLATINUM (Pt): Atomic number 78; atomic mass 195.09. Metal. White. m.p. 1,769°C; b.p. 3,827 ± 100°C. Discovered in 1735 by A. de Ulloa.
Sources: Often found free. Comparatively rare and costly. By-product in electrolysis of nickel ores.
Uses: Excellent catalyst. Platinum and its alloys used in magnets, electrical and electonic apparatus, chemical, dental, and medical instruments, textile machinery, weights and measures, and X-ray apparatus; in jewelry making, electroplating, and high-temperature processes; and as resistant coatings.

PLUTONIUM (Pu): Atomic number 94; atomic mass 242 (most stable isotope). Radioactive metal. Discovered in 1940 by G. Seaborg and associates.
Sources: Bombardment of uranium with deuterons.
Uses: Nuclear-reactor fuel.

POLONIUM (Po): Atomic number 84; atomic mass 210. Radioactive metal. Discovered in 1898 by M. and P. Curie.
Sources: Present in all uranium minerals. Neutron bombardment of bismuth.
Uses: A source of nuclear power, of radiation, and of neutrons.

POTASSIUM (K): Atomic number 19; atomic mass 39.102. Metal. Silver-white. m.p. 62.3°C; b.p. 770°C. Discovered in 1807 by H. Davy.
Sources: Seventh most abundant element in earth, forming 2½% of crust and smaller percentage of seawater. Not found free. Compounds are widely distributed in form of clays, micas, and feldspars.
Uses: Widely used as chemical reagent and in photoelectric cells. Liquid alloy of potassium and sodium serves as a coolant and in certain thermometers. Compounds used in photography, chemical manufacturing, bleaching, tanning, dyeing, cooking; in fertilizers, soaps, glass, matches, explosives, medicines. Synthetic radioactive isotope potassium-42 used in medical tracer studies.
Biological significance: Essential to all living things. Composes about 0.35% of human body. Performs vital function of nerve-impulse conduction.

PRASEODYMIUM (Pr): Atomic number 59; atomic mass 140.907. Metal. Yellow. m.p. 930°C; b.p. 3,020°C. Discovered in 1885 by C. von Welsbach.
Sources: Found with other rare earths in monazite, cerite, and allanite.
Uses: In phosphors, lasers, and masers; as core material for carbon arcs; as coloring agent in glazes and glass; as catalyst.

PROMETHIUM (Pm): Atomic number 61; atomic

mass 145. Radioactive metal. Silver-white. m.p. 1,160°C. Discovered in 1945 by J. Marinsky and L. Glendenin.

Sources: Fission of uranium and neutron bombardment of neodymium.

Uses: In nuclear-power generators and special semiconductor batteries.

PROTACTINIUM (Pa): Atomic number 91; atomic mass 231.0359. Radioactive metal. White. Discovered in 1917 by O. Hahn and L. Meitner and independently by F. Soddy and associates.

Sources: Uranium ores; irradiation of thorium-230.

RADIUM (Ra): Atomic number 88; atomic mass 226.0254. Radioactive metal. White. m.p. 700°C; b.p. 1,140°C. Discovered in 1898 by M. and P. Curie.

Sources: Extremely rare. Not found free. Chief sources are carnotite and pitchblende.

Uses: Chiefly as source of radiation in medical treatment of malignant growths and in industry. Neutron source when combined with beryllium.

RADON (Rn): Atomic number 86; atomic mass 222.02. Radioactive nonmetal; gas. f.p. −71°C; b.p. −61.8°C. Discovered in 1900 by E. Dorn.

Sources: Radioactive decay product of radium.

Uses: As tracer in leak detection, and in flow-rate measurement; in radiography and medical research; in treatment of malignant growths.

RHENIUM (Re): Atomic number 75; atomic mass 186.2. Metal. m.p. 3,180°C; b.p. 5,630°C. Discovered in 1925 by W. Noddack and associates.

Sources: Minerals columbite, tantalite, and wolframite. Also found in molybdenite.

Uses: As additive to tungsten- and molybdenum-based alloys; in electronic filaments, electrical contact materials, high-temperature thermocouples, igniters for flashbulbs, refractory metal, components for missiles; as catalysts; in electroplating.

RHODIUM (Rh): Atomic number 45, atomic mass 102.905. Metal. Silver-white. m.p. 1,966°C; b.p. 4,500°C. Discovered in 1803 by W. Wollaston.

Sources: Found together with platinum. Native in certain river sands.

Uses: Alloyed with platinum for high-temperature thermocouples, furnace windings, laboratory crucibles, electrical contacts, jewelry, optical instrument mirrors; in electroplating; as a catalyst.

RUBIDIUM (Rb): Atomic number 37; atomic mass 85.47. Metal. White. m.p. 38.5°C; b.p. 700°C. Discovered in 1861 by R. Bunsen and G. Kirchhoff.

Sources: Rare, occurring in various minerals. Small amount found in seawater.

Uses: In photocells, vacuum tubes, catalysts, engines of space vehicles.

RUTHENIUM (Ru): Atomic number 44; atomic mass 101.07. Metal. Gray. m.p. 2,450°C; b.p. 2,700°C. Discovered in 1844 by K. Klaus.

Sources: Found native with other members of the platinum group.

Uses: Hardener for platinum and palladium in jewelry; in electrical contact alloys, dental and medical instruments; in electroplating; as an experimental nitrogen-fixer.

SAMARIUM (Sm): Atomic number 62; atomic mass

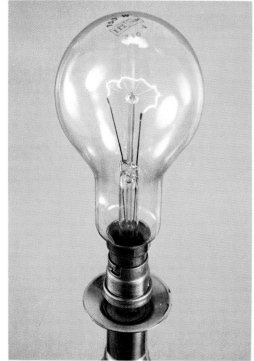

The chemical element tungsten is used in the manufacture of lightbulbs. Tungsten filaments are used in incandescent bulbs.

150.35. Radioactive metal. Gray-white. m.p. 1,072°C.; b.p. 1,900°C. Discovered in 1879 by F. Lecoq de Boisbaudran.

Sources: Found in very minute amounts in samarskite, cerite, and certain minerals occurring in Scandinavia.

Uses: As neutron absorber; in manufacture of laser crystals; in permanent-magnet research.

SCANDIUM (Sc): Atomic number 21; atomic mass 44.95. Metal. Silver-white. m.p. 1,539°C; b.p. 2,727°C. Discovered in 1879 by L. Nilson.

Sources: Obtained from various ores, including those of tin, tungsten, and beryllium.

SELENIUM (Se): Atomic number 34; atomic mass 78.96. Nonmetal. Red. m.p. 217°C; b.p. 685°C. Discovered in 1817 by J. Berzelius.

Source: Traces found in sulfur deposits.

Uses: Chiefly in glass and ceramic industries; also in photoelectric cells; as catalyst.

SILICON (Si): Atomic number 14; atomic mass 28.086. Nonmetal. Crystalline form, gray; amorphous form, brown; and a graphitelike form. m.p. 1,410°C; b.p. 2,355°C. Discovered in 1824 by J. Berzelius.

Sources: Second most abundant element in crust of earth; forms more than 25% of earth weight. Silica (silicon dioxide) is most abundant form. Present in vast array of rocks and minerals.

Uses: Elemental silicon is used primarily to make transistors and hard alloys with various metals, especially iron. Silicon carbide is a powerful abrasive. The oxide (in the form of quartz sand) and silicates find numerous applications, as in the manufacture of cement, concrete, ceramics, glass, silica gels, and water-glass. Silicones (organic silicon compounds) are plastics widely used as insulators, lubricants, binders, and cleansers.

Silicon also unites with hydrogen and the halogens to form industrially important substances.

Biological significance: Many lower animals secrete shells containing silicon.

SILVER (Ag): Atomic number 47; atomic mass 107.868. Metal. m.p. 960.5°C; b.p. 2,212°C. Known since ancient times.

Sources: Generally rare. Often free or combined in minerals and ores of gold, lead, copper, zinc, and nickel. Found in minerals argentite and cerargysite.

Uses: Once used chiefly for coinage and ornaments, now employed far more for industrial and related purposes, as the metal or in its various alloys and compounds; in medicine, dentistry, photography, steelmaking; in electrical and electronic devices, chemical and industrial equipment, mirrors, tableware, paints, solders, foil.

SODIUM (Na): Atomic number 11; atomic mass 22.9898. Metal. m.p. 97.5°C; b.p. 892°C. Discovered in 1807 by H. Davy.

Sources: Common element; sixth most abundant in earth's crust, of which it forms up to 3%. Very common in seawater. Not found free. Forms numerous mineral compounds: rock salt and other salts, feldspars, clays, amphiboles, hornblendes.

Uses: Element and its compounds are used in lamps, electric conductors, coolants, medicines, fertilizers, explosives, soaps and cleansers; in cookery, food manufacture, flavoring, and preservation, chemical production, papermaking, textile industry, petroleum industry, metallurgy, glassmaking.

Biological significance: Important in living matter. Forms 0.15% of human body. Vital in nerve impulse transmission.

Uranium is an important source of nuclear fuel for power plants. Here, the workmen examine uranium trioxide held in storage.

G. Hunter

STRONTIUM (Sr): Atomic number 38; atomic mass 87.62. Metal. Silver-white. m.p. 752°C; b.p. 1,380°C. Discovered in 1808 by H. Davy.

Sources: Common in earth's crust. Found in minerals strontianite, celestite, and brewsterite.

Uses: Metal itself used as oxygen eliminator in electron tubes and as fuel in some radioisotope-fueled thermoelectric generators. Compounds have numerous uses: in medicines, fluorescent lamps, luminous paints, fireworks and flares, lubricants, plastics; in optics, steelmaking, thickness gauges; glassmaking, chemical industry, sugar manufacture.

SULFUR (S): Atomic number 16; atomic mass 32.064. Nonmetal. Pale yellow. m.p. 112.8°C (rhombic); 119.0°C (monoclinic); b.p. 444.6°C. Known since ancient times.

Sources: Abundant in crust, more so in seawater. May occur free around volcanoes, hot springs, and salt deposits. Found in numerous minerals.

Uses: Sulfur and its compounds are useful in matches, explosives, pesticides, disinfectants, refrigerants, solvents, cleaners, bleaches, fertilizer, drugs, paints, dyes; in chemical analysis, rubber manufacture, photography, papermaking.

Biological significance: Essential component of living matter, especially proteins. Forms about 0.25% of human body.

TANTALUM (Ta): Atomic number 73; atomic mass 180.948. Metal. Blue-gray. m.p. 2,996°C; b.p. 5,425± 100°C. Discovered in 1802 by A. Ekeberg.

Sources: Not found free. Found chiefly in columbite or tantalite minerals.

Uses: As tough wire; in scientific, medical, dental, and chemical apparatus, cutting tooks, pen points, electrical and electronic devices, nuclear reactors, lenses, weights, aviation and spacecraft parts; and as catalyst in rubber manufacture.

TECHNETIUM (Tc): Atomic number 43, atomic mass 98.906. Metal. m.p. 2,200±50°C; b.p. 5,030°C. Discovered in 1937 by C. Perrier and E. Segre.

Sources: Bombardment of molybdenum with deuterons or neutrons. Uranium decay product.

Uses: Tracer studies.

TELLURIUM (Te): Atomic number 52; atomic mass 127.60. Nonmetal. White. m.p. 450°C; b.p. 990°C. Discovered in 1782 by F. Müller von Reichenstein.

Sources: Found in pure form and in compounds.

Uses: Coloring agent in glass. In certain alloys to give high electrical resistance.

TERBIUM (Tb): Atomic number 65; atomic mass 158.924. Metal. m.p. 1,356°C; b.p. 2,800°C. Discovered in 1843 by C. Mosander.

Sources: Reduction of the chloride with calcium.

Uses: Phosphor activator.

THALLIUM (Tl): Atomic number 81; atomic mass 204.37. Metal. Blue-white. m.p. 302°C; b.p. 1,457°C. Discovered in 1861 by W. Crookes.

Sources: Found in pyrites.

Uses: In alloys, photoelectric cells, rodenticides.

THORIUM (Th): Atomic number 90; atomic mass 232.038. Radioactive metal. m.p. 1,845°C. Discovered in 1828 by J. Berzelius.

Sources: Occurs in mineral thorite; obtained commercially from monazite sand.

Uses: Can be changed into a fissionable fuel in a breeder type of nuclear reactor. Thoria, the oxide, has been used in making incandescent glass mantles.

THULIUM (Tm): Atomic number 69; atomic mass 168.934. Radioactive metal. m.p. 1,550°C; b.p. 1,727°C. Discovered in 1879 by P. Cleve.

Sources: Occurs in very slight amounts in gadolinite, euxenite, xenotime, and others.

Uses: X-ray source.

TIN (Sn): Atomic number 50; atomic mass 118.69. Metal. Gray-white. m.p. 231.89°C; b.p. 2,260°C. Known since ancient times.

Sources: Rare, despite its seeming abundance and familiarity. Ore is usually cassiterite.

Uses: In familiar alloys brass, bronze, type metals, solder, pewter, fusible metals, bell metal. Tin and its compounds and alloys often used in foil, piping, plumbing, polishing powders; plating and coating.

TITANIUM (Ti): Atomic number 22; atomic mass 47.90. Metal. White. m.p. 1,675°C; b.p. 3,260°C. Discovered in 1791 by W. Gregor.

Sources: Found in igneous rocks and their sediments; as an oxide in titanates and many iron ores.

Uses: Steel alloys. Oxide used in white pigments.

TUNGSTEN (W): Atomic number 74; atomic mass 183.85. Metal. Gray or white. m.p. 3,370°C; b.p. 5,900°C. Discovered in 1783 by J. and F. de Elhuyar.

Sources: Generally rare. Obtained from minerals wolframite, scheelite, and others.

Uses: Metal and its alloys and compounds used in chemical industry and in laboratory, industrial, and medical equipment; cutting tools and abrasives; electrical devices; lamp, X-ray, and electronic elements; steel alloys; armor plate; munitions; needles.

URANIUM (U): Atomic number 92; atomic mass 238.93. Radioactive metal. m.p. 1,150°C; b.p. 3,818°C. Discovered in 1789 by M. Klaproth.

Sources: Occurs in a variety of minerals, especially pitchblende (uranite), carnotite, and other ores.

Uses: As nuclear fuel; in chemical analysis, photography; in pigments for glass and ceramics.

VANADIUM (V): Atomic number 23; atomic mass 50.942. Metal. m.p. 1,900±25°C; b.p. 3,000°C. Discovered in 1830 by N. Sefström.

Sources: Relatively rare, but found in minerals vanadinite and patronite.

Uses: In steel alloys, and as target material for X rays.

XENON (Xe): Atomic number 54; atomic mass 131.30. Nonmetal; gas. f.p. −112°C; b.p. −108.12°C. Discovered in 1898 by W. Ramsay and M. Travers.

Sources: Atmosphere; mineral springs.

Uses: High-speed photographic and vacuum tubes.

YTTERBIUM (Yb): Atomic number 70; atomic mass 173.04. Metal. m.p. 824°C; b.p. 1,427°C. Discovered in 1907 by G. Urbain and C. von Welsbach.

Sources: Gadolinite, xenotime, polycrase, and blomstrandine.

Uses: In lasers, as X-ray source.

YTTRIUM (Y): Atomic number 39; atomic mass 88.905. Metal. Gray. m.p. 1,490°C; b.p. 2,927°C. Discovered in 1794 by J. Gadolin.

Sources: Found in almost all rare-earth minerals.

Uses: In nuclear technology and in metal alloys.

ZINC (Zn): Atomic number 30; atomic mass 65.37. Metal. White. m.p. 419.5°C; b.p. 907°C. Known since ancient times.

Sources: Not too scarce in earth's crust; small amounts in seawater. Not found free; occurs in variety of minerals: calamine, sphalerite (chief ore), others.

Uses: In alloys brass, bronze, babbitt metals, solder, pewter, and others. Zinc and compounds used in coating metals, phosphors, medicines, and paint.

Biological significance: Vital to plant growth.

ZIRCONIUM (Zr): Atomic number 40; atomic mass 91.22. Metal. m.p. 1,857°C; b.p. 4,377°C. Discovered in 1789 by M. Klaproth.

Sources: Found in various igneous rocks; usually occurs as zirconium orthosilicate, or zircon.

Uses: Alloys used in steel manufacture; the oxides, in paints, lacquers, and abrasives. Zircon is a gemstone.

(ELEMENT 104): Atomic number 104; atomic mass 257 (uncertain). Radioactive metal. Discovered in 1969 by A. Ghiorso and associates and tentatively named rutherfordium. Soviet scientists proposed the name kurchatovium. Named unnilquadium in 1980.

(ELEMENT 105): Atomic number 105; atomic mass 260 (uncertain). Radioactive metal. Discovered in 1970 by A. Ghiorso and associates and tentatively named hahnium. Soviet scientists proposed the name nielsbohrium. Named unnilquintium in 1980.

(ELEMENT 106): Atomic number 106; atomic mass 263 (uncertain). Radioactive metal. Discovered in 1974 separately by U.S. and Soviet groups. Named unnilsexium in 1980.

(ELEMENT 107): Atomic number 107; three isotopes reported. Radioactive metal. A Soviet group led by G. N. Flerov reported isotopes of mass number 261 and 267 in 1976 and 1979, respectively; a group under P. J. Armbruster reported isotope 262 in 1981.

(ELEMENT 108): Atomic number 108; half-life of 2 ms. Radioactive metal. Synthesized and identified in 1984 in Darmstadt, Germany, by team led by P. J. Armbruster and G. Münzenberg. Unnamed as of yet.

The list of chemical elements is not complete. Physicists continue to turn up transuranium elements—elements with atomic numbers higher than uranium. These elements are radioactive and unstable. Most exist only briefly in laboratories, as, for example, element 109, a single atom of which was produced in 1982 and survived for less than five-thousandths of a second. In 1986, Soviet physicists claimed the discovery of element 110.

Scientists believe that elements of even higher atomic number might exist that have relatively long half-lives. Element 114 is one of the "superheavy" elements predicted to have such stability. It would appear in the periodic table in the same column as lead and should have similar properties.

INSIDE THE ATOM

For at least 2,500 years people have been puzzling about the nature of matter. We are still wondering. What are the tiniest particles to be found in matter? Is there a basic particle—a building block—so tiny that it cannot be broken up into still smaller bits?

In the fifth century B.C., the Greek philosopher Democritus suggested that all matter, of any kind whatsoever, can be ultimately broken down into tiny particles that cannot be further divided. These supposedly tiniest of all particles he called *atoms*. The word "atom" is a combination of two Greek words that mean "not" and "cut"; "atom" therefore means not capable of being cut. Democritus believed that, although the basic substance of all atoms was the same, atoms of different things—animals, say, or plants, or minerals—varied in size, shape, and weight.

At about the time that Democritus was introducing his theories about the atom, Empedocles, another Greek philosopher, was trying to explain the composition of matter in another way. He believed that all matter is made up of four primal, or basic, substances that he called "elements"—earth, air, fire, and water. Disagreeing with the atomic theory of Democritus, Empedocles held that there are four different kinds of atoms, corresponding to the elements.

Empedocles' idea received the support of the great Greek thinker Aristotle. Because of the prestige that Aristotle's support gave to Empedocles' theory, the ideas of Democritus and the other atomists of his time received scant attention for almost two thousand years.

In 1661, the British physicist and chemist Robert Boyle expressed, in *The Sceptical Chymist*, a surprisingly modern and correct conception of the elements. Boyle believed that the building blocks of most materials could be viewed as "certain primitive and simple, or perfectly unmingled, bodies." He called these bodies *elements*. Elements could be used to construct other substances, or *compounds*. Any compounds could ultimately be broken down into elements.

A number of scientists renewed their interest in Democritus' idea of the atom, but it was not until the nineteenth century that scientists began to devise experiments to determine the structure of the atom.

DALTON'S CONTRIBUTIONS

John Dalton, an English chemist and physicist, who lived from 1766 to 1844, was the first scientist to define the atom as we know it today—the smallest particle of an element that will behave chemically like that element. Dalton held that chemical reactions do not change the atoms themselves but only the way in which they are arranged. He thought that all the atoms of a given element are absolutely alike and differ from those of every other element. Today we know that atoms of the same ele-

Democritus believed that the basic substance of all atoms was the same, but that atoms differed in size, shape, and weight.

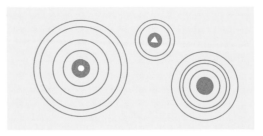

The English chemist John Dalton thought that the atoms of one element differed from those of another element.

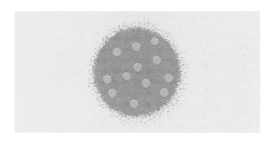

In the early 20th century J.J. Thomson proposed a model of atomic structure: negatively charged electrons embedded in a positively charged medium.

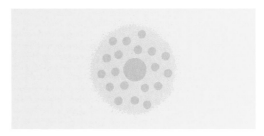

Lord Rutherford proposed that the positive charge is concentrated in a central core that is surrounded at some distance by an array of electrons.

ment may differ in weight, forming what are called *isotopes*. Dalton realized that atoms compose the larger particles of matter known as *molecules*, although he did not always distinguish very carefully between atoms and molecules.

A few scientists, however, wondered whether there might be even tinier particles.

THOMSON'S MODEL

Many years after Dalton's work, the English physicist J. J. Thomson began a series of cathode-tube experiments that changed scientists' thinking about the atom. In 1897, he reported that he had succeeded in tearing negatively electrified particles from atoms by applying electrical and magnetic forces. He came to the conclusion that "the atoms of the different chemical elements are different aggregations of ultimate particles of the same kind." Thomson called these particles corpuscles. Today we know them as *electrons*. They are essential parts of all atoms. Since the charge they carry is negative, we might speak of them as flecks of negative electricity.

By 1902 Thomson had proved that the electron is a universal and fundamental component of all matter. In 1906, Thomson advanced his theory of atomic structure. It showed the negatively-charged electrons embedded in a positively-charged medium.

LATER MODELS

Ernest Rutherford, a contemporary of Thomson's, proposed an alternate structure for the atom. Rutherford showed, indirectly, that the positive charge of the atom is con-

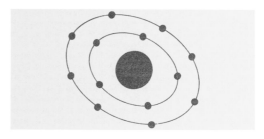

Later scientists developed an atomic model based on the idea that electrons encircled the nucleus in definite orbits, somewhat as the planets follow orbits around the sun.

centrated in a dense central core—the *nucleus*—and that electrons surrounded the nucleus at some distance away.

In 1913, the Danish physicist Niels Bohr analyzed the Rutherford model on the basis of later research. Bohr believed that the Rutherford model of the atom would be stable only if the electrons revolved around the atomic nucleus. Another scientist, Arnold Sommerfield, added that the orbit of the electrons would be elliptical.

The work of Rutherford, Bohr, and Sommerfield presented a neat package view of the atom. There are, however, problems with this view. Electrons do not behave as simple particles. Rather, they display characteristics known as wave properties. The Rutherford-Bohr-Sommerfield model could not account for this.

Several years later, based on the findings of several scientists, the German physicist Edwin Schrödinger proposed a different model of the atom—one that took into account the wave properties of electrons. According to this view, an atom has a positive charge located in a central region but the electrons do not occupy well-defined orbits around the central part. Rather, there is a "fuzzy" distribution of electrons around the nucleus. Only the probability of finding an electron at a certain distance from the nucleus can be determined, not the precise distance from the nucleus of any specific electron.

THE NUCLEUS

The nucleus, or center, of most atoms is very tiny, with a diameter of about 10^{-12} or 10^{-13} centimeters. The nuclei of all atoms—except hydrogen—have two major kinds of particles, known as *nucleons*. There are also many other kinds of particles to be found in atomic nuclei, some of which will be discussed later in the article, but it is the two major kinds that should be described first.

Electrons revolving around the nucleus at the same average distance are said to occupy the same shell. Below: the seven major shells. All of the shells except the K shell are also divided into two to four subshells.

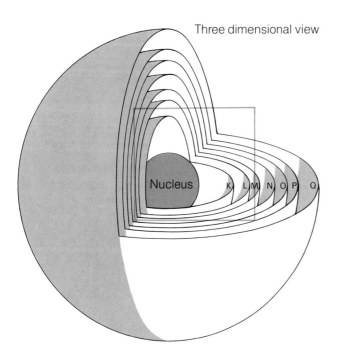

Three dimensional view

Nucleus   K L M N O P Q

One of the nuclear particles is the *proton*, which carries a positive charge. The mass of a proton is 1.007276 AMU. AMU stands for atomic mass unit. One AMU is equal to 1.6735×10^{-24} grams. A proton therefore weighs about 1.68567×10^{-24} grams.

The other nucleon is a *neutron*, which, as its name implies, carries no electrical charge. It is neutral. The mass of a neutron is 1.008665 AMU. It is slightly more massive than a proton.

In spite of the fact that the nucleons are so·tiny, most of the mass of an atom is in the nucleus. In one kilogram of any substance you can think of—iron, cork, gold, air—there are something like 600,000,000,000,000,000,000,000,000 nucleons. The protons and neutrons are held together within the nucleus by a force, known as the strong nuclear binding force. The overall shape of the nucleus is not spherical as was once thought. Rather it is not well-defined and is rather fuzzy.

ELECTRONS

The other important building block of the atom is the electron. It is much lighter than either the proton or the neutron. Its weight, or rather mass, is about 1,836 times less than that of the proton. The electron carries a negative charge that is equal to the charge of positive electricity carried by the proton.

SHELLS

Electrons revolving at the same average distance from the nucleus occupy the same *shell*. The hydrogen atom, with its lone electron, has only one shell, while the most complicated and heaviest of the atoms have seven shells. These shells are sometimes lettered or numbered for convenience. Chemists often refer to them, beginning with the one just outside the nucleus of the atom, as *K, L, M, N, O, P*, and *Q*. Only in atoms with very high atomic numbers do we find *P* and *Q*, however. The greatest number of electrons ever found in the inner shell is 2; in the second, 8; in the third, 18. The various shells, however, are not always filled to capacity.

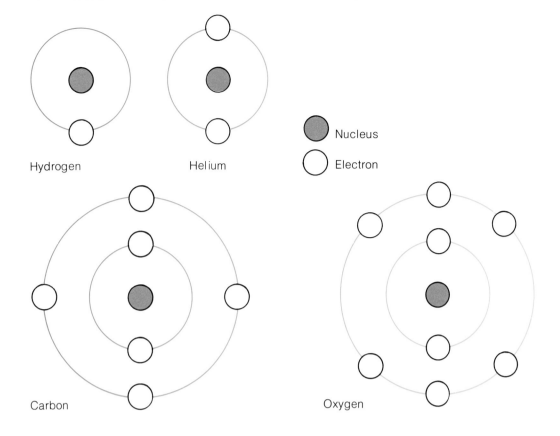

The basic structure of atoms of hydrogen, helium, carbon, and oxygen. Although it is now known that electrons do not occupy definite orbits as these models would imply, such representations provide chemists with a way to describe the properties of atoms and their valence, or ability to combine with other atoms.

Shells are not identical with the electron orbits, but rather are areas in which the orbits lie. With the exception of the innermost, or K, shell, the shells are divided into from two to four subshells. The L shell has two subshells, called s and $p;$ the M shell has three subshells — s, p, and $d;$ the other shells have four subshells — s, p, d, and f. As the atoms increase in mass, then, both shells and subshells increase in number and complexity.

Let us consider the shell arrangement of a few typical atoms. The single electron of the hydrogen atom moves within the limits of the first (K) shell. The helium atom has two electrons, both in the K shell. There are three electrons in the lithium atom. Two are in the K shell, one in the s subshell of the L shell. The neon atom has a total of ten electrons. It has two electrons in the K shell; the rest are in the L shell. The s subshell of the L shell contains two electrons; the p subshell, six.

The electrons in the outer shell are called *valence electrons*. *Valence* is a word that indicates the degree to which an atom can combine with others in chemical reactions. Only the electrons in the outer shell take part in these reactions.

ATOMIC WEIGHT AND NUMBER

Infinitesimally small as the atom is, it possesses weight. Of course we cannot weigh the atom directly. We measure atoms by indirect methods. We know, for example, that an oxygen atom weighs 0.000,000,000,000,000,000,000,026 grams. We call this the absolute weight of the oxygen atom, because we consider it by itself and not in relation to the weight of the other atoms.

The absolute weights of the various atoms are too small for practical use in chemical calculations. Hence, instead of concerning himself with absolute weights, the chemist uses a tried and true system of relative weights, called *atomic weights*. In this system we consider the weight of atom A only as it compares with the weight of atom B and atom C and all the rest.

We take an isotope of carbon as the basic unit of the system of atomic weights and give it the arbitrary value of 12.000. When we weigh equal volumes of carbon-12 and hydrogen, we find that the carbon is about 11.905 times as heavy as hydrogen. In our atomic scale, therefore, hydrogen has the value of 12.000 divided by 11.905, or about 1.008.

By appropriate methods, we can compute the ratios between the weights of all atoms. That is, we can determine the atomic weights of all elements. For example, the sulfur atom is 2.672 times heavier than the carbon atom. Its atomic weight, therefore, is 2.672 × 12, or 32.064.

Hydrogen is the lightest of the elements. Why do we not make the hydrogen atom the unit of our system of atomic weights? Why do we not give it the value 1.000 and rate the other relative weights accordingly? The answer is that, if we use carbon-12 as the unit, the relative weights of atoms are generally close to whole numbers. In many of our calculations involving atomic weights, we use the nearest whole number instead of using whole numbers plus fractions.

The weight of each atom is obviously the sum total of the weights of each of the subatomic particles (particles smaller than the atom) of which the atom consists. They are held within the atom by electrical, magnetic, and other kinds of attraction. When we say that they are held together, we do not mean that they are closely packed. On the contrary, they occupy very little of the space within the atom.

Atoms are listed, on the basis of their increasing complexity, by what is known as their *atomic number,* which is equal to the number of protons they possess. Hydrogen, the lightest of the elements, has atomic number 1; helium, the next heaviest, has atomic number 2; oxygen has atomic number 8; uranium, atomic number 92. In the majority of cases a higher atomic number goes hand in hand with increased weight. There are a few exceptions, however. For example, cobalt, with atomic number 27, is heavier than nickel, with atomic number 28.

Scientists have found that the atomic number of an element can tell them certain things about the properties of the element.

As we have pointed out, the number of

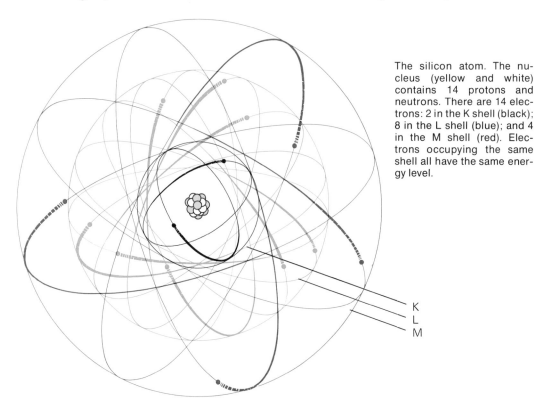

The silicon atom. The nucleus (yellow and white) contains 14 protons and neutrons. There are 14 electrons: 2 in the K shell (black); 8 in the L shell (blue); and 4 in the M shell (red). Electrons occupying the same shell all have the same energy level.

K
L
M

protons in an atomic nucleus is always equal to the atomic number. The number of neutrons in the nucleus can be determined by subtracting the atomic number from the atomic weight, reckoned in whole numbers.

OTHER SUBATOMIC PARTICLES

Protons, neutrons, and electrons are not by any means the only subatomic particles—that is to say, irreducible elementary particles smaller than the atom in size. Many more have been discovered and undoubtedly will be in the future. Several hundred kinds of subatomic particles are known today. These include different forms of a given particle. These forms differ in certain properties such as spin and charge.

A few of the subatomic particles, including electrons, protons, and neutrons, are stable. Most of them are unstable. They change, or "decay," into other particles of different types. Unstable particles are very short-lived. The positive *pi meson*, for example, has a life of 2.56×10^{-8} seconds. This means that, from the time it appears to the time it decays into other particles, it exists for $2.56 \times 1/100,000,000$ seconds. Other particles have even shorter lives.

The different particles are electrically positive $(+)$, negative $(-)$, or neutral (0). They generally show *spin*—that is, they rotate on their axes. Spin is one of their important identifying characteristics.

A remarkable feature of subatomic particles is that for each type there is, actually or theoretically, an "opposite number," or counter-form, known as an *antiparticle*. If a particle is positive or negative, its antiparticle has the opposite electric charge. The two differ in spin; that is, they rotate in opposite directions. In all other respects, they are the same. Certain neutral particles are identical with their antiparticles. When particles and their antiparticles collide, they are annihilated.

Perhaps the best-known antiparticle is the *positron*, or positive *antielectron*. Its existence was proved in 1932 by C. D. Anderson, of the California Institute of Technology. Since that time, a good deal of information about positrons has been acquired. It has been discovered, for exam-

LOSS OF MASS IN THE NUCLEUS OF AN ATOM

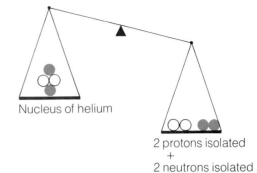

Nucleus of helium

2 protons isolated
+
2 neutrons isolated

Protons and neutrons have very little mass and some of that is lost when they join. The illustration above shows how two isolated protons and neutrons weigh more than two protons and neutrons combined. Some of their mass is lost as energy when they form a nucleus.

ple, that in some forms of radiation, the emitted particles are positrons.

Subatomic particles interact in several different ways. One type of interaction involves the light particles called *pi mesons*, or *pions*. It has been found that both protons and neutrons absorb and emit pi mesons rapidly and constantly. This so-called *pi-meson exchange* is the "cement" that holds together the nucleus.

There is also *electromagnetic interaction*, in which electrons and other light-mass particles emit and absorb *photons*. Electrons have several different energy states or *levels*. The less energy they possess, the lower their energy state. If they absorb photons, they jump to a higher energy state. When they return to the lower energy state, they emit photons. This interaction is weaker than interchange of pi mesons.

In a third kind of subatomic interaction, particles of light mass are emitted by slowly decaying heavier particles. Thus, when some of the heavy particles called *hyperons* decay, they emit photons. Others give off pi mesons. Pions ultimately decay into subatomic particles called *muons* (i.e., *mu mesons*) and *neutrinos*.

Our knowledge of the inside of the atom continues to grow almost daily through the research conducted in particle accelerators and colliders.

VALENCE

Everything in this world undergoes change. This is as true of living things as it is of nonliving things.

All changes may be divided into two classes—physical and chemical. *Physical changes* do not affect the molecules of which substances consist. For example, the molecule of water is made up of two atoms of hydrogen and one of oxygen. If water is heated and turned into steam or if it is frozen into ice, its molecule still consists of two hydrogen atoms and one oxygen atom. The steam will turn back into water when it cools. The ice will become water again when it melts. If we bend or crush a solid or grind it into powder, it undergoes only a physical change. Its molecules have not been affected.

A *chemical change* is entirely different. It involves the transformation of the molecules: the combination or the recombination of atoms. A different substance is formed. When the atoms of oxygen and hydrogen, both gases, combine chemically, they are transformed into water, a liquid. Mercuric oxide is a red solid. It is quite different from the liquid metal, mercury, and the colorless gas, oxygen, which make up its molecules. An even more startling transformation results when sodium, a highly reactive metal, and chlorine, a sharp-smelling, poisonous gas, combine chemically to form table salt.

COMBINING OF ATOMS

In this article we shall concern ourselves with chemical changes and particularly with the ways in which atoms combine to bring about such changes. At first glance there does not seem to be much rhyme or reason in these combinations. In a molecule of mercuric oxide (HgO), an atom of oxygen combines with an atom of mercury. The chemical formula HgO is made up of the symbols Hg, standing for "mercury," and O, standing for "oxygen." A small figure after and near the bottom of a symbol indicates the number of atoms. If no such figure is given, the number "one" is understood. For example, the formula for calcium carbonate—$CaCO_3$—shows that the calcium carbonate molecule is made up of one atom of calcium (Ca), one atom of carbon (C), and three atoms of oxygen (O).

The ammonia molecule (NH_3) consists of one atom of nitrogen plus three atoms of hydrogen. In ammonia, a single atom of one element has combined with three atoms of another. In the compound methane (CH_4), a carbon atom combines with four atoms of hydrogen. In other words, an atom may be linked, in a molecule, to a single atom or to two or more atoms.

A formula such as CH_4, which indicates the number of atoms of each kind in a molecule, is called a *molecular formula*. The name *structural formula* is given to any formula, such as those in Figure 1, that shows lines for the bonds

Fig. 1. Reading from left to right: the structural formulas for ammonia (NH_3), methane (CH_4), and benzene (C_6H_6).

Fig. 2. A diagram of an atom showing the seven shells that electrons can occupy around the nucleus, which itself is made up of particles.

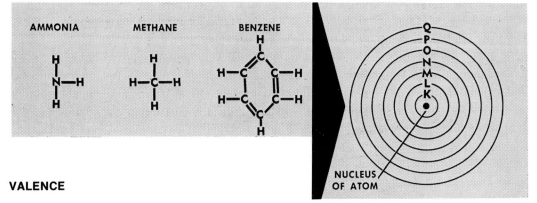

connecting one atom with another. In Figure 1, note that a nitrogen atom has three bonds (lines) and carbon four bonds, while hydrogen has only one. The number of bonds leading from an atom is called the *valence* of that atom. In the compounds shown in Figure 1, nitrogen has a valence of three, carbon a valence of four, and hydrogen only one. Valence, therefore, means the "combining power of an atom," expressed in terms of the number of hydrogen atoms with which it can combine. It is convenient to express valence in terms of hydrogen, since hydrogen always has a valence of one, and never more.

STRUCTURE OF AN ATOM

To understand why atoms combine, we must consider the structure of the atom. It consists of a core, or *nucleus,* made up of positive particles, called *protons,* neutral particles, known as *neutrons,* and other particles. The nucleus contains nearly all the weight of the atom. Revolving around the nucleus are a number of negatively charged particles, called *electrons.* They move around the atomic nucleus in a series of orbits, somewhat like a tiny version of the solar system, in which planets move around the sun.

Atoms are listed, on the basis of increasing complexity, by their *atomic number,* which always corresponds to the number of protons in the nucleus. In an electrically neutral atom, there are as many electrons orbiting around the nucleus as there are protons in the nucleus. Hydrogen, the lightest element, has atomic number 1. It has a single proton and a single electron. Oxygen, with atomic number 8, has eight protons and also eight electrons. The number of neutrons varies according to the atom.

Electrons revolving at the same average distance from the nucleus are said to occupy the same *shell.* A shell may contain from one to thirty-two electrons. The hydrogen atom, which has but one electron, has only one shell. Other atoms have two, or three or more; the most complicated ones have seven. Scientists refer to the seven shells, beginning with the one just outside of the nucleus, as the K, L, M, N, O, P, and Q shells (Figure 2). The maximum number of electrons in the K shell is two; in the L shell, eight; in the M shell, eighteen; and so on. The shells are not always filled to capacity. (In some cases the full quota of electrons for the M shell is eight.)

Let us consider the electron arrangement in a few representative atoms (Figure 3). Hydrogen has one electron, which is in the K shell. Helium has two electrons, filling the K shell to capacity. Boron has five electrons. Two of these fill the K shell; the other three are in the next shell —the L shell. This can hold eight electrons in all. Hence it is not filled in the case of boron; it lacks five electrons. Magnesium, with atomic number 12, has twelve electrons. Two of these fill the K shell, and eight the L shell. The remaining two are in the M shell. The electrons of the outer shell, in each case, are the only ones involved in chemical reactions.

Fig. 3. The electron arrangement in various atoms. The outer shells of hydrogen, boron, and magnesium are not filled. Whether atoms accept or donate electrons to form a complete outer shell is the basis of valence.

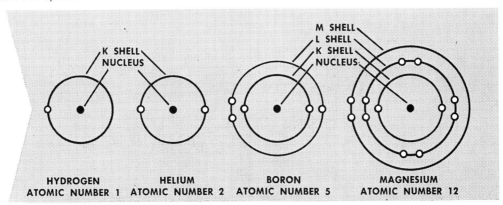

HYDROGEN
ATOMIC NUMBER 1

HELIUM
ATOMIC NUMBER 2

BORON
ATOMIC NUMBER 5

MAGNESIUM
ATOMIC NUMBER 12

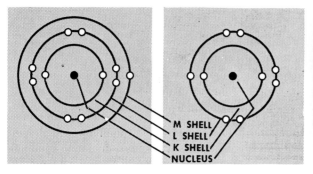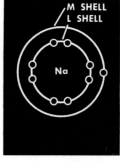

Fig. 4. This is the sodium atom, atomic number 11.
Fig. 5 This is the fluorine atom, atomic number 9.
These atoms are chemically active.

Fig. 6. This is a diagram of the two outer shells of sodium; the outermost shell has only one electron.
Fig. 7. This is the outermost shell of fluorine.

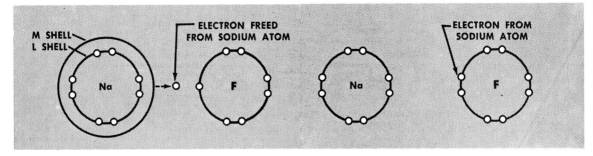

Fig. 8. The sodium atom (Na) shown above has lost the single electron in its outer (M) shell. It now has a total of ten electrons, instead of eleven.

Fig. 9. The electron from the M shell of the sodium atom has been taken up by the fluorine atom. This atom now has eight electrons in its outer shell.

THE INERT GASES

In certain atoms—helium, neon, argon, krypton, xenon, and radon, the so-called inert, or noble gases—the outer shells have their full quota of electrons. This is a very stable arrangement. It accounts for the fact that the inert gases almost never enter into chemical combinations. As for the other atoms, the outer shells are not filled to capacity. Such atoms can fill the full electron quota in their outer shells by gaining or losing electrons, or by sharing them. We say of such atoms that they are chemically active.

ELECTROVALENCE: ATOMS GAIN OR LOSE ELECTRONS

Let us see what happens when atoms combine by gaining or losing electrons—the type of chemical combination known as *electrovalence*.

The elements sodium and fluorine join together in this way. In Figure 4, we show an atom of sodium. It has eleven

positive charges on its nucleus, balanced by eleven electrons with one negative charge each. The innermost shell—the K shell—holds two electrons; the next, or L, shell holds eight. These shells are filled. The eleventh electron occupies the M shell all by itself.

Now, the nucleus attracts each electron with a force varying inversely as the square of the distance. This means that the two K electrons are held most tightly of all, while the lone electron in the M shell is attracted much less strongly than any of the others. It would not take much energy to remove this electron from the atom altogether.

Figure 5 shows the structure of a fluorine atom, with atomic number 9. It has nine electrons in all. The first two, as in the sodium atom, fill the K shell, closest to the nucleus. The other seven electrons go into the L shell; they do not quite fill it, since this shell can hold eight electrons.

Both the sodium and fluorine atoms are electrically balanced, with equal num-

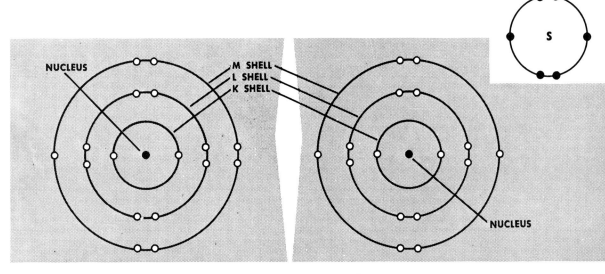

Fig. 10. In the diagram is shown the chlorine atom, atomic number 17. Chlorine has seven electrons in its valence, or outer, shell. Fig. 11. The sulfur atom is shown here. It is atomic number 16. In the inset at the upper right, only the valence shell is shown.

bers of positive and negative charges. They are capable of chemical activity because their outermost shells—the M shell of the sodium atom and the L shell of the fluorine atom—are not filled.

A chemist has various ways of representing atoms. Instead of showing the nucleus and all the shells of the sodium and fluorine atoms, as in Figures 4 and 5, he might show only the two outer shells of the sodium atom and the one outer shell of the fluorine atom, as in Figures 6 and 7. In

Figure 6, the dots represent the 8 electrons in the L shell and the one electron that is by itself in the M shell. The symbol for sodium Na, in the center, stands for the nucleus and the two electrons that are always in the K shell. The drawing of the fluorine atom in Figure 7 shows the seven electrons in the outer (L) shell. The symbol for fluorine, F, in the center represents the nucleus and the two electrons of the K shell.

The electron in the outer (M) shell of

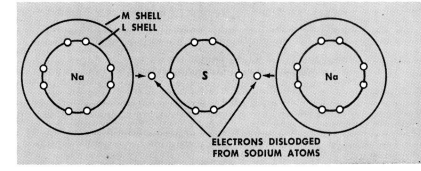

Fig. 12. Each of the two sodium atoms in the vicinity of the sulfur atom loses an electron in its outer (M) shell.

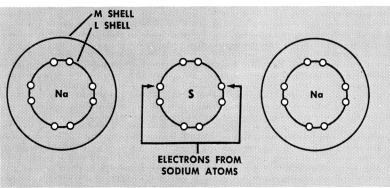

Fig. 13. The sulfur atom has captured an electron from each of the sodium atoms.

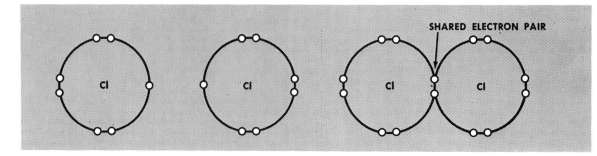

Fig. 14. Each of the chlorine atoms in this two-part diagram has seven electrons in the outer (M) shell. Fig. 15. The two atoms have combined in the diagram that is shown here. They now form a molecule by sharing a pair of electrons.

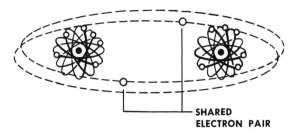

Fig. 16. These two chlorine atoms share a pair of electrons. Six inner electrons revolve around each of the nuclei. Two revolve around the nuclei of both atoms.

Fig. 17. Chlorine and fluorine atoms can also combine in the same manner as two chlorine atoms. This forms a compound called chlorine fluoride.

the sodium atom can be rather easily dislodged (Figure 8). As we have seen, it is comparatively far from the nucleus and is not attracted greatly by it. If the electron that a sodium atom has lost comes close to the fluorine atom, it will occupy the vacancy in the L shell (Figure 9).

An important transformation in the sodium and fluorine atoms has taken place. The sodium atom was electrically balanced formerly with eleven positive charges in the nucleus and eleven electrons. Now that it has lost an electron, it has eleven positive charges and only ten negative ones. Hence it has one net positive charge, indicated by a plus sign set to the right and near the top of the symbol for sodium: Na^+. The fluorine atom was also balanced to begin with, with its nine positive and nine negative charges. Since it has picked up an additional electron, it has ten negative charges as against nine positive ones. It has one net negative charge, indicated by

a minus sign: F^-. A charged atom, such as Na^+ or F^-, is called an *ion*. Ions differ greatly in their properties from the regular, uncharged atoms from which they have been derived.

We know that like electric charges repel, while unlike charges attract. The positive sodium ion and the negative fluorine ion will, therefore, attract one another. This *electrostatic attraction,* or *Coulomb attraction,* as it is called, is very strong. It is far more powerful than gravitational attraction. Hence, if there are a number of sodium and fluorine ions in a receptacle, say, they will be drawn together. They will arrange themselves so that each sodium ion will be as close to as many fluorine ions as possible, and each fluorine ion as close to as many sodium ions as possible.

Sodium chloride (NaCl), or table salt, is formed in the same way as sodium fluoride. Chlorine, atomic number 17, has complete K and L shells and a one-less-than-complete M shell (figure 10). It can take up an electron from a sodium atom to complete the outer shell, just as fluorine does.

Lithium, potassium, and cesium are metals that, like sodium, lose a single electron in combining chemically. Some metals (such as beryllium, magnesium, and calcium) lose two electrons; others (scandium, yttrium, barium), three; still others (titanium, zirconium, hafnium), four. When a magnesium atom, say, loses the two electrons of the outer shell and acquires two net positive charges, becoming an ion, two plus marks are added to its symbol: Mg^{++}.

We saw that the nonmetals fluorine and chlorine can each acquire a single electron to fill the outer shell. The nonmetal sulfur can take up two electrons when it combines with another atom. The sulfur atom is shown in Figure 11. You will note that its K and L shells are complete; its M shell contains six electrons and has room for two more.

Suppose that each of two sodium atoms in the vicinity of a sulfur atom loses the lone electron in its M shell (Figure 12). The sulfur atom can capture both of the free electrons (Figure 13). Each sodium atom, having lost an electron, now has a single net positive charge; it is now a sodium ion, Na^+. The sulfur atom has acquired two electrons. It now has a net negative charge of two; it has become a sulfur ion, S^{--}. The two positively charged sodium ions are attracted to the negatively charged sulfur ion, and the compound sodium sulfide, Na_2S, is the result. In this compound, sodium has a valence of one; sulfur, a valence of two.

An enormous number of compounds can be formed in the same way as those that we have just described—sodium fluoride (NaFl), sodium chloride (NaCl), and sodium sulfide (Na_2S). In each case an electron (or more than one electron) leaves one atom and is picked up by another atom; hence both atoms become electrically charged. Since they have different charges, they attract one another and combine. The resulting compound is called electrovalent. The bond that holds the ions together is known as *electrovalence* or *ionic valence*.

COVALENCE

Electrovalence is not the only kind of bond that unites atoms. Suppose two different atoms, such as those of chlorine or fluorine, have one vacancy in their outer shells. They can fill this vacancy by sharing a pair of electrons.

In Figure 14, we see two chlorine atoms, each with seven electrons in the M shell. They can combine as shown in Figure 15. Each atom now seems to have eight electrons in its outer shell, but only because two of the electrons are allowed to count twice, once for each atom. The orbits of the outer-shell electrons would look something like the one shown in Figure 16. You will note that six electrons are distributed around only one or the other of the two chlorine nuclei. Two electrons, however, are revolving around the nuclei of both of the chlorine atoms.

This type of bond is called *covalence*. Note that two electrons take part in one covalent bond. The two chlorine atoms linked together by a covalent bond in Figure 15 form the molecule of the gas chlorine (Cl_2). A chlorine atom and a fluorine atom can unite in a similar way. Both of these atoms, as we have seen, have seven atoms, out of a possible eight, in their outer shells. When they combine, as in Figure 17, they form the compound chlorine fluoride (ClF).

Nitrogen has a covalence of three: that is, it is bound to three atoms of the element with which it unites. Figure 18 shows a nitrogen atom and three hydrogen atoms in the uncombined state. The nitrogen atom has five electrons in its outer shell, while the hydrogen atoms have one electron each. Figure 19 shows the four atoms united in a covalent bond, to form the compound ammonia, NH_3. The nitrogen atom shares electron pairs with the three hydrogen atoms. Carbon generally has a covalence of four. When it combines with four chlorine atoms, each with a covalence of one, it forms carbon tetrachloride, CCl_4, a familiar dry-cleaning fluid (Figure 20). In all these diagrams, note that the electrons that are indicated as "shared" by two atoms really are distributed around both the atomic nuclei.

"IN-BETWEEN" CASES

The compounds we have described up to now have presented clear-cut cases of electrovalence or covalence. There are also a great number of intermediate cases.

For example, two atoms may share a pair of electrons, as in covalence, but as the two electrons orbit around the nuclei of both atoms, they may spend more time in the neighborhood of one of the nuclei. The atom favored in this way will acquire a partial negative charge. The other atom will have a partial positive charge. We would say of such a compound that it has "covalent bonds with partial ionic character." A good example is water.

The water molecule, H_2O, is made up, as we have seen, of two atoms of hydrogen and one of oxygen. Hydrogen has only one electron, which is in the K shell. Oxygen, with atomic number 8, has the usual two electrons in the K shell and six electrons

in the L shell. There are two vacancies, therefore, in this shell. The vacancies are filled as the oxygen atom combines with two hydrogen atoms. If all these atoms were linked together by covalent bonds, we could show the linkage as in Figure 21, or by the notation $H-O-H$. On the other hand, if the bonding were completely ionic, there would be two H^+ ions and one O^{--} ion, sharing two electrons and kept together by electrostatic attraction only. Neither way of indicating the molecule would be accurate.

For one thing, the hydrogen atoms do not form a straight line with the oxygen atom, but a bent line, making an angle of 105°. We know also that the water molecule is not neutral electrically. The oxygen atom has a slight negative charge; each of the two hydrogen atoms, a slight positive one. Figure 22 would give at least some idea of the water molecule. Note that the two shared electrons are shown nearer to the oxygen atom than to the hydrogen atoms, indicating that they spend more time in the vicinity of the oxygen atom as they orbit. The charges are in parentheses to show their partial nature.

Fig. 18. Left: uncombined hydrogen and nitrogen atoms. Fig. 19. Middle: nitrogen and hydrogen atoms combined to form ammonia (NH_3). Fig. 20. Right: carbon tetrachloride (CCl_4) with carbon and chlorine atoms combined.

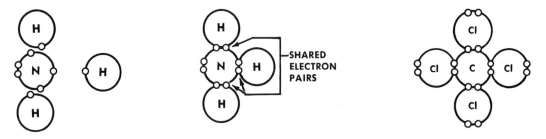

Fig. 21. The diagram at the left shows how the water molecule would look if its hydrogen and oxygen atoms were combined by covalence. Fig. 22. At right, the water molecule.

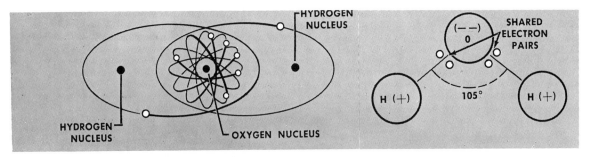

ISOTOPES

As we have already learned in other articles, an atom basically consists of a core, or *nucleus*, around which are distributed negative particles known as *electrons*. The nucleus itself is made up of positively charged particles known as *protons* and neutral particles called *neutrons*.

All the atoms of an element contain the same number of protons. This number is the element's *atomic number*. All atoms of an element do not, however, contain the same number of neutrons. When talking about these differences in the number of neutrons in an element's nuclei, we call these different atoms *isotopes* of that element. The isotopes of an element differ in *atomic mass*, or *atomic weight* — that is, in the total number of protons and neutrons in the nucleus.

STABLE OR UNSTABLE

When the ratio of neutrons to protons is within certain numerical limits, the isotopes are generally *stable*. That is, they do not change when left undisturbed. However, sometimes atoms will have too many neutrons for the number of protons, or vice versa. Such atoms are *unstable*. In such cases, the nucleus will undergo a spontaneous rearrangement, or readjustment, which will result in the release of energy. This energy will appear as radiations, which may be either electromagnetic or in the form of actual particles.

Isotopes that emit such radiations are said to be *radioactive*. The rearrangement that occurs is called *decay*. Radioactive isotopes are also sometimes called *radioisotopes*.

Isotopes are indicated by the name of the element, or its symbol, plus the mass number. Carbon-14, or C-14, therefore means the isotope of carbon with mass number 14. C-14 has six protons and eight neutrons.

HALF-LIFE

The stability of an isotope is measured by its half-life. The half-life of an isotope is the length of time that passes before one half of all the atoms originally in a given amount of the isotope decay. For some extremely unstable elements, the half-life may be only a few fractions of a second, while for other, more stable isotopes, the half-life is millions of years.

USES

Isotopes have become important tools in pure research, in industry, and in medicine. We can discuss only a few of the most widely used isotopes and their uses.

Isotopes of hydrogen. Hydrogen has two important isotopes: deuterium, with atomic mass 2; and tritium, with atomic mass 3. The supply of deuterium is nearly inexhaustible; it is easily obtained from seawater. Tritium can be synthesized from deuterium.

Deuterium and tritium are the fuels for a nuclear fusion reaction. In such a reaction, tremendous amounts of energy are released as the result of the fusion of two atoms. If a way can be found to control the reaction on a sustained basis, the energy released by the fusion of the atoms — deuterium and tritium being the two atoms — would solve all of the world's pressing energy problems.

Isotopes of carbon. The element carbon has several isotopes, some quite stable, some radioactive. Radioactive carbon-14 is the most widely used and perhaps the best known isotope. It is used as a tracer in biochemical and medical studies, as a radiation source for fine instruments, and as a measuring stick, or clock, to date geological and archaeological finds.

Isotopes of cobalt. Cobalt has several isotopes widely used in medical research and treatment. Cobalt-60 is used in the radiation treatment of cancer, in agricultural research and food processing, and in the testing of many household and industrial products.

Isotopes of other elements. Strontium-90 is used in geological research and as a radiation source in industry. Isotopes of several other elements, including sodium, iodine, and potassium, are used in medical tracer studies.

CHEMICAL REACTIONS

by Hilton A. Smith

Change is the keynote of all things, both living and nonliving, from the microscopic ameba to man, from the particles of dust dancing in a beam of sunlight to the proud monument built for the ages. Certain changes, with which we are all familiar, are spectacular and rapid. A blazing fire consumes many square kilometers of forest; a mountain slide sends thousands of metric tons of rocks crashing into the valley below. Other changes are slow and obscure. The restless waves of the ocean gradually carve away the land in one place. In another place, the winds just as gradually deposit many metric tons of dust and create new soil. Nothing escapes change.

All changes, great and small, swift and slow, may be divided into two classes — physical and chemical. A *physical change* does not cause any basic alteration in the molecules of which a given substance is composed. Consider, for instance, what happens when ice melts. The basic particle of ice is a molecule in which two atoms of hydrogen and one of oxygen are held together as a unit. If the ice is allowed to melt so that liquid water is formed, the water molecule remains unchanged: it still contains two atoms of hydrogen combined with one of oxygen. It remains unchanged, too, if the liquid water is heated so that it becomes steam. The liquefying of the solid ice and the vaporization of the liquid water are physical changes. They do not alter the composition of the water molecule.

Chemical change is something else again, for it involves the transformation of the molecules of a substance. We saw that only physical changes take place when ice melts or when water turns into steam. But we can bring about a chemical change if we pass a direct electric current through water between two electrodes. The water will gradually disappear, as hydrogen gas forms at one of the electrodes and oxygen gas at the other. In other words, the hydrogen and oxygen atoms that make up the water mole-cules will break apart and will become molecules of hydrogen and of oxygen.

NATURE OF CHEMICAL REACTIONS

Some chemical reactions are produced under natural conditions. Molds and bacteria penetrate a decaying log in a forest and cause the elements of which it is composed to combine with the oxygen of the air so that in time the log is utterly consumed. Natural chemical reactions also take place when iron rusts, or when peat is formed, or when food is digested in our bodies. Many chemical reactions, however, are deliberately brought about by people under conditions of their own choosing. It is a result of such reactions that coal and petroleum yield synthetic rubber, that the sticky substance called coal tar produces gorgeous dyes, that the stems and stalks of plants give rayon.

SYMBOLS, FORMULAS, AND EQUATIONS

Of course we can describe chemical reactions in everyday language. We can say that mercuric oxide, a substance whose basic particle is a molecule made up of an atom of mercury and an atom of oxygen, is transformed, when heated, into two other substances — mercury, whose molecule is made up of a single mercury atom, and oxygen, whose molecule is made up of two oxygen atoms. A chemist would indicate all this much more simply and directly by the following combination of letters and figures:

$$2HgO \rightarrow 2Hg + O_2$$

We find here (1) chemical symbols, (2) chemical formulas, and (3) a balanced equation.

To begin with, what do we mean by symbols? The answer is simple enough: a chemical symbol consists of a letter or of two letters representing a given chemical element. The letter C, for example, stands for carbon, the letter H for hydrogen, and

© Morton Beebe/The Image Bank

Fireworks are a spectacular example of a chemical reaction.

the letter N for nitrogen. A symbol, in other words, is an abbreviation.

If there were twenty-six or fewer chemical elements and each began with a different letter of the alphabet, we could use single letters as symbols for all of them. But as there are more than a hundred elements, we use combinations of two letters for the symbols of most of them.

There are four elements, for example, that began with H: hafnium, helium, holmium, and hydrogen. We indicate hydrogen by H; hafnium by Hf; helium by He; holmium by Ho. Similarly, B is the symbol for boron; Ba for barium; Be for beryllium; Bi for bismuth; Bk for berkelium; Br for bromine.

Some of the symbols may seem rather

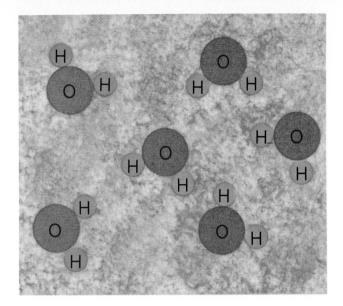

Diagram of water (H_2O) molecules in water vapor, which is a gas. In each molecule of water, two atoms of hydrogen are chemically combined with an oxygen atom.

odd. Why does Hg, for instance, stand for mercury; Fe for iron; Sn for tin? The reason is that the symbol of an element is sometimes the abbreviation of its Greek or Latin name. Hg stands for *hydroargyrum* ("liquid silver"), a slightly Latinized form of the Greek word for mercury. Fe represents the first two letters of *ferrum*, the Latin word for iron. Sn comes from the Latin *stannum*, meaning tin.

SYMBOLS STAND FOR VARIOUS THINGS

The symbols of the elements represent a sort of international shorthand system, used by the chemists of all nations. Obviously, a laboratory worker saves a good deal of time by writing H for hydrogen, Pt for platinum and U for uranium. But this is a special kind of shorthand, quite different from the ordinary kind. To the chemist H is not merely a convenient way of abbreviating the word hydrogen. It also represents an atom of hydrogen. O stands for an atom of oxygen, Cr for an atom of chromium, and so on. When a chemist adds a small figure after and near the bottom of a symbol he gives still another important detail—namely, the number of atoms. Thus O_2 stands for two atoms of oxygen; C_4 for four atoms of carbon. The "2" and the "4" are sometimes referred to as subscripts ("written under"). A chemist reading "O_2" and "O_4" aloud would say "O-two" and "O-four."

When two or more symbols are written together, it means that the atoms they represent are joined to form a molecule. H_2O means "a molecule consisting of two atoms of hydrogen plus one atom of oxygen"—a water molecule. CO_2 means "a molecule consisting of one atom of carbon plus two atoms of oxygen"—a carbon dioxide molecule. Since molecules are always made up of a definite number of atoms, we can use symbols to indicate any kind of molecule, no matter how complicated. The molecule of cane sugar, for example, is made up of 45 atoms—12 atoms of carbon, 22 of hydrogen, and 11 of oxygen. In the chemist's shorthand it is simply $C_{12}H_{22}O_{11}$. Of course, if a molecule is made up of the atoms of a single element, it may be written with a single symbol. A hydrogen molecule consists of two atoms of hydrogen, written H_2. A helium molecule consists of a single atom of helium, written He.

The symbols that we use to indicate the molecule of a substance are called the formula for that substance. The formula for water is H_2O; for cane sugar, $C_{12}H_{22}O_{11}$; for boric acid, H_3BO_3; for lye (sodium hydroxide), NaOH.

EXPRESSING REACTIONS

Let us see now how we can use symbols and formulas to express chemical reactions. Here is how we would set down the reaction in which mercuric oxide yields mercury and oxygen:

$$HgO \rightarrow Hg + O_2$$

A chemist would interpret the symbols in this way: "The molecule HgO, made up of one atom of mercury (Hg) plus one atom of oxygen (O), yields ($\rightarrow$) a molecule made up of a single atom of mercury (Hg) plus a molecule made up of two atoms of oxygen (O_2)."

We give the name *chemical equation* to the combination of symbols that indicates a chemical reaction. The substance that reacts—on the left-hand side of the arrow—is called the reactant. The result of the reaction is called the product. Thus HgO, above, is the reactant. Hg and O_2 are the products.

BALANCING EQUATIONS

A chemist would not be satisfied with the equation $HgO \rightarrow Hg + O_2$ because it is not balanced. Just what does that mean?

The law of the conservation of matter tells us that in the usual chemical reactions matter is never destroyed, though it may be transformed. Hence none of the atoms of a substance that undergoes chemical change is destroyed in the process, nor are any new atoms added to the product. There must be just as many hydrogen, or oxygen, or nitrogen atoms in the reactant as in the product, however much they may be shifted about in order to form new molecular patterns. If an equation fulfills this condition, we say that it is balanced.

What of the equation $HgO \rightarrow Hg + O_2$? One atom of mercury and one atom of oxygen are in the reactant. In the product we find one atom of mercury but two atoms of oxygen. Obviously when a single molecule of mercuric oxide decomposes into mercury and oxygen, only one atom of oxygen can be produced, for there is only one available.

To balance the equation, we put a 2 before HgO and another 2 before Hg. 2HgO, for example, stands for "2 molecules, each made up of one atom of mercury plus one atom of oxygen." The equation is now

$$2HgO \rightarrow 2Hg + O_2$$

In other words, "two molecules, each made up of an atom of mercury plus an atom of oxygen, yield two molecules of mercury, each made up of a single mercury atom, and one molecule of oxygen, made up of two oxygen atoms." We have now balanced the equation: there are two atoms of mercury and two of oxygen on the left-hand side, and the same number of the right-hand side.

You will note that in the product we wrote 2Hg, not Hg_2. The reason is that the mercury molecule, Hg, is made up of a sin-

| Two HgO molecules (each made up of one atom of mercury and one atom of oxygen) | are subjected to heat and yield | two Hg molecules (each made up of one atom of mercury) | plus | one oxygen molecule (made up of two atoms of oxygen) |

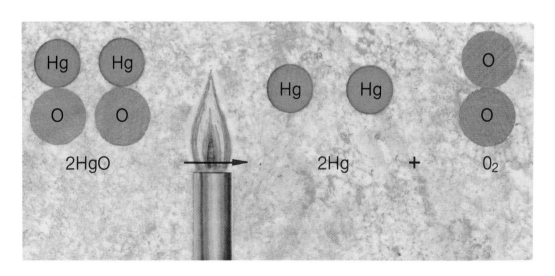

gle mercury atom. If we added a 2 subscript, we would really be saying that the molecule contains two atoms and that would be an error. Remember that we cannot change a molecule in balancing an equation. We can only change the *number* of molecules.

WHAT BALANCED EQUATIONS TELL US

Our equation

$$2HgO \rightarrow 2Hg + O_2$$

"Mercuric oxide yields mercury and oxygen" tells us that mercuric oxide decomposes in such a way that two molecules of mercury and one of oxygen are produced from the reaction of two mercuric oxide molecules. Now each of these molecules is very small and weighs very little. When we heat a test tube of red mercuric oxide, we are heating many millions of millions of tiny molecules of this material. Yet it does not matter how many molecules are involved in the reaction. The general relationship expressed by the equation $2HgO \rightarrow Hg + O_2$ still holds. Thus, if we heated exactly two million molecules of mercuric oxide, the product would be two million molecules of mercury plus a million molecules of oxygen.

The weights of elements are indicated by a system of relative weights called *atomic weights*. In this system, the weight of each element is given not in grams but in relation to the weights of the other elements. If we consult a list of atomic weights, we find that mercury has 200.6 atomic-weight units and oxygen 16. That is, an atom of mercury is about $12\frac{1}{2}$ times as heavy as an atom of oxygen. We can easily calculate the atomic-weight units in the reactant and in the product of the reaction, $2HgO \rightarrow 2Hg + O_2$.

This is how we go about it. We know that each mercuric-oxide molecule (HgO) is made up of one atom of mercury, weighing 200.6 atomic-weight units, plus one atom of oxygen, weighing 16 units — that is, 216.6 units in all. Each mercury molecule will contain 200.6 units and each oxygen molecule 32 units (2×16). If we substitute atomic-weight units for symbols, we have

$$2HgO \rightarrow 2Hg + O_2$$
$$2 \times (200.6 + 16) \rightarrow 2 \times 200.6 + 16 \times 2$$
$$2 \times 216.6 \rightarrow 2 \times 200.6 + 32$$
$$433.2 \rightarrow 401.2 + 32$$

Thus our equation tells us that 433.2 atomic-weight units of mercuric oxide will produce 401.2 atomic-weight units of mercury and 32 atomic-weight units of oxygen.

The equation $2HgO \rightarrow 2Hg + O_2$ gives us the atoms and molecules and the weight relationships involved in a particular reaction. However, it does not yield any information about the conditions under which the reaction will take place. If we stored mercuric oxide in a bottle under ordinary conditions, it would continue to be mercuric oxide indefinitely. We would have to apply heat to the substance to make it decompose into mercury and oxygen. We could indicate that fact by writing the word "heat" above the arrow, thus:

$$2HgO \overset{heat}{\rightarrow} 2Hg + O_2.$$

"Mercuric oxide when heated yields mercury plus oxygen."

SPEED OF REACTIONS

Chemical equations do not show how rapidly a given reaction takes place. Some chemical processes proceed very quickly. In the cylinders of an automobile engine, vaporized gasoline and air (that is, the oxygen in air) react almost instantaneously when fired by an electric spark. Other chemical reactions are very slow. Water, portland cement, and sand, blended together in suitable proportions, react to form concrete, but it takes weeks or months.

The principal factors that influence the speed of a chemical reaction, apart from the very nature of the molecules involved, are (1) the temperature, (2) the degree of contact between the reacting materials, (3) the concentration of these materials, and (4) the presence of substances called *catalysts*.

The effect of temperature is an obvious factor. Practically all chemical processes are speeded up by an increase in temperature. All molecules are constantly in motion except at the extreme temperature called

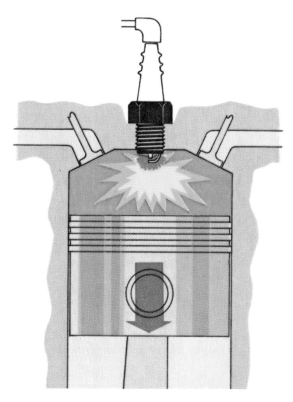

A rapid chemical reaction occurs above. Vaporized gasoline and air in an auto cylinder react almost instantly when ignited by an electric spark. The explosion produced pushes the piston down in the cylinder. Below: spreading a mixture of portland cement, sand, and water, which will react to form concrete. In this case, the reaction will be complete only after weeks or even months have elapsed.

©E. Allen McGee, Inc.

absolute zero, corresponding to −273.16° Celsius. As the temperature of a reaction mixture rises, the molecules of reacting substances move with greater speed than before. They collide more often and more vigorously. More and more molecules are jarred so effectively that their atoms are released to form new patterns.

The chemist is generally able to control the temperature at which reactions take place in the laboratory. Sometimes, however, reactions are speeded up by temperature rises that are not due to any human agency. You may have heard of piles of greasy rags, or oily wastes, or fermented straw catching fire of themselves through spontaneous combustion. This is what happens. All these materials undergo oxidation — that is, combination with the oxygen of the air — even under ordinary room temperatures. As a result heat is released. Since the substances in question are poor conductors, they retain a good deal of heat. This raises the temperature, and as the temperature rises oxidation goes on at a more rapid rate. At last combustible gases are expelled, and by this time the temperature is so high that the materials burst into flame.

The degree of contact between reacting molecules also influences greatly the speed of reaction. The grinding up of large solid particles into very fine ones speeds up a reaction because it causes contact over a vastly increased area. Let us take the fuel coal as an example. Coal burns because the carbon of the coal reacts vigorously with the oxygen of the air at a high temperature. It takes time to set fire to large chunks of coal, because the amount of surface which is exposed to oxygen is limited. But powdered coal offers an almost infinite number of points of contact. If such fuel is blown out of a nozzle, so that the particles will not pack together, it burns almost as rapidly as liquid fuel.

It is because particles of coal dust, sawdust, and grain dust are so finely divided that dust is an ever-present danger in coal mines, sawmill factories, and grain elevators. The total surface exposed to the oxygen of the air is so great that a mere spark may cause a terrible explosion. Even aluminum and bronze dust will be ignited and burn up at an explosive rate under such conditions.

CATALYSTS

Even when the temperature is high and when reacting substances are in intimate contact, some reactions will take place slowly. In such cases outside materials, not involved in the reaction, will sometimes speed up the chemical change. These outside materials are called catalysts. They are generally solids, but they may be liquids or gases. They modify the rate of a reaction without themselves undergoing any permanent change. Different catalysts are used to modify the rate of different reactions. Living cells contain natural catalysts, called enzymes, which make possible many of the chemical changes occurring within the cells.

The chemist makes frequent use of catalysts. Sometimes he adds them in small quantities to reacting materials. Thus he combines finely divided nickel with cottonseed oil so that the oil will react with hydrogen gas to form the solid fats sold as shortening or used in the manufacture of soap. In other cases, the catalyst is present in a bed over which the reacting materials are passed. A mixture of air and sulfur dioxide, passed over a catalyst consisting of finely divided platinum, reacts rapidly to produce sulfur trioxide.

REVERSIBLE REACTIONS AND EQUILIBRIUM

The chemical equation

$$2HgO \xrightarrow{\text{heat}} 2Hg + O_2$$

tells us that mercuric oxide when heated yields mercury and oxygen, but it does not show how complete the reaction is. If the mercuric oxide is heated in a closed tube, a twofold reaction will take place at one and the same time. As the mercuric oxide decomposes into mercury and oxygen, the mercury and oxygen will react to form mercuric oxide:

$$2Hg + O_2 \rightarrow 2HgO$$

Right: a greatly simplified diagram of the Haber process for the production of ammonia. A mixture of nitrogen gas and hydrogen gas is compressed and then led into a synthesis chamber. Here, under high temperature and pressure and with the use of a catalyst, the nitrogen and hydrogen are converted into ammonia. A reverse reaction also takes place. Some of the newly formed ammonia decomposes to form nitrogen and hydrogen. When the state of equilibrium is reached, only about one fourth of the nitrogen and hydrogen has been converted into ammonia. The mixture of gases—ammonia and unconverted nitrogen and hydrogen—is led to a condenser, where it is cooled. The ammonia gas is converted here into a liquid and is led to a storage tank. The nitrogen and hydrogen gases pass to a compressor. After they have been compressed, they again make their way to the synthesis chamber.

Naturally, if this continues, the mercuric oxide will never decompose completely. A chemist would call this a *reversible reaction.*

We indicate a reversible reaction by using two arrows in the equation. One of them shows the decomposition or forward process. The other shows the recombination of the products: that is, the reverse process. This is how we would show that mercuric oxide produces mercury and oxygen, while at the same time mercury and oxygen yield mercuric oxide:

$$2HgO \underset{}{\overset{\text{heat}}{\rightleftharpoons}} 2Hg + O_2$$

Let us follow through with this reversible reaction. Let us suppose that we begin by heating mercuric oxide (HgO) in a closed tube that is evacuated (that is, from which the air has been removed). At first, there is neither mercury nor oxygen in the tube and therefore the reaction can only proceed in the forward direction. As mercury and oxygen are formed, however, the reverse reaction commences. It is very slow in the beginning since only a little mercury and oxygen are present. These substances accumulate, however, as decomposition continues. Since increased concentration increases the speed of a reaction, the rate at which the mercury and the oxygen recombine to form mercuric oxide slowly increases. It continues to increase until the rates of the two opposing reactions are the same. That is, the mercuric oxide is being reformed at precisely the same speed

THE HABER PROCESS

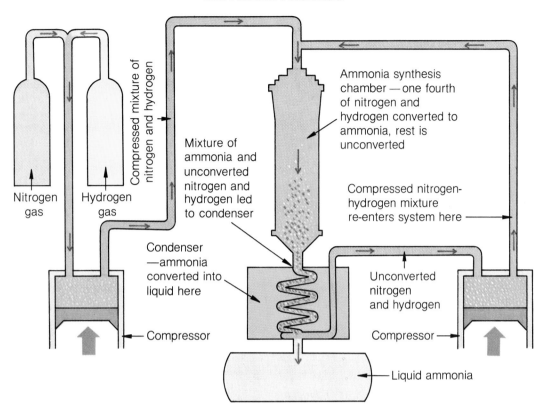

Compressed mixture of nitrogen and hydrogen

Nitrogen gas

Hydrogen gas

Mixture of ammonia and unconverted nitrogen and hydrogen led to condenser

Ammonia synthesis chamber — one fourth of nitrogen and hydrogen converted to ammonia, rest is unconverted

Compressed nitrogen-hydrogen mixture re-enters system here

Condenser —ammonia converted into liquid here

Unconverted nitrogen and hydrogen

Compressor

Compressor

Liquid ammonia

as it is being decomposed. A reversible reaction at this stage is said to be in chemical *equilibrium*.

In some cases the reaction mixture at equilibrium will consist mainly of reactants, in other cases of products. The exact ratio depends on the nature of the chemicals involved. This ratio is also affected by certain variable factors. One of these is the temperature at which the reaction is carried out. Another is the concentrations or the pressures of the reactants and the products. By controlling the variable factors, the chemist can modify the ratio between reactants and products at equilibrium.

We can carry a given reversible reaction to completion by entirely removing one or more of the products from the reaction mixture. We have seen that the gas oxygen is one of the two products formed by the decomposition of mercuric oxide. If we allow this gas to escape from the reaction mixture into the atmosphere, it will be unable to unite with the mercury that is also

formed as a product. Therefore, mercuric oxide will completely decompose into oxygen and mercury whenever the reaction is carried out in an open tube.

A TYPICAL REACTION

To show how the chemist analyzes chemical reactions and then modifies the factors that govern them, let us examine the Haber process, in which nitrogen and hydrogen are made to react in order to produce ammonia. This process, named after Fritz Haber, the German chemist who discovered it, is extremely important. The product, ammonia, is in great demand, particularly in the manufacture of fertilizers, explosives, refrigerants, and certain types of rayon. The reacting materials, nitrogen and hydrogen, are cheap and readily available. Because of the Haber process, an important chemical can be manufactured cheaply and in great quantities.

We can represent the chemical reaction by the following equation:

$$N_2 + H_2 \rightarrow NH_3$$

Nitrogen plus hydrogen yields ammonia.

The reaction is not balanced as it stands, for there are two atoms of nitrogen in the reactant, but only one in the product. Also, there are two atoms of hydrogen in the reactant, but three in the product. The fact that there are two atoms of nitrogen in the reactant means that there must be two ammonia molecules in the product:

$$N_2 + H_2 \rightarrow 2NH_3$$

Still the reaction remains unbalanced, for the six atoms of hydrogen in the two ammonia molecules will require six atoms of hydrogen—that is, three molecules of hydrogen—on the left-hand side. By adding the figure 3 before the symbol for the hydrogen molecule in the reactant, the equation is completely balanced, thus:

$$N_2 + 3H_2 \rightarrow 2NH_3$$

"A molecule of nitrogen plus three molecules of hydrogen yields two molecules of ammonia."

The atomic weight of nitrogen is 14; of hydrogen, 1. Making the appropriate substitutions in the equation, we have:

$$N_2 \qquad + 3H_2 \qquad \rightarrow 2NH_3$$
$$2 \times 14 + 3 \times (2 \times 1) \rightarrow 2 \times (14 + [3 \times 1])$$
$$28 \qquad + 6 \qquad \rightarrow 34$$

That is, 28 atomic-weight units of nitrogen will combine with 6 atomic-weight units of hydrogen to form 34 atomic-weight units of ammonia. We have seen that the same proportion would hold, no matter what weight units we used. Hence, if we combined 28 kilograms of nitrogen with 6 kilograms of hydrogen, the yield would be 34 kilograms of ammonia.

The reaction $N_2 + 3H_2 \rightarrow 2NH_3$, unfortunately, goes on too slowly at ordinary temperatures and pressures to be of any use. To speed up the reaction, the temperature is increased to about 500° Celsius and the pressure is raised to between 500 and 1,000 times that of the atmosphere. Furthermore, a catalyst, generally iron, is employed. Under such conditions the reaction is fairly rapid.

This reaction is reversible. That is, some of the ammonia that is produced decomposes to form nitrogen and hydrogen. The high pressures that are used cause an increase in the yield of ammonia. Higher temperatures increase the rates of both the direct and reverse reactions, but that of the reverse reaction is increased more. When the state of equilibrium is reached, only about one-fourth of the nitrogen and hydrogen has been converted into ammonia. In order to get further production, it is necessary to lead the mixture of gases—ammonia, nitrogen, and hydrogen—from the heavy steel bomb in which the reaction is taking place. After the ammonia has been extracted from the mixture, the nitrogen and hydrogen are reheated and recompressed in order to form still more ammonia as a product.

Thus we can modify the conditions under which reactions are produced. And we can transform a chemical reaction from a natural event, such as the rusting of a metal, into one of our mightiest tools.

In industrial chemistry, the control of a particular reaction is sometimes not as important as the choice of a proper combination of reactions. For example, in the 1860s a Belgian chemist, Ernest Solvay, devised a method for producing sodium carbonate, or soda ash, an important chemical in the glass and chemical industries. His process was valuable for making efficient use of ammonia, which was very expensive in the nineteenth century. The Solvay process is still the main one used in industry for making soda ash.

The steps in the process are too complicated to describe in detail. In effect, the soda ash is produced by combining ammonia, carbon dioxide, and a salt solution under carefully controlled conditions. The resulting sodium bicarbonate is converted to sodium carbonate by means of another reaction. The product, ammonium chloride, undergoes a further reaction, as well, to produce ammonia gas. This ammonia is then re-used to produce more soda ash. By devising such series of reactions, chemists have added greatly to the industrial power of the modern world.

CATALYSIS

by Anthony Standen

You may have heard a person spoken of as a "catalyst" at a party. The meaning is that he or she is able to enliven the party by getting strangers to converse freely with one another. Without the "catalyst," it might have taken quite a long time for these people to get to know one another. Perhaps they would never have become acquainted. This use of the word "catalyst" is borrowed from chemistry. A *catalyst*, in the language of chemistry, is a substance that can alter the speed of a chemical reaction. It can bring about reactions that might otherwise not have taken place at all. But the catalyst itself is not used up. This process is called *catalysis*.

To show how catalysis works, let us consider the familiar cigarette lighter. The fluid that one puts in an ordinary lighter does not burst into flame spontaneously when it comes in contact with the oxygen of the air. It must first be heated to quite a high temperature by a spark, produced by friction as a revolving metal wheel comes in contact with a "flint" that is made of a mixture of various rare-earth metals. Once the fluid is heated, it can react with oxygen and burn. In so doing, it produces enough heat to raise some more fluid to a high enough temperature so that the burning can go on. But certain lighters can work not by means of a spark but by means of catalysis. The catalyst in this case is a so-called igniter, consisting of platinum wires attached to platinum sponge. A wick in a "wicked chamber" is saturated with methanol (wood alcohol). When not in use, the chamber is covered with a cap. If the cap is removed and the igniter is inserted in the chamber, the platinum sponge absorbs the methanol gas fumes and heats the platinum wires. The fumes are ignited, producing a flame. The reacting substances — methanol and the oxygen of the air — are converted into something chemically different. The platinum serving as the catalyst is not used up.

Catalysis was discovered quite early in the nineteenth century. It was given its name in 1836 by the great Danish chemist Joens Jakob Berzelius, who supposed that it represented a special force in nature, different from any other. It is now known that this is not the case. Catalysis operates within the framework of chemical laws.

HOW DO CATALYSTS WORK?

It is not possible in this short article to provide a full answer to the question: "How do catalysts work?" The experts themselves, indeed, disagree in the higher reaches of catalysis research. But at least we can give some idea of catalytic action and show how it follows the accepted laws of chemistry.

Let us consider, first of all, what happens when we mix hydrogen gas and oxygen gas. We know that when heat is applied to the mixture, the hydrogen and oxygen molecules will react violently and that there will be an explosion. The product of this reaction is water. Water is H_2O, a compound in which two atoms of hydrogen are combined with one atom of oxygen. But suppose we keep the hydrogen-oxygen mixture at room temperature. Nothing will happen. Why?

The fact is that molecules must have a certain level of energy before they can react. At room temperature, some of the hydrogen and oxygen molecules in our mixture have more energy than others. But none of them has enough to react with the other. But when we heat the hydrogen-oxygen mixture, we raise the average energy level of all the molecules. Some of the hydrogen and oxygen molecules will now be able to react with one another and will combine chemically. In doing so, they will release a large amount of heat. This will heat the rest of the mixture, so that more of the molecules will have enough energy to combine chemically. The reaction will proceed so rapidly that, as we pointed out, an explosion will take place.

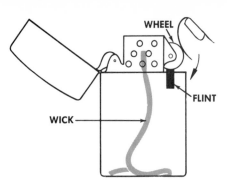

Above: a friction-wheel cigarette lighter. In the cutaway photo at the right, we see the lighter in operation. The sparks that result as the wheel revolves rapidly against the flint ignite the wick and produce a flame.

These three photographs illustrate how one type of catalytic lighter works. *A* shows the device when not in use, with a double chamber covered by a double cap. The larger chamber contains a wick saturated with methanol. At the base of the smaller cap is an igniter, consisting of platinum wires and a platinum sponge. To work the lighter, the igniter is inserted part way into the chamber containing the wick (*B*). As it is withdrawn, a flame is produced (*C*).

The significant thing about catalysts is that they can bring about a reaction in a case like this even when the *reactants* — the reacting substances — are at room temperature. They do this in a roundabout but effective way, which we shall now examine.

Let us first consider a generalized reaction in which catalysis plays no part. Suppose that two substances, A and B, react to produce the substance AB:

$$A + B \rightarrow AB \qquad (1)$$

"A reacts with (+) B and yields (→) AB."

We shall assume that this reaction will take place only when heat is applied to A and B. We shall assume, too, that the reaction does not take place at room temperature because none of the molecules of A and B has enough energy to combine chemically with another.

Suppose now that we use a piece of platinum as a catalyst. When a molecule of A comes in contact with platinum, it is often adsorbed (attached) to the platinum, forming the compound PtA, Pt being the chemical symbol for platinum:

$$Pt + A \rightarrow PtA \qquad (2)$$

"Platinum (Pt) reacts with A and yields PtA."

We shall assume that only a low energy level is required for this reaction. Hence it can take place readily at room temperature.

Molecules of B will also be adsorbed on the surface of the platinum, forming the compound PtB:

$$Pt + B \rightarrow PtB \qquad (3)$$

"Platinum reacts with B and yields PtB."

This would lead to the situation represented in the diagram on this page. The surface of the platinum is now covered with A's and B's, forming the compounds PtA and PtB, respectively, with the platinum.

Now another thing may happen. If an A molecule and a B molecule are adsorbed at points very close together, the following reaction may occur:

$$PtA + PtB \rightarrow 2Pt + AB \qquad (4)$$

"A molecule of PtA reacts with a molecule of PtB and yields 2 molecules of platinum and a molecule of AB."

We have now obtained the desired reaction product, the compound AB, without having had to heat A and B. The platinum catalyst has been regenerated and is now ready to serve in other catalytic reactions.

Let us sum up. To obtain AB directly by the reaction of A and B, we would have to raise the energy of A and B by applying heat. If we use platinum as a catalyst, as indicated in equations (2), (3), and (4), the route is roundabout, but the reaction resulting in the compound AB can take place at room temperature.

There are many different kinds of catalysts. Some speed up chemical reactions, while others slow them down. There are *homogeneous catalytic systems* and *heter-ogeneous catalytic systems*. In homogeneous systems, the catalyst and the reacting substances are in the same state of matter. Thus, as we shall see, gaseous oxides of nitrogen act as catalysts in the reaction in which the gases sulfur dioxide and oxygen combine to form sulfur trioxide. In heterogeneous catalytic systems, the catalyst and the reacting substances are in different states. For example, in the cigarette lighter mentioned above, the catalyst, platinum, is a solid. The reacting substances – the fuel in vapor form and the oxygen – are gases. Most catalytic reactions are heterogeneous.

Not all catalysis takes place at moderate temperatures. Sometimes high temperatures must be provided. But even at these high temperatures, there will often be no reaction at all unless a catalyst is employed.

Actually, catalysis is a more complicated procedure than we have indicated here. However, certain basic principles apply to all catalysts. For one thing, they all provide an easier path, from the viewpoint of the energy required to set the reaction going, than would otherwise be available. What is just as important, they are not used up themselves in the reactions that they promote.

CATALYSIS IN INDUSTRY

Catalysts are extremely important in industry. They were applied in industrial processes many years before their nature was understood. They have been used increasingly and with more highly developed techniques in the present century. It would take many pages simply to list the industrial processes in which catalysis plays a part. Let us consider a few examples.

Catalysts play a key role in the preparation of the important chemical sulfuric acid, which is manufactured in greater quantities than any other chemical. The

Diagram showing how substances *A* and *B* react through the catalytic action of platinum.

| A | B | A | B A | A | B | A B |

PLATINUM

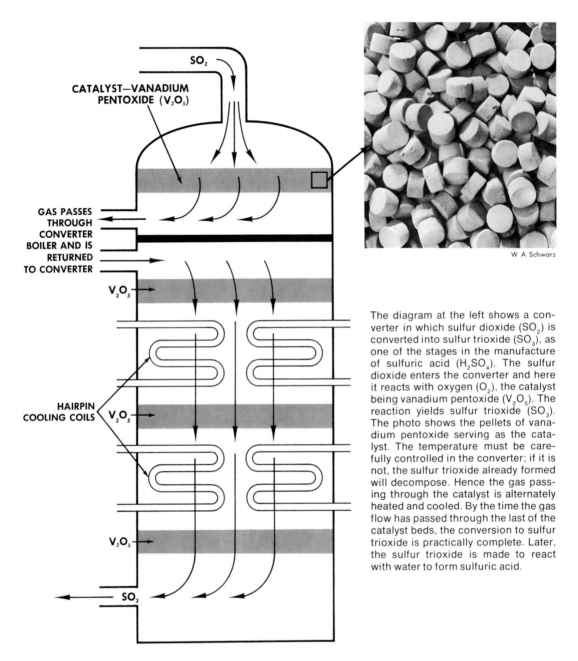

SO₂

CATALYST—VANADIUM
PENTOXIDE (V₂O₅)

GAS PASSES
THROUGH
CONVERTER
BOILER AND IS
RETURNED
TO CONVERTER

V₂O₅

HAIRPIN
COOLING COILS V₂O₅

V₂O₅

SO₃

W A Schwarz

The diagram at the left shows a converter in which sulfur dioxide (SO_2) is converted into sulfur trioxide (SO_3), as one of the stages in the manufacture of sulfuric acid (H_2SO_4). The sulfur dioxide enters the converter and here it reacts with oxygen (O_2), the catalyst being vanadium pentoxide (V_2O_5). The reaction yields sulfur trioxide (SO_3). The photo shows the pellets of vanadium pentoxide serving as the catalyst. The temperature must be carefully controlled in the converter; if it is not, the sulfur trioxide already formed will decompose. Hence the gas passing through the catalyst is alternately heated and cooled. By the time the gas flow has passed through the last of the catalyst beds, the conversion to sulfur trioxide is practically complete. Later, the sulfur trioxide is made to react with water to form sulfuric acid.

acid is prepared in two different ways, each involving the use of a catalyst. In both methods, sulfur is first oxidized to form sulfur dioxide (SO_2). The problem then is to oxidize sulfur dioxide to sulfur trioxide (SO_3). In the lead-chamber process (so called because the reactions take place in large lead chambers), oxides of nitrogen serve as the catalyst. The contact process uses a solid catalyst, either platinum or a compound of vanadium. In either case, once sulfur trioxide is obtained, it reacts with water to form sulfuric acid (H_2SO_4). The three basic reactions in the preparation of sulfuric acid are given in the equations that follow:

$$S + O_2 \rightarrow SO_2 \qquad (1)$$
"A molecule of sulfur (S) reacts with a molecule of oxygen (O_2) and yields a molecule of sulfur dioxide (SO_2)."

$$2SO_2 + O_2 \xrightarrow{\text{catalyst}} 2SO_3 \qquad (2)$$
"Two molecules of sulfur dioxide ($2SO_2$) react with a molecule of oxygen (O_2), in the presence of a catalyst, and yield 2 molecules of sulfur trioxide ($2SO_3$)."

$$SO_3 + H_2O \rightarrow H_2SO_4 \qquad (3)$$
"A molecule of sulfur trioxide (SO_3) reacts with water (H_2O) and yields sulfuric acid (H_2SO_4)."

The lead-chamber process was developed in the eighteenth century; the contact process, in the nineteenth.

Ammonia, NH_3, another key chemical, is synthesized from nitrogen and hydrogen. Not only do these two elements not combine at room temperature but no appreciable reaction will take place even if they are subjected to great heat. Catalysis must be employed. A compressed mixture of nitrogen and hydrogen, heated to about 500° Celsius, is passed over a catalyst, which is generally iron. The catalyst is preferably "promoted," or made more active, by the addition of small amounts of aluminum oxide and potassium oxide. A *promoter* may be thought of as a catalyst for a catalyst. This process for making ammonia is called the Haber process.

Ammonia, in turn, can be converted by the use of a catalyst to nitric acid, HNO_3, which gives rise to a large family of products—explosives, fertilizers, and dyes, among others. The catalyst in this case is platinum, perhaps mixed with another metal closely related to platinum.

Today's petroleum industry is dependent upon catalysis for the production of gasoline. At the beginning of this century, in the very early days of the automobile industry, enough gasoline for the few cars on the road could be made by simply distilling petroleum. As more and more cars were manufactured, the demand for gasoline increased so enormously that other means of producing the fuel had to be developed. *Thermal cracking,* in which large hydrocarbon molecules were reduced to smaller ones, depended only on the action of heat. Even the added quantities of gasoline produced by this method did not suffice. It was necessary to develop other processes involving the use of catalysts. One of these processes, *catalytic cracking,* is one in which large molecules are broken up. Other catalytic processes, such as *alkylation,* involve building up small molecules into larger ones. The catalysts used for these processes include acids as well as oxides of aluminum, silicon, and chromium.

Another very important use of catalysis is in the conversion of liquid oils, such as cottonseed oil and soybean oil, into semisolid fats that can be used as margarine or for other purposes. The conversion is brought about by the process called *hydrogenation.* It involves adding hydrogen atoms to the molecules of the liquid oils. The liquid is heated with hydrogen under pressure. The catalyst used is finely divided nickel.

Catalysis is helping to solve a serious problem that has confronted the automobile industry in recent years. The exhaust systems of cars have been the target of much criticism. Unfortunately, an automobile usually does not completely burn up its fuel, gasoline. It discharges into the air through the exhaust system very small quantities of unsaturated hydrocarbons. These react with compounds of nitrogen and oxygen, formed in the atmosphere by direct combination through the action of lightning. The products of such reactions are the chemicals that are among the most disagreeable and dangerous features of smog. A catalytic converter, installed in the exhaust system of automobiles, oxidizes and destroys the unsaturated hydrocarbons in the exhaust.

CATALYSIS IN LIFE PROCESSES— ENZYMES

In the instances of catalysis that we have discussed up to now, the catalysts are inorganic substances and on the whole they are chemically rather simple. There is another large group of catalysts—the *enzymes* —which are complex, organic substances.

They are produced by living cells of animals and plants, and they catalyze the reactions upon which life depends. Without enzymes, these reactions would either not take place at all or they would proceed so slowly as to be quite useless to the organism.

Enzymes are proteins. They contain large numbers of amino acids, strung together like the beads in a necklace. Each amino acid consists of a carbon atom to which are attached (1) an amino group, NH_2, made up of nitrogen and hydrogen; (2) a carboxyl group, COOH, made up of carbon, oxygen, and hydrogen; (3) a hydrogen atom, H; and (4) any one of various other groups, represented by the letter R. The structural formula of a typical amino acid is:

$$R-\overset{\displaystyle NH_2}{\underset{\displaystyle H}{C}}-COOH$$

The amino-acid links in the enzyme chain are very numerous. Enzymes do not require high temperatures to bring about their effects. As a matter of fact, they lose their potency if subjected to too great heat.

Enzymes play a vital role in digestion. They are responsible for the chemical changes that take place in the digestive tract. Each enzyme acts on a specific kind of food. Thus the enzymes known as the proteases act on proteins; the amylases, on starch; the lipases, on fats; the amidases, on amides; the cellulases, on cellulose; the maltases, on the sugar known as maltose. Enzymes generally work in groups. Each enzyme carries out a single step in a series of reactions that may involve 25 enzymes or more.

Already many hundreds of enzymes have been isolated and described. Digestion is not the only life process in which they serve. They play an important part in the conduction of nerve impulses, in the clotting of the blood, and in the contraction of muscle tissue. They are vital in photosynthesis—the manufacture of food in plants through the energy supplied by light.

Enzymes have served man in industrial operations since ancient times, though of

Catalytic action is involved in the leavening, or raising, of bread. Yeast is added to the dough. The sugar glucose in the dough is decomposed to ethyl alcohol and carbon dioxide, the catalyst being zymase, an enzyme produced by yeast. At left, dough with yeast has been put in a pan. At right, the dough after four hours. It has expanded due to the formation of carbon dioxide bubbles.

W. A. Schwarz

The action of the enzyme known as rennin is vital in the making of cheese. Rennin curdles milk to produce solid cheese and liquid whey.

Swiss National Tourist Office

course the nature of their action was not understood until quite recently. Winemaking, breadmaking, and cheesemaking are examples of industrial processes in which enzymes play an essential part in the production of the final product.

In making wine, use is made of an enzyme, zymase, produced by the microorganisms called yeasts. Zymase acts as a catalyst in the decomposition of the sugar glucose to ethyl alcohol and carbon dioxide:

$$C_6H_{12}O_6 \xrightarrow[\text{by zymase}]{\text{catalyzed}} 2C_2H_5OH + 2CO_2$$

"A molecule of sugar ($C_6H_{12}O_6$), catalyzed by zymase, yields 2 molecules of ethyl alcohol ($2C_2H_5OH$) and 2 molecules of carbon dioxide (CO_2)."

The same reaction is involved in the leavening of bread. In this case, however, the carbon dioxide gas is the important product of the reaction. The gas forms bubbles in the dough and causes it to "rise." The many little holes in bread attest to the action of these bubbles.

According to an ancient legend, the art of cheesemaking was discovered accidentally. An Arab merchant, so the story goes, set out on a journey and put his supply of milk in a pouch made of a sheep's stomach. By journey's end, the milk had separated into liquid whey and solid curd.

Regardless of the question of whether or not there is any truth in this story, it is a fact that from early times the curdling of milk to form curds was promoted by a substance obtained from the stomach of ruminants — a substance called rennet. We know now that the curdling action of rennet is due to an enzyme known as rennin. Rennet is still used in cheesemaking.

METALS AND NONMETALS

by W. H. Slabaugh

The hundred or so chemical elements, both those occurring in nature and those created by man, may be placed in two broad classes—metals and nonmetals. Generally speaking, the typical *metal* conducts electricity and heat. It shows lustrous surfaces, usually with white, or so-called metallic, luster, It is ductile, or capable of being drawn out into long, thin strands, and malleable, or capable of being shaped by hammering or by pressing between rolls. *Nonmetals,* by way of contrast, are poor conductors of electricity and heat, are often colored, and, when in solid form, are brittle. A comparatively small number of elements, called *metalloids,* do not fit too well in either of these two classes. They display properties of both metals and nonmetals. Arsenic and antimony are examples of metalloids.

Every classification of the chemical elements has its exceptions, and this is true of the metals-and-nonmetals classification. For example, the metals gold, silver, and copper conduct electricity extremely well but, unlike most metals, they are colored. Aluminum is as ductile and malleable as one might expect of a metal. Yet at high temperatures it becomes so brittle that when it is hammered, it breaks into small

Metals are durable, pliable materials which can be molded, pressed, or hammered into desired forms. They have been fashioned into many fine pieces of art. At left: Benvenuto Cellini's gold saltcellar. Above: bronze cat, a relic of ancient Egypt.

Tétrel — musée du Louvre, Paris

fragments. As a matter of fact, that is how powdered aluminum is obtained. Some metals are extraordinarily malleable. Gold, for example, can be hammered into sheets so thin that light can pass through them. Other metals are not malleable at all. Thus chromium is so hard even at room temperature that it shatters whenever it is hammered.

Metals are solid at room temperature, with the exception of mercury, which is a liquid. At this temperature, practically all the nonmetals are either solid (sulfur, silicon, carbon, and so on) or gaseous (oxygen, hydrogen, fluorine, and so on). Bromine is the only liquid nonmetallic element.

Some of the metals and nonmetals are familiar to everyone. The names of others would be recognized only by persons with a certain amount of training in chemistry or metallurgy. In the table on page 78, we give a list of some of the more important and familiar metals and nonmetals to be found in the earth's crust, together with a few of their uses and special properties.

You will note several things in examining the table. In the first place, most of the elements listed here are metals. As a matter of fact, metals are far more numerous than nonmetals. Of the sixty most common elements, less than a fifth are nonmetals. Another thing that may strike you is that some of the most important metals and nonmetals included there are quite rare in nature. Copper, for example, is familiar to all of us and plays an important role in civilization. Yet there is a hundred times more titanium than copper in the earth's crust. Fortunately for us, copper has been naturally concentrated in certain ores, such as chalcocite and cuprite, and it is relatively easy to obtain the metal from these ores.

STRUCTURE OF METALS AND
    NONMETALS

Metals differ from nonmetals in the nature of the bonds between atoms. This accounts in large part for the differences between metallic and nonmetallic substances.

Conductivity. In a metal, the electrons in the outer shell of the atom are loosely held and can take part in the electron flow that we call *electric current.* That is why metals conduct electricity. The metals with the most loosely held electrons, such as silver and copper, make the best electrical conductors.

We may think of a metal as made up of (1) "loose" outer-shell electrons, free to move from one atom to another, and of (2) ions, fixed in the crystalline structure of the metal. There are always the same relative number of outer-shell electrons and ions in the metal, even when electric current is flowing through it. The reason is that the same number of electrons are added to one end of the conductor as are removed from the other end. Hence the composition of the metal remains unchanged.

At extremely low temperatures, some metals become better and better conductors. Ultimately, they can develop into *superconductors*. Superconductors permit electric current to flow through them with no resistance. For example, when mercury is in the shape of a closed loop, or "doughnut," at 4° Celsius above absolute zero, it will carry an electric current for many weeks. When the metal is warmed to a slightly higher temperature, the electric current immediately drops to zero because the metal develops a small but effective resistance to it.

The best electrical conductors are also excellent conductors of heat. It is believed, therefore, that heat conductivity, as well as electrical conductivity, results from the flow of mobile electrons.

In contrast to the electrons in metals, the electrons in nonmetals, in the solid state, are firmly fixed. A particular atom is often attached to its neighbors by sharing valence electrons in what is called a covalent bond. Let us consider the structure of diamond, which is a form of carbon. The carbon atom of diamond has four valence electrons in the outer shell. (See *A* in Figure 1.) Every one of these electrons is paired with an electron from a neighboring carbon atom. This electron pair is shared by the two atoms in question (*B* in Figure 1). Since there are four of these neighboring atoms in all, each carbon atom of diamond has eight valence electrons—four of its own and one from each of its four neighbors (*C* in the diagram). The four neighboring atoms are grouped around the carbon atom in such a way as to form the geometric pattern called a tetrahedron (*D* in the diagram). This tetrahedral pattern is continued throughout the diamond crystal. The tight bonding between the carbon atoms of diamond accounts for the fact that diamond does not conduct electricity: there are no loosely held electrons available for this purpose.

Luster. The luster of metals as compared to the lack of luster of nonmetals can also be explained on the basis of loose electrons as against fixed electrons. When light strikes a polished metal surface, all of its wavelengths are reflected because of the "looseness" of the valence electrons. Hence the color of the reflected light is not altered.

A nonmetal, on the contrary, because

COMMON METALS AND NONMETALS IN THE EARTH'S CRUST		
Metals	Percentage by weight in earth's crust	Uses and special properties
Aluminum	8.13	Utensils, foil, conductors, alloys
Iron	5.00	Steels, cast iron
Magnesium	2.09	Flash bulbs, lightweight alloys
Titanium	0.440	Aircraft alloys
Manganese	0.100	Alloys
Chromium	0.020	Plating, alloys
Nickel	0.0080	Alloys, coinage
Zinc	0.0065	Alloys, galvanizing
Copper	0.0045	Conductors, alloys, coinage
Lead	0.0015	Batteries, sheet
Tin	0.0003	Tin plate for tin cans, alloys
Silver	0.00001	Plate, alloys, coinage, jewelry
Gold	0.0000003	Plate, alloys, jewelry
Nonmetals		
Oxygen	46.6	Steel manufacture, welding
Silicon	27.7	Alloys (with iron), computer chips
Phosphorus	0.118	Matches
Sulfur	0.052	Chemicals, paper, insecticides
Carbon	0.032	Black pigment, lubricant
Chlorine	0.031	Bleach, disinfectant, water purification

Fig. 1 THE CARBON ATOMS OF DIAMOND

Shared electron pair

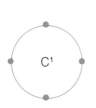 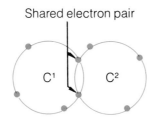 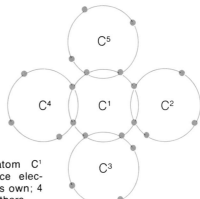

A. The above carbon atom, C¹, has 4 electrons in its outer shell.

B. Carbon atom C¹ shares an electron pair with adjoining carbon atom C².

C. Carbon atom C¹ has 8 valence electrons—4 of its own; 4 shared with others.

of the fixity of its electrons, cannot re-emit the light waves so readily. Hence most nonmetals alter the color of light that falls upon them by absorbing certain colors without re-emitting them. Sulfur is yellow, for example, because its electrons are capable of both absorbing and reflecting only yellow. The other colors are absorbed and are not reflected. Carbon in the form of graphite is black because it absorbs all the colors of light and re-emits none.

Ductility and Malleability. Atomic structure accounts for the differences in ductility and malleability between metals and nonmetals. The arrangement of atoms in a typical metal may be altered merely by displacing a layer or row of metal atoms. We can displace the atoms along *slip planes,* or the planes along which the atoms of a crystal move when subjected to a force, and change the shape of a metal without appreciably changing its structure. Most nonmetallic solids, however, are made up of molecules that cannot easily slip past each other, or else, as in the case of diamond, the atoms are firmly attached to one another and great force is required to displace them. Hence nonmetallic solids are brittle and cannot be shaped by hammering or rolling.

SOME CAN CHANGE

Changes in atomic structural arrangements may cause metals to become nonmetals. Thus common tin, or beta-tin, is metallic in the usual sense. It is malleable; it conducts electricity; it is lustrous white. At

a temperature above 13.2° Celsius, each atom of common tin has six near neighbors in its crystalline structure. Below 13.2° Celsius, however, the metal changes its structure to a crystalline form like that of diamond. Each atom now has four near neighbors and is covalently bonded to them. This form of tin, called alpha-tin, is nonmetallic. It does not conduct electricity, and it is gray in color.

The rate of conversion from metallic tin to the nonmetallic gray tin that we call alpha-tin is quite slow at temperatures slightly below 13.2° Celsius. It becomes much more rapid when the temperature reaches −50° Celsius. At such low temperatures, the metal may disintegrate—a condition known variously as *tin pest, tin plague,* or *tin disease.* There have been some striking cases of this condition in the past. Thus the tin organ pipes at Zeitz, in Germany, crumbled to dust because of the cold in 1851. Tin blocks disintegrated on a wholesale scale in Russia in the course of the bitter winter of 1867–68.

This characteristic of tin, of course, rules out its use wherever low temperatures are involved, as for example in air-conditioning equipment. It should be pointed out, however, that when the metal is in thin layers, as in tinplate, or when it is alloyed with other metals, it does not undergo the change described above.

Certain elements may be changed from the nonmetallic state to the metallic. At extremely high pressures and temperatures, the covalent structures of the nonmetals germanium, silicon, and diamond may be converted to metallic structures that conduct electricity. Thus diamond becomes an electrical conductor at a pressure 600,000 times greater than normal atmospheric pressure and a temperature of 1,000° Celsius. Under these conditions, the 4-neighbor structure of diamond changes to a 6-neighbor structure. It is conceivable that even under certain natural conditions, as for instance within the interior of the earth, elements ordinarily classed as nonmetals may act like metals.

A FEW TYPICAL NONMETALS

The two crystalline forms of carbon. The element carbon occurs free in nature in two crystalline forms—diamond and graphite. We have already described one of these—diamond—as a structure involving a 4-neighbor arrangement, held together with covalent bonds. The other form of carbon—graphite—has a 3-neighbor structure in which the carbon atoms lie in distinct layers or sheets. The covalent bonds in graphite are such that one of the electrons in the carbon atom is quite mobile. Hence graphite conducts electricity, but only in directions parallel to the layers of carbon

D. The carbon atoms C^2, C^3, C^4, C^5 are grouped around carbon atom C^1 so as to form a tetrahedron. This pattern is continued throughout the entire diamond crystal.

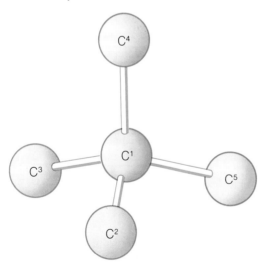

atoms. In directions at right angles to the layers, graphite is an insulator. (Fig. 2.)

The differing structures of diamond and of graphite help explain the differences in their uses. Diamond, as everyone knows, is an extremely hard substance. This hardness is due not only to the geometric arrangement of the atoms but also to the very short distances between atoms in the diamond crystal. Few atoms are smaller than those of carbon, and few chemical bonds are shorter than those in diamond. Because of its hardness, durability, and refraction of light, diamond is a favorite ornament. It is also used as a grinding material.

In graphite, the layers are about 3.4 Å (angstroms—a unit equivalent to a hundred-millionth of a centimeter) apart. This is a not inconsiderable distance when we deal with atoms. Consequently, these layers are only weakly attracted to each other and they can move over one another easily. We may compare a stack of graphite layers to a deck of playing cards. The cards lie in separate planes and can slide over one another. Yet the fifty-two cards, stacked one on top of the other, make up a deck of definite size and shape. In like manner, the different graphite layers, set one on the other, constitute a crystal of graphite. Because of its layered arrangement, graphite is soft. It serves very well as a lubricant.

Graphite can be converted to diamond, but only under extreme conditions of temperature and pressure. Diamonds can be made more easily by chemical processes from carbon-containing compounds such as methane. They are much like natural ones except for size and gem quality. They serve just as well as natural diamonds for industrial uses, such as cutting and polishing hard materials.

Silicon. Silicon belongs to the same family of elements as carbon, and it has the same 4-neighbor structure as diamond. The silicon-silicon bonds in this arrangement are somewhat longer (and consequently somewhat weaker) than those in diamond. As a result, silicon is not so hard a substance as diamond.

Sulfur. As a nonmetal, sulfur is quite different from diamond. It usually exists in

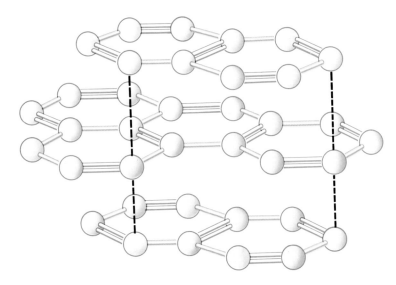

Fig. 2. The crystalline structure of graphite. Each circle represents a single carbon atom; each atom is joined to three of its neighbors. Three layers of atoms are illustrated here.

the form of small molecules rather than as a giant molecule – that is, one with the same atomic pattern endlessly repeated throughout the crystal.

At room temperature, ordinary sulfur is a yellow solid, consisting of molecules made up of 8 atoms arranged in a ring so as to form a sort of crown. When the solid is heated, it undergoes some striking changes. Upon melting, it forms a straw-colored liquid, still containing 8-atom molecules in the shape of rings. Since these rings slide over one another, the liquid flows easily. But when the sulfur is heated to a higher temperature, the sulfur rings open up and the ends join to form long chains of sulfur atoms. These chains become entangled and as a result the liquid, which is now dark, becomes viscous – that is, does not flow readily. At 444.6° Celsius, the boiling point of sulfur, the long chains break down into smaller ones, and the liquid again flows easily.

If the hot liquid is now poured into cold water, a rubbery form of sulfur, called plastic sulfur, is formed. It is *amorphous;* that is, it has no definable crystalline pattern. If, however, it is left standing at room temperature, it gradually forms small crystals of sulfur, composed of S_8 molecules.

A FEW TYPICAL METALS

Iron. This metal shows striking structural changes when it is heated to certain temperatures. At room temperature, the metal has a crystalline structure in which each atom has 8 neighbors. When it is heated to 910° Celsius, the structure becomes more compact, each atom having 12 neighbors. If the metal now cools slowly, the structure reverts to the original 8-neighbor form.

Permanent changes may be brought about in iron by dissolving carbon in it with the application of heat. If the iron-carbon system, a form of steel, is permitted to cool slowly, it becomes soft and malleable. If it is cooled rapidly, or tempered, the metal may become hard and brittle, or else tough, depending upon the rate of cooling.

Aluminum. This metal is reactive chemically. It combines with oxygen when it is exposed to the outer air. If this reaction continued indefinitely, the aluminum would be seriously weakened and at last would disintegrate. Are we to assume, then, that this would happen to the aluminum wings, fuselage, and rudder of an airplane made of aluminum? Fortunately the answer is "No." It is true that the surface of aluminum oxidizes even more rapidly than does that of iron. However, the outer aluminum-oxide layer formed on the surface of the metal adheres tightly to the metal beneath and forms an impervious coating for it. Consequently, we may use aluminum for structural purposes without fear of its high chemical activity except when it is exposed to alkalis, in which the oxide layer is quite soluble. It is for this reason that an aluminum teakettle should never be cleaned with household lye. The aluminum-oxide coating would react vigorously and dissolve in the lye solution. On the other hand, many acids, including concentrated nitric acid, have no effect on aluminum, because the aluminum-oxide layer is impervious to them.

Copper is not very active chemically. It slowly oxidizes in the atmosphere to form a gray-green "rust" which is a basic copper carbonate. This coating is called *patina*. Patina grows thicker and thicker as the metal continues to be exposed to the outer air. Microscopic growth rings, analogous to those in growing trees, are displayed by patinas in copper. No connection has ever been discovered between these rings and climatic or weather conditions.

Below is a microsection of the crystal structure of commercial aluminum. If hammered at high temperatures it will break into small fragments.

Alcan

Lead. Lead is a historic metal. It was used in large quantities by wealthy Romans for pipes and domestic utensils. The Romans called the metal *plumbum,* a word from which our words *"plumbing"* and *"plumber"* are derived. The chemical symbol for lead — Pb — is derived, of course, from *plumbum.*

When lead is absorbed into the human body, its poisonous qualities are related to the amount absorbed. Lead poisoning can develop very slowly, with the initial symptoms hard to recognize. A significant number of cases have been found among children who eat chips of lead-containing paint peeling off the walls of deteriorating buildings. Old lead water pipes also present a hazard because minute amounts of lead get into the drinking water and can have a slow cumulative poisoning effect. Although the use of lead in such products as paints and water pipes is no longer permitted, the metal is still valuable. It is used in storage batteries and as a shield against high-energy radiation.

Lead alloys are important in industry. To make the machining of metals such as steel, bronze, and brass more effective, lead alloys are added. They serve as bearing metals and as solders. The metal type used in printing is an alloy of lead, tin, and antimony. The tin makes the molten metal more fluid in the mold and produces finer detail in the finished type. The antimony makes the type harder.

DIFFERENCES IN CHEMICAL PROPERTIES

The chemical elements that tend to lose electrons — that is, to be *oxidized* — upon reacting with other elements are generally metals. On the other hand, the elements that gain electrons — are *reduced* — in chemical reactions or that share electrons with other elements are generally classified as nonmetals.

Reactions of metals. When a metal is in the gaseous state, a measure of its tendency to lose electrons and form ions is *ionization energy* — that is, the energy required to pull an electron from the atom. For sodium the ionization energy is expressed as follows:

Na → Na⁺ + e⁻ 5.14 electron volts

This is not so mysterious as it may seem. It means simply: "The sodium atom (Na) yields (→) a sodium ion with one positive charge (Na⁺) plus an electron (e⁻), when it is subjected to 5.14 electron volts." An electron volt is a unit of energy equivalent to the energy gained by an electron in passing through a potential difference of one volt. Or, putting it somewhat differently, "when a force of 5.14 electron volts is applied to a sodium atom, it loses an electron and becomes a sodium ion with one positive charge." The number of positive charges of the ion is equivalent to the number of electrons it loses.

Within a group of similar elements, such as those which make up the alkali family, the ionization energy decreases as the atoms of the elements become larger. By way of example, here are four of the alkali elements. We give the size of the atom and the ionization energy required to pull electrons from the atom:

Name of element	Atomic diameter	Ionization energy
Na (sodium)	3.7 Å (angstroms)	5.14 ev (electron volts)
K (potassium)	4.6 Å	4.34 ev
Rb (rubidium)	4.9 Å	4.18 ev
Cs (cesium)	5.3 Å	3.89 ev

Obviously, of the four atoms in the table above, cesium loses its valence electrons most easily; sodium, least easily.

Various factors have to be reckoned with in predicting the chemical activity of the different groups of metals in the periodic table. In addition to ionization energy, one must consider the number of electrons a metal atom loses. Some may lose up to three. One must also take into account the diameter of the atom, because the smaller the atom, the more difficult it is to extract electrons from it. The existence of such varying factors accounts for the unusually high reactivity of the alkali metals and the unusually low reactivity of such metals as gold, platinum, and silver.

In aqueous solutions (solutions with

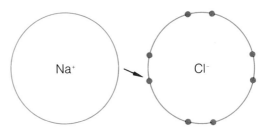

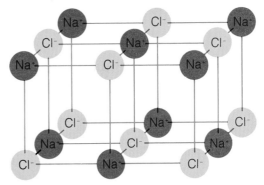

Fig. 3. Above: when sodium and chlorine atoms interact, the single electron in the outer shell of the sodium atom is captured by the chlorine atom, which now has 8 electrons in its outer shell. The sodium atom has become a sodium ion with a single positive charge (Na⁺). The chlorine atom is now a chlorine ion with one negative charge (Cl⁻). Below: the sodium ions and chlorine ions attract one another and form a crystal of table salt (sodium chloride).

water as the solvent), or under most natural conditions where moisture is present, the chemical reactivity of metals is most conveniently summarized in the *electrochemical*, or *electromotive*, *series*. This consists of a list of metals, arranged in the order of their tendency to undergo oxidation in aqueous solutions. By oxidation we mean here the loss of electrons by the atom, with the resulting formation of an ion, an atom with an electric charge. For example, when lithium undergoes oxidation, it loses one electron, and the resulting lithium ion has one positive charge:

$$Li \rightarrow Li^+ + e^-$$

"Lithium (Li) yields a lithium ion with one positive charge (Li⁺) plus one electron (e⁻)."

When lead (Pb) is oxidized, it loses two electrons, and the resulting ion has two positive charges:

$$Pb \rightarrow Pb^{++} + 2e^-$$

As we pointed out, the number of positive charges in the ion corresponds to the number of electrons lost by the atom.

Here is a representative list of metals arranged in their appropriate place in the electrochemical series:

$$Li \text{ (lithium)} \longrightarrow Li^+ + e^-$$
$$Ca \text{ (calcium)} \longrightarrow Ca^{++} + 2e^-$$
$$Na \text{ (sodium)} \longrightarrow Na^+ + e^-$$
$$Mg \text{ (magnesium)} \rightarrow Mg^{++} + 2e^-$$
$$Al \text{ (aluminum)} \longrightarrow Al^{+++} + 3e^-$$
$$Fe \text{ (iron)} \longrightarrow Fe^{++} + 2e^-$$
$$Ni \text{ (nickel)} \longrightarrow Ni^{++} + 2e^-$$
$$Pb \text{ (lead)} \longrightarrow Pb^{++} + 2e^-$$
$$Cu \text{ (copper)} \longrightarrow Cu^{++} + 2e^-$$
$$Ag \text{ (silver)} \longrightarrow Ag^+ + e^-$$
$$Au \text{ (gold)} \longrightarrow Au^+ + e^-$$

At the top of the list are the metals most easily oxidized; at the bottom, those hardest to oxidize. The metals most easily oxidized are those hardest to reduce.

Metals generally form compounds with many nonmetals. Thus they combine with oxygen, sulfur, and chlorine to form oxides, sulfides, and chlorides, respectively.

Reactions of nonmetals. Nonmetals enter into chemical reactions by gaining electrons—reduction—or by sharing their electrons with other elements.

Chlorine, for example, is an extremely active nonmetal and reacts vigorously by gaining an electron:

$$Cl + e^- \longrightarrow Cl^-$$

"The neutral atom chlorine (Cl) gains an electron (e^-) and becomes a chloride ion with a single negative charge (Cl^-)."

A nonmetal's tendency to gain electrons is called *electron affinity.* It is a measure of the energy released when the nonmetallic atom combines with an electron. For chlorine it is 3.78 electron volts.

When sodium reacts with chlorine, the transfer of electrons from the sodium to the chlorine accounts for the reaction:

$$Na + Cl \longrightarrow Na^+ + Cl^-$$

"When sodium (Na) reacts with chlorine (Cl), they yield a positive sodium ion (Na^+) and a negative chloride ion (Cl^-)."

The resulting positive sodium ions and negative chloride ions attract one another and form a crystal of table salt (NaCl: sodium chloride). This crystal is a collection of equal numbers of sodium ions and chloride ions, held together by the mutual attraction of positively charged particles and

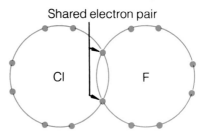

Fig. 4. When chlorine reacts with fluorine, forming the compound chlorine fluoride (ClF), they share an electron pair. This is an example of covalence.

negatively charged particles. (See Fig 3.) This sort of chemical union is called *ionic valence,* or *electrovalence.* The energy of this interaction of ions is called *lattice energy.* The net energy (the combination of ionization energy, electron affinity, and lattice energy) determines the stability of the compound.

In some cases, as we have pointed out, nonmetals form chemical combinations by sharing electron pairs—a form of chemical combination called *covalence.* A familiar example of covalence is the binding together of chlorine and fluorine atoms. The chlorine (Cl) atom has seven valence electrons in its outer shell. So has the fluorine (F) atom. When these two nonmetals combine, they form the compound chlorine fluoride (ClF). The chlorine and the fluorine atoms in this compound now share an electron pair. (Fig. 4)

METAL ORES

Only a few metals, such as gold, silver, and copper, occur as free metals in nature. Most metals are combined with other elements. The ease with which they may be obtained from their ores is closely related to the position of the metal in the electrochemical series described above.

The metals at the top of the series,

such as lithium and calcium, are the hardest to reduce. They can be recovered from compounds only with the most vigorous type of chemical reaction—*electrolysis*. That is how magnesium is obtained from magnesium chloride:

$$MgCl_2 \xrightarrow{\text{electrolysis}} Mg + Cl_2$$

"A molecule of magnesium chloride ($MgCl_2$), subjected to electrolysis, yields a molecule of magnesium (Mg) and a molecule of chlorine (Cl_2)."

At the bottom of the electrochemical series are the metals that are the easiest to reduce. They can be obtained merely by heating their oxide or sulfide ores. Gold is won from gold oxide in that way:

$$2Au_2O_3 + heat \longrightarrow 4Au + 3O_2$$

"Two molecules of gold oxide ($2Au_2O_3$) are subjected to heat and yield 4 molecules of gold (4Au) and 3 molecules of oxygen ($3O_2$)."

In many common metals that are some distance from either end of the electrochemical series, the free metal can be obtained by heating its ore with coke (a form of carbon), a mild reducing agent. That is how zinc is obtained from the compound zinc oxide:

$$ZnO + C \longrightarrow Zn + CO$$

"A molecule of zinc oxide (ZnO), heated with coke (C), yields a molecule of zinc (Zn) and a molecule of carbon monoxide (CO)."

Further treatment is usually required in order to remove certain impurities from the metals thus obtained.

FORMATION OF ALLOYS

If the common metals were used only in the pure form, modern industry would be severely handicapped. By mixing metals with other metals or with nonmetals, we obtain metallic systems, called *alloys*, that offer a wide variety of useful properties.

Some alloys are simply solutions of metals. They are made by melting two or more metals in almost any desired ratio and then letting the mixture solidify. The metals whose atoms are of nearly the same size are most likely to form solution alloys. Copper and silver form solution alloys with each other; so do silver and gold, iron and nickel, and sodium and potassium.

Two metals of considerable difference in chemical properties may combine in an alloy to form chemical compounds. Thus copper and aluminum form the compounds Cu_3Al and Cu_9Al_4. The ratio of the combining metals is not always what we would expect.

Some outstanding alloys are produced by mixing certain nonmetals with metals. Thus the addition of small percentages of carbon or silicon to iron produces various types of steel, with different properties.

The so-called *interstitial alloys* form an interesting group chemically. They result when the interstices, or gaps, between the closely packed elements of a metal are occupied by the much smaller atoms of another element—metal or nonmetal. For ordinary close-packed metallic systems, the maximum ratio of parent metal to interstitial element is 1 to 1. Among interstitial alloys we sometimes find this maximum ratio, as in the case of TiC (titanium carbide), TiN (titanium nitride), and WC (tungsten carbide). Some of these substances, such as tungsten carbide, are extremely hard and are used for machine-cutting tools that must remain sharp even at red heat.

In some metals, such as palladium, not only are the interstitial holes large enough to accommodate small atoms, such as those of hydrogen, but these small atoms are so loosely attached to their metallic neighbors that they migrate from hole to hole. For example, if hydrogen gas (H_2) is placed in a palladium cylinder, the hydrogen molecules will first form hydrogen atoms (H). The latter will trickle through the interstitial holes of the palladium. Finally emerging on the outside of the cylinder, they will combine to form hydrogen molecules. Obviously, hydrogen should never be stored in a palladium cylinder, since it will leak out. Fortunately, other metals are impervious to hydrogen.

U.P.I.

Washington State Department of Commerce and
Economic Development

WATER

by A. B. Garrett

Water is probably the most abundant as well as the most important compound upon the earth. We find it in the oceans that cover some three-quarters of the earth's surface; in rivers and lakes; and, in frozen form, in the huge ice caps of the Arctic and Antarctic regions. It is present in the soil and, as a vapor, in the atmosphere. It makes up a large percentage of plant and animal tissue.

The action of water is the principal erosive force on the earth's surface and it is responsible for much of the variety of the landscape. The earth would be an entirely different kind of place if there were no water in the soil and air, for under such conditions life could not possibly survive. Truly our world is a "water planet."

Water plays a vital role in our daily lives. We drink it and cleanse ourselves with it. We use it to cook our food and to cool our engines. Water serves as a vast highway for our ships. The farmer uses it to grow his crops and to sustain his animals. It is indispensable in a vast number of industrial operations.

The energy derived by the earth from the sun provides temperatures high enough so that water exists most commonly as a liquid. Yet, as you know, water is not restricted to the liquid state. Under the proper conditions, it can pass from solid to liquid to gas to liquid to solid and so on. In the case of certain celestial bodies, such as the moon and the planet Mercury, when water was converted to the gaseous state, it escaped into outer space because the force of gravity was not strong enough to retain it. The gravitational force of the earth is sufficiently strong to reduce to a tiny percentage the escape of water vapor from the atmosphere.

MOLECULAR STRUCTURE OF WATER

Water has the chemical formula H_2O, meaning that in every water molecule two atoms of hydrogen are attached to a single atom of oxygen. The three atoms are bonded together in such a way as to form a triangular structure. If straight lines were drawn from the hydrogen atoms so that they would intersect at the oxygen atom, the resulting angle would be about 105° (Figure 1). The bonds in this case are

Jones and Laughlin Steel   U.P.I.

covalent. This means that the oxygen atom shares electrons with the two hydrogen atoms. The oxygen atom has a slightly greater attraction for electrons than the hydrogen atoms do. Hence the shared electron pairs will be a little closer to the oxygen atom than to the hydrogen atoms. As a result, the oxygen end of the water molecule will have a slight negative charge; the hydrogen ends, a slight positive charge. A molecule of this type, with one part negatively charged and another positively charged, is called a *polar molecule*. Since the bonds between the atoms of the water molecule are both polar and covalent, they are often called *polar covalent*. This gives water unusual properties, as we shall see.

A striking characteristic of water molecules is that in the liquid state they are generally joined together to form linkages of two or more molecules. This is because of the polar character of the water molecule. One of the two positively charged hydrogen atoms of such a molecule will link itself to the negatively charged oxygen end of another molecule. Perhaps the other hydrogen end of the first molecule will attach itself to the oxygen end of a third molecule (Figure 2). The hydrogen bridge formed in this way between the oxygen ends of two different water molecules is called a *hydrogen bond*.

U.P.I.

The structurally simple compound H_2O is indispensable to all of us. Besides being vital to industry, agriculture, and recreation, water is a basic constituent of living organisms and many minerals. Water is also responsible for many chemical and physical processes that occur in the universe. The photos on these two pages give some idea of the many uses of water.

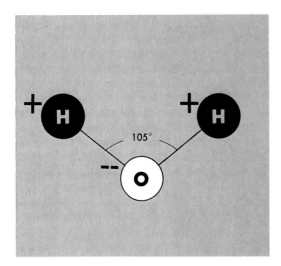

Fig. 1. In the water molecule, an oxygen atom (O) is bonded to two atoms of hydrogen (H). The oxygen and hydrogen ends of the molecules are electrically charged, as here.

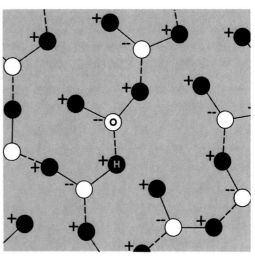

Fig. 2. A typical linkage of water molecules. A positively charged hydrogen end of one molecule is linked to the negatively charged oxygen end of another molecule.

IMPORTANT PHYSICAL PROPERTIES

Water as a solvent. Water is an excellent solvent because of the polar character of the water molecule. It can attach itself readily to either positive or negative ions, surrounding them and promoting the solvent effect. At the same time, water has a high *dielectric,* or insulating power, for the electric forces existing between ions. Hence it tends to keep electrically charged ions apart, thus spreading the solvent effect over a wide area.

The dissolving power of water is one of its most important properties. Various substances in the earth's crust go into solution in rain or in standing or flowing water. Huge rocks can gradually be broken down in this way. Minerals dissolved in water are absorbed by plants and are used to make new tissue. The food manufactured by a plant in the process of photosynthesis is transported, dissolved in water, throughout the plant. In the human body, dissolved food is carried to the cells in the bloodstream, which is made up mostly of water.

The expansion of water upon freezing. Most substances contract when they solidify from the liquid state. Water, however, expands considerably when it freezes. This remarkable quality is due to the linkage between molecules because of hydrogen bonding. As the temperature becomes lower, and the molecules become more sluggish, there will be more and more hydrogen bonding. Finally when freezing takes place—at O° Celsius—there will be a pattern of molecular linkages joining together all the molecules and forming a network structure. The molecules of ice, therefore, will occupy more space and will be less dense than those of liquid water.

The expansion of water as it freezes has various important consequences. For one thing, the water that makes its way between cracks in rock expands upon freezing and exerts great pressure. This gradually causes fragments of rock to break off. Thus freezing water becomes an effective erosive agent.

Because the ice that forms on a body of water—a river, say, or a pond—is lighter than the water that has not yet been frozen, it stays on the surface instead of sinking to the bottom. The ice, therefore, will collect on the surface and the water will freeze from the surface downward. The surface ice tends to insulate the liquid water underneath it from the effects of freezing temperatures. The thicker the ice becomes, the greater its insulating force will be. Hence unless a body of water is very shallow, it may never be frozen over entirely. It is for this reason that aquatic plant and animal life can survive freezing temperatures. The fact that freezing takes place only on the surface of bodies of water has an important effect on the climate.

The high specific heat of water. The *specific heat* of a substance is the heat required to raise the temperature of a gram of the substance 1° Celsius. Because so much energy is required to break hydrogen bonds as the temperature is raised, a relatively large amount of heat will be necessary to raise the temperature of a gram of water 1°. The specific heat of water is higher than that of any other substance, and therefore it is used as a measuring stick. It is given as 1. The specific heat of other substances is given as a fraction of 1.

Because of its high specific heat, water can absorb large amounts of heat with relatively little change in temperature, compared to other substances. That is why it is used in the cooling system of automobiles. Water absorbs the heat produced by the explosions in the cylinders of the engine. It is cooled as it passes through the radiator and it is then returned to the water jacket of the engine to absorb more heat.

Since so much heat is required to raise the temperature of water, it is often used in heating systems. The water in a hot-water heating system, for example, absorbs a large amount of heat in the furnace and gives off this heat as it passes through the radiators in the different rooms of the house.

The high boiling and freezing points of water. The linking of water molecules by hydrogen bonds has an effect upon its *boiling* and *freezing points*. It requires added energy to break down the linked molecules of water as the temperature is raised to the boiling point, when the liquid will pass to the gaseous state. Therefore the boiling point of water — 100° Celsius — is much higher than we would expect.

When a liquid becomes a solid, its molecules are combined to form a crystal lattice. Because so many water molecules are already joined together even in the liquid state, the water will become solid — that is, will freeze — at a higher temperature than we would expect. The freezing point of water, as we have seen, is 0° Celsius.

Above the freezing point many of the water-molecule linkages will break up and as a result the liquid will become denser until the temperature reaches the 4° Celsius mark. Above this reading, the liquid will begin to expand because of the increase in the vibration of the molecules with increasing temperatures. Most substances expand with increasing temperatures when they go above the 0° Celsius mark.

The high heats of vaporization and fusion. A certain amount of heat is required to change one gram of liquid into a gas without any change in temperature. This is called the *heat of vaporization*. For water the heat of vaporization is 539.6 calories per gram at its boiling point of 100° Celsius.

When a body of water freezes, the ice collects on the surface, and the water freezes from the surface downward. The surface ice tends to insulate the liquid water below from freezing temperatures. Hence living aquatic things, like the whale shown here, can survive freezing temperatures.

A certain amount of heat is absorbed by one gram of a solid when it changes into a liquid with no change in temperature. This is known as the *heat of fusion.* For water it is 79.6 calories per gram.

Both the heat of vaporization and the heat of fusion are higher for water than for most other substances. This has a pronounced effect upon the climate of the land areas near large bodies of water, as well as upon wind direction and wind velocity near such bodies of water.

IMPORTANT CHEMICAL PROPERTIES

The stability of water at room temperature. We generally consider that water is very stable at room temperature. Actually its stability will depend upon the kind of substance with which it is mixed. Thus gold, at room temperature, will not react with water at all. Iron will react with water very slowly to form rust. But when potassium comes in contact with water at this temperature, there will be an explosive reaction.

When water promotes combustion. At room temperature, the bonds between the oxygen atom and the hydrogen atoms of the water molecule are strong. A few of the water molecules will split, however, at this temperature. One hydrogen atom of the molecule will break away, becoming a hydrogen ion with a positive charge (H^+). The hydrogen ion does not exist by itself. It always combines with one or more water molecules to form a *hydronium ion,* the simplest type being H_3O^+. The remaining hydrogen atom will remain attached to the oxygen atom of the molecule. This oxygen and hydrogen will now form the negative ion OH^-, which is called a *hydroxyl ion.* At 25° Celsius, one out of every 55×10^7 molecules of water will undergo this change.

As the temperature is raised, the vibration of the atoms that make up water molecules gradually increases until the bonding energy between oxygen and hydrogen atoms will no longer be strong enough to keep them together. The molecules will now begin to dissociate, or break apart, to free oxygen atoms and hydrogen atoms. At 2,000° Celsius about 2 per cent of the water

molecules are dissociated. Since it can now supply free oxygen, water becomes an oxidizing agent. That is why metals such as magnesium, aluminum, and iron burn readily in water vapor at temperatures above 2,000° Celsius.

Some reactions of water. When water combines with certain compounds, it forms acids. Thus it reacts with carbon dioxide, CO_2, to produce carbonic acid, H_2CO_3. When water reacts with certain other compounds, it forms bases. For example, calcium oxide, or quicklime, CaO, combines with water to form the base calcium hydroxide, or slaked lime, which has the formula $Ca(OH)_2$. Water reacts with some salts, such as ammonium chloride, NH_4Cl, to form acid solutions. With other salts, such as sodium carbonate, Na_2CO_3, it reacts to form basic solutions. In the case of a few salts, the only effect of water appears to be to split these salts up into ions. Thus when sodium chloride, $NaCl$, is added to water, the sodium ions (Na^+) are separated from the chloride ions (Cl^-). The resulting solution is a good conductor of electricity.

The formation of hydrates. Water molecules attach themselves directly to certain substances to form compounds called *hydrates.* Thus when crystals of copper sulfate, $CuSO_4$, are formed in water, five molecules of water are attached to the two ions of copper sulfate—Cu^{++} and SO_4^{--}. The resulting hydrated copper sulfate has the formula $CuSO_4 \cdot 5H_2O$. The water in such a crystal formation is called *water of hydration.* If the copper sulfate crystals are heated, the water is driven off and the copper sulfate is converted into a powder called *anhydrous* copper sulfate. Anhydrous means "without water."

Water as a promoter of chemical reactions. Certain substances that would otherwise not react at all will combine chemically when water is added. For example, chemical reactions in plant or animal tissue take place only in the presence of water. Iron combines with the oxygen of the air to form iron oxide, or rust, only because water is present in the air in the form of water vapor. Various explanations have been offered for the role of water in

Phelps Dodge Industries

Hydrated copper sulfate crystals, which are blue in color, are shown in the left-hand photo. When the crystals are treated with controlled heat, like that in an electric oven, the water contained in the crystals is driven out, and a powder called anhydrous copper sulfate is obtained (right-hand photo). If the latter substance is allowed to stand in air, it will combine with water vapor from the air, and blue crystals of hydrated copper sulfate will again be formed.

furthering chemical reactions. In many cases, it serves as a medium in which reacting substances can be brought in contact more readily. In some cases, it may be a catalyst.

HEAVY WATER

As is pointed out elsewhere in this set, there are generally several different kinds of atoms in a given element—all of them alike chemically but differing in atomic weight. These different types of atoms are known as the *isotopes* of the element. Ordinary water is made up of the commonest isotope of hydrogen, with mass number 1, and the commonest isotope of oxygen, with mass number 16. However, in a small number of water molecules, each of the hydrogen atoms is an isotope with mass number 2. The name deuterium is given to such an isotope. It has the symbol D. Water containing deuterium instead of the common hydrogen isotope is called heavy water and is indicated by the formula D_2O.

There are other kinds of heavy water. Oxygen may be combined in the water molecule with hydrogen of mass number 3, known as tritium (symbol T). Again, the oxygen in the water molecule may be an isotope with atomic mass 17 or 18 instead of the commonest isotope of oxygen with atomic mass 16.

Deuterium oxide is widely used for research purposes in nuclear chemistry. It is also used as a moderator to control fission reactions in certain kinds of nuclear reactors. Tritium oxide has also been used in research and it is one of the components of the hydrogen bomb. Water containing the heavier isotopes of oxygen has served as a "geologic thermometer," to analyze variations of temperature in past geologic ages.

TRANSFORMATIONS OF THE WATER MOLECULE

The lowest theoretical temperature is absolute zero, representing the point at which all molecular motion ceases. It is indicated on the Kelvin scale as 0° Kelvin, and it corresponds to −273.16° Celsius. The Kelvin scale corresponds to the centigrade scale except that it begins with absolute zero (0°K) and has no negative temperatures. The highest known temperature, reached in some of the very hot stars, is about 5,000,000,000° Kelvin. It will be interesting to see what happens to a water molecule or its parts as it passes over this vast temperature range. The table on page 92 traces the changes of a water molecule.

SUBSTANCES FOUND IN WATER

Pure water is seldom found in nature. For one thing, water is such an excellent solvent that it generally contains dissolved materials in greater or lesser amounts. Salts dissolved in seawater make up something like 35 parts per thousand on the average. By far the most abundant of these salts is sodium chloride, or common salt. There are also considerable quantities of magnesium chloride, magnesium sulfate, calcium sulfate, and other salts. There are far less dissolved salts in freshwater lakes and streams that are fed chiefly by precipitation (rain, snow, sleet, and so on) and small streams, but they are present nevertheless.

	History of Water from 0° K to Stellar Temperatures*	
0° K	Water molecules are arranged in regular crystal lattice. Little or no molecular motion.
100° K	Water molecules are vibrating about fixed positions in crystal lattice.
273° K	Water molecules in the crystal are vibrating so rapidly that appreciable numbers of them break away to form a liquid or a gas.
373° K	Water molecules are moving rapidly in all directions; may be in either gaseous or liquid state depending upon pressure.
700° K	All water molecules are in gaseous state regardless of pressure. Atoms of the molecule are vibrating vigorously; atoms ionize.
2,000° K	Atoms are vibrating so vigorously that they cause appreciable numbers of the molecules to dissociate into atoms; atoms ionize.
5,000° K	Water molecules are dissociated into atoms and the atoms have so much energy that they are ionized. (No known substances can exist in the solid state at this temperature.)
10,000° K	All molecules of matter are dissociated into atoms and the atoms are highly ionized.
1,000,000° K	All atoms of matter are ionized; many are stripped to bare nuclei. All matter is converted to gaseous mixture of electrons, nuclei, and ions.
10,000,000° K	Atomic nuclei undergo fusion.
5,000,000,000° K	Complex atomic nuclei probably break up.

*This table has been adapted, with permission, from Garrett, Richardson, Montague: Chemistry, revised edition, Ginn and Company, 1966.

Gases are also dissolved in water. When air goes in solution in water, the oxygen it contains is used by fishes and other forms of aquatic life for respiration. Nitrogen, carbon dioxide, and the other gases found in the air are also dissolved in water.

When water sinks into the earth, it dissolves minerals contained in the soil and rocks. If the water penetrates limestone formations, which are composed mainly of calcium carbonate, $CaCO_3$, a chemical reaction takes place involving calcium carbonate, water, and the carbon dioxide dissolved in water. The product of this reaction is calcium bicarbonate, $Ca(HCO_3)_2$, which is readily dissolved in water. In humid regions, so much of the original calcium carbonate is converted to calcium bi-

carbonate and then dissolved in water that huge caverns are formed.

PURIFICATION OF THE WATER SUPPLY

Certain purification processes are going on all the time in nature. Organic wastes are diluted as they flow along streams and are acted on by the oxygen dissolved in these streams. A considerable proportion of the substances in suspension in the water, such as sediment, is precipitated. Water is aerated as it plunges down waterfalls or rapids. Groundwater is filtered as it sinks through the soil. Small fish and other forms of aquatic life eat the algae that are often found in water.

But natural purification methods such as these rarely suffice except in unspoiled surroundings. In most cases it would not be possible to use safely the water found in nature without some form of treatment. Water that is to be drunk should not contain harmful bacteria or protozoans or dissolved substances that would make it dangerous to health or give it a bad taste or odor. Water to be used for cleansing purposes should be free of minerals that might impede its cleansing action. The water that is to serve in industry has to be treated to remove most of the salts that would cause corrosion or that would precipitate in tanks or in the pipes of boilers or that would in any other way interfere with industrial processes. For certain industrial purposes and in chemical laboratories, the water must be especially free from impurities, since these would affect chemical reactions in which water is not just the medium but is rather one of the reacting substances.

Various methods are used for the purification of water. Surface waters are purified in a series of steps involving filtration, aeration, coagulation, chlorination, and other processes. Groundwater is treated by several chemical processes to remove its dissolved minerals. The particular method that is selected will depend on the uses to which the water will be put, the purity required, the quantity to be purified, and the cost. For a discussion of water treatment, see the article "The Water Supply" in this volume.

Acids, bases, and salts are importants parts of everyday life. Most of the medicines in this pharmacist's dispensing area can be classed as acids, bases, or salts.

ACIDS, BASES, AND SALTS

by Anthony Standen

The substances known as *acids, bases,* and *salts* are among the most important with which the chemist deals. They are widespread in nature, in industry, and in the home. Sulfuric acid plays a major part in many industrial processes; so does nitric acid. Carbonic acid is particularly familiar to us in the form of soda water. Hydrochloric acid, which is present in small quantities in our stomachs, is a vital factor in the digestive process. Among the more familiar bases are lye, or caustic soda, and ammonia water. By far the best-known salt is sodium chloride, or common table salt. It is extensively used in industry and in the home. It is present in the water of the ocean and in our own blood stream. Other well-known salts are Epsom salts and baking soda, or sodium bicarbonate.

PROPERTIES

It is easy to describe acids in terms of what they do. Chemists have been familiar with the properties of these substances for many years. Among other things, they are sour and they corrode metals. For example, the acetic acid found in vinegar is responsi-
ble for the tart taste that imparts flavor to salads and other foods. It is easy to show that acetic acid corrodes metals. If you leave a steel knife immersed in vinegar, the knife will quickly corrode. The acetic acid in vinegar is only moderately strong, but at full strength the acid blisters the skin. Sulfuric acid is so corrosive that it is seriously damaging to the human skin. In some very weak acids, the properties of sourness and corrosion are hardly noticeable, and these substances may even be used as medicines. Thus acetylsalicylic acid (better known as aspirin) is a valuable drug, used to relieve pain. Boric acid is used in ointments and in eye washes.

Strong bases, too, are usually quite corrosive, and they are apt to feel "soapy" when dissolved in water. In fact, bases—especially sodium hydroxide—are used in the manufacture of soaps. When a base is mixed with an acid in the right proportions, the two substances neutralize one another. They react to form a salt, a substance with quite different properties from those of acids or bases. Salts do not taste sour, nor do they feel "soapy" in solutions.

H. J. Heinz   Armour-Dial, Inc.   Morton Salt Co.

Familiar examples of an acid, a base, and a salt. Left to right: vinegar (acetic acid); ammonia (a base); and table salt (sodium chloride).

ACIDS

Just what is an acid? We can understand the answer to that question most clearly in terms of the modern theories concerning the structure of the atom and the electrical nature of chemical forces.

The hydrogen atom is much the simplest of all the atoms, for it has only one proton and one electron. Suppose now that it loses its electron. It then has only a proton, with a positive charge of electricity. It is no longer neutral electrically. It has become an ion—an electrically charged particle—with a single positive charge. The hydrogen ion is indicated by the symbol H^+. The H stands for hydrogen. The plus sign means that the ion has a positive charge. Since the hydrogen ion consists only of a proton, the two terms "hydrogen ion" and "proton" may be used interchangeably.

Now that we know what a hydrogen ion is, we can understand one of the most widely accepted definitions of an acid. According to this definition, an acid is any substance having in its molecule a hydrogen atom that can separate from the rest of the molecule as a hydrogen ion. In other words, all acids are sources of hydrogen ions, or protons.

Let us see what happens when hydrochloric acid (hydrogen chloride, HCl) dissolves in water (H_2O). Theoretically, the HCl would split up into two ions—a hydrogen ion, H^+, and a chloride ion, Cl^-. The minus sign indicates that this ion would be negative.

In actual fact, the hydrogen ion, H^+, does not exist by itself in solution in water. It always combines with a water molecule, H_2O, as follows:

$$H^+ + H_2O \rightarrow H_3O^+$$

H_3O^+ is an ion, since it has an electrical charge (a positive charge). It is called a *hydronium ion*. A hydronium ion, then, is nothing but a hydrogen ion that is combined with a molecule of water. It can easily split up again into a molecule of water and a hydrogen ion. It will do so, as a matter of fact, when another substance reacts with the ion.

When hydrogen chloride, therefore, is dissolved in water, it becomes a source of hydronium (H_3O^+) ions, and the latter can then react with other substances. That is why, in the presence of water, hydrogen chloride is an acid. The name "ionization" is given to the process of ion formation of which this is an example.

We can now write the full equation for the reaction that takes place when hydrochloric acid is dissolved in water. It is:

$$HCl + H_2O \rightarrow H_3O^+ + Cl^-$$

BASES

Like acids, bases ionize, or split up into ions, when they are dissolved in water. As we have seen, a typical base is lye (sodium hydroxide), which has the formula NaOH. It is composed of three atoms, one each of sodium (Na), oxygen (O), and hydrogen (H). The oxygen and the hydrogen remain together. When NaOH ionizes, therefore, it is broken up into a positive sodium ion, Na^+, and a negative ion, OH^-, called *hydroxyl* and composed of an oxygen atom and a hydrogen atom. The hydroxyl ion has one negative charge, which is carried by both atoms acting together as a unit. We can write the equation for the ionization of sodium hydroxide:

$$NaOH \rightarrow Na^+ + OH^-$$

A solution of sodium hydroxide in water, therefore, has both sodium ions and hydroxyl ions. It is characteristic of all bases that their solutions in water contain hydroxyl ions.

SALT FORMATION

Let us see what happens when an acid and a base react, neutralizing one another so that the product—a salt—has neither acidic nor basic properties. The hydrogen ion, H^+, of the acid and the hydroxyl ion, OH^-, of the base, upon reacting with one another, form water:

$$H^+ + OH^- \rightarrow H_2O$$

Or, rather, since the hydrogen ion is carried on a water molecule in the form of a hydronium ion, H_3O^+:

$$H_3O^+ + OH^- \rightarrow 2H_2O \text{ (two water}$$
$$\text{molecules)}$$

The hydrogen ion derived from acids

causes their sour taste. The hydroxyl ion derived from bases causes their "soapy" feel. But since nothing but water is formed when the hydrogen and hydroxyl ions react, these properties disappear.

The last equation, concentrating on the hydronium ion and the hydroxyl ion, did not show all the ions that are present when an acid and a base, dissolved in water, react with one another. Suppose that we mix hydrochloric acid with lye, or sodium hydroxide, which, as we have seen, is a base. The hydrochloric acid solution has chloride ions Cl^-, in addition to hydronium ions, H_3O^+. The sodium hydroxide solution has sodium ions, Na^+, as well as hydroxyl ions, OH^-. We can write the full equation:

$$H_3O^+ + Cl^- + Na^+ +$$
$$OH^- \rightarrow 2H_2O + Na^+ + Cl^-$$

The right-hand side of the equation, giving the product of the reaction, shows nothing but water, sodium ions, and chloride ions. Now when common salt, NaCl, dissolves in water, it forms sodium ions and chloride ions:

$$NaCl \rightarrow Na^+ + Cl^-$$

We can see, therefore, that when we treat hydrochloric acid with sodium hydroxide in water, the product is simply a solution of common salt in water. If we evaporate the water, the salt will be left behind as a white solid.

Another example of a salt is potassium nitrate, KNO_3, a colorless solid which, mixed with charcoal and a little sulfur, makes gunpowder. The potassium nitrate molecule consists of one atom of potassium (K), one atom of nitrogen (N), and three atoms of oxygen (O_3). Potassium nitrate is formed from nitric acid, HNO_3, and the base called potassium hydroxide, KOH,

Beckman Instruments, Inc.

A modern, digital pH meter. The instrument measures the pH of the solution in the beaker at the left. Then it gives the reading on the large dial.

which is very similar to sodium hydroxide. Nitric acid dissolves in water, forming the positive hydronium ion, H_3O^+, and the negative nitrate ion, NO_3^-:

$$HNO_3 + H_2O \rightarrow H_3O^+ + NO_3^-$$

Potassium hydroxide forms the ions K^+ and OH^- in solution. When nitric acid and potassium hydroxide solutions are mixed, the reaction is:

$$H_3O^+ + NO_3^- + K^+ +$$
$$OH^- \rightarrow K^+ + NO_3^- + 2H_2O$$

When the solution is evaporated, potassium nitrate, KNO_3, is left as a colorless solid.

From the last two examples we can see that neutralization involves replacing the hydrogen ions of an acid with the metal ions (such as sodium, potassium, or magnesium) of a base, and that a salt is a product formed by just such a replacement. Some acids have two hydrogen atoms. In that case either one or both of the hydrogen atoms can be replaced by an atom, or atoms, of a metal. Thus sulfuric acid, H_2SO_4, can form two series of salts. In one series—the acid sulfates—only one hydrogen atom is replaced, giving $NaHSO_4$ or $KHSO_4$. In the other series—the normal sulfates—both hydrogen atoms are replaced, giving Na_2SO_4 or K_2SO_4. In some cases, a single atom of a metal can replace two atoms of hydrogen. The salt magnesium sulfate, $MgSO_4$, is derived from sulfuric acid, H_2SO_4. One atom of magnesium has taken the place of both hydrogen atoms of H_2SO_4. In phosphoric acid, H_3PO_4, no less than three hydrogen atoms can be replaced.

Acids, bases, and salts all split up into ions when in solution in water. Water itself ionizes, although only to a very slight extent. It forms H^+ and OH^- ions. Then the H^+ ion combines with a water molecule, forming a hydronium ion, H_3O^+.

$$H_2O + H_2O \rightleftarrows H_3O^+ + OH^-$$

You will note that there are two arrows in the above equation. The small arrow, going from left to right, indicates that starting with water only a very small amount of H_3O^+ and OH^- is formed. Much more than 99 per cent of the water does not become ionized. The arrow going the other way indicates that if more than very small amounts of H_3O^+ and OH^- are present together (as when an acid and a base are mixed), these ions react with one another to form water. The fact that the arrow is large shows that this reaction goes almost to completion: only an extremely small amount of H_3O^+ and OH^- will be left.

pH

In ordinary water, there are as many H_3O^+ ions as OH^- ions, The quantity of each is exceedingly small. It is convenient to think of the concentration of H_3O^+ in terms of H^+, since each H^+ is carried on a water molecule, H_2O. It has been calculated that the concentration of H^+ in water is 0.0000001 gram per liter. This number can be written as 10^{-7}. The number written above the line, $^{-7}$, is called the exponent. The minus sign indicates that the figure 1 is to the right of the decimal point. The figure 7 indicates that it is seven places to the right.

In some solutions there is a comparatively high concentration of H^+ ions—typical acid ions—and only a slight concentration of OH^- ions—typical basic ions. Such solutions are acid. In other solutions the situation is reversed: there are many more OH^- ions than there are H^+ ions. Such solutions are *basic*, or *alkaline*. "Alkali" means "base." Since there are as many H^+ ions as OH^- ions in ordinary water, this liquid is neutral.

If a solution had a H^+ concentration of 0.1 gram per liter, this would be 1,000,000 times higher than the H^+ concentration in pure water: the solution would be strongly acid. If the H^+ concentration were 0.0001 grams per liter, it would be 1,000 times higher than the H^+ concentration in water. This solution would also be decidedly acid, though not so much so as the first one. We can express the acidity of a solution, therefore, by giving its H^+ concentration. We could write this as 10^{-1} and 10^{-4}, respectively, in the case of the two solutions we have just mentioned.

A convenient way of indicating the H^+ concentration in a solution is to forget the ten in a figure like 10^{-1} or 10^{-4} and simply give the exponent without the minus sign. This is called the pH of the solution. A solution with a H^+ concentration of 10^{-4} grams per liter would have a pH value of 4. A neutral solution, or pure water, has a pH of 7. A pH of 6 is very slightly acid, a pH of 1 very acid.

The pH system can be applied to basic solutions too. A pH of 8 would mean a hydrogen ion concentration of 10^{-8}, or 0.00000001 grams per liter. This is only one tenth of the concentration of the hydrogen ion, H^+, in pure water, and would imply that the OH^- concentration would be correspondingly greater. Such a solution would be slightly alkaline. The higher the pH, the more basic, or alkaline, a solution is. The degree of acidity of alkalinity of any solution, ranging from 1 to 14, can be most conveniently expressed by the pH system.

It is a matter of great importance in many fields of science and industry to be able to determine the acidity of a solution and express it in terms of pH. Chemists, biologists, bacteriologists, agricultural experts, and many others are interested in the pH of the particular solutions with which they work. For example, the physiologist is often concerned with the acidity of various body fluids. The juices of the stomach are strongly acidic, having a pH of from 1.0 to 3.0. The pH of the blood ranges from 7.3 to 7.5. This means that the blood is very slightly alkaline. If the pH of the blood were to fall to 7.0, exactly neutral, or rise to 7.7, slightly more alkaline than normal, the results might be fatal. Fortunately, the body has systems of *buffers* to combat excessive acidity or alkalinity. These are substances that are able to remove, by chemical action, an excess of either H^+ or OH^-.

INDICATORS

The chemist can determine the pH of a solution in various ways. One of the simplest systems involves the use of substances called *indicators*. When these come into contact with acidic or basic solutions, they change color at different points on the pH scale. A familiar indicator is litmus paper. This changes color at a pH of nearly

pH INDICATOR CHART

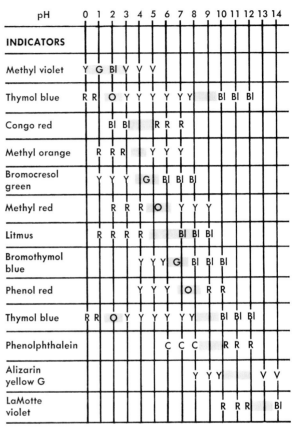

pH	0	1	2	3	4	5	6	7	8	9	10	11	12	13	14
INDICATORS															
Methyl violet	Y	G	Bl	V	V	V									
Thymol blue	R	R	O	Y	Y	Y	Y	Y	Y	Y	Bl	Bl	Bl		
Congo red				Bl	Bl		R	R	R						
Methyl orange		R	R	R	Y	Y	Y								
Bromocresol green		Y	Y	Y	G	Bl	Bl	Bl							
Methyl red				R	R	R	O	Y	Y	Y					
Litmus			R	R	R	R		Bl	Bl	Bl					
Bromothymol blue						Y	Y	Y	G	Bl	Bl	Bl			
Phenol red							Y	Y	Y	O	R	R			
Thymol blue	R	R	O	Y	Y	Y	Y	Y	Y	Y	Bl	Bl	Bl		
Phenolphthalein								C	C	C	R	R	R		
Alizarin yellow G										Y	Y	Y		V	V
LaMotte violet												R	R	R	Bl

Color code:

Bl — blue	R — red
C — colorless	V — violet
G — green	Y — yellow
O — orange	

The zoned bars indicate color transitions of the indicators.

7, which is convenient, because 7 is the value for neutrality. Other indicators change at different pH values. By testing a given solution with a number of different indicators, one can determine its pH with enough accuracy for many purposes.

The table on this page shows a number of indicators and the place on the pH scale at which they change color. With these indicators it is not at all difficult to find the pH of any solution you care to test. Try some of the solution with thymol blue. Suppose this indicator turns yellow. This means that the pH is between 3 and 8. Next, test some of the solution with methyl red. If the indicator turns red, this shows that the pH is less than 4.5. Obviously, the pH of the solution must be between 3 and 4.5. Similarly, for any solution, you can narrow the pH range by using several of the indicators shown in the table.

TITRATION

To indicate the total amount of acid or base in a solution, the chemist uses the process called *titration*. A certain amount of the unknown solution is measured out into a beaker, and a few drops of an indicator are added. If the solution contains a base, an acid solution whose concentration is known is dropped into it from a burette, a graduated glass tube provided with a small aperture and a stopcock. After a certain amount of acid has been added the indicator begins to turn color, showing that the solution in the beaker is now neutral. The indicator chosen must be one that changes at a pH of 7. The volume of the acid solution required to neutralize the base is read off from the burette scale, and the chemist can then calculate the concentration of the base. If the unknown solution is acidic, an alkaline solution of known concentration is delivered from the burette.

USES OF ACIDS AND BASES

Sulfuric acid is used in greater quantities than any other chemical. Its sales offer an excellent index of business conditions. It enters into the manufacture of explosives, dyestuffs, and drugs. It is used in oil and sugar refining and in the preparation of fertilizers. Hydrochloric acid serves in the manufacture of glue, gelatin, and dextrose, and in cleaning metals. Doctors prescribe very dilute hydrochloric acid for certain cases of indigestion. Nitric acid is used in large quantities in the manufacture of fertilizers and of explosives and also in the preparation of dyes and plastics. Acetic acid is a good solvent for certain organic substances. It plays an important part in the preparation of cellulose acetate, which is used as a plastic and in the manufacture of acetate fiber. Hydrofluoric acid is used

in making refrigerants and certain new plastics. It has the useful property of being able to etch glass.

Bases are also very useful to man. Sodium hydroxide is used in the rayon, film, soap, paper, and petroleum industries, and in many others. Hydrated lime (calcium hydroxide) serves to raise the pH of soils and to make insecticides and many other chemicals. Ammonium hydroxide is another important base. It is employed as a raw material for making many important compounds, including the ones that are employed in fertilizers.

USES OF SALTS

Common salt, sodium chloride, is undoubtedly the most valuable to man of all the salts. It is used to prepare various compounds of the metal sodium. It is also the principal source of the gas chlorine, which is used for bleaching, for purifying water, and in many plastics. Salt is also an important preservative of meat, fish, hides, skins, and various other products. The salt called sodium bicarbonate ($NaHCO_3$) is an essential ingredient in baking powders. Epsom salts (magnesium sulfate, $MgSO_4 \cdot 7H_2O$) and calomel (mercurous chloride, Hg_2Cl_2) have medicinal uses. Gypsum (calcium sulfate, $CaSO_4 \cdot 2H_2O$) yields plaster of Paris and is also used in the manufacture of gypsum wallboard. Hypo, (or sodium thiosulfate, $Na_2S_2O_3 \cdot 5H_2O$) plays an important part in photography as a fixing agent. A list of some salts of potassium, alone, provides a good idea of the wide-ranging uses of salts. For example, potassium carbonate (K_2CO_3) is used in making glass, dyes, and some soaps. Potassium bromide (KBr) is used in making photographic paper, plates, and developers. Potassium nitrate (KNO_3) is used in the production of gunpowder, while potassium chloride (KCl) and sulfate (K_2SO_4) are used in making fertilizers. Pure monopotassium tartrate ($K_2H_4C_4O_6$), called cream of tartar, is used in the manufacture of baking powder.

These are but a few examples of the ways in which acids, bases, and salts serve us. They have innumerable uses.

SOLUTIONS

by James S. Coles

When a person puts sugar in a cup of coffee and stirs it vigorously, the sugar apparently disappears without leaving a trace. What really happens is that when the sugar is dissolved, its particles are so tiny and so uniformly distributed in the coffee that they cannot be seen. The chemist calls a mixture of this sort a *solution*. The chemist could say that we have made a solution of sugar or that sugar is in solution in the coffee. The coffee itself is a solution, made up primarily of water.

We are most familiar with solutions that consist of a solid dissolved in a liquid, but this is only one kind of solution. Water, in the form of water vapor, may be dissolved in air. Gas may be dissolved in water, as in carbonated drinks. One solid may be dissolved in another, as in monel metal, which is an alloy of nickel, copper, and other elements. A true solution is formed when two or more solid, or liquid, or gaseous substances are mixed, without undergoing a chemical change, so that you can no longer distinguish between them.

Solutions are to be found all about us and they play an exceedingly important part in the lives of living things. The air we breathe is a solution of different gases, such as oxygen, nitrogen, and water vapor. Ocean water is a solution formed by salts and other substances dissolved in water. The roots of plants absorb food from the soil solution. Most of the food we eat is broken down into water-soluble materials in the process of digestion. Otherwise it could not be carried in the blood to the tissues of the body. Many familiar substances are solutions. Gasoline is a solution; so are steel, glass, and mouthwash.

SOLVENTS AND SOLUTES

In every solution, there must be a *solvent*, or dissolving substance, and a *solute*, or dissolved substance. Let us assume that the components of a solution are not in the same physical state—that one, say, is a liquid, while the other is a solid. In that case, the solvent is generally the substance that exists in the same state as the resulting solution. For example, when sugar, which is a solid, is dissolved in a liquid, such as water, the sugar is the solute and the water is the solvent. When water evaporates in the air, the water is the solute and the air is the solvent.

The components of a solution may be in the same physical state: solid, or liquid, or gaseous. In that case, the substance that is present in greater quantity is called the

This beaker has been filled with warm tap water, saturated with dissolved gas (air). The water has become warmer, and some of the dissolved gas is leaving the solution, forming bubbles as it goes out.

W. A. Schwarz

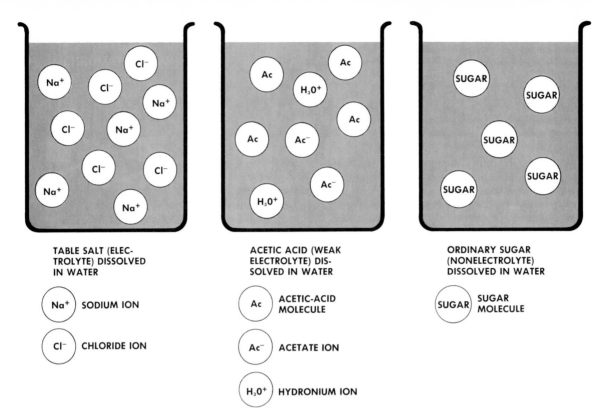

What happens when three different types of solutes are dissolved in water. The solutes are table salt, or sodium chloride (NaCl), an electrolyte; acetic acid ($C_2H_4O_2$), a weak electrolyte; and ordinary sugar, or sucrose ($C_{12}H_{22}O_{11}$), a non-electrolyte. For an explanation, see the text. We have indicated acetic acid by the symbol Ac. The acetate ion (we call it Ac⁻) has the formula $C_2H_3O_2^-$. The hydronium ion—symbol H_3O^+—consists of a hydrogen ion H^+, combined with a water molecule (H_2O).

solvent. For instance, water and alcohol can be mixed to form a solution in any desired proportion. If ten parts of alcohol are dissolved in ninety parts of water, the alcohol is the solute and the water the solvent. If ten parts of water are dissolved in ninety parts of alcohol, water is the solute; alcohol, the solvent.

When a solute, such as sugar or salt, goes into solution, individual particles of the solute break loose and move about freely in the solvent. The solute particles of a sugar such as sucrose are individual molecules. In the case of table salt, or sodium chloride—a chemical compound of sodium and chlorine—the solute particles are smaller than molecules. They consist of positively charged sodium atoms, called sodium ions, and negatively charged chlorine atoms, called chloride ions.

When some solutes are dissolved, their particles are mixtures of molecules and ions. White vinegar, which is a solution of acetic acid in water, is a good example. There are three different kinds of solute particles in vinegar: acetic-acid molecules, positively charged hydronium ions, and negatively charged acetate ions.

If the individual solute particles moving about in a solution are positively and negatively charged ions, the solution is able to conduct an electric current and the solute is called an *electrolyte*. If the solute particles represent a mixture of ions and molecules, the solution can conduct only a weak electric current and the solute material is known as a weak electrolyte. If the solute particles consist entirely of molecules, the solution cannot conduct electricity and the solute is called a *nonelectrolyte*. Table salt is an electrolyte. Acetic acid is a weak electrolyte. Sugar is a nonelectrolyte.

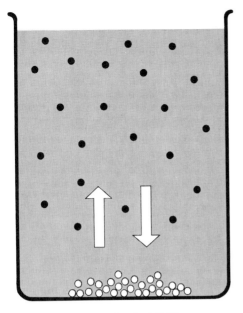

○ UNDISSOLVED SUGAR
MOLECULES IN SOLID STATE

● DISSOLVED SUGAR
MOLECULES IN SOLUTION

Diagram of a saturated solution. The double arrow indicates that the rate at which sugar dissolves has become equal to the rate at which dissolved sugar crystallizes.

A solution is said to be *dilute* when only a small amount of solute is dissolved in a large quantity of solvent. The sap of a maple tree, for example, represents a dilute solution of sugar. If a large quantity of solute is dissolved, we have a *concentrated* solution. A good example of a concentrated solution is the thick simple syrup of sugar and water that is mixed with various flavorings to make the soft-drink syrups at soda fountains.

SATURATED SOLUTIONS

When a lump of sugar is placed in a glass containing water, the sugar is all solid sugar at first, and the water is all water. Almost immediately, however, the sugar molecules begin to leave the solid lump and move about freely in the water, forming a solution. In the beginning there will be only a few sugar molecules in solution in the water, and the solution will be very dilute. Occasionally one of the sugar molecules in solution will strike the solid sugar and may remain attached to it. This will happen more often when the solution becomes more concentrated. The more sugar molecules there are in solution, the greater the possibility that one of them will collide with the solid.

The number of sugar molecules that leave the solid and go into solution in each second per square centimeter of surface remains constant, provided the temperature of the solution does not change. On the other hand, the number of sugar molecules that leave the solution and are deposited on the solid sugar in each second will increase as the number of sugar molecules in a given volume of solution increases: that is, as the concentration increases.

In time, the number of sugar molecules going into solution in each second will be just equal to the number of sugar molecules leaving the solution. When this condition is reached, there will be no further increase in the concentration. The sugar in the solid lump will then be in equilibrium with the sugar in solution. We will then have what is called a *saturated solution*.

CRYSTAL FORMATION

We can produce crystals of a given substance from a saturated solution of that substance. First we carefully filter out undissolved matter from the saturated solution. The temperature is then gradually lowered, causing crystals to separate from the solution. Another way to produce crystals is to let the solvent evaporate, instead of lowering the temperature.

If the cooling or evaporation is slow, the first small crystals that form will gradually increase in size as more solute separates from the solution. If the cooling or evaporation is rapid, there will not be enough time for the crystals that first separated to grow. Instead, many new crystals will form as the solute continues to leave the solution.

It is possible to grow a single large crystal of a substance such as alum. First we allow a carefully filtered saturated solution of alum to stand for a few days in a flat, open container in a room kept at uniform temperature. A crop of small crystals will form. We select a particularly likely specimen as a seed crystal and suspend it by means of a fine thread in a fresh portion of saturated alum solution. As this solution slowly evaporates, the solute that separates from the solvent will be deposited on the seed crystal attached to the string, and this crystal will grow to large dimensions. The temperature must be kept uniform if good specimens are to be grown.

This diagram shows what happens when cold tap water, saturated with dissolved gas (air) is left standing in a warm room. As the water becomes warmer, some of the dissolved gas leaves the solution, forming bubbles. The arrows indicate that equilibrium has been reached. The rate at which air enters the solution has become equal to the rate at which it leaves the solution.

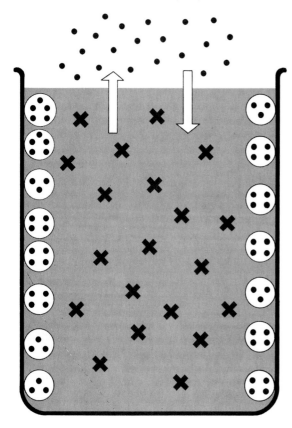

◯ BUBBLE OF GAS (AIR)

• UNDISSOLVED GAS MOLECULE

✕ GAS MOLECULE DISSOLVED IN WATER

The concentration of a solute in a given solvent that will produce saturation at a specified temperature is known as the *solubility* of the solute. Chemists usually define solubility as the number of grams of solute that will dissolve in 100 grams of solvent at a given temperature. For example, the solubility of table salt is 35.7 grams in 100 grams of water when the temperature of the solvent is 0° Celsius. It is 39.8 grams at 100° Celsius.

Some substances form saturated solutions when only a small amount of the solute is present. For example, a saturated solution of lime (calcium hydroxide) in water contains only 0.165 grams of lime in 100 grams of water at 20° Celsius. Even a saturated solution of limewater, therefore, is very dilute. On the other hand, when sugar (sucrose) is dissolved in water, the saturated solution will contain more sugar than water—about 200 grams of sugar for each 100 grams of water (at room temperature). That means that a sugar solution may be quite concentrated and still not be saturated.

TEMPERATURE AND SOLUBILITY

Temperature has an important effect upon the solubility of substances. Usually the higher the temperature, the greater the solubility of a solid or liquid. In the case of gases, the higher the temperature, the less soluble the gas, for fewer gas molecules occupy a given volume of solvent.

Difference in temperature accounts for the bubbles that appear on the sides of a glass or pitcher of water that has been standing for some time in a warm room. The cold water that we draw from the faucet contains a certain quantity of dissolved air. Gases are less soluble in warm water than in cold water. Therefore, if the cold water is saturated, or almost saturated, with dissolved air as it comes from the faucet, some of this dissolved gas must leave the solution when the water becomes warmer. The gas separates from the solution by forming small bubbles, which usually appear at tiny imperfections on the inner surface of the container. Once these bubbles are produced, an equilibrium is established between the gas in the bubbles and the gas dissolved in the solution. Any other gas molecules leaving the solution go for the most part into the bubbles already formed, thus enlarging them. The air that is dissolved in the surface layer of water establishes its equilibrium with the air above the surface by releasing some of the dissolved gas to the atmosphere.

VAPOR PRESSURE IN SOLUTIONS

Vapor pressure is an important factor in solutions. To understand what we mean by vapor pressure, let us imagine that we have a quantity of water in a closed container. Water molecules are escaping from the liquid into the confined space above it, through the process of *evaporation*. They have become molecules of water vapor. The chemical composition of the water remains unchanged. At the same time, molecules of the gaseous water vapor will strike the surface of the water, will be held to it by attractive forces known as van der Waals' forces, and will become molecules of liquid water—a process called *condensation*.

As evaporation goes on, the density (and therefore the pressure) of the gas above the surface of the water will increase until finally there will be an equilibrium between the rate of evaporation and the rate of condensation. The gas pressure at which this equilibrium exists in a closed container and at a given temperature is called the *vapor pressure* of the water. Of course, the gas above the surface of the water is water vapor.

The adjustment of the equilibrium between a liquid and a gas accounts for some striking effects. It explains, for instance, what happens when you open a bottle of ordinary soda water, which consists of carbon dioxide dissolved in water. Not only do you hear a fizzing sound, but you see a great many bubbles forming and rising to the surface. Yet before the bottle is opened there are no bubbles at all.

As long as the bottle is tightly sealed by its cap, the carbon dioxide in the space above the liquid in the neck of the bottle is under pressure. The carbon-dioxide gas dissolved in the water is in equilibrium with this gas under pressure. As soon as the cap is removed, gas escapes from the bottle and the pressure is reduced. There is now more gas dissolved in the liquid than is required to maintain equilibrium under the new condition of reduced pressure. To restore equilibrium, the excess gas rises to the surface in the form of bubbles. As it leaves the solution, it produces a fizz.

BOILING AND FREEZING POINTS

The boiling point of a solution is higher than that of the pure solvent it contains. On the other hand, the freezing point of a solution is lower than that of the solvent. If we drive an automobile in a climate where the temperature is below freezing in the winter, we add an antifreeze, such as ethylene glycol or alcohol, to lower the freezing point of the liquid in the radiator. If a street is icy and we wish to melt the ice, we sprinkle salt upon it. The salt is dissolved in the small amount of liquid water present on the surface of ice, forming a salt solution. Since the freezing point of the salt solution is lower than that of water, the ice will melt and go into solution. Of course, if the outside temperature is below the temperature at which even a very concentrated salt solution freezes, the ice will not melt when salt is spread upon it.

MISCIBLE AND IMMISCIBLE SOLUTIONS

When a liquid is dissolved in another liquid, they may be totally *miscible*. That is, they will mix completely with one another, regardless of the proportions of the two substances. Alcohol and water are totally miscible. No matter how much alcohol we put in a given amount of water, no matter how much water we put in a given amount of alcohol, the two liquids will dissolve completely in one another and form a single solution.

However, some liquids will not mix. They are said to be *immiscible*. Water and oil are immiscible: oil will not dissolve in water, nor will water dissolve in oil. A great many liquids are partially miscible. They will mix only to a certain extent. Gasoline and alcohol are partially miscible.

A. Salt is spread on a snow-covered street in Evanston, Illinois, in a study of the effects of salt in winter street maintenance. B. The salt application has now begun to take effect. A part of the snow has melted and the pavement appears here and there. C. Still later, the snow has melted almost to the bare pavement.

Salt Institute

A

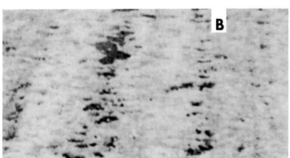

B

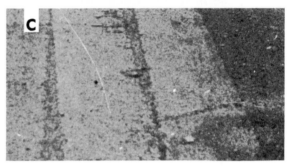

C

OSMOSIS

One of the most remarkable phenomena involving solutions is *osmosis*. It represents the passage, or *diffusion*, of solvent molecules through a semipermeable membrane from a pure solvent into a solution or from a less concentrated solution into one that is more concentrated. The semipermeable membrane in question allows molecules of one kind to pass through but not molecules of another kind. Various explanations have been offered for this selective effect. Some believe that the membrane acts as a sieve, that it has holes that will permit only comparatively small molecules to pass through. Others think the phenomenon is caused by adsorption. In some cases, it may be due to electrical charges on the surface of the membrane. The charges may repel certain ions and attract others. Whatever the underlying causes of osmosis, its effects are clear enough. So is the pressure it produces—*osmotic pressure*.

Here is an example of osmosis at work. We place a semipermeable membrane across the bottom of a U-shaped tube. This membrane will allow water to pass through, but not sugar. We then pour water into the arms of the U-tube. Some of the water molecules will pass through the membrane from the right side to the left. Others will pass from the left side to the right. If the temperature and pressure of the water on both sides are equal, the rate at which molecules move from left to right will equal that at which they move from right to left. The two sides will be at equilibrium.

We dissolve enough sugar in the water in the right arm of the U-tube so that there is one sugar molecule for every nine water molecules. Only nine out of every ten molecules on the right-hand side of the membrane now can pass through it. Hence the rate at which water molecules move from right to left is only 90 per cent of the original rate. But the rate at which the water molecules move through the membrane from left to right has not changed, since the temperature and pressure have not changed.

This means that more water is passing from left to right than from right to left. The quantity of liquid on the right increases therefore, and it will continue to increase until the rate of passage of the molecules through the membrane in the two directions is equal. This equilibrium is brought about in the following way. As the height of the water in the right arm of the U-tube becomes greater and the height in the left arm becomes less, the pressure of the water on the membrane from the left will decrease and the pressure of the solution on the membrane from the right will increase. As the pressure in the right increases, the rate at which the solution molecules strike the membrane increases. Hence the number of water molecules passing through from right to left increases. At the same time, decreasing pressure from the left will lower the rate of water molecules passing through from left to right. Ultimately there will again be equilibrium across the membrane.

All plant and animal life depends upon osmosis for its very existence. The cell, which is the basic element of living things, is provided with a semipermeable membrane. This membrane is impermeable to many of the materials within the cell, but permeable to the water that surrounds it as well as to the gases and nutritive substances contained in the water. Hence, by the process of osmosis, the cell can draw in not only water but also oxygen and nutritive elements. Wastes also pass through the semipermeable membrane by osmosis from the cell into the surrounding liquid.

A very delicate balance in osmotic pressure must be kept in the case of cells. It is for this reason that whenever your physician injects a quantity of liquid in the body the liquid is brought up to a certain concentration by the addition of salts. The concentration has to equal that of the blood or of the interior of the blood cells: about 0.9 per cent by weight. A 0.9-per-cent solution like this is called a *physiological salt solution*. If the injected liquid were too dilute, fluid would pass through the membranes surrounding the blood cells into the cells themselves, causing them to swell and perhaps burst. If, on the other hand, the solution were too concentrated, fluid would leave the blood cells and pass into the solution, causing the cells to wither.

SUPERSATURATED SOLUTIONS

Among the most interesting of all solutions are those called *supersaturated*. They have a greater proportion of solute than is contained in a saturated solution. It is not difficult to prepare supersaturated solutions if one is careful, and many fascinating experiments can be performed with them.

If we dissolve ordinary photographic hypo (sodium thiosulfate) in boiling water, we need about 428 grams of hypo per 100 grams of water to form a saturated solution. If the solution were cooled in contact with a little hypo, some hypo would come out of the solution in the form of crystals as the temperature became lower, since the solubility of the substance would decrease with falling temperatures.

Suppose that we carefully poured off the clear solution, while it was still at the boiling point, into a carefully cleaned bottle of glass and covered the top with clean paper or cotton so that no dust or lint would fall into the liquid. It would be possible to cool the solution without having any of the solute come out of it in the form of crystals. We could then have a supersaturated solution at room temperature – about 20° Celsius. It would contain 428 grams of hypo for every 100 grams of water. A saturated solution at this temperature would contain only 84.5 grams of hypo for every 100 grams of water. In other words there will be an excess of 343.5 grams of hypo per 100 grams of water in solution.

This excess will come out of solution if there is any solid hypo with which it can establish equilibrium. If we put even the tiniest hypo crystal into the supersaturated solution, ions of the hypo in solution will begin to attach themselves to the crystal. It is true that the solid hypo will begin to dissolve at the same time. But the rate at which this takes place will be much less than the rate at which the dissolved hypo

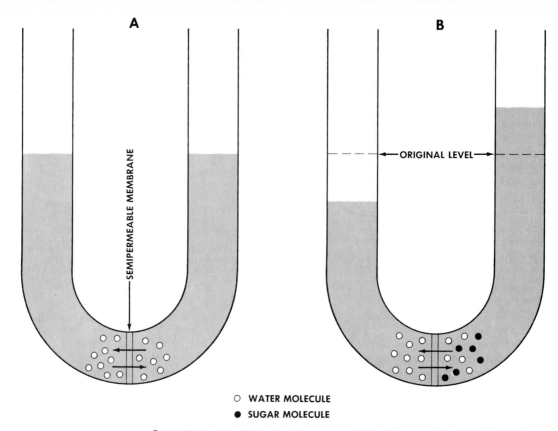

A

B

ORIGINAL LEVEL

SEMIPERMEABLE MEMBRANE

○ WATER MOLECULE
● SUGAR MOLECULE

Osmosis at work. Water has been poured into the arms of a U-tube (A); then sugar has been dissolved in the right arm of the tube (B.) A description of the experiment is given in the text.

will crystallize. Hence there will be a net increase in the amount of solid hypo.

Finally, as the concentration of the hypo in solution decreases, the rate at which it crystallizes out becomes equal to the rate at which solid hypo goes into solution. Equilibrium is now established between the solid and the saturated solution. The amount of hypo that separates from the solution in the process of coming to equilibrium will be the excess amount present in the supersaturated solution—that is, 343.5 grams.

It is not always necessary to put a crystal of solid solute in a supersaturated solution in order to cause crystallization of the excess solute. The same result can be obtained by jarring the container or rubbing the inside with a glass rod or other object. This introduces tiny particles around which crystals of solute will then form rapidly and precipitate out of the solution. Some substances crystallize so easily that it is almost impossible for them to form supersaturated solutions.

MOLAR AND NORMAL SOLUTIONS

Chemists frequently use two other terms to descibe the concentrations of solutions. A *molar solution* is a solution that holds one gram molecular weight of a solute in 1,000 grams of the solvent. A *gram molecular weight* of a substance is the sum of the atomic weights, in grams, of the atoms in the molecule. The gram molecular weight of sulfuric acid, H_2SO_4, is $(2 \times 1) + 32 + (4 \times 16)$, or 98 grams, 1, 32, and 16 being the atomic weights of hydrogen, sulfur, and oxygen, respectively. A one molar solution of sulfuric acid would, therefore, contain 98 grams of the acid.

A *normal solution* contains one gram equivalent weight of a solute in 1,000 milliliters of solvent. A *gram equivalent weight* is the molecular weight divided by the number of active hydrogen atoms or hydroxyl groups in a molecule of the substance. In the case of sulfuric acid, the gram equivalent weight is 98 divided by 2, or 49 grams. A normal solution of sulfuric acid would, therefore, contain 49 grams of the acid.

COLLOIDS

by A. B. Garrett

The substances making up what we call matter fall roughly into three main divisions on the basis of size. On one hand, there are the things that are large enough to be seen by the human eye, with or without the aid of an optical microscope. At the other extreme, there are ordinary molecules. The still smaller building blocks of which gases, liquids, and solids consist are atoms, which often combine to form molecules; and ions, which are molecules or atoms with an electrical charge. Finally, there are certain particles too small to be made out by an optical microscope but larger than ordinary molecules, or atoms, or ions. The intermediate particles in question are responsible for a vast number of phenomena, including red sunsets, smog, the green color of water that has collected in a gravel pit, the semirigid structure of jellies, and the color of certain gems, stones, and glasses.

CERTAIN SIZE PARTICLES

Until the nineteenth century, these intermediate particles were unknown to scientists. One of the first to pick up their trail was a Scots chemist, Thomas Graham, who had been studying the diffusion of different types of particles through membranes. In a paper that he published in 1861 in the *Philosophical Transactions,* Graham pointed out that certain substances, like salt and sugar, can be readily dissolved and diffused, or passed, through the pores of a parchment membrane immersed in water. Other dissolved substances, such as glue, gelatin, and gums, will diffuse very slowly or not at all. He concluded that if a substance in solution has been reduced to its molecules or its ions, it can diffuse readily. It cannot do so, however, if it is made up of clumps of molecules.

Since the materials that readily diffuse when they are in solution all form crystals when they become solids, Graham called them *crystalloids.* He gave the name of *col-*

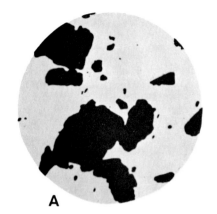

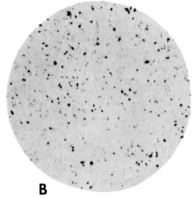

A, above, is a photomicrograph of the finest available powdered graphite, enlarged 700 times. B is a photomicrograph of graphite particles in the colloidal size range, also enlarged 700 times.

Apparatus used by T. Graham to demonstrate the existence of colloidal matter.

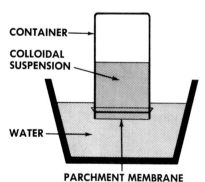

CONTAINER

COLLOIDAL SUSPENSION

WATER

PARCHMENT MEMBRANE

loids, or gluelike materials, to the glue and other substances that do not pass readily through membranes when in solution. Graham assumed that only the substances that he called colloids were "gluelike." The word "colloid" comes from the Greek *kolla,* meaning "glue."

We know today that crystalline materials can be prepared in such a manner that, like colloids, they too will not diffuse readily through a parchment filter. We have found that a better distinguishing characteristic of a colloidal particle is its size rather than its gluelike property. The word colloid today is applied to any material containing particles that fall within a certain range of dimensions.

Scientists have noted that the smallest dimension that can be made out with an ordinary microscope is about 1/1,000 of a millimeter, or 1 micron (written μ). The largest particle that can pass readily through a parchment membrane is roughly 1/1,000,000 of a millimeter, or 1 millimicron (written mμ), in size. These two dimensions — 1 micron and 1 millimicron — have been established as the limits within which colloidal particles are to be found. The particles in question cannot be made out with a microscope, nor will they diffuse through a parchment membrane.

A particle may have colloidal properties if one, two, or all three of its dimensions are in the colloidal range. If only one dimension is in this range, the particle is sheetlike. The particle is threadlike if two dimensions are in this range. When all three dimensions are colloidal in size, the particle shows most of the characteristics of colloidal particles. Among the colloids are polymers, such as rubber, plastics, and synthetic fibers. We also include certain very large molecules — proteins and carbohydrates — which are found in plant and animal tissues.

A distinct branch of chemistry, called *colloid chemistry,* is devoted to the study of colloids. The colloid chemist is concerned with the composition and the properties of colloidal particles and the substances in which they are to be found. The colloidal properties of the materials used in industry are examined and the knowledge gained is used to create new industrial products and to improve old ones.

COLLOIDAL SYSTEMS

The colloidal particles and the medium, or material, in which the particles are dispersed or suspended is called a *colloidal system.* The particles in question may be the grains of a solid, the bubbles of a gas, or the droplets of a liquid. The *disperse medium* — the medium in which the particles are dispersed, or scattered — may be a solid, a liquid, or a gas. The table below shows the various possibilities.

We use the terms *sol, gel,* and *emulsion* for some common types of colloidal systems. A *sol* is usually a solid dispersed in a liquid, but it may also be a gas or a liquid in a liquid. A *gel,* or jelly, is a colloidal system in which the long colloidal particles

Colloidal particles	Medium in which the colloidal particles are scattered	Examples
Solid	Gas	Smokes and dust, consisting of solid particles suspended in air.
Solid	Liquid	Finely divided metals suspended in liquids.
Solid	Solid	Finely divided gold in ruby glass.
Liquid	Gas	Clouds, fogs, mists; these consist of water droplets suspended in air.
Liquid	Liquid	Emulsions: dispersions of fine particles of a liquid in another liquid.
Liquid	Solid	Liquid droplets in minerals, like opal.
Gas	Gas	This type of colloidal system *cannot* occur.
Gas	Liquid	Foams, froths; these are bubbles of air suspended in water.
Gas	Solid	Bubbles of gas, of colloidal dimensions, occurring in certain minerals, like meerschaum.

A beam of light passed through a pure solution, such as that in the first beaker, produces no visible path. However, a beam of light passed through a solution containing colloidal particles, such as that in the second beaker, produces a bright path of light— a phenomenon known as the Tyndall effect.

form a branchlike or "brush-heap" type of structure, in which the liquid is trapped in little pools. Under certain conditions, sols can become gels and gels can be changed into sols. An *emulsion* is usually a liquid dispersed in a liquid. It consists of very minute drops of the first liquid scattered through the other one.

CHARACTERISTIC PROPERTIES

An important property of colloidal particles is the one discovered by Graham — namely, their failure to diffuse readily through a parchment membrane. Since that time, a number of other characteristic properties have been noted. (1) The particles in several types of colloidal systems are in constant, random motion. (2) A sharp path of light is produced when a beam of light is passed through many colloidal systems. (3) Many colloidal particles carry electric charges. (4) Many colloidal systems remain remarkably stable. (5) Some of them show characteristic colors. (6) Colloidal particles have marked adsorption properties. These properties are not all limited to particles in the colloidal size range but they are very pronounced in these particles.

CONSTANT, RANDOM MOTION

What causes the particles in a colloidal suspension to remain dispersed throughout the medium instead of precipitating, or settling out? One reason is that they are constantly bumping into one another as the result of a constant, random motion that is called the *Brownian movement.* It is named after the Scots scientist Robert Brown, who

first observed it in 1828 while examining, under the microscope, fine grains of pollen suspended in water. He noticed that the grains kept up a curious, zigzag motion which mystified him. Modern science has provided an explanation. Grains of pollen or other colloidal particles suspended in a liquid medium move because they are constantly bombarded by the molecules of the medium. The collisions impart very considerable and erratic motions to the colloidal particles, even though the latter are much heavier than the molecules striking them.

PATH OF LIGHT PRODUCED

Many liquids under ordinary circumstances appear to be clear and even colorless. Yet in reality they may contain a good deal of finely suspended matter not removable by ordinary filtration. It is only necessary to pass a strong beam of light through the liquid in a dark room. If there is no suspended matter, the beam of light has no visible path through the liquid. But if colloidal particles are present, light is reflected by these particles, and all of them become visible sources of light. We then see a bright path of light through the liquid. This is called the *Tyndall effect,* or *Tyndall beam.* It was named after the 19th century English scientist John Tyndall, who first studied it in detail. Pure solutions and solvents may show only a very faint Tyndall beam in the absence of colloidal matter.

The German scientists H. Siedentopf and R. Zsigmondy made use of the Tyndall beam in developing their ultramicroscope in 1903, which proved to be very useful in

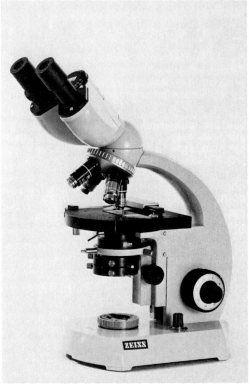

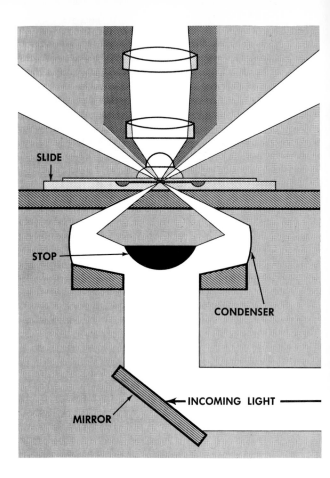

Colloidal particles suspended in a liquid may be studied with a dark-field microscope, which causes the object that is being viewed to appear bright against a dark background. The dark-field microscope shown at the left, above, consists of a regular inclined binocular model with a condenser, for concentrating the light beam, mounted under the stage, the small platform upon which an object is placed for examination. The diagram shows how the dark-field microscope works. A drop of the liquid that is to be examined is set on the central portion of a slide and a cover glass is put on the latter. Excess liquid flows into a ringlike groove in the slide, leaving a thin film in the central part where the observation will be made. An opaque stop in the center of the condenser keeps direct light from striking the slide from below. Parallel rays of light, reflected from the mirror, are also reflected from the curved bottom of the stop and the curved inner wall of the condenser. They focus, in the form of a hollow cone of light, in the liquid that is being examined. The colloidal particles contained in the liquid are then made visible by the illumination.

the study of colloidal systems. This device consisted of a small cell or glass to hold the colloidal dispersion, a powerful source of light to send a pencil of rays through the cell, and a good microscope with which to look at the colloidal dispersion at right angles to the beam of light. The minute colloidal particles were seen in the ultramicroscope as bright specks of light moving in

zigzag paths. The ultramicroscope has been quite generally replaced by the dark-field microscope in the study of colloidal systems.

ELECTRIC CHARGES

A considerable number of colloidal particles carry electrical charges. It appears probable that one way in which the particles become charged is by attracting to their surfaces some simple ions already existing in the solution in which they occur. If an electrical field is set up, the imprisoned ions are attracted to the electrode having the opposite electrical charge, dragging the whole colloidal clump along.

HIGH STABILITY

The high stability of many colloidal systems is due to the combined effect of particle bombardment, or Brownian movement, and to the fact that in such stable systems the particles carry like electrical charges and thus constantly repel one another. The force of gravity is not strong

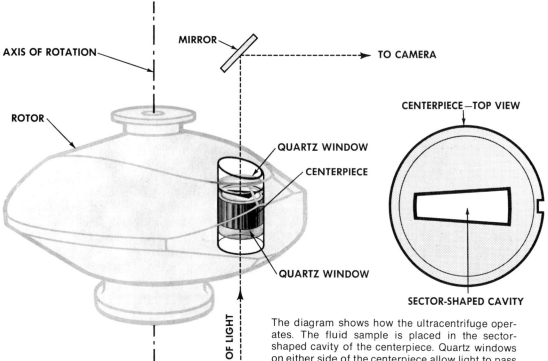

AXIS OF ROTATION

ROTOR

MIRROR

TO CAMERA

QUARTZ WINDOW

CENTERPIECE

QUARTZ WINDOW

PATH OF LIGHT

CENTERPIECE—TOP VIEW

SECTOR-SHAPED CAVITY

Diagrams adapted from materials supplied by Beckman Instruments, Inc., Spinaco Div.

enough to overcome these two effects. Hence the particles do not precipitate, or settle.

The inherent stability of a colloidal system is weakened if the electrically charged particles lose their charge. For example, if a solution containing ions with a positive charge is added to a colloidal system in which the particles have a negative charge, the positive ions will seek out the negative ions and the charge on the latter will be neutralized. The force of gravity may then cause them to settle. Again, if an electric field is set up in a colloidal suspension, the charged colloidal particles will move over to the electrode with the opposite charge. This is called *electrophoresis*. As the charged particles come in contact with the electrode, they lose their charge and they then precipitate.

Often a colloidal system will be produced in a solution containing ions which will gradually neutralize the electrical charge on the colloidal particles. It is possible to separate the ions from the colloidal particles in such a solution by causing the former to diffuse through a membrane. This

The diagram shows how the ultracentrifuge operates. The fluid sample is placed in the sector-shaped cavity of the centerpiece. Quartz windows on either side of the centerpiece allow light to pass through the sample and to a mirror above it. The light is then reflected from the mirror to a camera lens. As the rotor, driven by an electric motor, spins at speeds up to 60,000 revolutions per minute, great centrifugal force is applied to the fluid sample. The camera records the behavior of solute molecules or of suspended particles in the sample. The right-hand diagram gives a view of the centerpiece as seen from above, showing the sector-shaped cavity in which the sample is put. In the photograph below, a rotor is being attached to a motor-driven shaft, which will hold it in place during the spinning operation. The rotor is driven by an electric motor, not shown here.

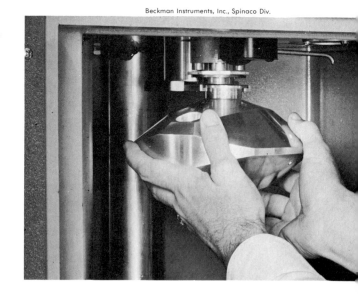

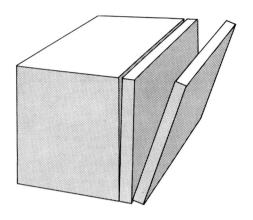
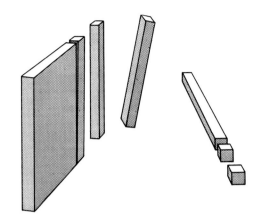

The smaller a substance is, the more surface it has in comparison with its volume. When the cube shown above is sliced, split, and cut into small cubes, there is many times as much surface as there was when there was only a single cube. Yet the volume has remained the same. Colloidal particles also have a vast amount of surface in relation to their volume.

method of purifying a colloidal system is called *dialysis*.

In studying colloidal systems, the colloid chemist often has occasion to cause the colloidal particles to precipitate by using a device called an *ultracentrifuge*. It places a force many times that of gravity on the particles and causes them to settle. From data on the rate of settling in a centrifugal field, calculations can be made concerning the weight and sometimes the size of the colloidal particles.

COLORS OF COLLOIDAL SYSTEMS

Many of the colors that we see around us are due to the scattering of light from colloidal particles. The particular colors depend on the size of the particles, their texture, and the direction in which they are oriented, or turned. For example, the color of certain bird feathers results from such scattering of light. In these feathers, a great many air bubbles of colloidal size are suspended in a solid medium. These bubbles scatter blue light but transmit red light, which is absorbed by the dark background formed by the body of the bird. As a result, the feathers appear to be blue.

ADSORPTION PROPERTIES

By *adsorption*, the chemist means the adherence of atoms, molecules, or ions of any kind to the surface of a solid. The particles of which a solid is composed are held together by attractive forces operating from particle to particle within the body of the solid. At the surface, however, there are free, unbalanced forces of attraction. These will hold an adsorbed layer which is frequently not more than one molecule deep. The tightness with which the adsorbed layer is held depends on the nature of the solid and of the adsorbed substance. The attraction may be relatively weak or may resemble the stronger forces of chemical bonds. The quantity of substance adsorbed depends largely on the surface area of the material.

Colloidal particles have a vast amount of surface compared to their mass. We can illustrate this point by systematically cutting up a cube of any substance with an edge-length of 1 centimeter. The cube with which we begin has 12 edges, each of them 1 centimeter in length. It has 8 corner points. It has a total surface area of 6 square centimeters. Its volume is 1 cubic centimeter. If we cut this cube into smaller cubes with an edge-length of 1 millimeter, we will produce 1,000 cubes with a surface area of 60 square centimeters. If we cut the original cube into cubes 0.5 micron or 500 millimicrons thick (that is, in the size range

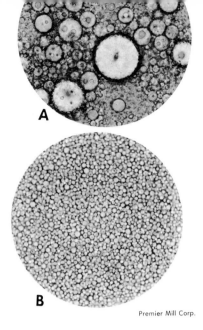

A

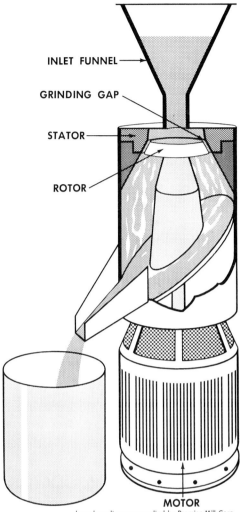

B

Premier Mill Corp.

INLET FUNNEL

GRINDING GAP

STATOR

ROTOR

MOTOR

based on diagram supplied by Premier Mill Corp.

The colloid mill shown in this diagram breaks down comparatively coarse particles into finer ones of colloidal size. The product to be broken down is poured into the inlet funnel and is fed to a rapidly spinning rotor, set within a matching stationary part, called a stator. The distance between rotor and stator is adjustable. As the material comes in contact with the rotor, it is flung out to the edge by centrifugal force. This force continues to push it out and down through the closely set space between rotor and stator. As the material whirls around, it is subjected to a great many eddies of tremendous force; as a result, the particles are broken up. The product is flung out and passes out of the colloid mill through a suitable opening. At the left we see how a colloid mill serves. *A* shows an ordinary oil-in-water emulsion, magnified 1,000 times. *B* shows the same emulsion after it has passed through a colloid mill.

of colloidal particles), we will produce the amazing number of 8,000,000,000,000 cubes, with a total of 96,000,000,000,000 edges and 64,000,000,000,000 corner points. The overall surface of these cubes will reach the total of 120,000 square centimeters. There is exactly 20,000 times more surface now than when the substance was in the form of a single cube. Yet the combined volume of all these tiny cubes is the same as that of the single cube with which we started — 1 cubic centimeter. The vast increase in surface area and in the number of corners and edges makes adsorption an important property of colloids.

PREPARATION OF COLLOIDAL
    PARTICLES

In bringing about various desired chemical reactions, chemists often have occasion to prepare particles of colloidal dimensions. They can do this in several ways.

They can break down comparatively coarse particles into particles of colloidal size. This is called *peptization,* from the Greek word *peptein,* meaning "to digest." One way to peptize materials is to grind them in what is known as a *colloid mill.* If coarse particles are held together by weak molecular forces, they can be broken up into smaller particles if one puts them in a liquid for which the particles have great affinity — that is, to which they are strongly attracted. Thus latex particles can be reduced to colloidal dimensions if they are immersed in benzene.

The chemist can also produce particles

of colloidal size by building up smaller particles until they reach the colloidal range of dimensions. Substances that have been reduced to their molecules or ions in a solution can be made, under certain conditions, to form tiny crystals of colloidal size.

Such, then, are the colloidal particles and the systems in which they occur. These systems are involved in many living processes and other processes set up or modified by man.

COLLOIDAL PHENOMENA IN LIVING THINGS

All living particles are based on matter in the colloidal state. The cells of which living things consist are made up mainly of colloidal particles suspended in water. The vast surface area offered by these particles accounts for the rapid rate at which reactions occur within cells. If the particles gather together in clusters and settle out in a cell, that cell dies.

The clotting of blood is a colloidal phenomenon. Colloid particles of a protein called fibrinogen are suspended in the fluid part of the blood. Ordinarily they form a sol. But when the blood vessel is injured or when blood escapes through a wound, the fibrinogen is transformed by special mechanisms within the body into a gel—a blood clot that helps to prevent loss of blood.

Muscles are a colloid gel. They may be compared to the gelatin, in gel form, that we eat for dessert. If you shake gelatin too vigorously, it will gradually lose its water content and will shrink. If it is kept undisturbed, it will readsorb the water in time and will regain its original condition. Much the same thing happens in the case of our muscles. We constantly agitate them during the day. As a result, they too are gradually dehydrated. That is, they lose their water content. With rest, however, they return to their former state.

Certain colors of living things are due to colloidal particles. The color of the iris of the eye in some cases depends on the way in which the colloidal particles are dispersed, or scattered. As you know, certain animals, like the chameleon, are able to adapt their color to their surroundings.

They do so by altering the degree of dispersion of the colloidal particles in the skin.

COLLOIDS IN THE NON-LIVING WORLD

The colloids play an important part in the world of inanimate, or non-living, nature. They account, among other things, for much of the color that we see about us. The blue of the sky and the red of sunsets are both derived from the scattering of light from particles of dust in the atmosphere.

The formation of deltas at the mouths of rivers is a colloidal phenomenon. The silt of such rivers contains many coarse particles and also others of colloidal size. The coarse particles gradually settle out as the river flows along its course. The colloidal particles, however, remain in suspension as the current carries them along, since they have the same electrical charge and therefore repel one another. The river now approaches the ocean. The salts dissolved in seawater consist of ions. As the ions come in contact with the colloidal particles in the river, they neutralize the electric charge upon them. The particles no longer repel one another. They merge, and the large combined particles drop to the bottom of the river mouth. In time, they give rise to vast delta systems, like those of the Nile and the Mississippi.

COLLOIDAL PHENOMENA IN FOODS

Foods offer a great many examples of colloidal phenomena. The curdling of milk is one of them. The particles of casein, a protein that occurs in milk, are negatively charged. When the sugar in milk ferments, lactic acid is produced. This contains positive hydrogen ions. The hydrogen ions are attracted to the negative particles of casein and the latter are neutralized. They no longer repel one another but coagulate, and thus curds are formed.

The preparation of mayonnaise dressing is based upon a colloidal phenomenon. When oil droplets are suspended in a vinegar medium, the two liquids soon separate. The oil is collected in one layer; the vinegar in another. To make mayonnaise dressing, egg yolk is added. This serves as an emulsifying agent or, as it is sometimes called, a

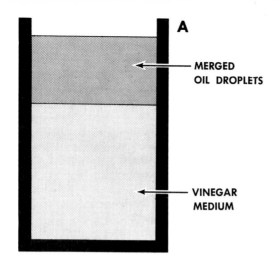

A

MERGED OIL DROPLETS

VINEGAR MEDIUM

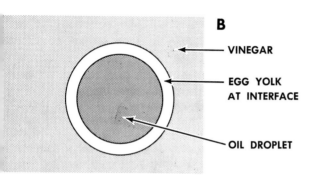

B

VINEGAR

EGG YOLK AT INTERFACE

OIL DROPLET

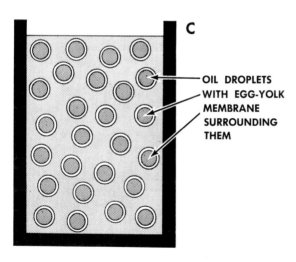

C

OIL DROPLETS WITH EGG-YOLK MEMBRANE SURROUNDING THEM

The preparation of mayonnaise dressing. In *A*, above, oil droplets have been added to a vinegar medium. These droplets have flowed together and, being lighter than the vinegar, rise to the top of the container. Thus we have two layers of liquid, one above the other. If egg yolk is added *(B)*, it concentrates on the interface (boundary) between a given oil droplet and the vinegar. The egg yolk forms a membrane around each oil droplet. The droplets can no longer merge; as a consequence, they now remain suspended throughout the vinegar medium *(C)*.

protective colloid. It concentrates in the interface, or boundary, between the liquid of the colloidal particle (the oil) and the liquid in which the colloidal particle is suspended—that is, the vinegar. The egg yolk forms a membrane around each oil droplet, and thus prevents the droplets from merging.

We change the nature of a food in a somewhat similar fashion when we prepare homogenized milk. In ordinary milk, the globules of fat that make up the cream are larger than colloidal particles. They have a lower specific gravity than the water and the proteins contained in the milk. Therefore, if we let milk stand, the butterfat globules will rise and concentrate at the surface. To prevent this from taking place, we break up the globules by squirting the milk under high pressure against a plate of agate, or a hard metal disk, or through a nozzle equipped with a special type of diaphragm. Each of the fat globules is broken up into a number of smaller ones of colloidal size. These become coated with the proteins in the milk, which concentrate at the interface and serve as protective colloids. The protein concentration, together with the Brownian movement, will prevent the globules of butterfat from flowing together as before.

COLLOIDS IN INDUSTRIAL PROCESSES

Colloidal phenomena play an important part in a vast number of industrial processes. In many cases, these go back hundreds or even thousands of years, but only since the development of colloid science have we been able to understand them. In other instances, entire industries owe their rise to the development of colloid science.

The ceramic industry, devoted to the manufacture of pottery, tiles, brick, dishes, and various other articles, draws its raw materials from different clays, which are a good example of matter in the colloidal state. All the clay particles have the same kind of electric charge and repel one another. Because they are so closely packed, they cannot move freely but maintain the same relative position. If a force is exerted

Actual photograph of bubbles produced in the flotation process. Each bubble is carrying a load of mineral. The flotation process is described in the text.

H. Rush Spedden

The process of *flotation* by which minerals are separated from their ores, involves particles of colloidal size and adsorption. First, the ore is ground to a fine dust and is mixed with water. A small amount of carefully selected oily material is then put into the water. This oil will be adsorbed only on the surface of the desired mineral particles. The mixture is now agitated and frothy suds are produced. As you know, oil is lighter than water. The oily part of the suds carries the mineral particles upward with it. The impure materials in the ore, which have not absorbed any oil, remain at the bottom. The froth that has accumulated at the surface is then skimmed off, and the mineral is obtained.

upon the clay, the particles slip by each other, thus permitting the shaping of the vase or plate or other object that is being fashioned by the potter. When the force is removed, the particles retain their new position, since the same electrical forces as before act upon them.

The plastics industry offers various examples of colloidal phenomena. In the preparation of rayon, for example, cellulose, the framework of plants, is transformed into a colloidal sol by the use of appropriate solvents. It is then squirted through a spinneret into a coagulating bath and a thread of artificial fiber is obtained.

Demonstration showing one of the many uses of activated charcoal. The solution shown in *A* has unwanted color. Activated charcoal is added to the solution *(A)*, which is poured into a filter *(B)*. The result is a clear solution *(C)*. The unwanted color in the solution has been removed by the adsorbing action of the activated charcoal.

The extensive soap and synthetic-detergent industry provides an excellent example of colloidal particles at work. The

Atlas Chemical Industries

A

B

C

An electrostatic precipitator, or scrubber, is designed to capture dust and fumes escaping from furnaces. This device can eliminate much of the air pollution over industrial areas. Instead of discharging particles and fumes into the air, the electrostatic precipitator captures and collects them and then dumps them in a suitable spot. The diagram shows how a precipitator works. The gas, laden with dust particles, passes between a series of collecting plates, between which a high-voltage electrostatic field is maintained. The field ionizes the gas, and the gas ions cause the particles to acquire an electric charge. The field forces the charged particles to the collecting plates, from which they will later be dislodged, dropping into hoppers. Only two plates are shown here; actually there are many of them.

Diagram adapted from material supplied by Research-Cottrell, Inc.

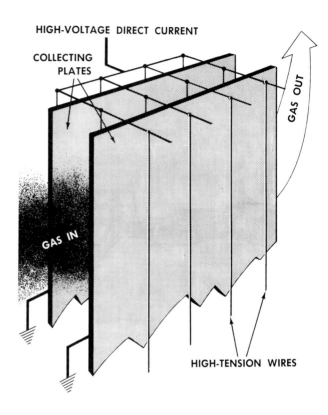

HIGH-VOLTAGE DIRECT CURRENT

COLLECTING PLATES

GAS OUT

GAS IN

HIGH-TENSION WIRES

colloidal suspension of soap in the water removes dirt. The paper industry offers another good example. Sheet formation, coating, glossing, and printing are based on the colloidal properties of cellulose fibers, glue, and inks. The photographic industry utilizes gelatin and colloidal silver salts as the light-sensitive ingredients of the film or photographic plate.

Adsorption plays an important part in various industrial processes. Activated carbon is used widely in industry for the recovery of volatile solvents—those which readily evaporate. The vapor is passed through a column or tower stacked with activated carbon, and the vapor particles are adsorbed on the surface. The saturated adsorber is now heated and the solvent that has concentrated upon its surface is removed. Activated carbon is also used in gas masks. Poisonous gases are adsorbed on the surface of the carbon while the oxygen and nitrogen contained in the air pass through freely to the lungs.

Among the unfortunate by-products of industry is the smoke that escapes from the chimney flues of many manufacturing plants. Such smoke is always annoying and it may be extremely injurious to both animal and plant life. Fortunately, colloid science has made it possible to deal with this nuisance.

In the first decade of the twentieth century, a U.S. chemist, Frederick Gardner Cottrell, developed a device for precipitating by electrical means the particles contained in smoke. This device, which is called the *Cottrell electrostatic precipitator,* has been widely adopted. In one type of precipitator (see diagram), there are a number of parallel plates, housed within a shell and set approximately 30 centimeters apart. Wires running parallel to the plates are set midway between them. As smoke enters the precipitator, a powerful electric charge is passed through the wire and an electric field is set up. The gas is ionized, and the ions cause the particles in the smoke to be charged. Since the plates, which are in another circuit, have the opposite charge, they attract the charged particles, which collect upon the plates. The particles are dislodged from the plates by rapping or scraping, which is done either by hand or automatically. Then the particles drop into hoppers and are removed. The Cottrell process does away with bothersome fumes. Furthermore, it recovers valuable materials, such as metallic oxides, which would otherwise make their way to the outer air.

CRYSTALS

Exquisitely fashioned snowflakes, flawless diamonds with glittering facets, the almost perfect cubes of salt grains — these objects are not random and unrelated forms. They are all fine examples of *crystals* — bodies with a pattern of flat surfaces that meet at definite angles. The world is full of such crystalline patterns, for almost all inorganic, or nonliving, substances in the solid state form crystals. We find them in ice, snow, sugar, salt, and sulfur; in metals like gold, silver, copper, iron, and mercury; in precious stones like zircon, emerald, topaz, and sapphire.

Certain nonliving substances, like glass and obsidian, do not form crystals. Their inner structure is haphazard and shapeless, as is their outer structure, unless it is modified. We call such solids *amorphous* (from the Greek "without form"). Amorphous inorganic solids are not numerous. The great majority of inorganic solids assume some definite crystalline pattern or other.

INNER STRUCTURE

The external differences between crystals are based on differences in internal structure. The particles of matter within a crystal are arranged in a framework called a crystal *lattice*. Scientists deduced the existence of the lattice many years ago, but it was not until X rays were used in the study of solids that definite proof was obtained. X rays reveal the lattice by causing a definite shadow pattern to be cast when they pass through it.

There are three types of structural units in crystal lattices: (1) molecules, (2) ions, or electrically charged molecules and atoms, and (3) atoms.

CRYSTALS MADE UP OF MOLECULES

In substances like ice, iodine, and solid carbon dioxide (dry ice), the structural units of the crystal lattice are small molecules. These are held together by rather weak forces, partly gravitational and partly electrical. There is much space between the molecules, and the crystals are comparatively light in weight.

In general, crystals in which small molecules are the structural units have low melting points. They are good insulators, and they are comparatively soft. In some cases the bonds between the molecules are so weak that the solid will change into a gas without first becoming a liquid. This is what happens in the case of dry ice. We call this phenomenon *sublimation*.

Some crystals consist of larger molecules. Chrysotile asbestos is a good example of a substance that forms crystals from larger molecules. The crystals take the form of long, parallel fibers.

IONIC CRYSTALS

In salts, the unit making up the crystal is an ion, which, as we pointed out, is an

Marking an uncut diamond with ink to show how it should be cut. Like most inorganic substances, diamonds form crystalline patterns.

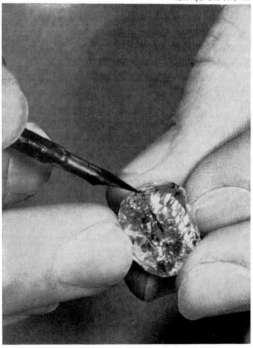

electrically charged molecule or atom. Let us recall that each atom has a nucleus or central core made up chiefly of protons, each with a positive electrical charge, and neutrons, which have no charge. Around this central core electrons are distributed. Each electron has a negative charge. Ordinarily the charge on an atom is neutral; that is to say, there will be as many negative charges as there are positive charges. If an atom loses an electron, it has one excess positive charge. It becomes a positive ion, or *cation*. If an atom gains an electron, it has one excess negative charge. It becomes a negative ion, or *anion*.

Let us see what happens when sodium, normally a metal, and chlorine, normally a gas, react to form the solid called sodium chloride (NaCl), which is simply table salt. Each sodium atom transfers an electron to a chlorine atom. As a result, the sodium atom becomes a positive ion since it now has an excess positive charge. Each chlorine atom, on the other hand, acquires a single excess negative charge. It is now a negative ion.

Since ions with unlike charges attract each other, the chlorine ions will attract the sodium ions, but they will hold off the other chlorine ions since ions with like charges repel each other. As a result of the attraction between the oppositely charged particles, each chlorine ion will surround itself with six sodium ions. Each sodium ion will surround itself with six chlorine ions. The lattice pattern that results is shown in Figure 1. This pattern will be repeated throughout the crystal. In the drawing we have represented the sodium and chlorine ions as being a considerable distance apart, in order to show the interlacing pattern clearly. Actually the ions are closely packed together in the crystal lattice, as shown in Figure 2.

This particular arrangement is common when the ions are of about the same size. If the ions vary in size, it may be easier for them to fit together in a lattice of the type shown in Figure 3. Here the central ion is equally distant from eight other ions of opposite charge. The compound cesium chloride (CsCl) forms this sort of crystal.

Substances that have the ionic type of lattice have moderate insulating properties and high melting points. They are hard, but they can be split along definite lines.

CRYSTALS MADE UP OF NEUTRAL ATOMS

In metals, the atom is the structural unit in the formation of a crystal. The atoms may be thought of as spheres having the same diameter and packed together as closely as possible. To illustrate one arrangement, let us imagine that fifteen billiard balls are racked up to form the base, or foundation layer, of the pyramidal structure shown in Figure 4. Six more are set on top of the first layer of balls. Then another ball is placed on the second layer. This shows the closest packing possible in a cube. Iron, lead, gold, silver, and aluminum assume this kind of pattern. There are several other arrangements of atoms in metallic crystal lattices. Lattices of this kind are opaque. They have moderate hardness, high melting points, and are the best conductors of heat and electricity. These qualities make metals very useful.

PROPERTIES OF CRYSTALS

The variations in internal structure shown by different crystals have a direct

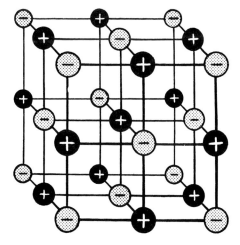

1. How the ions are arranged in the crystal lattice of sodium chloride (table salt). The + circles represent positive sodium ions; the − circles represent negative chlorine ions.

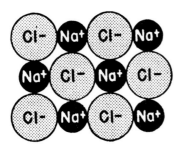

2. Positive sodium ions (Na⁺) and negative chlorine ions (C1⁻), closely packed together in the crystal lattice of table salt.

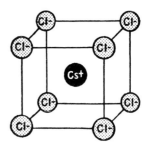

3. The crystal pattern of cesium chloride. The central ion (Cs⁺) is equally distant from the others.

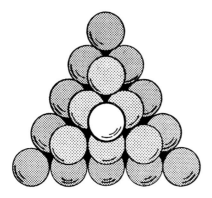

4. How atoms (viewed from above) are packed in the crystal lattice of various metals, including iron, lead, gold, silver and aluminum.

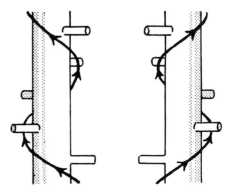

5. Spiral arrangement of the molecules of right-handed and left-handed silicon-dioxide crystals. The right-handed crystal is a mirror image of the left-handed one.

bearing upon their properties. Different crystals have different *lines of cleavage*— lines along which they split most readily. They conduct heat at different rates. They react differently to magnetic and electrical forces.

A striking example of the different properties that crystals can exhibit, depending on their internal structure, is provided by the element carbon. There are two crystalline forms in which carbon occurs as an element in nature. One is graphite, a black, opaque substance. The other is diamond. They are so different in appearance that a person unaware of their chemistry would not think they were made up of the same element. Their differences result from variations in their crystalline structure.

In the graphite crystal, atoms of car-bon are joined together in flat, hexagonal plates. The plates lie parallel to one another (see Figure 6). The bonds between the layers of plates are indicated by dotted lines in Figure 6. These bonds are weak, in comparison with the bonds between the atoms within the hexagonal plates themselves. Hence one layer slips easily over the layer beneath or above it. This property accounts for graphite's value as a lubricant.

Diamond, like graphite, is made up of carbon atoms. In the diamond crystal, however, each atom of carbon is bonded to four neighboring carbon atoms, which are grouped about it at equal distances. In Figure 7, for example, the carbon atom A is bonded to carbon atoms B, C, D, and E. Atoms B, C, D, and E are each bonded to four other carbon atoms in the same way.

The bonds between all of the atoms in the diamond crystal are of equal strength. The result is a very rigid formation. In fact, diamond is the hardest substance known, and it is very difficult to cut. Diamond also has a high melting point. It is a good insulating material, and it is transparent.

POLARIZATION OF LIGHT

A few crystals, like those of the mineral Iceland spar, have a particularly interesting property—that of allowing only light-waves that vibrate in parallel planes to pass through them. This effect is called *plane-polarized light.* To understand what is involved, let us try to pass a knife blade between the pages of a closed book. This will be possible only if the knife blade is held parallel to the pages. The book in this case would correspond to the Iceland spar crystal. The knife would correspond to one of the parallel planes in which the light would vibrate.

Other crystals have a rather peculiar effect on plane-polarized light. If such light is allowed to pass through a selected crystal of quartz, the plane of polarized light is twisted to the right through an angle. Other quartz crystals will turn the plane to the same angle to the left. Crystals of the first type are called right-handed. Those of the second type are called left-handed.

X-ray analysis has shown that the molecules are arranged in spirals in such crystals, as shown in the diagram of the right-handed and left-handed silicon dioxide (SiO_2) crystals in Figure 5. These crystal "twins" have the same relation to each other as an object and its reflection in a mirror. They have the same chemical composition, but the structural arrangement that makes one right-handed and the other left-handed also makes their chemical behavior different.

The fact that different crystals will rotate the plane of polarized light in different directions forms a reliable means of identifying certain substances. For example, sugars in solution will rotate the planes of light through different angles. Therefore, the angle of rotation can be used to identify each sugar in question.

EXTERNAL STRUCTURE

Our knowledge of the inner structure of crystals is of comparatively recent date. The outer structure of crystals, however, has been carefully studied for several hundred years. *Crystallography,* the study of the external characteristics of crystals, is a highly technical subject and it takes years to become expert in it. In the following brief account we shall give some idea of the principles upon which it is based.

Over a hundred years ago scientists adopted a convenient method for classifying crystals on the basis of their external forms. The method is still used to identify minerals occurring in nature as well as chemical compounds of many different kinds. According to this classification, all crystals fall into thirty-two different classes

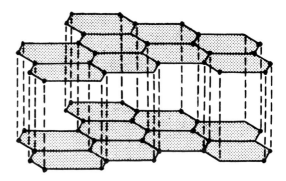

6. Crystals of graphite, made up of carbon atoms, consist of parallel layers of flat, joined hexagonal plates. The bonds between layers are shown by dotted lines in the diagram.

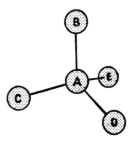

7. How the atoms of carbon are grouped in diamond. *A, B, C, D* and *E* are carbon atoms.

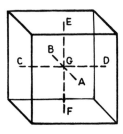

8. How the three axes of a cube meet.

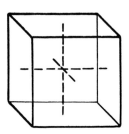

9. Isometric or cubic system. Left: sodium-chloride crystal; right: pyrite tetrahedrite.

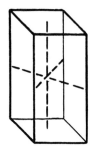

10. Tetragonal system.

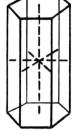

11. Hexagonal system.

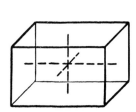

12. Orthorhombic system.

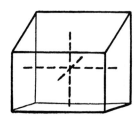

13. Monoclinic system.

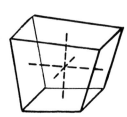

14. Triclinic system.

from the viewpoint of external structure. The thirty-two classes, in turn, are grouped in six primary divisions, called *crystal systems*. These systems are based on the arrangement of the axes—imaginary lines which intersect at a point within the crystal.

To show more clearly what we mean by axes, let us examine a crystal that has the form of a cube (Figure 8). This crystal has six faces, consisting of squares with the same dimensions. If you look carefully at the cube, you will see that there are only three directions in which the edges can run. Each axis will be parallel to one of these directions.

Let us drill a hole through the center of the cube on one face, at A, so that it will come out in the middle of the opposite face, at B. This hole, AB, will represent one axis of the cube. It will be parallel to one set of edges. Let us now drill CD and EF, also axes of the cube, in the same way. The three axes meet at the point G.

As we shall see, the characteristic axis arrangement is the same for all the crystals within a given system, however simple or

complicated they may be. Thus, in Figure 9 the sodium-chloride crystal and the far more complicated pyrite-tetrahedrite crystal belong to the same system because the arrangement of their axes is the same. In the following description of the six crystal systems, we shall discuss only the simplest form within each group.

THE SIX CRYSTAL SYSTEMS

The first of these groups is called the *isometric* (equal measure) or cubic system (Figure 9). In this system are included the crystals that form cubes or modifications of the cube shape, such as octahedrons (eight-faced solids) and dodecahedrons (twelve-faced solids). These forms have three axes of equal length set at right angles to one another. Among the substances forming isometric crystals are salt, diamond, fluorspar, galena, gold, silver, garnet, iron, copper, lead, and certain forms of silicon and phosphorus.

The second group is the *tetragonal* (four-sided) system. In the simplest form of this system (Figure 10), the four sides are equal rectangles. The top and bottom are equal squares. There are two axes of equal length and a third axis that is different in length. All three axes meet at right angles to one another. Zircon, chalcopyrite, scheelite, wulfenite, and tin all form crystals of this general pattern.

The third group is the *hexagonal,* or six-sided system, illustrated in Figure 11. The crystal shown here has six sides. The top and bottom surfaces are at right angles to these sides. There are three horizontal axes, all in the same plane, set at an angle of sixty degrees to one another. There is also a fourth axis set at right angles to the first three. It may be shorter or longer than the other axes. In the hexagonal system we find most of the gem minerals, including sapphire, ruby, emerald, aquamarine, and tourmaline. The crystals of ice and snow fall in this group. So do those of the elements arsenic, antimony, and bismuth.

The fourth group is the *orthorhombic* system. In the representative crystal shown in Figure 12, all six faces are rectangles that meet at right angles. There are three axes of unequal length set at right angles to one another. This system includes sulfur, topaz, chrysoberyl, calamine, and benzidine.

The fifth group is the *monoclinic* (one-slant) system. In the form shown in Figure 13, only the sides meet at right angles to one another. The top and bottom surfaces are tilted from the perpendicular. Two of the axes meet at right angles, but the third is tilted. Gypsum belongs to this system. So do kunzite, borax, naphthalene, and cane sugar.

In the sixth crystal system, the *triclinic* (three-slant), none of the faces meet at right angles. All three axes are inclined at different angles to one another (Figure 14). Turquoise and the feldspars belong to this group.

It is difficult to find a perfect crystal bounded by absolutely smooth plane faces. Unless the surfaces are formed under exceptionally favorable circumstances, they will show various irregularities. Sometimes the crystalline pattern will be so obscure that only the most careful measurements will reveal it. The faces are often marked by grooves, or pits, or small outgrowths. Sometimes they are curved from strain. Occasionally, too, many small crystals combine erratically. That is why uncut stones so often appear to be dull and valueless. They do not reveal their full beauty until they pass through the hands of the jewelry cutter. Yet nature can produce crystals lovelier by far than any man-cut gems.

Horne — John Innes Institute

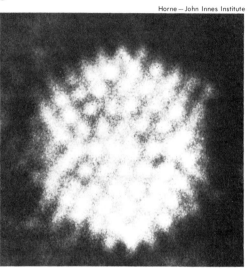

The remarkable photograph above, taken with an electron microscope, shows the crystals of a virus. It has been enlarged 750,000 times. The illustration below shows the geometric precision of the structure of the crystal.

Horne — John Innes Institute

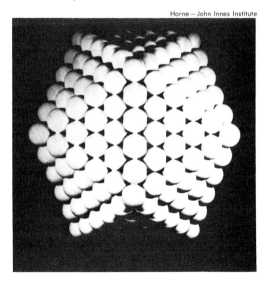

CHROMATOGRAPHY

by W. Thomas Lippincott

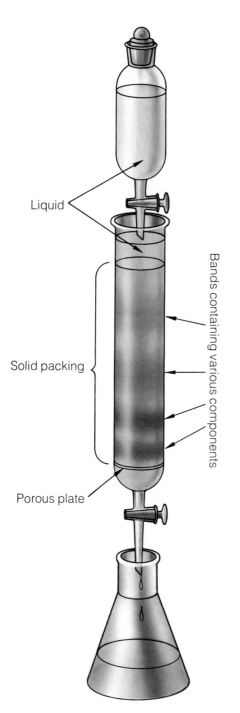

Liquid

Solid packing

Porous plate

Bands containing various components

The chemist often has occasion to separate and identify the different components that make up a chemical mixture. One of the most important tools in this type of analysis is the technique, or series of techniques, known as *chromatography*.

The name "chromatography" comes from two Greek words meaning "color recording." In the earliest application of chromatography by the Russian botanist Michael Tswett in the first decade of the twentieth century, different components were separated as distinct pigments. That accounts for the name "chromatography," given to the technique by Tswett. In certain kinds of modern chromatography, as we shall see, components are not distinguished on the basis of color.

The basis of chromatography is differential migration—that is, the flow of molecules at different rates from a starting point. The sample that is to be analyzed is in a moving phase. It may be either a liquid or a gas. It is passed through a medium that is stationary. This medium may be either a solid or else a liquid that is held by a solid. Since each of the components of the sample flows at a distinctive rate past or through the stationary medium, the components can be effectively separated from one another.

There are various kinds of chromatography. In this article, we shall consider three of the most important types: (1) liquid-column chromatography; (2) paper chromatography; and (3) vapor chromatography, also known as gas chromatography.

LIQUID-COLUMN CHROMATOGRAPHY

For this type of chemical analysis, a special column is prepared, as shown in Figure 1. It consists of a glass tube which has a porous plate set across its lower end and which is filled with a specially prepared powdered solid—generally silica gel or alumina. Both these solids have a large

Fig. 1. Chromatography column

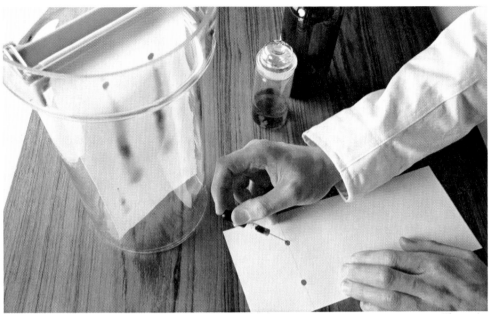

Demonstration of the dye separation technique for paper chromatography.

adsorptive capacity. They will adsorb as much as 70 per cent water by weight without appearing to be wet. By adsorption is meant the adherence of atoms, ions, or molecules to the surface of a solid in a thin layer. This layer is sometimes not more than one molecule deep.

The silica gel or alumina that is to be used in the tube is first carefully purified and dried. Then it is mixed with about 50 per cent alcohol or water. The liquid is adsorbed on the surface of the powder, which is then put in the glass tube. The alcohol or water adsorbed on the silica gel or alumina is known as the *stationary liquid*. We could also say that it is in the stationary phase.

The sample that is to be analyzed is dissolved in a liquid that does not mix with the stationary liquid. A chemist would say that it is immiscible with the stationary liquid. A part of the sample solution is then dropped onto the column and is allowed to flow past the stationary liquid. The particular components of the sample mixture that are most soluble in the stationary liquid will be held nearer the top of the column. The components that are least soluble will be carried downward with the moving solution and will be held nearer the bottom of the column. Usually it is necessary to add more of the sample solution to the column to bring about a satisfactory separation of components.

In this way, the sample mixture can be separated into a series of bands, moving down the column with varying speeds. If the bands are not colored, they can sometimes be made visible by adding a suitable indicator. If we continue to add sample solution to the column, the bands can be washed off the column and collected separately.

The process of separating the mixture into a series of bands on the column is known as *development*. The resulting pattern of bands is called a *chromatogram*. The name *elution* is given to the procedure of washing the bands off the column.

An example of the use of liquid-column chromatography is the separation of mixtures of organic acids. In this case, silica gel is used as the powdered solid. Methanol, adsorbed on the surface of the silica gel, is the liquid in the stationary phase. The acids to be analyzed are dissolved in petroleum ether. The sample solution is poured onto the column and additional petroleum ether is added slowly to develop the chromatogram. The bands may be made visible by using an indicator such as bromcresol green. If elution of the various bands is desired, even more petroleum ether must be dropped onto the column.

PAPER CHROMATOGRAPHY

It is often convenient to use filter paper, rather than silica gel or alumina, as the solid support for the stationary liquid. This is known as paper chromatography.

In one type of paper chromatography, a drop of a solution containing the mixture that is to be separated is placed near an edge or corner of the paper. The solvent is then allowed to evaporate, leaving a spot containing the mixture. The strip of filter paper is now suspended in a tall container so that the end of the paper near the spot dips into a liquid. The latter consists of a mixture of water and an organic solvent, such as butanol or collidine, which is partly miscible with water.

The liquid starts climbing up the strip of paper through capillary action. As it moves upward, it reaches the spot containing the mixture that is to be separated. Passing beyond the spot, the liquid carries the components of the mixture with it. These components move along the paper at different rates, depending on their relative solubility in water. When the advancing liquid front reaches the top of the filter paper, the latter is removed from the container, dried, and sprayed with an indicator. This should reveal a number of spots along the strip. Each spot on the strip represents a different component or combination of components. One can identify each component by comparing its rate of flow, called Rf, with that of known samples. The formula for the rate of flow is as follows:

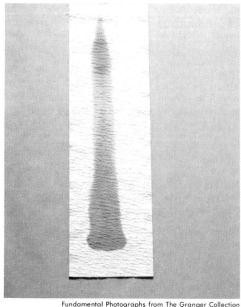

Components of crimson lake dye are separated by paper chromatography. Top, dye is placed at bottom of filter strip within a beaker. Middle, dye is carried upward by weak acetic acid solution. Bottom, dried chromatograph—dye has separated into three different colors.

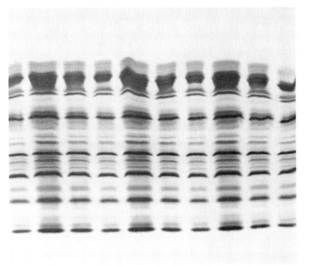

Print of an isoelectric focusing experiment, another chromatographic technique. Materials such as proteins travel through a medium of continuously changing pH in an electric field. Components stop at point where their net electric charge is rendered zero by the pH level.

$$Rf = \frac{\text{distance moved by the component}}{\text{distance moved by the liquid front}}$$

In some cases, the location of the components of the unknown sample is compared with the location of the components of a known sample.

Sometimes a two-dimensional chromatogram will provide a better separation of components. In this case, a large square of filter paper, perhaps 45 centimeters on a side, is used. The solvent containing the mixture to be separated is placed near one corner of the sheet and is allowed to evaporate, leaving a spot. The paper is then suspended so that one edge near the spot dips into a liquid consisting of one kind of solution. The liquid moves up the sheet by capillary action, carrying the components of the mixture sample up along with it. When the liquid reaches the top of the paper, the sheet is removed and dried.

The paper is now turned through 90°. An edge adjacent to the edge that was first dipped is suspended in a different solution, in which the Rf values are different. This will cause components of the spots that were separated in the first operation to move across the paper in a direction at right angles to the first movement. Thus a further separation of components will be brought about.

Paper chromatography is generally used in qualitative analysis—that is, in determining the kinds and not the quantities of components. There are various methods, however, for actually measuring quantities in paper chromatography. For example, the processed paper can be placed in a spectrophotometer, which measures the relative intensities of light in different parts of a spectrum. The transmission of light of particular wavelengths can then be measured.

GAS CHROMATOGRAPHY

When a gaseous mixture is passed over a solid or an immobilized liquid, it is usually possible to separate the components of the mixture. This technique, known as vapor chromatography, gas chromatography, or most often, gas-liquid chromatography, is one of the most widely used separation procedures. It is possible to operate the apparatus at temperatures up to 420° Celsius or even higher. Any stable substance having an appreciable vapor pressure at such temperatures can be separated from a mixture through the use of vapor chromatography. This technique has been employed effectively in a wide variety of studies, from the analysis of orange juice to the identification of polluting substances in the atmosphere.

A typical vapor chromatograph is shown in Figure 2A. The basic part of the apparatus is a column folded or coiled on itself and filled with a porous granular solid, whose pores are saturated with a liquid having a high boiling point. The sample gas, mixed with another gas, called the carrier gas, is made to pass through the column. The latter is usually a tube of glass or metal 1¼ to 3 meters long and 60 millimeters in diameter. It is enclosed in an oven equipped such that the apparatus may be used at different temperatures. In front of the intake end of the column (see Figure 2A) is a carrier gas injector, a flowmeter to monitor the flow rate of gas, and a sample injector.

The carrier gas, which in most cases is helium, is used to keep the sample gas moving through the column. When the mixture of carrier and sample gas leaves the column, it passes through a detector. It is then either passed to the outer air or else collected for further use.

The presence of the different components of the gaseous mixture being analyzed is detected by examining the exit gas—the gas that has passed through the column. The exit gas consists largely of the carrier gas, but at any time it may contain one of the sample mixture components.

In a typical gas analysis, the component least tightly held by the column packing passes out first. Under ideal conditions, the next component appears in the exit gas only after all of the first component has been carried away.

The most common method for detecting the presence of the various gaseous components in the exit gas involves the thermal (heat) conductivity of the components. A thin wire or a special resistor is placed in the path of the exit gas and is heated as electric current is passed through it. The resistance of the wire or thermistor, continuously measured, will vary with its temperature. Heat will be conducted away more rapidly by some gases, more slowly by others, and the resistance will change.

The resistance, recorded on a strip of paper (see Figure 2B), identifies the components of the exit gas.

Other methods for detecting gaseous components are used, such as making measurements of the components' density, their response to magnetism, or their absorption spectra.

In one method, the exit gas is ionized by alpha or beta particles from a radioactive source. The ions formed in this way are analyzed. A special technique can be used to isolate each component of the sample.

Vapor chromatography can be used to make a qualitative analysis of the sample being examined. A quantitative analysis can be obtained by measuring and comparing the areas under the different peaks in the chromatograph record (Figure 2B).

Other chromatographic techniques include the popular high performance liquid chromatography (HPLC), which uses high pressures to pump a mobile liquid phase and offers high resolution and sensitivity; gel permeation chromatography relies on filtering action to separate materials based on molecular size.

Fig. 2. Below left (A) is the diagram of a vapor chromatograph. Below right (B) we see a vapor chromatograph record of a number of methyl esters of fatty acids. The peaks and valleys in the record indicate the changing resistance of a wire or thermistor as the exit gas moves past it. The peaks in the record indicate various components of the original gaseous sample.

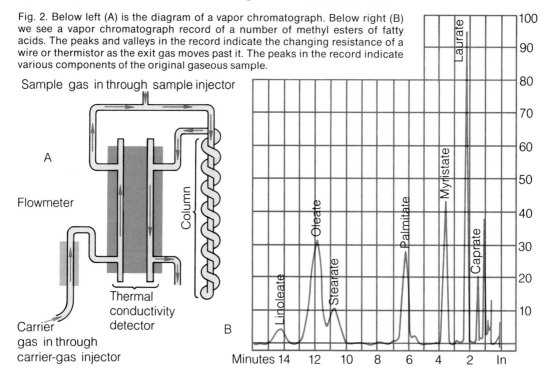

Young students of chemistry begin by studying simple organic compounds.

ORGANIC CHEMISTRY

by Donald C. Gregg

When chemistry first became a bona fide science, in the latter part of the eighteenth century, a number of chemical compounds were obtained from plants and animals. Scientists called them *organic substances* because they were derived either from living or formerly living organisms. The branch of chemistry devoted to their study became known as *organic chemistry*. The name *inorganic chemistry* was applied to the study of the chemical compounds obtained from minerals.

Chemists soon recognized that organic compounds derived from plants and animals had one thing in common: they all contained carbon atoms. Because of this fact, organic chemistry was also called the chemistry of the compounds of carbon.

The early chemists spent most of their time studying the acids, bases, and salts. They were interested chiefly in the substances obtained from, or related to, minerals. It was much easier to work with such inorganic compounds, since most of them were soluble in water or in the common acids. Usually their reactions were quite rapid. It was not too difficult to understand these reactions, or at least to explain them logically by using the theories that were then believed to be correct. Chemists could prepare many inorganic compounds by using very simple methods.

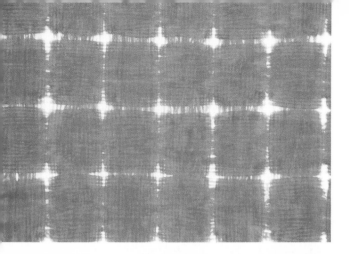

All three photos: Runk/Schoenberger from Grant Heilman

Organic chemists have developed many compounds that can be woven into cloth. Three common synthetic fibers are shown above. Top: nylon weave, magnified 50 times. Middle: rayon weave, magnified 50 times. Bottom: polyester weave, magnified 50 times.

A "VITAL FORCE?"

Organic compounds seemed to be quite different in nature and behavior. Many of them reacted with most common inorganic substances very slowly, or not at all. Heat played an important part in the preparation of inorganic compounds, but it usually caused organic compounds to decompose. As we pointed out, chemists could make inorganic compounds in the laboratory, but apparently it was impossible to create an organic compound. Chemists wondered how such compounds were produced in plants and animals.

The Swedish scientist Joens Jakob Berzelius was one of the foremost authorities in chemistry in the nineteenth century. This learned man made no effort to explain the behavior of the organic compounds. He simply stated that since they were derived from plant or animal organisms, their formation was probably due to some "vital force," a force totally lacking in inorganic material. It would be impossible for man, he maintained, to duplicate this vital force. Hence, nobody would ever be able to prepare an organic compound.

In 1828, a young German chemist, Friedrich Woehler, succeeded in doing what Berzelius had declared was impossible. He heated an inorganic compound, ammonium cyanate, and produced urea, an organic compound that occurs in urine. Chemists now began to doubt the existence of Berzelius' vital force. They theorized that perhaps other organic compounds might be prepared in the laboratory.

SYNTHETIC COMPOUNDS ARE COMMON

In the years that followed, chemists succeeded in producing many such compounds. More and more researchers devoted themselves to organic chemistry, and the number of synthetic, or man-made, compounds mounted steadily. In a few decades they far outnumbered the inorganic compounds.

UNUSUAL CHEMISTRY OF CARBON

Today we know that the basic laws that govern the formation and behavior of inorganic compounds also apply to the compounds of carbon. However, we still study organic compounds separately for several reasons. For one thing, carbon atoms have the capacity (which is almost unique) of producing chains and rings of atoms as they combine with each other and with other atoms. Then too, the organic compounds as a group are much less stable than the inorganic, so that chemists must use different techniques in working with them.

Why do we still call the compounds of carbon "organic" when we realize that many of them are man-made and not found in any living organism? The chief reason is that the compounds derived from nature and those produced in the laboratory are closely related to one another chemically. It has always seemed best, therefore, to give the name of organic compounds to most of the compounds of carbon whether they are of natural origin or produced in the laboratory. The carbon compounds that are not considered organic are the carbon oxides, carbonates, and cyanides. These are considered inorganic.

Hundreds of organic compounds are synthesized or discovered every year. It would be impossible to overemphasize their importance. Plants and animals are composed mostly of organic compounds. So is most human food, derived directly or indirectly from living things. Many fabrics are produced from organic materials (such as wool, cotton, and linen) obtained from plants or animals. Many, too, consist of synthetic organic materials (rayon, nylon, Dacron, and the like), developed in the laboratory in an attempt to duplicate natural fibers or to improve upon them. Plastics, rubber, paper, and wood are all organic substances. Petroleum, from which we obtain fuels and many industrial products, is a mixture of carbon compounds. So is coal tar, which yields drugs, perfumes, flavorings, and other useful products.

STRUCTURE OF THE CARBON ATOM

Each atom has a certain number of negatively charged particles, called elec-trons, revolving around a central core in a number of shells. There are two such shells in the carbon atom. The first shell has two electrons and the second has four. Only the electrons in the outer shell are involved in chemical changes.

This shell, with its four electrons, is not filled to capacity. There is room for four more electrons. The carbon atom can share electrons with other atoms in order to fill up the gap—a type of combination called *covalence*. When a carbon atom combines with four hydrogen atoms, sharing four pairs of electrons, we say it has a covalence, or combining power, of four. There are so very few cases in which a carbon atom has a covalence of three that we may leave them out of consideration for the purposes of this article. We shall assume that in all simple organic compounds, such as we shall take up, each carbon atom has four covalent bonds.

In order to have a clear idea of the carbon atom, we may picture the four valence electrons as being equidistant from each other and from the nucleus of the atom. We can visualize this best if we think of them as located at the corners of a regular tetrahedron. A regular tetrahedron is a figure or solid with four faces, consisting of equal triangles—triangles whose sides are all equal. Of course, it must always be remembered that

Highly simplified diagram of the carbon atom. Six electrons revolve around the nucleus in two shells. The inner shell has only two electrons; the outer shell has four electrons.

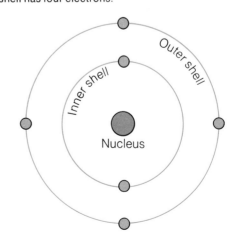

Nucleus

Inner shell

Outer shell

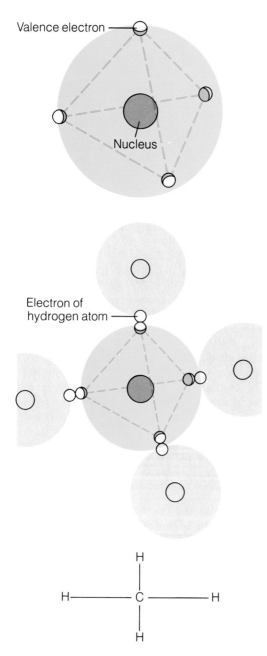

Valence electron

Nucleus

Electron of
hydrogen atom

H
|
H————C————H
|
H

Top: nucleus and outer shell of a carbon atom. Note that the four valence, or outer, electrons are equal distances from the nucleus and from each other. Middle: model of a carbon atom combined with four hydrogen atoms to form methane, CH_4. Four electrons, each from a hydrogen atom, have been added to the carbon atom. Bottom: a two-dimensional structural formula for the methane molecule. The long dashes represent bonds between the carbon atom and the hydrogen atoms.

electrons are in constant movement in space. However, for our purposes the tetrahedral arrangement is a reasonable representation.

The simplest organic compound is methane, in which one atom of carbon, represented by C, is combined with four atoms of hydrogen, represented by H. The formula for this compound is CH_4. A formula such as this is called a *compositional formula*. It indicates the kinds and numbers of atoms in a compound.

A space model of CH_4 is shown in the diagram on page 132. Four electrons, each from a hydrogen atom, which has but one electron, have been added to the outer shell of the carbon atom. This shell, therefore, now has its full quota of eight electrons.

On page 132, we also show a simplified projection of this model: a carbon atom is shown linked to four hydrogen atoms. This is not a true representation, of course. For one thing, it is two-dimensional, while a real molecule has three dimensions. However, chemists represent organic compounds in this simple manner because it is convenient and gives at least some idea of the actual three-dimensional structure. Such pictures of the way in which atoms are joined together are called *structural formulas*.

METHANE, OR PARAFFIN, SERIES

Methane, CH_4, is called a *hydrocarbon* because it is made up only of atoms of hydrogen and carbon. There are a great number of other hydrocarbons, all containing only carbon and hydrogen atoms. They occur as gases, liquids, or solids. They are found in the gases formed by heating coal or wood in a closed vessel. Petroleum and natural gas are made up almost entirely of hydrocarbons.

Methane is a member of the so-called methane series of hydrocarbons. Each compound in the series has the general formula C_nH_{2n+2}. If there is only one carbon atom in a compound of the methane series, n is equal to 1. If there are two carbon atoms, n is equal to 2, and so on. In the case of methane, for example, since there is only one carbon atom, n equals 1. Substituting 1 for n in the formula C_nH_{2n+2}, we have $C_1H_{(2 \times 1)+2}$, or CH_4. In the hydrocarbons of the methane series, the quota of electrons for the outer shells of the carbon atoms is filled. Hence, these atoms cannot take on any more hydrogen atoms. In other words, they are saturated with hydrogen atoms. For this reason they are called *saturated hydrocarbons.*

The compounds of the methane series are sometimes known as *paraffin hydrocarbons.* The word paraffin comes from the Latin *parum affinis,* meaning "having very little affinity." The name paraffin hydrocarbons is quite appropriate, since the compounds of the methane series have little affinity with other atoms: that is, little tendency to join them in chemical combinations. The members of the methane series are also called *alkanes.*

Methane, the first of the methane series, is a gas occurring in natural gas. It is sometimes called marsh gas because it is found in marshes, where it is formed by the decay of vegetable matter. Sometimes it collects in coal mines. Miners called it fire damp. When methane is mixed with the oxygen contained in air, it is explosive. Many disastrous accidents in mines have been due to the ignition of such a mixture. Methane is a common source of hydrogen. It is also used in the manufacture of carbon black.

The second member of the methane series is ethane, which has the compositional formula C_2H_6. It is a colorless, odorless gas, found in coal gas and natural gas. Its structural formula is as follows:

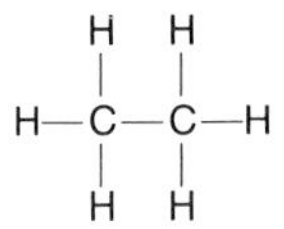

H   H
| |
H—C—C—H
| |
H   H

Each link in the above formula, of course, indicates a covalent bond. Whenever we write a structural formula such as this, we must check it to make sure that the four covalent bonds of the carbon atom are accounted for. Note that in each case a carbon atom is linked to three hydrogen atoms and also to the other carbon atom.

The structural formula shown above can also be written as CH_3CH_3. This is called a *simplified structural formula.* It indicates that a carbon atom (C), which is connected to three hydrogen atoms (H_3), is linked to another carbon atom (C), also connected to three hydrogen atoms (H_3).

Propane is the third member of the methane series. It has the compositional formula C_3H_8 and the structural formula:

H   H   H
| | |
H—C—C—C—H
| | |
H   H   H

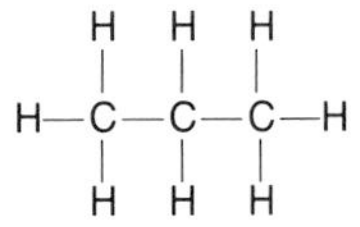

The simplified structural formula of propane is $CH_3CH_2CH_3$. This means that a carbon atom (C), connected to three hydrogen atoms (H_3), is linked to another carbon atom (C), joined to two hydrogen atoms (H_2). The second carbon atom is linked in its turn to a third one (C), joined to three hydrogen atoms (H_3). Compressed propane is used as a fuel gas.

The fourth member of the methane series is butane, with the compositional formula C_4H_{10}. Since we have so many building blocks, in the form of atoms, to work with in butane, we might suspect that the different carbon and hydrogen atoms can be combined in more than one way. This is actually the case. One structure has the normal straight-chain form:

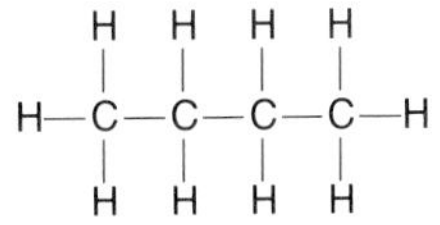

H   H   H   H
| | | |
H—C—C—C—C—H
| | | |
H   H   H   H

The other has a chain of three carbon atoms, with a carbon attached to the middle carbon of the chain as follows:

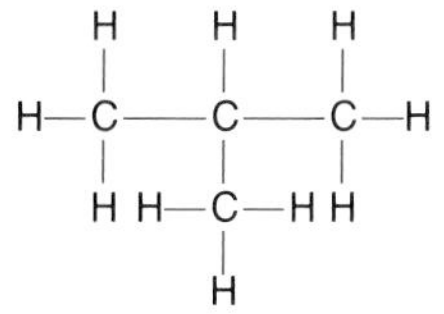

H   H   H
| | |
H—C———C———C—H
| | |
H   H—C—H H
        |
        H

This forms what is called a *branched-chain configuration.* It shows a chain with a branch leading from one part of it.

The two butane structures shown above represent two different compounds, each having the compositional formula

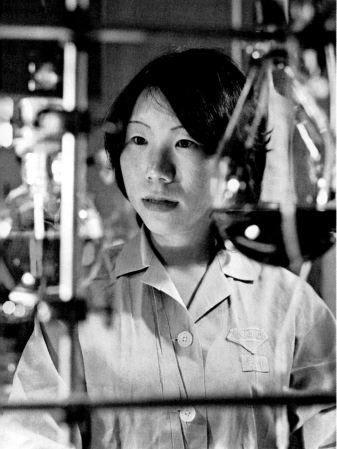

This technician is working in a lab that produces polyethylene. Polyethylene is widely used as a wrap for packaging food and other items.

J. Alex Langley, DPI

C_4H_{10}. The compound with the straight-chain structure is called normal butane or *N*-butane. The branched-chain compound is known as isobutane. They have different physical properties and chemical behavior. *N*-butane is a major source of butadiene, the principal chemical in some kinds of synthetic rubber. Isobutane is used in the manufacture of certain superior grades of motor fuel.

Two or more compounds with the same compositional formula but with different structural formulas are called *isomers*. *N*-butane is the straight-chain isomer of butane. Isobutane is the branched-chain isomer. There are only two isomers in the case of butane. As we shall see, other compounds have a great many isomers. The existence of isomers is a common and significant phenomenon in chemistry.

In the fifth member of the methane series—pentane—five carbon atoms are combined with twelve hydrogen atoms. The compositional formula is C_5H_{12}. There are three different isomers of pentane, each with a different structural formula: *n*-pentane, isopentane, and neopentane. Here are the structural formulas:

a) n-pentane

b) isopentane

c) neopentane

Other members of the methane series include hexane (C_6H_{14}), heptane (C_7H_{16}), and octane (C_8H_{18}). The number of possible isomers increases as the number of carbon atoms in the molecule increases. For $C_{10}H_{22}$ there are at least 75 possible isomers; for $C_{20}H_{42}$, at least 366,319. Only a few of the latter are known. Obviously, because of the existence of isomers, the number of possible organic compounds reaches almost astronomical proportions.

The lighter members of the methane series of hydrocarbons, containing less than five carbons per molecule, are gases. Those having from five to fifteen carbons are usually liquids. Those with more than fifteen carbons are solids.

The paraffins are constituents of natural gas and petroleum. Natural gas is mostly methane with some of the other gaseous paraffin hydrocarbons. Crude oil consists of a mixture of many different liquid and solid hydrocarbons. In gasoline, there is also a mixture of hydrocarbons, generally ranging from C_6H_{14} to $C_{12}H_{26}$. The yield of gasoline from a given amount of crude oil may be increased by building up hydrocarbons containing less than six carbons to produce molecules with from six to twelve carbons. It can also be increased by breaking down the hydrocarbons that have more than twelve carbon atoms. The latter process is known as *cracking.*

Pure paraffin hydrocarbons are colorless, almost odorless, and inflammable. They are very insoluble in water and they do not form ions in water solutions. When they undergo complete combustion, they yield only carbon dioxide (CO_2) and water. It is rather difficult, however, to bring about this type of combustion. Hence, carbon and carbon monoxide are often produced in addition to carbon dioxide and water. For example, when gasoline is burned in the motor of a car, carbon may be deposited in the cylinders and carbon monoxide (CO), a poisonous gas, may issue from the exhaust pipe.

Certain saturated hydrocarbons have a ring structure. These compounds are called *cyclic.* We can form such a structure by joining the ends of a chain together, provided there are at least three carbon atoms. Cyclopropane, with the compositional formula C_3H_6, is a good example of a ring formation. This substance, which is used as an inhalation anesthetic, has the following structural formula:

J. Alex Langley, DPI

Chemical plants like this one manufacture many compounds which are widely used in industry for the production of synthetic materials.

Saturated hydrocarbons with a ring structure are called *cycloparaffins.* Cyclobutane (C_4H_8), cyclopentane (C_5H_{10}), and cyclohexane (C_6H_{12}) are cycloparaffins.

UNSATURATED HYDROCARBONS

In all the carbon compounds we have discussed thus far, there is only one bond between any two carbon atoms. There is more than one such bond in certain hydrocarbons. For instance, in the substance called ethylene, C_2H_4, there are two links between the two carbon atoms in its molecule, as the structural formula shows:

When there is more than one bond between carbon atoms, the molecule is not saturated, since other atoms could be added to it.

For example, instead of the double carbon link in ethylene, we might substitute a pair of hydrogen atoms, enclosed in squares in the following formula:

This is exactly what happens when ethylene is treated with hydrogen gas and a suitable catalyst. The new substance produced in this reaction is ethane gas (C_2H_6).

Substances such as ethylene, in which carbon atoms are joined by double bonds, are called *unsaturated,* because they are capable of taking up more hydrogen atoms or other atoms. They are quite reactive.

In some unsaturated hydrocarbons, there are three bonds between carbon atoms. To give a familiar example, acetylene (C_2H_2) has the following structural formula:

$$H—C≡C—H$$

Acetylene is widely used for oxyacetylene welding and cutting and in the preparation of many important organic compounds, such as acetic acid and acetone. It is the first member of a series of hydrocarbons with triple bonds between carbon atoms — the *acetylene series.*

The unsaturated hydrocarbons play an important part in the rubber industry and many others. The basic unit of rubber is isoprene (C_5H_8), an unsaturated hydrocarbon with the following structure:

Rubber is a *polymer.* This means that it is made up of a great number of identical small units. This is how two of the many isoprene units in rubber would be linked (the double bonds shift position):

Isoprene unit when polymerized   Isoprene unit when polymerized

SOME DERIVATIVES OF THE PARAFFIN HYDROCARBONS

The alkyl halides. The alkyl halides are simple compounds derived from the paraffin hydrocarbons. In them, a halogen atom takes the place of a hydrogen atom. If a chlorine atom (Cl) is substituted for a hydrogen atom (H) in methane, CH_4, we obtain the alkyl halide known as methyl chloride, CH_3Cl, as the following structural formulas show:

Methane   Methyl chloride

Methyl chloride has been used extensively as a refrigerant.

Two other familiar alkyl halides are ethyl choride (C_2H_5Cl) and ethyl bromide (C_2H_5Br):

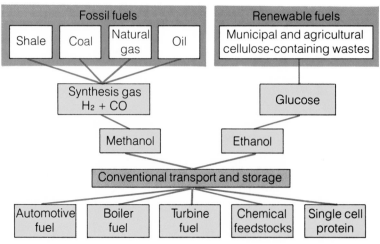

ALTERNATIVE LIQUID FUELS
Sources and uses

Chart showing possible ways of producing and using alternative fuels. As the availability of fossil fuels diminishes and costs rise, more research is being devoted to a search for substitute fuels.

Courtesy of Hoffman—La Roche

Antibiotics, produced in chemical laboratories (left), are organic compounds used in medicine. Anesthetics (below) are often derived from organic substances.

Shostal

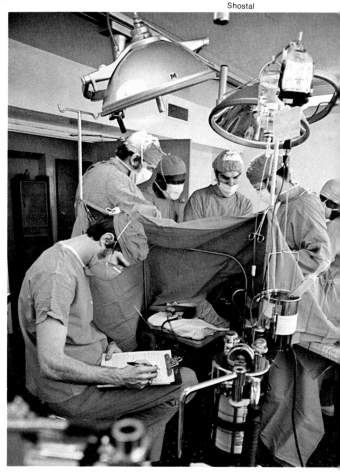

H—C—C—Cl   H—C—C—Br
Ethyl chloride   Ethyl bromide

Both these substances have been used as anesthetics.

The typical alkyl halide is quite reactive, as the halogen atom in its molecule is held rather loosely. This atom may be removed completely or it may be replaced by certain atoms or groups of atoms.

The alcohols and ethers. In the compounds called *alcohols,* a hydroxyl group, OH, or more than one such group, is substituted for one of the hydrogen atoms in a paraffin hydrocarbon. When a hydrogen atom of methane (CH_4) is replaced in this way, we have methyl alcohol (CH_3OH):

H—C—H   H—C—OH
Methane   Methyl alcohol

Methyl alcohol, often known as wood alcohol, can be made by the destructive distillation of wood. It can also be prepared from carbon monoxide and hydrogen (CO and H_2). Methyl alcohol is a colorless liquid

Courtesy of PPG Industries

Ethylene glycol, a dihydric organic alcohol, is used as an antifreeze in automobile engines.

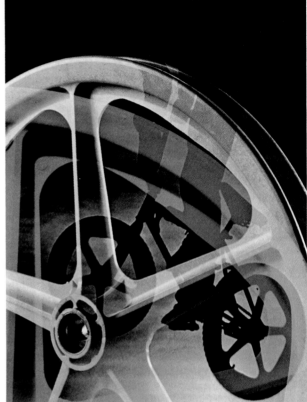

Courtesy of DuPont

This wheel is made of nylon, a durable synthetic fiber.

with an odor suggesting that of wine. When taken internally, it can bring about blindness or cause death. Methyl alcohol is used to denature ethyl alcohol, or grain alcohol, making it unfit for human consumption. It is also useful as a solvent in the manufacture of varnishes and shellacs. Quantities are employed in the preparation of formaldehyde, a preservative and disinfectant.

The hydrocarbon ethane (C_2H_6), as we have seen, has the structural formula

$$H-\underset{\underset{H}{|}}{\overset{\overset{H}{|}}{C}}-\underset{\underset{H}{|}}{\overset{\overset{H}{|}}{C}}-H$$

Replacing a hydrogen atom in this molecule by a hydroxyl group, OH, we have ethyl alcohol (C_2H_5OH), or common alcohol, with the structural formula

$$H-\underset{\underset{H}{|}}{\overset{\overset{H}{|}}{C}}-\underset{\underset{H}{|}}{\overset{\overset{H}{|}}{C}}-OH$$

Ethyl alcohol, also called grain alcohol, is produced by the fermentation of sugars, and is an ingredient in alcoholic beverages and many medicinal preparations. It has many important industrial uses. Among other things, it is an important solvent in the preparation of varnishes, shellacs, and enamels.

In some alcohols, two or more hydroxyl (OH) groups are attached to carbon atoms. An alcohol with two such groups is called *dihydric*. Ethylene glycol, C_2H_4 $(OH)_2$, is a typical dihydric alcohol. It has the following molecular structure:

$$\begin{array}{ccc} H & & H \\ | & & | \\ H-C & - & C-H \\ | & & | \\ OH & & OH \end{array}$$

It is used as an antifreeze in automobile radiators and as a solvent.

The alcohol glycerol, or glycerine, $C_3H_5(OH)_3$, has three OH groups:

$$\begin{array}{ccccc} H & & H & & H \\ | & & | & & | \\ H-C & - & C & - & C-H \\ | & & | & & | \\ OH & & OH & & OH \end{array}$$

This substance is a useful solvent and serves in the manufacture of nitroglycerin and dynamite.

The *ethers* are obtained by eliminating water from alcohols. The most common ether is diethyl ether, $(C_2H_5)_2O$, with the structural formula

$$\begin{array}{ccccccc} H & H & & H & H \\ | & | & & | & | \\ H-C & -C & -O- & C & -C-H \\ | & | & & | & | \\ H & H & & H & H \end{array}$$

It may be prepared from ethyl alcohol, with concentrated sulfuric acid serving as the dehydrating (water-removing) agent. Diethyl ether serves as an anesthetic.

The aldehydes, ketones, and organic acids. When simple alcohols are oxidized (combined with oxygen), aldehydes, ketones, or organic acids are produced.

The *aldehydes* contain the group

$$\begin{array}{c} H \\ | \\ R-C=O, \end{array}$$

where R represents an *alkyl group*. The general formula of an *alkyl group* is C_nH_{2n+1}. A simple aldehyde is formaldehyde (HCHO), which can be prepared by passing the vapor of methyl alcohol over a heated metal catalyst. In formaldehyde, hydrogen replaces the alkyl group. The structural formula is

$$\begin{array}{c} H \\ | \\ H-C=O \end{array}$$

Trained organic chemists in a well-equipped laboratory perform a variety of tests to determine the identity of an organic compound.

Formaldehyde, a gas, is used in the manufacture of plastics, dyes, and drugs. When it is dissolved in water, it serves as a potent disinfectant and antiseptic.

In the *ketones,* we always find the group $\overset{O}{\underset{}{\|}}$. Dimethyl ketone, or acetone, $R-C-R'$ $(CH_3)_2CO$, is the most important of the ketones. It is widely used as a solvent. Its structural formula is

$$H-\overset{\overset{H}{|}}{\underset{\underset{H}{|}}{C}}-\overset{\overset{O}{\|}}{C}-\overset{\overset{H}{|}}{\underset{\underset{H}{|}}{C}}-H$$

Organic acids contain the *carboxyl group,* $O=\overset{}{\underset{|}{C}}-OH$. The simplest of these acids is formic acid (CH_2O_2). The name is derived from the Latin word *formica,* meaning "ant". The acid was formerly prepared by distilling red ants. A much more effective method of preparation is to oxidize methyl alcohol in the presence of a platinum catalyst. Formic acid is a colorless liquid with an unpleasant odor. It is used, among other things, to dye cloth. The structural formula of the acid is

$$O=\overset{}{\underset{\underset{H}{|}}{C}}-OH$$

Other important organic acids are acetic acid, lactic acid, tartaric acid, oleic acid, and stearic acid.

The esters. The *esters* are the most important derivatives of the organic acids. They are prepared by heating an organic acid with an alcohol in the presence of a small quantity of concentrated sulfuric acid. (Esters can also be prepared from inorganic acids.) For example, when acetic acid reacts with ethyl alcohol, it forms an ester called ethyl acetate $(CH_3CO_2C_2H_5)$, which is used in synthesizing various organic compounds. Its molecular structure is

$$H-\overset{\overset{H}{|}}{\underset{\underset{H}{|}}{C}}-\overset{\overset{}{\|}}{\underset{\underset{O}{\|}}{C}}-O-\overset{\overset{H}{|}}{\underset{\underset{H}{|}}{C}}-\overset{\overset{H}{|}}{\underset{\underset{H}{|}}{C}}-H$$

AROMATIC HYDROCARBONS

Chemists were able to obtain certain hydrocarbons in large quantities by the distillation of coal tar long before petroleum became available in vast amounts. Since these coal-tar hydrocarbons had rather marked odors, in contrast to the practically odorless paraffins, they received the name of *aromatic hydrocarbons.* They have since become exceedingly important in industry and medicine. They are used in the manufacture of explosives, synthetic dyes, perfumes, photographic developers, drugs, and many other products.

The simplest aromatic hydrocarbon is benzene, C_6H_6. This substance was first obtained by Michael Faraday by the distillation of coal. For a number of years chemists were puzzled about the structure of the benzene molecule. They wondered how six atoms of carbon could be combined with six atoms of hydrogen. The German chemist, August Friedrich Kekule, solved the problem. He pointed out that the six carbons formed a ring, with single bonds between carbon atoms alternating with double bonds, as follows:

$$\text{H}-\overset{\overset{\text{H}}{|}}{\underset{}{\text{C}}} \quad \text{(benzene ring structure)}$$

Since the double bonds actually shift about, organic chemists generally represent the benzene molecule as a hexagon:

In derivatives of benzene, one or more of the hydrogen atoms are replaced by other atoms or groups. To indicate the structural formulas of such derivatives, a hexagon may be used, with the substituted group at one of the corners. For example, chlorobenzene, C_6H_5Cl, can be represented

Plastics are very moldable organic materials. Here, a plastic product is worked into shape at an industrial plant.

J. Alex Langley, DPI

either by the complete structural formula or by a hexagon, with a chlorine atom (Cl) at a corner, as follows:

A well-known aromatic carbon is trinitrotoluene, $C_6H_2(CH_3)(NO_2)_3$, popularly known as TNT. It is used as an explosive. It has the structural formula

CH_3, known as methyl, is a *radical,* or

group that remains unchanged through most chemical reactions. It has replaced one of the hydrogens in the benzene ring. Three other hydrogens have been replaced by NO_2, the nitro group.

ORGANIC COMPOUNDS IN LIVING MATTER

The processes of life take place in a jellylike substance, the protoplasm, which is found in all living cells. It is made up mostly of water (85 to 90 per cent). It also contains a small percentage of minerals and 10 per cent or more of the organic substances known as proteins, fats, and carbohydrates. Let us examine briefly the chemical structure of these substances.

The proteins. The proteins are among the most complex of all molecules. Their molecular weights sometimes come to many millions. The human body has hundreds of different proteins, with special structures that enable them to perform different tasks. All these proteins contain nitrogen, carbon, hydrogen, and oxygen, and often various other elements, including iron, sulfur, copper, and phosphorus.

Proteins can be broken down chemically into substances called *amino acids*. There are a number of different kinds of such acids. All contain an *amino group*, $-NH_2$, consisting of nitrogen and hydrogen, and also a carboxyl group, $-COOH$, made up of carbon, oxygen, and hydrogen. The simplest amino acid is glycine, $CH_2(NH_2)COOH$, with the structural formula:

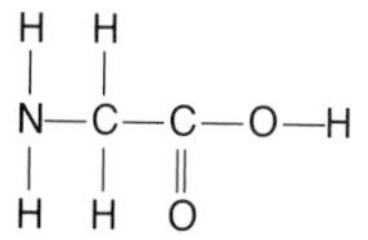

```
    H   H
    |   |
N — C — C — O — H
    |   |   ‖
    H   H   O
```

More than twenty amino acids in all have been obtained by breaking down plant and animal proteins. They are essential to life. Among the other amino acids are alanine, valine, leucine, isoleucine, threonine, methionine, phenylalanine, tryptophan, lysine, and histidine. The proteins consist of long chains of these acids, stretched out in zigzag form, or coiled in a spiral, or compactly folded together.

The fats. The natural fats are made up of carbon, hydrogen, and oxygen atoms. They are esters of the alcohol glycerol and a fatty organic acid. Among the fatty acids are palmitic and stearic acids, which are saturated, and oleic and linoleic acids, which are not. The saturated fatty acids yield solids (at ordinary temperatures), when they react with glycerol; the unsaturated ones, liquids. Generally the fats occurring in nature are mixtures of solid and liquid fats. For example, the solid fats palmitin (from palmitic acid) and stearin (from stearic acid) make up about two fifths of hog lard. About three fifths of hog lard is a liquid fat, olein (from oleic acid).

The carbohydrates. The carbohydrates contain three chemical elements — carbon, hydrogen, and oxygen. The general formula for all of them is $C_xH_{2y}O_y$. In the case of the carbohydrate d-glucose, x is equal to 6 and so is y. Hence the compositional formula for this compound is $C_6H_{12}O_6$. Note that in every carbohydrate there are always twice as many hydrogen atoms as there are oxygen atoms, just as in the water molecule (H_2O). The d-glucose molecule has the following structure:

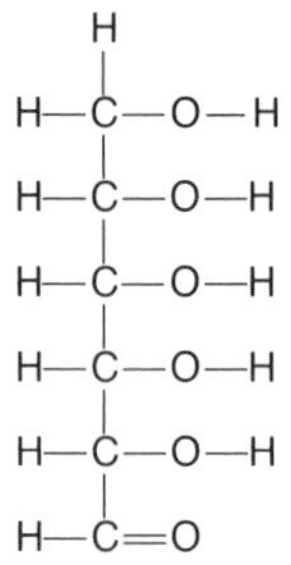

```
        H
        |
H — C — O — H
    |
H — C — O — H
    |
H — C — O — H
    |
H — C — O — H
    |
H — C — O — H
    |
H — C = O
```

The carbohydrates are called *saccharides,* or compounds made up of sugars. The simplest are the *monosaccharides,* which are also known as the *simple sugars.* Glucose, mentioned above, is a monosaccharide. Fructose and galactose, which have the same compositional formula as glucose ($C_6H_{12}O_6$), are also monosaccharides. When two monosaccharides combine chemically, they form a disaccharide, or double saccharide. The general formula for this type is $C_{12}H_{22}O_{11}$. The three most important disaccharides are sucrose (table sugar), lactose (milk sugar), and maltose (malt sugar). When such sugars are in solution, they can enter into equilibrium with ring forms of the same molecules. The ring forms of hexose (six-carbon) sugars such as glucose are known as pyranoses.

The more complex carbohydrates are known as polysaccharides. They are made up of combinations of monosaccharides and their molecules are very large. They include starch, glycogen, and cellulose, all of which have the general formula ($C_6H_{10}O_5$). The polysaccharide starch is found chiefly in the seeds or tubers of plants and serves as a food reserve. Glycogen occurs principally in the internal organs and blood of animals; it furnishes most of the fuel required for cellular activity. Cellulose forms the cell walls and woody structure of plants.

In this short article we have been able to give only the briefest sort of introduction to the vast field of organic chemistry. However, you have learned something about the significance of molecular structure and the intimate relationships existing between organic compounds.

ELECTROCHEMISTRY

by A. B. Garrett

When you turn on the lights of an automobile or close the switch of a flashlight, chemical reactions in an electric battery produce an electric current. If you use an electric current to charge a storage battery, chemical changes occur in the battery as the result of the current passing through the battery solution. The chemical reactions that produce an electric current and those produced by an electric current are the subject matter of *electrochemistry*.

In all the electrochemical reactions that we consider, atoms, molecules, and ions and oxidation-and-reduction reactions are involved. To understand the meaning of these terms, let us briefly consider atomic structure. The atom of any element contains a tiny nucleus composed primarily of protons and neutrons. This nucleus is surrounded by electrons in various energy levels, or shells. Only the electrons in the outer shell—the so-called *valence electrons*—are involved in chemical reactions.

An atom has as many protons as it has electrons and hence is electrically neutral. However, many atoms, especially those of metals, give up their outside electrons in chemical reactions to other atoms, especially to those of nonmetals. When this happens, the atom that has lost electrons has more protons than electrons and is no longer electrically neutral. It has become a charged particle, called an *ion*. The atom to which this ion has lost electrons now has more electrons than protons. It too no longer has an equal balance of positive and negative charges. It too has become an ion. The ionic charge is positive if electrons have been given off by an atom, because the resulting ion has more protons than electrons. The charge is negative if electrons have been added to an atom, because in this case the ion that is formed has more electrons than protons.

The process of giving off electrons is called *oxidation*. That of taking on electrons is known as *reduction*. The two processes in a given reaction always occur together. After all, if the atom of a metal loses electrons, it must always give them up to some other atom, and as a result that atom will gain electrons.

Oxidation originally meant only "combining with oxygen." Reduction meant only "removal of oxygen." Thus, when mercury combines with oxygen to produce mercuric oxide, the mercury is said to be oxidized. When mercuric oxide is heated, producing mercury and oxygen, the mercuric oxide is said to be reduced. The terms oxidation and reduction can still be used referring to such reactions, since electrons are lost and gained in the processes.

Let us consider a typical example of this electron loss-and-gain process. The electrically neutral atom of sodium, a metal, has 11 protons and 11 electrons, with a single electron in the outer shell. The atom of chlorine, a nonmetal, has 17 protons and 17 electrons, with 7 electrons in the outer shell. If sodium atoms and chlorine atoms

The carbon atom shown below has 6 protons *(P)* and 6 neutrons*(N)* in its nucleus. It has 2 electrons *(E)* in its inner shell and 4 in its outer shell, making a total of 6. The atom has as many protons, representing positive electric charges, as it has electrons, representing negative charges; so it is electrically neutral.

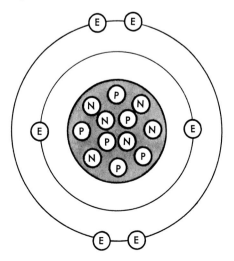

are brought together, they react spontaneously to form sodium chloride.

A sodium atom involved in the reaction loses the single electron in its outer shell to a chlorine atom. As a result, each sodium atom has only 10 electrons but still has 11 protons. Therefore it becomes a positively charged ion, with one excess positive electric charge. This is indicated by adding a single plus sign to the symbol for sodium, thus: Na^+. The chlorine atom now has 18 electrons and 17 protons. It has become a negatively charged ion, called a chloride ion, with one excess negative electric charge. It is written as Cl^-.

If an atom loses two electrons, giving it two more positive charges than it has negative charges, two plus signs are added. Thus the positively charged nickel ion, which has lost two electrons, is written as Ni^{++}. The negatively charged oxygen ion, with two more negative charges than positive charges, is indicated by the symbol $O^=$.

In our example, the sodium atom, having lost an electron, has undergone oxidation. The chlorine atom, having taken on an electron, has undergone reduction:

$$Na + Cl \rightarrow Na^+ + Cl^-$$

This is the chemist's shorthand for the statement that "a sodium atom (Na) reacting with (+) a chlorine atom (Cl) yields ($\rightarrow$) a positive sodium ion (Na^+) and a negative chloride ion (Cl^-)."

The reaction could be reversed if we had a device that would "pump" the extra electrons off the chloride ions and put them back onto the sodium ions. As we shall see, this is actually done with electric batteries or electric generators, which are really electron "pumps."

ELECTROLYSIS REACTIONS

We have just pointed out that sodium and chlorine yield sodium ions and chloride ions, respectively, when they react. The positive sodium ions are attracted to the negative chloride ions. They combine in a crystal structure shown on page 145, and they form the substance called sodium chloride (NaCl), which is the chemical name of the familiar table salt. If crystals of sodium chloride are dissolved in water, they will form free sodium and chloride ions. The resulting solution will conduct electricity.

Any substance that becomes a conductor of electricity when it is dissolved in water or some other liquid is called an *electrolyte*. Salts (including sodium chloride), acids, and bases are all electrolytes. Substances that do not conduct electricity

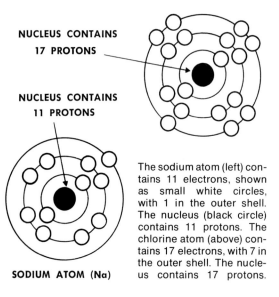

CHLORINE ATOM (Cl)

NUCLEUS CONTAINS 17 PROTONS

NUCLEUS CONTAINS 11 PROTONS

The sodium atom (left) contains 11 electrons, shown as small white circles, with 1 in the outer shell. The nucleus (black circle) contains 11 protons. The chlorine atom (above) contains 17 electrons, with 7 in the outer shell. The nucleus contains 17 protons.

SODIUM ATOM (Na)

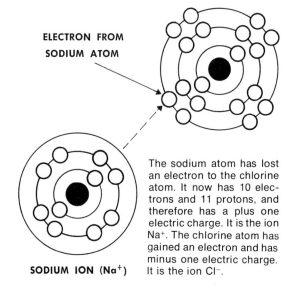

CHLORIDE ION (Cl$^-$)

ELECTRON FROM SODIUM ATOM

The sodium atom has lost an electron to the chlorine atom. It now has 10 electrons and 11 protons, and therefore has a plus one electric charge. It is the ion Na^+. The chlorine atom has gained an electron and has minus one electric charge. It is the ion Cl^-.

SODIUM ION (Na$^+$)

when they are dissolved are known as *nonelectrolytes*. Sugar ($C_{12}H_{22}O_{11}$) is a familiar nonelectrolyte.

Suppose that we put two carbon rods in a jar containing molten sodium chloride and that we then connect an electric battery to the carbon rods. These rods will serve as *electrodes*, or *electrical terminals*. One of them will be the negative electrode, or *cathode* — the electrode where reduction occurs. The other electrode is the positive electrode, or *anode*. This is the electrode where oxidation occurs. The negative electrode is connected to the negative pole of the battery. The positive electrode is connected to the positive pole.

As current passes through the fused salt, metallic sodium will be produced at the negative electrode, and chlorine gas at the positive electrode. The molten sodium chloride has been decomposed into sodium and chlorine as the result of the passage of electricity through the fused salt. We call such decomposition *electrolysis*, from two Greek words meaning "breaking down by means of electricity." The chemical reaction that takes place is called an electrolysis reaction. Only direct current, such as is supplied by a battery or a direct-current generator, can be used to bring about such a reaction.

Let us consider what happens when a solution of copper chloride is electrolyzed — that is, undergoes electrolysis. (See diagram on page 146.) The solution contains positive copper ions (Cu^{++}) and negative chlorine ions (Cl^-). The two electrodes are strips of platinum, which take no part in the reaction. When a battery is connected to the platinum electrodes, it draws electrons from the positive platinum electrode — the anode — and pumps a surplus of electrons to the negative platinum electrode — the cathode. The latter is often described as electron-rich: the positive electrode is described as electron-poor.

The negative chlorine ions (Cl^-) in the solution are attracted to the positive electrode, or anode. When they reach this electrode, they give up one electron each to it, and they become neutral chlorine atoms (Cl). They have now lost their electric charge or, as we might also say, they have been discharged. Two chlorine atoms team up to form a chlorine molecule (Cl_2). The many chlorine molecules formed in this way escape from the anode as bubbles of chlorine gas. The change that takes place at this electrode is an oxidation because the chloride ions have lost electrons. The equation representing this change is

$$2Cl^- \rightarrow Cl_2 + 2e$$

Two negatively charged chloride ions ($2Cl^-$) become a chlorine molecule (Cl_2) after they have contributed an electron apiece, or two in all ($2e$), to the anode.

At the same time, the positive copper ions (Cu^{++}) have been attracted to the negative electrode, or cathode. They pick up two electrons each from that electron-rich electrode, and become neutral copper atoms, as indicated by the equation

$$Cu^{++} + 2e \rightarrow Cu$$

A positively charged copper ion (Cu^{++}) picks up two electrons ($2e$) from the cathode and becomes an electrically neutral copper atom (Cu). The copper ions are reduced in this process, since they acquire electrons.

The electrons that accumulate at the anode are pushed through the external cir-

Crystal lattice of sodium chloride (table salt). The lattice is made up of chloride ions having a negative charge (Cl^-) and sodium ions with a positive charge (Na^+).

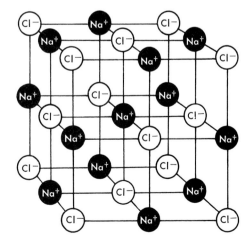

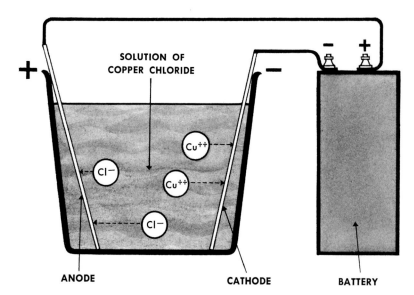

SOLUTION OF COPPER CHLORIDE

+

−

− +

ANODE

CATHODE

BATTERY

Electrolysis of a solution of copper chloride. The solution contains positive copper ions (Cu^{++}) and negative chloride ions (Cl^-). The positive copper ions are attracted to the negative electrode (cathode), pick up 2 electrons each and become copper atoms (Cu). The negative chloride ions are attracted to the positive electrode (anode), give up an electron each and become chlorine atoms (Cl). The neutral copper atoms are plated on the cathode. The neutral chlorine atoms team up in pairs to form chlorine molecules (Cl_2), and they escape from the anode in the form of chlorine-gas bubbles.

cuit by the electric battery. And so the anode remains electron-poor. A fresh supply of electrons is pumped by the battery to the cathode, which remains electron-rich.

Thus the conduction of the current through the solution has been made possible by the ions. One set, the Cl^- ions, race through the solution and unload their electrons onto the positive electrode. The other set of ions, the Cu^{++} ions, rush to the cathode and pick up electrons, thus relieving the electron pressure at that point. This process can go on until the carrier ions are all used up. Then the electrolysis generally stops. It can continue only if the voltage of the current is so high that the molecules of the solvent (the liquid in which the ions have been dissolved out) are decomposed. They can then both supply and take on electrons.

If the electrodes are properly constructed, the chlorine gas can be collected as it bubbles away from the positive electrode. The copper atoms are deposited, or plated out, on the negative electrode. Suppose we electrolyze the solution until all the copper ions have been discharged. If we then weigh the copper that has been plated out on the cathode, we can tell exactly how much copper was in the solution we used. This method of quantitative analysis is called *electrochemical analysis*. It can be

applied only to metals that can be easily and completely discharged by an electric current.

In the electrolysis reaction we have just described, the electrodes themselves took no part in the reaction. They simply acted as stations where traveling ions could pick up or give up electrons. In certain cases, however, the electrodes are involved in electrolysis reactions. To illustrate, let us again work with a copper-chloride solution. This time, however, the anode is to be of copper. When the current is turned on, as we have seen, the battery always draws electrons from the anode. If the latter is of platinum, the electrons are derived from the chloride ions (Cl^-) in the solution, each one giving up an electron as it reaches the anode. However, if the latter consists of copper, it is easier for the atoms of copper in the anode to give up electrons than it is for the anode to capture electrons from the chloride ions in the solution. Because of this, no chlorine gas is formed at the anode. Instead, the anode dissolves and the copper atoms become copper ions, as follows:

$$Cu - 2e \rightarrow Cu^{++}$$

A copper atom (Cu) gives up 2 electrons (2e) and becomes a copper ion (Cu^{++}).

The electrons yielded up by the copper atoms are pumped to the battery. The copper ions go into the solution. They are attracted to the cathode, where they pick up electrons and become copper atoms.

FARADAY'S LAWS

There are various other types of electrolysis reactions, all based on the discharge of ions at cathodes and anodes. One of the first men to work with such reactions was the great 19th-century English scientist Michael Faraday. He discovered the following laws of electrolysis, known as *Faraday's laws:*

(1) The weight of a substance deposited at an electrode in electrolysis is directly proportional to the quantity of electricity passed through the cell.

(2) The weights of different substances produced by the same quantity of electricity are proportional to the equivalent weights of the substances. The equivalent weight is the atomic weight divided by the valence number.

APPLICATIONS

It would take pages merely to list the many applications of electrolytic reactions in modern industry. We shall mention only a few in the following paragraphs.

Electroplating. We all know that much of the silverware used in homes is not solid silver, but is made from steel or copper that has a coating of silver metal on it. The coating, or plating, of silver is put on by electrolysis reactions in the process called *electroplating.*

To coat a metal object, such as a spoon, with silver, it is placed in a solution of silver cyanide. The spoon is used as the cathode. It is attached to the negative wire of an electric generator. The anode may be a piece of platinum or perhaps a bar of silver. When the current is turned on, positive silver ions (Ag^+) are attracted to the spoon that serves as the negative electrode and are discharged on it. Thus they cover it with a coating of metallic silver. Many objects are plated in this way.

Formerly, the accepted method of coating steel with tin was to dip steel sheets into molten tin, so that they became covered with a thin layer of the latter metal.

This is a wasteful process because the tin coating is much thicker than is needed. Nowadays, the preferred method of coating is by electrolysis. In this case, the steel sheet is the cathode, and the plating tanks contain solutions of tin salts. The tin ion, called the stannous ion (Sn^{++}), is attracted to the steel serving as the cathode and is plated out on it. After the tin has been deposited on both sides of the steel sheet, the latter is run through an oven which melts the coating so that it will form a smooth covering.

Steel and other metals can also be coated with such metals as nickel, chromium, and gold. Nickel plating provides effective protection from atmospheric corrosion. A very thin layer of chromium is applied as a final layer on the shiny parts of automobiles and various household appliances. Many articles of jewelry are plated with gold. In inexpensive articles, the plating may be only 0.00005 millimeters thick.

Metal plates for printing purposes are produced by the electroplating process called *electrotyping.*

Preparing hydrogen and oxygen by electrolysis. When an electric current is passed through water, the water is decomposed into the gases hydrogen and oxygen. This method of preparing the two gases in question is a very important one.

Water always contains a certain number of hydrogen ions (H^+), which are combined with water molecules (H_2O) to form hydronium ions (H_3O^+). It also has an equal number of hydroxyl ions (OH^-). Pure water, however, is a very poor conductor of an electric current because the concentration of ions is low. If we dissolve sodium hydroxide (NaOH) in water, forming sodium ions (Na^+) and hydroxyl ions (OH^-), the solution conducts current very well.

Two electrons (2e) react with two water molecules ($2H_2O$) to produce a molecule of hydrogen (H_2) and two hydroxyl ions ($2OH^-$):

$$2H_2O + 2e \rightarrow H_2 + 2OH^-$$

The hydrogen molecules formed in this way bubble off as hydrogen gas. At the anode, the hydroxyl ions give up an electron each.

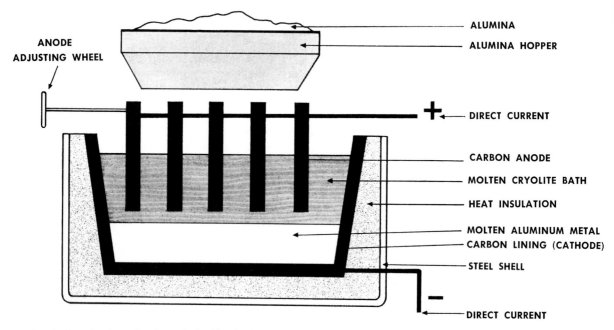

ANODE
ADJUSTING WHEEL

ALUMINA

ALUMINA HOPPER

DIRECT CURRENT

CARBON ANODE

MOLTEN CRYOLITE BATH

HEAT INSULATION

MOLTEN ALUMINUM METAL

CARBON LINING (CATHODE)

STEEL SHELL

DIRECT CURRENT

Producing aluminum by electrolysis. Alumina (aluminum oxide, Al_2O_3), derived from bauxite, passes from a hopper to a bath of molten cryolite (Na_3AlF_6) and is dissolved in it. The bath is contained in a cell with an outer steel lining and an inner lining of carbon, which serves as the cathode in the electrolytic reaction. A set of carbon anodes, adjusted as required by an adjusting wheel, dips into the bath. When electric current is passed through the bath, the dissolved alumina is converted into oxygen and molten aluminum. The oxygen is deposited on the carbon anode, which is gradually consumed. The molten aluminum is heavier than the cryolite in the bath, and therefore it sinks to the bottom of the cell. It is drawn off at intervals, and is charged into furnaces or cast into pigs. We have shown a single cell here. In an aluminum smelting works, a number of cells (or pots) are arranged in series.

Four hydroxyl ions, losing electrons in this way, decompose into a molecule of oxygen gas and two molecules of water:

$$4OH^- - 4e \rightarrow O_2 + 2H_2O$$

The gaseous oxygen then bubbles out at the anode.

The electrolysis of brine. When electric current is passed through a strong solution of brine (water saturated with common salt), three valuable industrial products are produced. These are hydrogen gas, chlorine gas, and sodium hydroxide, also called lye or caustic soda.

When the sodium chloride goes into solution, as we have seen, it yields free sodium ions (Na^+) and chloride ions (Cl^-). At the negative electrode, the hydrogen ion (H^+) obtainable from water molecules has a greater attraction for electrons than does

the sodium ion (Na^+). Hence hydrogen gas (H_2) is formed from water, leaving hydroxyl ions (OH^-). At the positive electrode, chloride ions (Cl^-) will give up their electrons more readily than will the oxygen ions ($O^=$) obtainable from water molecules. Hence chlorine gas (Cl_2) is formed at this electrode.

Hydroxyl ions and sodium ions are left in the solution after the electrolysis operation has been completed. The solution is then evaporated and sodium hydroxide ($NaOH$) is obtained as a white salt. It is used in large quantities in the petroleum, paper, textile, rayon, and rubber industries.

The manufacture of aluminum. Electrolysis plays a very important part in the manufacture of aluminum. When alumina, or aluminum oxide, Al_2O_3 is dissolved in a bath of molten cryolite and kept at a tem-

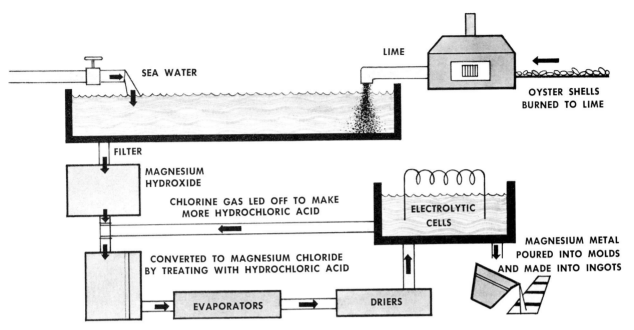

Producing magnesium from sea water by electrolysis. First, the water is pumped into huge tanks, where it is mixed with lime that has been obtained by roasting oyster shells. The magnesium in the sea water is converted by the lime into magnesium hydroxide, or milk of magnesia, which is filtered out. This is treated with hydrochloric acid (obtained from the reaction of chlorine with natural gas) and produces a solution of magnesium chloride. The water contained in it is evaporated as the solution passes through evaporators and driers. The dry magnesium chloride then goes to a series of electrolytic cells, where it is broken up into chlorine gas and molten magnesium. The gas is led off and is used to make more hydrochloric acid, while the magnesium is poured into molds and made into ingots. Since sea water contains about 0.13 per cent magnesium, it offers an almost unlimited supply of this valuable metal.

perature of 1,000° Celsius, it forms aluminum ions (Al^{+++}) and oxygen ions ($O^=$). The container is lined on the inside with carbon, which serves as the cathode. The anodes, which dip into the solution, are also made of carbon. Oxygen ions are attracted to the anode, where they lose electrons and become oxygen atoms. These atoms pair up and escape as bubbles of oxygen gas:

$$O^= - 2e \rightarrow O; \; 2O \rightarrow O_2 \uparrow$$

At the cathode, the aluminum ions gain electrons and form aluminum metal:

$$Al^{+++} + 3e \rightarrow Al$$

The temperature of the solution is higher than the melting point of aluminum. Therefore, the metal collects as a liquid. The molten aluminum is drawn off at intervals and is poured into molds. The cryolite is not affected by the electrolysis reactions.

VOLTAIC CELLS (BATTERIES)

A *voltaic cell* is a device by which chemical energy is changed into electric energy. It is often called an *electric battery*. It should be pointed out that this term also applies to devices, such as solar batteries, that generate current through other than chemical means. In the following pages, we shall use the word "battery" to refer to the devices that generate electricity through the release of chemical energy.

There are two general types of such batteries: *primary* and *secondary*. The primary battery cannot be readily recharged after its chemical energy has been converted to electric energy. It is called a primary battery because, before the electric generator was developed in the 1870s, it was used to charge other batteries, which were known as secondary batteries. The term "secondary battery" is still used in the sense of a voltaic cell that is recharged at intervals. This means that active chemicals within it are restored to their original condition as electricity is passed through the device. The secondary battery is also called a storage battery because it can be used to store electricity, which can then be applied whenever it is required.

PRIMARY CELLS

The primary battery in most common use today is the *Leclanché day cell,* named after the 19th-century French physicist Georges Leclanché, who first developed it. We show a cross section of a familiar type of dry cell on this page. It has a zinc case, which serves as the negative pole, or electrode. The positive pole is a rod of carbon set in the center of the cell and embedded in

a black mixture of manganese dioxide (MnO_2) and carbon particles. The electrolyte consists of a solution of ammonium chloride and zinc chloride combined with corn starch or other materials to form a paste between the zinc electrode and the black mixture. The battery is sealed at the top.

When the cell is in use, atoms of the zinc in the outer case are oxidized, giving up electrons and forming zinc ions (Zn^{++}):

$$Zn - 2e \rightarrow Zn^{++}$$

The electrons lost by the zinc atoms then flow through the outer electric circuit, supplying electric energy. They re-enter the battery at the carbon rod that serves as the positive pole. The manganese dioxide, which has a greater electron-attracting power than does zinc, is then partially reduced. This type of cell gives about 1.5 volts. By connecting many dry cells in se-

A typical dry cell. The zinc case serves as the negative electrode. The positive electrode consists of a rod of carbon set in the center of the cell and embedded in a black mixture of manganese dioxide and carbon particles. The electrolyte, which is in the form of a paste, is set between the black mixture and the zinc electrode. The case is sealed at the top.

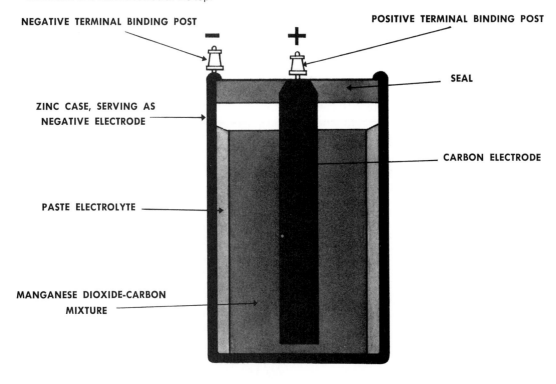

NEGATIVE TERMINAL BINDING POST

POSITIVE TERMINAL BINDING POST

SEAL

ZINC CASE, SERVING AS NEGATIVE ELECTRODE

CARBON ELECTRODE

PASTE ELECTROLYTE

MANGANESE DIOXIDE-CARBON MIXTURE

ries, one can produce a rather high voltage.

Another type of dry cell—the mercury cell—uses zinc as the negative electrode, mercury as the positive electrode, and potassium hydroxide as the electrolyte. This cell gives about 1.3 volts. It is used in hearing aids and transistorized radios, among other things.

The air cell is another primary cell, in limited use. It is similar to the Leclanché cell, but in place of manganese dioxide, it uses the oxygen of the air. In this cell, the positive electrode is porous, and atmospheric oxygen diffuses through it. The oxygen molecules pick up electrons to form oxygen ions. Air cells serve in certain railroad systems.

SECONDARY OR STORAGE BATTERIES

The lead-acid storage battery is the oldest type and it is still the most widely used. The kind that we find in automobiles consists of six identical units called *cells*. They can deliver 2 volts each, or a total of 12. Each cell is made up of a number of alternating vertical positive and negative plates, separated by thin layers of wood, hard rubber, or plastic, called separators. All the positive plates are connected to a unit called a strap or lug. All the negative plates are connected to another strap. The straps are attached to the terminal posts, positive and negative. Each of the plates is in the form of a grid, having a number of pockets. Those of the positive plate are filled with a paste of lead dioxide (PbO_2). Those of the negative plate are filled with spongy lead (Pb). All plates are immersed in a solution of dilute sulfuric acid—the electrolyte.

When the battery is in use, the lead atoms in the negative plates lose electrons and become lead ions (Pb^{++}). The electrons freed in this way flow through the outer cir-

A lead-acid storage battery, used extensively in power boats. It consists of three cells, which can deliver 2 volts each, or a total of 6 volts. Each cell is made up of alternating positive and negative plates, kept apart by separators. Positive plates are connected to one post strap; negative plates to another. The straps are attached to terminal posts.

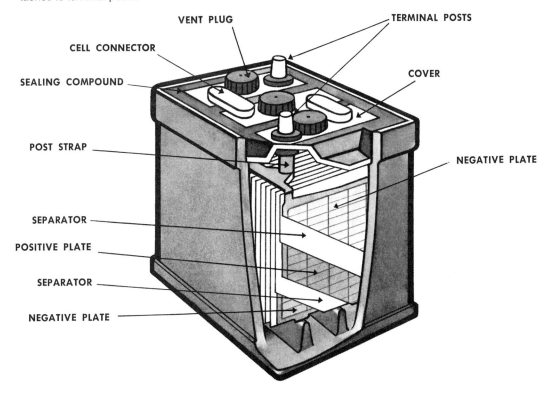

VENT PLUG

TERMINAL POSTS

CELL CONNECTOR

COVER

SEALING COMPOUND

NEGATIVE PLATE

POST STRAP

SEPARATOR

POSITIVE PLATE

SEPARATOR

NEGATIVE PLATE

cuit, as in the case of voltaic cells. The lead ions combine with sulfate ions ($SO_4^=$) in the solution and form a white coating of lead sulfate ($PbSO_4$) on the negative plates:

$$Pb^{++} + SO_4^= \rightarrow PbSO_4$$

When the electrons that make up the electric current arrive at the positive pole, they are picked up by the lead dioxide. With the help of hydrogen ions in the solution, they change the lead dioxide to lead ion (Pb^{++}) and water (H_2O). The lead ion at once reacts with sulfuric acid in the solution to form water and lead sulfate that forms a white coating on the positive plate.

Hence, as a cell discharges, its plates become coated with lead sulfate, and it becomes less and less efficient as a source of electric energy. It is recharged and restored to its former condition by being connected to an outside current. In the automobile, this current is supplied by the generator. Electrolysis takes place in each cell, as current flows through it in a direction opposite to that followed by the current when the cell is in use. The lead sulfate at the negative pole is changed to lead, and that at the positive pole to lead dioxide.

Two other widely used types of storage batteries are the Edison battery and the nickel-cadmium battery. The Edison battery uses a solution of potassium hydroxide (KOH) as the electrolyte; iron for the negative electrode; and nickel dioxide (NiO_2) for the positive electrode. The battery gives 1.35 volts per cell on full charge. The voltage drops gradually as discharge takes place. Edison batteries are not harmed appreciably by overcharge or by standing for a long time uncharged. The nickel-cadmium battery is similar but uses cadmium for the negative electrode.

CORROSION

The study of electrochemical systems has cast a good deal of light on corrosion — the chemical reaction between a metal and its surroundings or between different particles within a metal. Corrosion is often a kind of electrochemical reaction, involving an oxidation-reduction process.

One type of corrosion results from impurities in a metal. These impurities serve as electrodes. They bring about a type of corrosion known as *local cell action*. For example, if a sheet of zinc contains impurities of copper or carbon, each one of the copper or carbon particles within the zinc can serve as a cathode if the metal is covered with a solution that can serve as an electrolyte. The zinc will act as the other electrode, and as a result will be pitted or else more or less evenly corroded.

Another type of corrosion occurs in the case of very pure metals on which oxygen molecules of the air are unevenly adsorbed. If an acidic layer of moisture is present, the oxygen molecules will pick up electrons to form oxygen ions. The pure metal will act as an air cell and corrode.

Water pipes often corrode when two dissimilar metals, such as copper and iron, are in contact or otherwise connected electrically, producing an electrochemical couple. Occasionally corrosion occurs as the result of electrolysis produced by electrical conductors grounded through water pipes. The pipes serve as electrodes taking part in the electrolysis reaction. They are corroded as their molecules gradually yield up electrons and become ions.

Corrosion of metals can be prevented or reduced in a number of ways. While some impurities can greatly speed up the rate of corrosion by electrochemical action, other trace impurities can have an equal but opposite effect of slowing down the corrosion rate. Thus, various alloys have much greater corrosion resistance than the pure metals that go into such alloys. An example of this is the addition of aluminum to brass to increase the corrosion resistance of the brass. Metals may also be electroplated onto other metals for this purpose. Nickel and chromium are often used.

Another method of reducing corrosion is known as *cathodic protection*. This involves the use of a "sacrificial" anode such as aluminum or zinc. The metal to be protected is connected electrically to this anode, which is then used up by the corrosion action. This technique protects buried gas and water lines.

PHYSICS

Fundamental Photographs

*A drop of water hits water in a tank.
A wall of water develops at the point
of impact and concentric circular
ripples radiate outward. The science
of physics includes the study of
such phenomena — impact, motion, and
wave formation.*

SIMPLE MACHINES

by Alexander Joseph

Most people, perhaps, think of a machine as a complicated device with hundreds of moving parts and powered by some sort of engine or by electricity. As a matter of fact, even such simple tools as a hammer and a screwdriver are also machines. To a scientist, a *machine* is any object that allows us to do work with less effort than if we did the job with nothing but our own bodies. You may have to strain in lifting a heavy weight by hand, but you can often push the weight up an inclined plane, such as a ramp, very easily. The amount of work you do in both cases is equal, but the inclined plane permits you to use less effort. Therefore the inclined plane is a machine.

If we were to take apart a complicated machine, such as a typewriter, we would find it made up a number of simpler elements. There are actually just six of these basic elements, which we call *simple machines*. These are the *lever,* the *inclined plane,* the *wedge,* the *screw,* the *wheel and axle,* and the *pulley.*

What do we mean by saying that a machine allows us to use less effort in doing a given amount of work? First of all, we must see what a scientist means when he speaks of *effort* and *work.* No object moves without being pushed or pulled in some way. In the language of physics a push or pull is called a *force. Effort* means the same thing as force.

DOING WORK

If you braced yourself against a very large building and pushed with all your might, you would not succeed in moving it. This is because your force is not great enough to overcome the resisting force of the building. You are certainly exerting effort, but a physicist would say that you are not doing work. For a physicist, *work* is performed only when a force results in moving a body. To move a body, and thus to do work, we must exert a steady force that is greater than the resisting force of the

body. The resisting force of the body is the force, such as gravity, friction, or inertia, that prevents the body from being moved. We must also exert this force through a definite distance.

To find the amount of work we do in performing any task, we multiply our force by the distance through which it acts. Suppose a person pushes with 200 newtons of force against a piano and continues to push while the piano moves 10 meters. That person performs 2,000 newton-meters of work — 200 newtons of force times 10 meters. (A *newton* is the unit of force needed to move a body whose mass is one kilogram at an acceleration of one meter per second every second.)

LAW OF MACHINES

Mechanical work is a form of energy. As such, it cannot be destroyed nor can it be created out of nothing. A simple machine has no energy of its own and cannot do work by itself. It is only when we perform work upon it that it will perform work, in turn, upon some other object. The machine never adds to the sum total of work that is accomplished, since that would mean creating energy out of nothing. The *law of machines* states this fact in more scientific language: it tells us that the *work output* of a machine is always equal to the *work input.* In reality, every machine has a certain amount of friction. Since some of the work input is wasted in overcoming this friction, the work output is a little less than the work input.

Once we understand this law, we can see why machines multiply our strength, or force. We know that the work input and output of a machine must be the same, and that each of these is equal to a force multiplied by a distance. Let us suppose that a force of 100 newtons acts through a distance of 1 meter, and another force, of 1 newton, acts through a distance of 100 meters. Each force is doing the same

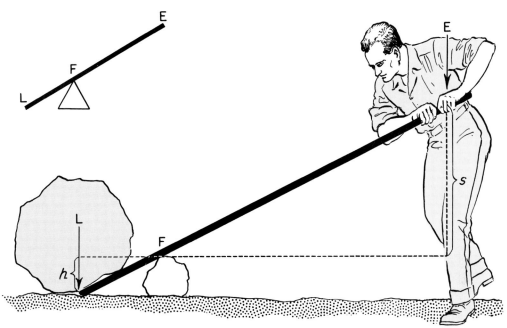

Fig. 1. First-class lever.

amount of work, but one is 100 times larger than the other.

We can use a small force at one end of a simple machine to move an object against a much greater force, such as that of gravity, at the other end of the machine, if the object does not move as far as we do. The ratio between the resisting force on the object that is moved and the force that is applied to this particular object is called the *mechanical advantage*.

THREE CLASSES OF LEVERS

The first of the simple machines is the lever, which is merely a rigid bar, capable of turning about one point, or axis. One of the most common levers is the crowbar, a long iron or steel bar that is used to move heavy objects or to pry things apart. Figure 1 shows a man lifting a boulder with a 2-meter crowbar. One end is placed under the load L and the bar is braced, at a point near the load end, against a small rock that serves as a pivot. The point F, resting against the pivot, is called the *fulcrum*. The man applies his effort E to the upper end of the crowbar. This kind of arrangement, with the fulcrum, or pivot point, between

the load and the effort is called a *lever of the first class*.

We can divide the lever into two parts: the *effort arm,* extending from the fulcrum to the effort end, and the *load arm,* from the fulcrum to the load end. Let us suppose, in this example, that the effort arm is 2.5 meters long and the load arm is 0.5 meter, and that the weight, or downward force, of the boulder is 1,000 newtons. How much force must the man use in order to lift this weight?

If he pushes his end of the crowbar down through a certain distance s (the ef-

Fig. 2. Second-class lever.

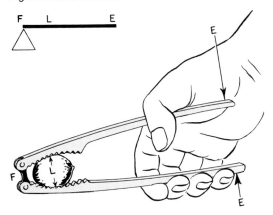

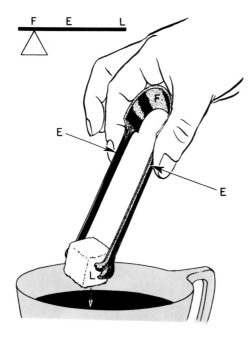

Fig. 3. Third-class lever.

fort distance), the crowbar will raise the boulder through the distance h (the load distance). The man will be doing $E \times s$ newton-meters of work, which will be equal to the $L \times h$ newton-meters of work done by the machine. From the length of the lever arms we can see that s is 5 times longer than h; therefore, the effort will be 5 times smaller than the load. The man will have to exert a force of only 200 newtons to raise the 1,000-newton boulder. The mechanical advantage of this crowbar is 1,000 divided by 200, or 5. Applying effort to the long arm, you could always move a load 5 times greater than the effort.

The mechanical advantage of every simple machine is found in the same way. We just divide the load by the effort required to move it, or we divide the effort distance by the load distance, which amounts to the same thing. In the case of levers, we can also divide the effort arm by the load arm. In the above example, 2.5m ÷ 0.5m = 5.

A pair of scissors is a combination of two first-class levers, attached to a common fulcrum. No doubt you have noticed that it is much easier to cut something with the scissors if you place the material close to the pivot point. When you do this, you are shortening the load arm while the effort arm remains the same. Thus you increase the mechanical advantage and the load requires less effort.

In some levers, the load is placed between the fulcrum and the effort: these are known as *levers of the second class*. The ordinary nutcracker shown in Figure 2 is a common example. Actually, it is made up of two second-class levers (the upper and lower parts of the nutcracker in Figure 2), whose fulcrums are joined together. The load is the nut, which resists being crushed. We apply the effort by squeezing the handles together. In this case, and for all second-class levers, the effort arm consists of the entire length of the lever. The load arm of each lever is the distance from the nut to the fulcrum. Since the effort arm is longer than the load arm, the nutcracker has a mechanical advantage of more than one.

A familiar second-class lever is the rowboat oar. At first glance it might seem that an oar should be a first-class lever. See if you can find out why it really belongs to the second class. As a clue, remember that the purpose of the oar is to move the boat, and not the river.

Grasping a lump of sugar with a pair of tongs, as shown in Figure 3, involves the use of the lever. Tongs and tweezers belong to the *third class of levers*—here the effort is applied between the load and the fulcrum. The fulcrum (F) is located where the two branches of the tongs join, and the load is placed at the opposite end, between the jaws. Each branch is itself a third-class lever. In third-class levers, it is the load arm that takes up the entire length.

Since the effort is always applied somewhere in the middle, the effort arm is always shorter than the load arm. This means that the mechanical advantage of a third-class lever is always less than one. In effect, we must use more pressure on the tongs to keep the cube from slipping than we would have to apply if we grasped the cube directly with our fingers. This can sometimes be very useful. Tweezers, for example, are ideal for handling stamps, in-

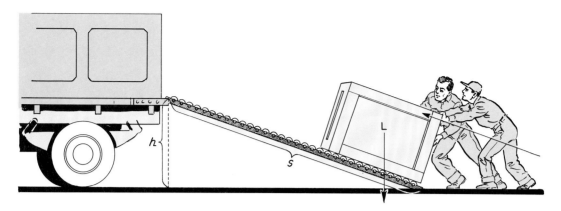

Fig. 4. Inclined plane.

sects, and other delicate objects because they lessen the crushing force of our fingers.

THE INCLINED PLANE

In Figure 4 the workmen are pushing a 2,000 newton crate up a loading conveyor into a truck. This conveyor is nothing more than an inclined plane with roller wheels on it for reducing friction. An inclined plane lessens the effort required to lift a load. If the men tried to lift the crate to the height h by themselves, they would have to apply 2,000 newtons of force—an almost impossible feat. If h were 2 meters, they would be doing 4,000 newton-meters of work.

Using the inclined plane they do the same amount of work, but their effort is applied over the longer distance s, the length of the plane. If this length were 8 meters, they would need only 500 newtons of force between them to do the work. Dividing the load of 2,000 newtons by the effort of 500 newtons, we find a mechanical advantage of 4. For any inclined plane, the mechanical advantage is the length of the incline divided by its vertical height at the top.

Railroad terminals often use inclined planes, in the form of gradually sloping ramps, instead of stairs. In mountainous country, roads are constructed in long, gradual slopes so that automobiles may ascend the mountains without imposing too great a strain on their motors.

THE WEDGE

The simple machine called the *wedge* is really a small inclined plane. Instead of pushing a load up the incline, the entire wedge is driven under, or into, the load, as shown in Figure 5. The wedge is ideal for separating two objects that are held together by a great force. For the boulder and the earth this force is gravity. For the two halves of the log the force is the rigid strength of the wood itself.

When using a wedge, the effort is applied to the head, or thick end, and the load

Fig. 5. Wedges—really inclined planes.

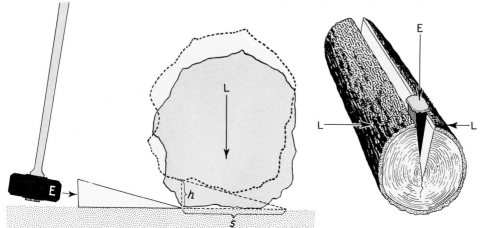

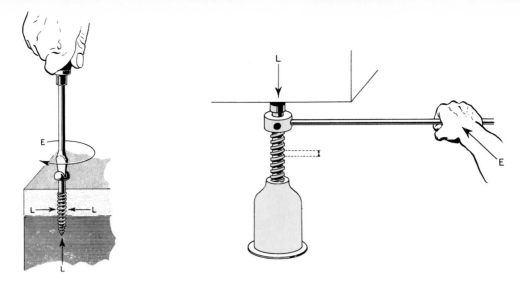

Fig. 6. Wood screw.

Fig. 7. Jackscrew.

acts against the long sloping side. If the hammer drives the wedge completely under the load, its effort has moved through a distance *s,* equal to the length of the wedge. At the same time, the load will be raised through the distance *h,* the thickness of the head. Dividing the effort distance by the load distance gives us the mechanical advantage. Thus, the mechanical advantage of a wedge is its length divided by its greatest thickness. A wedge 12 centimeters long with a 2-centimeter head would have a mechanical advantage of 6. It would have the same mechanical advantage if it were 18 centimeters long and had a 3-centimeter head.

THE SCREW

It may seem strange to think of a screw as an inclined plane, but you can prove this for yourself. If you follow the thread of a screw with a pointed pencil, you will see that it is a plane, constantly curving upward around a central shaft. In wood screws (Figure 6) the central shaft is tapered. In most machine screws and bolts the shaft is a straight cylinder. Screws are used widely to apply tremendous force with very little effort.

Figure 7 shows a jackscrew lifting a load. The screw is turned by applying effort to the end of a long handle. Each time the handle makes one revolution, the screw does likewise. Thus, the effort moves through the circumference of a circle with a radius equal to the length of the handle, while the screw, with the load, moves up-

ward through the distance from one thread to the next, or the *pitch* of the screw. If the handle were 60 centimeters and the pitch were 0.8 centimeters, this would be equivalent to an effort distance of about 377 centimeters and a load distance of 0.8 centimeters or a mechanical advantage of about 471.

We can easily see how much greater this advantage is than that of any practical lever or wedge. With this jackscrew, an 8-newton effort could left a weight of 3,768 newtons. Screws are designed with even higher mechanical advantages. The screw principle is applied in all types of vises and machine tools to exert maximum force.

Fig. 8. Wheel and axle.

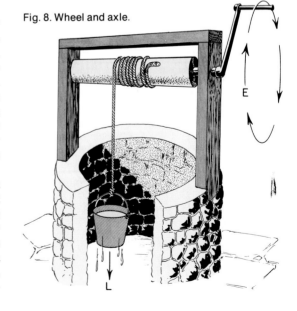

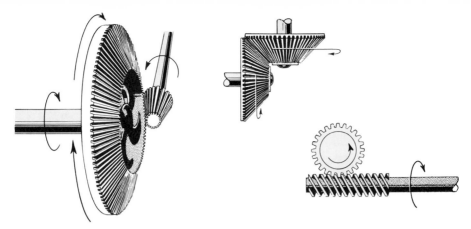

Fig. 9. Examples of gears.

THE WHEEL AND AXLE

In an old-fashioned well shown in Figure 8, the heavy water bucket is raised by turning a large crank attached to a wooden axle. As the axle turns, the rope winds around it and thus lifts the bucket. This is an example of another simple machine—the wheel and axle. As the arm of the crank turns, it moves in a full circle. A wheel, turned by a rope that winds around its outer rim, could easily be used in place of the crank.

When we apply effort to the wheel and turn it through a complete circle, the axle also makes one revolution, taking up a length of rope equal to its circumference. The axle radius is small compared to the radius of the wheel—let us say 5 centimeters as against 20—and the circumferences of the axle and wheel are in the same proportion as their radii. The circumference of the wheel is the effort distance and that of the axle is the load distance. Thus the mechanical advantage is the radius of the wheel divided by the radius of the axle. In this case, the mechanical advantage would be 4. We could lift a bucket of water weighing 400 newtons with an effort of 100 newtons on the crank.

A wheel with teeth cut into its outer rim is called a *gear*. When the teeth of two gears are meshed together, one gear can turn the other. Each of the gears is connected to a shaft, one supplying effort and the other moving a load. Thus, a pair of gears accomplishes the same thing as a wheel and axle.

The mechanical advantage of a pair of gears depends on the number of teeth in each gear. This is called the *gear ratio*. If a gear with 4 teeth is connected to one with 20 teeth, the gear ratio is 5 to 1. The smaller gear turns 5 times for every revolution of the larger one. If effort is applied to the smaller gear the mechanical advantage of the pair will be 5. A 400-newton force on the smaller gear will move a load of 2,000 newtons on the larger one. If we apply the effort to the larger gear, the mechanical advantage becomes $\frac{1}{5}$. A 400-newton force now would move a load of only 80 newtons on the smaller gear, but the load would move 5 times as fast as the effort.

Gears are vital elements in almost all modern machinery. As we have seen, a set of gears can be used to increase mechanical advantage, to increase speed, or to do both at different times. Teeth can be cut in a variety of shapes for different purposes (see Figure 9), and the gears can be arranged to change the direction of motion. An automobile uses gears in almost every conceivable way. The steering mechanism uses a worm gear to apply force for angling the front wheels. The transmission and differential carry the turning force of the engine's flywheel to the rear wheels by means of gears in some cars. By shifting these gears, we alter the ratio of engine speed to gear speed. Thus we supply more or less mechanical advantage for starting, climbing hills, or speeding over level ground.

PULLEYS

Another simple machine with many applications is the pulley. A homemaker pulls on a clothesline looped around two fixed pulleys. The dentist uses a system of pulleys to transmit the motion of an electric motor to the drill that he applies to your teeth. A derrick employs a *tackle,* or a combination of ropes and pulleys, rigged at

Fig. 10. Simple pulley.

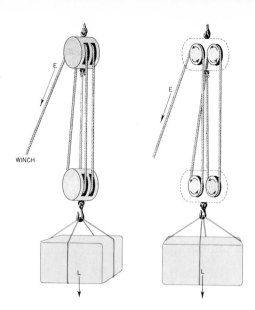

Fig. 11. Block and tackle.

the end of a *spar* or beam.

The simplest kind of pulley is shown in Figure 10—just a rope slung over a grooved wheel and attached to a load. When the rope is pulled, the wheel turns but does not move up or down. This single, fixed pulley is useful because it is more convenient to pull down a rope than to pull up a load. The pulley allows us to change the direction of our effort. However, this kind of pulley has no mechanical advantage. To raise the load one meter, we have to pull down the rope one meter.

In order to have a mechanical advantage there must be at least one pulley that can move with the load. The block and tackle shown in Figure 11 is an example of this kind of pulley system. Here, the rope passes around four pulley wheels: two of them are in the upper block, which is fixed, and two are in the lower block, which moves up and down with the load. One end of the rope is anchored to the upper block. The load is supported, then, by four lengths of rope. If we were to raise the load one fourth of a meter, each length of rope would have to be shortened by one fourth of a meter, making a total of 1 meter for the whole rope. In other words, the effort end of the rope must be pulled through a distance of 1 meter, and the effort will be 4 times smaller than the load—the mechanical advantage is 4.

For all pulley systems the mechanical advantage is equal to the number of ropes

(that is, lengths of a single rope) that are supporting the load. Theoretically, there is no limit to the load that can be moved by a block and tackle type of pulley. We could undoubtedly make a block with a thousand pulley wheels that could be used to raise a very heavy load. Such a device would, however, be too heavy and clumsy to be practical.

SIMPLE MACHINES COMBINED

In complex machines, we find almost every kind of simple machine. In the automobile, for example, a complicated linkage of levers with many joints and pivot points allows us to move the heavy gears in the transmission by making slight motions with the gear shift. The hand brake, clutch pedal, and foot-brake pedal are also parts of lever systems that supply great mechanical advantage. The heart of the engine is the crankshaft, which is simply an axle with a number of cranks built into it. It not only converts the up-and-down force of the pistons into turning force, but also adds a mechanical advantage for moving the entire load of the automobile. Gears, of course, may be used in the transmission. They are used to synchronize the crankshaft with the camshaft and the distributor. By means of belts and pulleys, the crankshaft turns the water pump, fan, and electrical generator. These are but a few examples of the simple machines that combined with one another form part of the automobile.

Sir Isaac Newton—in many ways the father of modern physical sciences. He invented the calculus, described motion and gravity, and made many contributions to astronomy.

National Portrait Gallery, London

ISAAC NEWTON
by Aaron B. Lerner

Isaac Newton may be said to have set the universe in order for the age of science. For centuries to come, his theory of universal gravitation sufficiently described the laws of motion of the earth, the moon, the sun, and all of the other heavenly bodies. In mathematics he was one of the co-developers of the calculus and also made other fundamental contributions to that science. He also founded the science of physical optics and invented the reflecting telescope.

The esteem in which Newton has been held can be gathered from a famous couplet by the 18th century English poet Alexander Pope: "Nature and nature's laws lay hid in night. God said, 'Let Newton be!' and all was light."

Newton himself had no false modesty, but he said that if he had achieved anything,

it was only because he stood on the shoulders of giants. And there was much truth in this. What Newton did was to gather together in his mind the work of other great scientists and create a clear, logical, all-embracing system. He is justly honored as one of the greatist scientists in history.

EARLY LIFE

Isaac Newton was born in Woolsthorpe, England, on December 25, 1642. His early childhood was not happy. His father died before he was born. Premature, Newton remained small and weak until he reached school age. Before he was two, his mother remarried and turned him over to his grandmother to raise on a farm. He attended school in Skillington and Stoke during those years.

His mother reclaimed Newton when he was 14, with the intent of making him a farmer — something in which he had no interest at all. Instead he spent his time in solitary thinking and reading. He also enjoyed handiwork and making gadgets and models. His mother relented, and at 18, Newton entered Cambridge University's Trinity College. He did well there but was mainly involved in his own projects: observing comet paths, studying the refraction of light, and developing the general binomial theorem — of major importance in mathematics.

A GREAT CREATIVE PERIOD

For about 18 months the university was shut down because of an outbreak of bubonic plague. Newton spent that time at his farm in Woolsthorpe. It was a period of amazing creativity. As Newton later wrote, "In those days I was in the prime of my age for invention, and minded mathematics and physics more than at any time since."

In those few months Newton developed both the differential and integral calculus, independently of the German mathematician Leibniz. The calculus was absolutely essential to the development of modern physics. It is the branch of mathematics that deals with the study of continuously changing quantities. Without the calculus, Newton could not have worked out his theory of gravitation, since it involves the mathematics of bodies in motion — that is, bodies continuously changing their position in space.

In this period, Newton also showed that the colors of the spectrum can be obtained by diffracting white light with a prism, and can then be recombined with another prism to produce white light again. He speculated on the nature of light, and ended up with an outlook that in some ways resembles the modern one — that light behaves both as if it travels as particles and in waves. In the course of his work on light, he designed the first reflecting telescope.

Finally, Newton began to work out his gravitational theory, coming to the conclusion that gravitational forces work at a distance between bodies in space. The story that his musings on gravitation were set off by the sight of an apple falling comes from Newton himself.

Newton was 24 when he returned to the university. A professor of mathematics, Isaac Barrow, had earlier recognized Newton's genius. It is generally thought that Barrow resigned his professorship so that Newton could take his place as Lucasian professor of mathematics, at the age of 26. He remained at the university until he was 53 years old. His work in mathematics and optics had already made him a famous man, and he was elected a member of the Royal Society at the age of 30.

THE *PRINCIPIA*

In 1684, while Newton was at the university, he was visited by the English astronomer, Edmund Halley, who wished to consult with him on the forces that control planetary orbits. Specifically, he wanted to know what sort of orbit a planet would follow under the influence of a force that varies inversely with the square of the distance between the planet and the sun. A number of scientists had been working on this problem, but none of them had been able to come up with an answer.

Newton immediately replied that the planet would follow an elliptical orbit. When asked how he knew this, he said that he had worked it out, but he could not find the calculation among his papers. Halley urged him to write it down again. A first paper, *De motu corporum (On the Motions of Bodies)*, was soon presented to the Royal Society. It represented a portion of what was to become Newton's most famous work, *Philosophiae naturalis principia mathematica (Mathematical Principles of Natural Philosophy)*, better known simply as *Principia*.

The book was written in a surprisingly short time and was published on July 5, 1687. Halley must be given great credit for his efforts in getting the book out. Although not wealthy, he paid for its publication. He also steered the touchy Newton through scientific squabbles with contemporaries such as the still more touchy Robert Hooke, who claimed equal credit for New-

Using prisms, Newton was able to decompose light and analyze the nature of light rays.

ton's insights into gravitation but lacked the mathematical ability to work them out.

The *Principia* is one of the greatest scientific books ever written. It is divided into three major sections. The first sets out the three laws of motion and various laws of force. The second takes up motions in different kinds of fluids. The third and most important section presents Newton's theory of universal gravitation. It shows how this force accounts for all motions, whether of bodies on the earth — including the water motions we call tides — or of bodies in space. Simply put, he had accounted for the laws of motion that control the universe.

NEWTON THE MAN

For the rest of his life, Newton continued to receive awards and honors for his achievements in science. He produced some further work and improved and added to his *Principia,* but nothing equaled his earlier achievements. He retained his mathematical powers, however, and quickly solved difficult problems that were presented to him. In 1696 he was appointed Warden of the Mint, and he became Master of the Mint three years later. He retained that post until his death, in Kensington, on March 20, 1727, at the age of 84.

Newton was a solitary and often a difficult man. During his long stay at Cambridge he worked alone or with one assistant. He did not do much teaching, and he had only a very few friends. He slept little and ate little and was nearly always working. He spent his days reading, writing, making notes, pacing back and forth in his room, going for walks, doing experiments, forgetting about meals, avoiding naps — and always, always concentrating on his projects. His relaxation came in switching back and forth between theoretical physics, mathematics, and — there was a mystical side to Newton — alchemy and theology. His contemporaries said that he seldom laughed or even smiled.

Newton's health as an adult was excellent and he was clever with his hands. However, he had at least two nervous breakdowns, and he lost his temper easily. He engaged in several long scientific battles with fellow scientists over his theories, including Hooke and Leibniz and the English astronomer John Flamsteed. In such battles Newton could be without mercy. The philosopher John Locke, a friend, wrote that "Newton was a difficult man to deal with and a little too apt to raise in himself suspicions where there was no ground." It should be said, however, that justice was often on Newton's side and that he could deal quite patiently with other scientists in trying to explain his theories.

All of those conflicts are long since over. The achievements of Isaac Newton remain. Among the galaxy of scientists, his star shines very brightly.

GRAVITATION

by Louis M. Heil

One of the first things a very young child learns is that any object (including himself) will fall to the ground if it is free to fall. He will note, later, that any object tends to bend in the direction in which it would fall if it were dropped. For example, a rope that is hung between two vertical poles will sag.

NEWTON'S LAW OF GRAVITATION

Observations such as these led the ancients to believe that the earth is the natural resting place of all objects. They held that a ball dropped from a height will fall to the ground and that a rope hung between two poles will sag because the ball and the rope are trying to get to their natural resting places. Scientists now have a much more satisfactory explanation. It was worked out in the seventeenth century by the great English philosopher and mathematician Sir Isaac Newton.

Newton was trying to find out why Mars, Venus, Jupiter, and the other planets of our solar system keep moving in very definite orbits around the sun. We can compare the paths they take to the path described by an object that a boy swings at the end of a string. There is one important difference, however. There is a mechanical connection—the string—between the boy's hand and the object that he is whirling. Obviously, there is no mechanical connection between the sun and the planets. Newton came to the conclusion that a planet keeps revolving around the sun because the two bodies attract one another. This type of action-at-a-distance he called *gravitation*.

He made a careful analysis of the action-at-a-distance, and he developed his famous *law of universal gravitation*. According to this law, every particle in the universe attracts every other particle with a force which is proportional to the product of their masses and inversely proportional to the square of the distance between the various particles.

Let us give a simple illustration. Suppose that two bodies are set at a given distance from each other. We assume that one body has a mass of 4 metric tons and the other a mass of 8 metric tons and that the gravitational force acting on each is 1 newton. The product of the two masses is 32 (4×8). We now double the mass of the first body, leaving the distance between the two bodies unchanged. Since the first body now has a mass of 8 metric tons, the product of the two masses is 64 (8×8). This product is twice the former one. Therefore the gravitational force between the two bodies is doubled. That means that it is increased from 1 newton to 2 newtons.

We now consider again the two original masses of 4 and 8 metric tons. Suppose we reduce the distance between the centers of the two bodies by $1/2$. Since the gravitational force between the bodies is inversely proportional to the square of the distance between them, this force will now be increased 4 times ($2/1$ squared). It will be 4 newtons. If the distance between the two bodies was doubled, the force would be reduced to $1/4$ its original value ($1/2$ squared). It would therefore be $1/4$ newton.

It is important to note that the force resulting from gravitational attraction acts on both bodies involved. The earth, whose mass is much greater than that of the moon, attracts the moon through the force of gravitation. But the moon also attracts the earth. We assume that both the earth and the moon are moving about a common center (Figure 1). This is much closer to the center of the earth than to the center of the moon because of the difference in their masses. Actually, as the diagram shows, the common center around which both the earth and the moon move lies within the earth.

UNIVERSAL GRAVITATION

Gravitational attraction applies to all masses, regardless of where they are located. It explains why a baseball falls to the ground when it is dropped from the hand, why a rope sags when it is slung between two poles, and why planets keep whirling in their orbits around the sun. We are able to analyze the forces involved with such accuracy that we can confidently predict the motions of the planets.

It was because the orbit of a planet did not follow exactly the law of gravitation that another, hitherto unknown, planet was discovered. For many years the orbit described by Uranus had puzzled astronomers, because this planet should have followed a somewhat different course if the law of gravitation held true. Two astronomers—the Englishman John Couch Adams and the Frenchman Urbain-Jean-Joseph Leverrier—found the true cause. Uranus apparently was attracted not only by the sun but by some other heavenly body, hitherto unknown. In the 1840s, the two astronomers, working independently, calculated the orbit of the unknown body, applying the law of gravitation. In 1846, a German astronomer, Johann Gottfried Galle, discovered this body—the planet now called Neptune—in the position predicted by Leverrier. Pluto, lying beyond the orbit of Neptune, was discovered in much the same way.

THE CAVENDISH EXPERIMENT

The law of universal gravitation was accepted at first by scientists because it seemed to give the right answers. Then an 18th-century English scientist, Henry Cavendish, put the law to the test in a famous laboratory experiment.

His apparatus (Figure 2) consisted of a torsion balance capable of supporting a large mass. He suspended two massive lead balls, M_1 and M_2 in the diagram, from the ends of the balance. Then he brought up an equally massive lead ball, M_3, until it was close to M_2. If the law of gravitation held true, the attraction between M_2 and M_3 should make M_2 move slightly, thus causing the suspended wire to be twisted somewhat. This is exactly what happened. Cavendish then changed the masses M_2 and M_3 as well as the distance between their centers. He found that M_2 and M_3 attracted each other in accordance with Newton's law of universal gravitation. Experiments like the one performed by Cavendish have been carried out by other scientists since his time.

Albert Einstein showed that Newton's law does not apply exactly when we deal with speeds approaching the speed of light (over 300,000 kilometers per second). In such cases, we must use the calculations of the relativity theory. However, for calculations that do not involve such fantastic speeds, Newton's law still holds.

The force of attraction between the earth and a baseball dropped from the hand is so strong because the mass of the earth is so very great. Every particle on the earth acts on every particle of an object on or above the surface of the earth. The result of

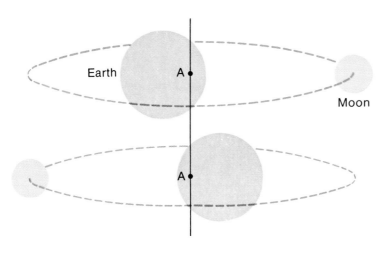

Fig. 1. The earth and moon, drawn to each other by gravitational force, move around a common center (A), located within the earth.

all these little forces is a single large force directed toward the center of our planet.

However, the force of attraction between two ordinary masses of about equal size is exceedingly small. Suppose that a one-kilogram iron object is located so that its center is one meter from the center of another one-kilogram iron object. The gravitational attraction acting on both masses is about 0.0000000000667 newtons. This would be approximately the weight of a very small cube of iron with an edge about equal to the thickness of a sheet of paper.

ACCELERATION DUE TO GRAVITY

When the earth attracts a certain quantity of sugar, say, the result is always a definite force, which we call weight. Suppose this force is one newton. In that case, we say that the sugar weighs one newton at the surface of the earth. Weight is not the same thing as mass. *Mass* is the quantity of a substance—the amount of matter it contains. The *weight* of a body is the force acting upon it due to the gravitational attraction of the earth.

This attraction brings about a downward acceleration (a constant increase in the rate of speed) of all objects toward the center of the earth. The acceleration in question is called *gravity*. It is the same for all objects, regardless of their mass, at a given point on the earth's surface, if air resistance is small. If the air resistance is considerable, it will affect light objects more than heavy ones. The acceleration of the lighter objects will not be quite so great. The acceleration due to gravity is called g. It is approximately 9.8 meters per second per second. This means that if an object is accelerated toward the earth's center, it will travel 9.8 meters per second faster at the end of the first second than at the beginning of the first second. It will travel 9.8 meters per second faster at the end of the second second than at the beginning of the second second and so on.

The words "gravity" and "gravitation" are sometimes used as if they meant the same thing. Strictly speaking, gravitation refers to the acceleration of any two objects in the universe toward each other, while gravity, as we have just seen, represents the

Fig. 2. The Cavendish gravity experiment. Two lead balls were suspended at the opposite ends of a torsion balance. When a third ball, M_3, was brought close to M_2, a mutual attraction caused M_2 to approach M_3, causing the bar to turn.

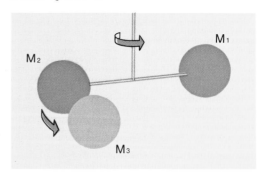

Fig. 3. The pull of gravity is not so strong at the mountaintop, A, as it is in the plain, B. The reason is that A is farther away from the center of gravity of the earth than B.

Fig. 4. The plumb line shown in the diagram does not point exactly to the earth's center of gravity because it is attracted to A, which is a dense part of the earth's crust.

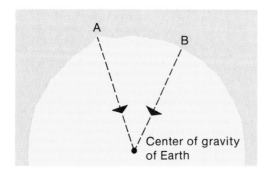

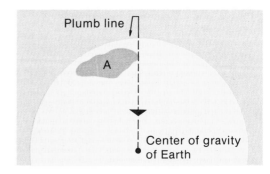

gravitational acceleration toward the center of the earth. The word "gravity" has also been applied to the gravitational acceleration toward the center of other celestial bodies. Thus we can speak of the force of gravity on the planet Mars.

FACTORS AFFECTING g

The value of g has been accurately measured at many different places on the earth's surface and has been found to vary, as shown in the table on this page. The variation is due to several factors. One of these is distance from the center of the earth. The nearer an object is to this point, the greater the earth's attraction for it and the greater the value of g. If two places are at approximately the same latitude, the one at a higher altitude will show a lower value of g than the other one (Figure 3).

Another factor affecting the value of g is the uneven distribution of the mass on the earth, particularly in the crust, which contains very dense substances in some locations and less dense substances in others. There are certain places where the bob, or weight, at the end of a plumb line does not point toward the exact center of the earth because of the attraction exercised by dense masses in a nearby area (Figure 4).

The weight of an object on the earth's surface will differ, therefore, from place to place, because the gravitational attraction varies from place to place. Yet the mass of an object never changes. A standard mass of one kilogram and a balance (Figure 5) can be used to measure one kilogram of mass anywhere on the earth's surface, or for that matter on the surface of any planet. The difference between the two ends of the balance is so small that there is no difference in the gravitational attraction upon them. This is because the earth's radius is much longer than the balance arm.

A standard one-kilogram mass may be used with a balance in Cuba to measure a one-kilogram mass of sugar. Suppose this same one-kilogram standard and the kilogram of sugar are taken to Iceland. They will be in equilibrium on a balance located there even though the actual force acting on

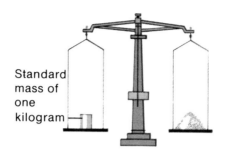

Fig. 5. To measure out a quantity of sugar with a mass of one kilogram, we put a standard mass of one kilogram on one pan of a balance and some sugar on the other pan. If the two pans balance, the mass of the sugar is equal to one kilogram.

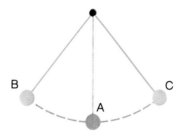

Fig. 6. A simple pendulum. As the bob, A, swings to and fro between B and C, it is acted on by the force of gravity.

both the standard and the sugar in Iceland will be greater than the force acting on both in Cuba.

MEASURING g

An accurate measurement of g, the acceleration due to gravity, can be made

Values of g, the acceleration due to gravity

Place	Value of g
(centimeters per second per second)	
Cambridge, Massachusetts	980.398
Denver, Colorado	979.609
Eagle City, Alaska	982.183
Galveston, Texas	979.272
Greenwich, England	981.188
Honolulu, Hawaii	978.946
Madras, India	978.281
New Orleans, Louisiana	979.324
Ponta Delgada, Azores	980.143
Reykjavik, Iceland	982.273

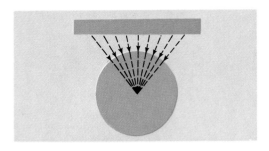

Fig. 7. If the bar shown in the diagram above and the earth were comparable in size, the gravitational forces acting on the bar, all pointing downward toward the center of gravity of the earth, would differ markedly in direction.

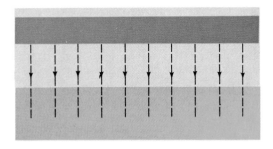

Fig. 8. Here the bar and the earth's surface (of which only a few centimeters are shown) are drawn to exact scale. The center of gravity of the earth is so far from the bar that the gravitational forces shown are practically parallel.

with a pendulum. A simple pendulum (Figure 6) consists of a small sphere, called a bob, suspended from a fixed support by means of a long, thin thread. The bob is set moving by a slight sideways push. As it swings upward it will be acted on by the force of gravity until finally its upward course will be halted. Then the force of gravity will cause the bob to swing downward in the opposite direction to the first one. Its momentum will carry it past the starting point and then upward. Again it will be slowed down by gravity until its upward course will be halted, and it will swing downward in the opposite direction, as before.

The path the pendulum describes as it moves from one high point of its course to the other and then back again to the first one is called a *complete vibration*. The time required for a complete vibration is known as the *period*. Of course the length of the period will be affected by the force of gravity. The stronger the gravitational pull, the shorter the path of the pendulum and the shorter the period will be. If we know the period and the length of the pendulum, we can calculate the pull of gravity—that is, the value of g.

If we are using a simple pendulum, such as the one we described, the margin of error would be about one per cent. Using more complicated pendulums, scientists are able to determine the value of g to six significant figures. If, for example, we calculate the value of g at a given place as

982.183 centimeters per second per second, the six significant figures would be 9, 8, 2, 1, 8, and 3.

Why should scientists want to make such very accurate measurements of g? For one thing, you will recall that the attraction depends on the mass of the particles on the earth as well as their distance from the object on which they exert a gravitational force. Hence, an accurate measurement of g gives definite clues concerning the density (mass per unit volume) of the substances immediately below the earth's surface. These clues are useful to the geologist in determining how movements of the earth's crust take place.

CENTER OF GRAVITY

The gravitational force acting on each bit of mass of a substance, as the earth attracts it, is generally directed toward the center of the earth, as we have seen. Suppose the earth were comparable in size to a horizontal bar, such as the one shown in Figure 7. In that case, the gravitational forces affecting the different parts of the bar would differ in direction, as shown in the diagram. But since the earth is so very much larger than the iron bar, the forces of attraction are to all intents and purposes parallel as shown in Figure 8.

In the case of the horizontal bar or any other object, for that matter, there is always one point at which the combined forces of gravity will be centered. This point is located at what is called the *center of gravity*.

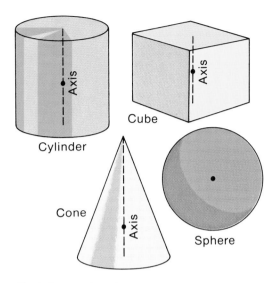

Fig. 9. Center of gravity of four types of solid bodies having regular geometrical shapes.

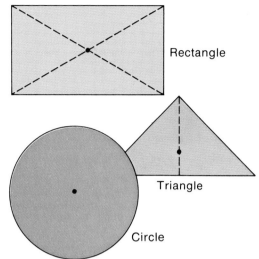

Fig. 10. Center of gravity of three kinds of regularly shaped areas in the form of sheets.

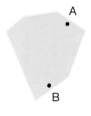

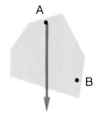

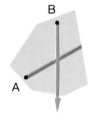

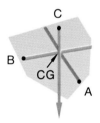

Fig. 11. To find the center of gravity of the piece of cardboard shown above, we punch holes A and B at different places near the edge.

The cardboard is suspended at A from a nail. A vertical line, drawn with a plumb line guide, will pass through the center of gravity.

We now suspend the cardboard at B and draw the vertical line from B. The center of gravity is where the lines from A and B cross.

We make another hole at C and hang the cardboard at this point. A vertical line from C will likewise pass through the center of gravity.

We may think of it as the point where all of the weight of the body is concentrated.

In the case of solid bodies with regular geometric shapes, the center of gravity is always at the geometric center of the body—that is, if the density is the same throughout. We show some bodies of this kind in Figure 9. Note that the center of gravity of a cube or of a sphere is at the exact center of these solids. For a cone, it is on the axis and at a point a fourth of the way from the base of the cone to its vertex, or tip.

If a body is in the form of a sheet, the center of gravity corresponds to the center of the area (Figure 10). If the sheet is rectangular in shape, the center of gravity is located at the point where the two diagonals intersect. In the case of a triangle, it is one third of the distance from the middle point of any of the three sides to the opposite point. For a circular sheet, it is at the center of the circle.

Even if a body is irregular in shape, the center of gravity can be determined. Suppose we want to locate it in the case of the piece of cardboard shown in Figure 11. First we punch two holes, A and B, at different points along the edge of the cardboard. Then we hang the cardboard at point

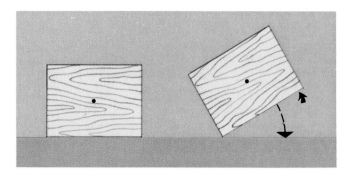

Fig. 12. The left-hand block of wood is in a state of stable equilibrium. If we lift up one end of it somewhat (right-hand drawing), the center of gravity will be raised. When we let go of this wooden block, it will return to its former position.

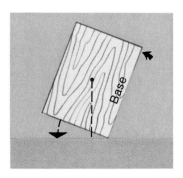

Fig. 13. If we were to tip a block of wood so that a vertical line from its center of gravity were to fall outside of the base of the block, it would fall over.

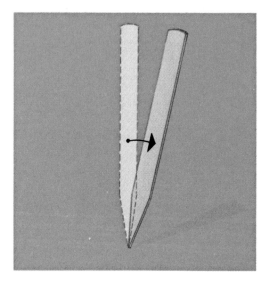

Fig. 14. If we tip the long stick at the left, we lower its center of gravity and as a result it will fall. Before it is tipped, the stick is said to be in unstable equilibrium.

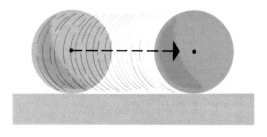

Fig. 15. If we roll a ball along a table, the center of gravity will be neither lowered nor raised. The ball is said to be in neutral equilibrium.

A from a nail driven into a post, say. When the cardboard comes to rest, its center of gravity will be located on a vertical line extending from the point of support. To find the vertical line, we hang a plumb line from the same nail. Using it as a guide, we draw the vertical line on the cardboard. Then we hang up the cardboard at point B and repeat our procedure. The vertical line from B will cross the one from A at the center of gravity. We can prove that this is so by making another hole, C, in the cardboard and suspending the cardboard at point C. A vertical line from C will pass through the center of gravity.

THREE TYPES OF EQUILIBRIUM

The location of a body's center of gravity will determine how stable it is—that is, to what extent it resists any effort to disturb its equilibrium, or balance. Generally speaking, there are three states of equilibrium. They are known as stable, unstable, and neutral.

A body is said to be in *stable equilibrium* if we raise its center of gravity when we tip it or lift one end. If we release the body, it will fall back to its former position. A square block of wood resting on the floor (Figure 12) is a good example of a body that is in stable equilibrium. If we lift up one end of the block slightly, the weight of the body, concentrated at the center of gravity, will make it return to its original position. However, if we were to tip the block so that a vertical line from the center of gravity would fall outside the base of the block (Figure 13), it would tip over. It would be in a state of stable equilibrium in its new position. To keep a body as stable as possible, it is advisable to provide a wide base and a low center of gravity. This is done in designing automobiles, boats, furniture, and the like.

If we lower the center of gravity of a body when we tip it, the body is in *unstable equilibrium*. Suppose we set a pointed stick in a vertical position on a horizontal table, as shown in Figure 14. Any slight tipping of the stick will lower the center of gravity and will make it fall over suddenly.

If the center of gravity of a body is neither raised nor lowered when the body is displaced, we say that the body is in a state of *neutral equilibrium*. A ball on a table (Figure 15) is in a state of neutral equilibrium, because any force that moves the ball sideways neither lowers nor raises the center of gravity. It will continue to remain in equilibrium, no matter how we move it along the table.

GRAVITATIONAL ATTRACTION ON THE MOON

Does a traveler on the moon find that the same forces of gravitational attraction hold for the moon as for the earth? If he drops a ball from a height on the moon, does it fall to the moon's surface? If he strings a heavy wire between two high places, does it sag downward? The answer is "Yes." However, a traveler from the earth would find that the forces of gravitational attraction would not be nearly so great on the moon as on the earth, because the mass of the earth is so much greater.

The acceleration due to gravity at the surface of the moon would be about 2.1 meters per second per second, as against 9.8 meters per second per second on the earth's surface. In 3 seconds, an object dropped from a height on the earth would travel a total distance of 44 meters. The same object dropped from a height on the moon would fall a total distance of only about 9.5 meters in the same period of time.

Because of the lessened gravitational force on the moon, our traveler is able to perform feats that would be quite impossible on the earth. He is able to carry with ease objects that on the earth would tax the powers of a champion weight-lifter. The world's record for the high jump is about two meters. Even without any kind of athletic training, a traveler on the moon would be able to clear the seven-meter mark without any great effort.

As television images of the astronauts on the moon showed, moving about in a lessened gravitational field actually takes some getting used to. The astronauts had to remain much more aware of their center of gravity than we do on the earth, since it was easier to get off balance as they maneuvered.

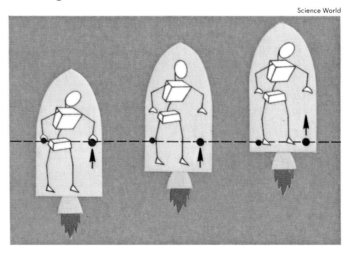

Experiments carried out in an accelerated field have the same results as experiments performed in a gravitational field. If a man inside an accelerating spaceship drops two marbles—one light and one heavy—at the same time, he observes that they hit the floor of the spaceship at the same time—just as they would if gravity were operating within the spaceship.

MOTION

by Ferdinand L. Singer

From one point of view, nothing is ever at rest. It is true that the chair on which you sit, the table on which you write, and the bookcase by the wall are not moving toward you or away from you. But the earth on which they rest revolves around the sun. The sun itself moves constantly in space.

It would be very confusing if we were to take into account all these different kinds of motion in discussing, say, the movement of an automobile on a city street. For most purposes it suffices to consider motion only with respect to the earth. We can define motion as change of position in relation to certain fixed points. For example, an automobile moves past a road sign; a hit baseball moves out from home plate. The road sign and home plate would be the fixed points of reference.

If the size of a moving body is negligibly small compared to the size of its path, we need not consider the motions of the individual parts that compose the body. When we discuss the motion of an automobile as it rounds a curve, we need not take into account the movements of the valves, or pistons, or other parts of the car. A body such as an automobile, considered only as a self-contained unit, is called a *particle*. We

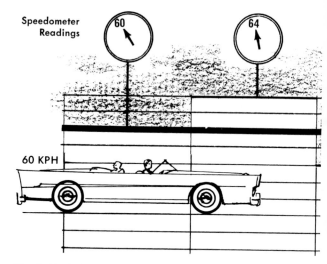

Fig. 1. Variations in the speed of an automobile traveling at an average speed of 60 kilometers per hour.

can think of the earth as a particle when we trace its orbit around the sun. We can disregard the motions of the automobiles that are traveling along its surface as it whirls through space.

DIFFERENT KINDS OF MOTION

There are various kinds of motion. To begin with, there is *rectilinear motion*, or motion in a straight line, as when an automobile travels along a straight road. When we go around a curve in a car, our motion becomes *curvilinear*. We may travel at *constant speed* by traveling over equal dis-

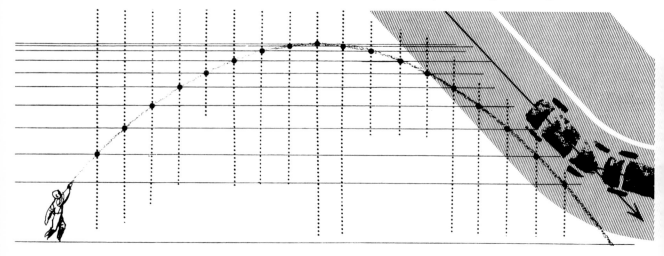

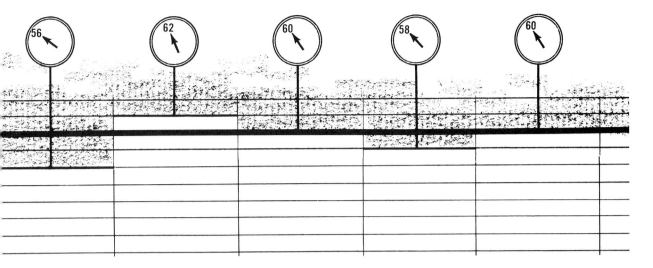

tances in equal intervals of time, or we may let our speed vary. If our speed increases, we say that we are *accelerating*. If we slow down, we are *decelerating*.

The simplest type of motion is rectilinear motion at constant speed. Suppose that an automobile covers 2 kilometers in two minutes. If its speed has been constant, it will have traveled at the rate of 60 kilometers an hour. Actually, try as he may, no driver would be able to keep the speedometer needle exactly at the 60 kilometers-per-hour mark. The needle will sometimes go above and sometimes below this mark. When we say, therefore, that an automobile travels over a certain stretch of road at 60 kilometers per hour, we really mean that its

average speed is 60 kilometers per hour. (See Figure 1.) The average speed is equivalent to the distance divided by the elapsed time.

CHANGES IN SPEED

Changes in speed are described in terms of the rate at which we change our speed. If an automobile increases its speed by 4 kilometers per hour for every second of time, we say that it has an acceleration of 4 kilometers per hour per second. Starting from rest, the car will be traveling at 4 kilometers per hour at the end of 1 second, at 8 kilometers per hour at the end of 2 seconds, at 12 kilometers per hour at the end of 3 seconds, and so on. This car is increasing

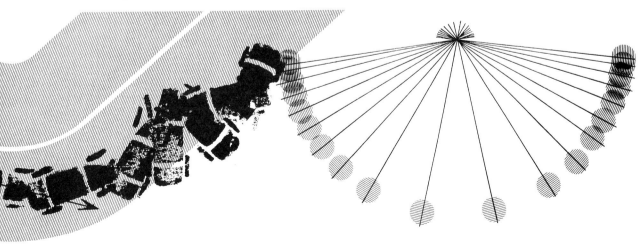

its speed at a constant rate. For this reason it is said to have motion together with a constant rate of acceleration.

The average speed in a uniformly accelerated motion is the sum of the initial speed and the final speed divided by 2. This relationship also applies to objects which come to rest while losing speed at a constant rate. The rate at which speed is reduced is called the *deceleration,* or *negative acceleration.* We sometimes give the name *positive acceleration* to acceleration involving an increase in velocity.

Freely falling bodies move with constant acceleration. The force bringing this about is the gravitational attraction of the earth. The force of gravitation differs at various locations. It is stronger, for example, at the poles than at the equator.

The gravitational attraction of the earth causes all objects to fall with an acceleration of about 9.8 meters per second per second. The exact rate will depend on the gravitational force at a given part of the earth's surface. Suppose we drop a baseball from the top of a skyscraper (Figure 2). At the end of 1 second, the ball will have attained a velocity of 9.8 meters per second. At the end of 2 seconds, the velocity will be 19.6 meters per second, at the end of 3 seconds, 29.4 meters per second, and so on. In this case, of course, we have positive acceleration. Actually, the velocity in each case would not be quite so great as we have indicated, because the air resistance would hold the baseball back by a slight amount.

If an object is cast straight up in the air, the force of gravity will decelerate it (decrease its velocity) at the rate of 9.8 meters per second per second. Suppose a girl throws a baseball straight up in the air at a velocity of 29.4 meters per second. At the end of the first second, the ball will be traveling at the velocity of 19.6 (29.4-9.8) meters per second. At the end of the second second, the velocity will be reduced to 9.8 (19.6-9.8) meters per second, while at the end of the third second the velocity will be zero. This is a good example of deceleration, or negative acceleration. Of course the ball will not remain motionless in the air after its velocity has been reduced to zero. The force of gravity will pull it down so that it will fall with an acceleration of 9.8 meters per second per second.

At one time it was thought that the speed of fall depended on the weight of an object. The great Italian scientist and mathematician Galileo Galilei, who lived during the period 1564 to 1642, is said to have proved, in a famous experiment, that this was not the case. He caused several objects of different sizes and materials to be dropped from the Leaning Tower of Pisa at the same instant of time. According to the traditional account, observers saw these objects hit the ground at the same time. Some modern authorities are rather skeptical about the story. They point out that the

Fig. 2. A baseball, dropped from the top of a skyscraper, falls with a constant acceleration of 9.8 meters per second per second.

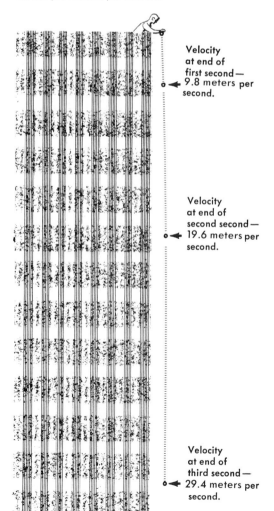

Velocity at end of first second — 9.8 meters per second.

Velocity at end of second second — 19.6 meters per second.

Velocity at end of third second — 29.4 meters per second.

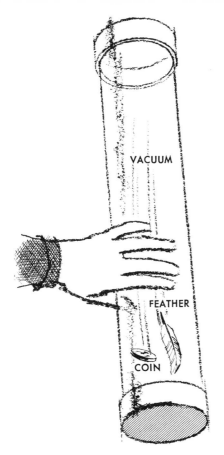

Fig. 3. When air is pumped out of the tube, creating a vacuum, the coin and the feather will fall at the same rate.

Fig. 4. When the car starts to go around the curve, it will change direction from A to B. It will have to travel a greater distance than arc ab. To maintain a given speed, there will have to be an acceleration. Actually, the car changes directions much more often than shown here.

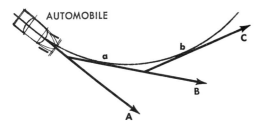

Fig. 5. A simple pendulum. The time that is required for it to swing back and forth once is known as the period. The extent of the pendulum's swing is called the amplitude.

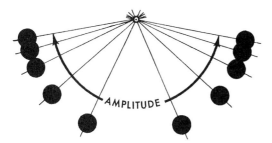

resistance of the air would affect the objects as they fell, and that the lighter objects would be held back most by this resistance.

However, we can prove, by a familiar experiment, that the weight of an object has nothing to do with the rate at which it falls. We put a feather and a coin in a long glass tube, which is made airtight (Figure 3). When the air is pumped out of the tube and the tube is suddenly inverted, the feather and the coin will be seen to fall at the same rate. When air is admitted to the tube the coin will fall more rapidly, because the resistance of the air will affect the feather much more than the heavier coin.

SPEED AND VELOCITY

We often use "speed" and "velocity" interchangeably, and sometimes we are justified in doing so. However, speed is not always the same thing as velocity. Strictly speaking, speed measures the rate at which we travel, while *velocity* involves not only speed but also direction. In *rectilinear motion*, or motion in a straight line, velocity and speed are practically synonymous, since only one direction is involved. In motion along a curve, however, the direction of the velocity is always different from the actual path along which the moving object travels.

To show what we mean, let us assume that a car (Figure 4) is traveling along a straight road in the direction A at the rate of 80 kilometers an hour. Suddenly it comes to a curve. The car will now be headed toward a new direction, B. It will really cover a greater distance in going from a to b, on the curved path, than the arc ab. To maintain the speed of 80 kilometers per hour, it will have to travel faster. In other words, there will be an acceleration. The same thing will happen when the car changes its direction from B to C. Each time the car changes its direction, there will be an acceleration. The direction of the car at any one time will be at a tangent to the curved path that the car will follow — that is, the direction will touch the curve at one point and then will diverge from it.

The velocity, then, is always changing as a vehicle moves in a curved path. If the

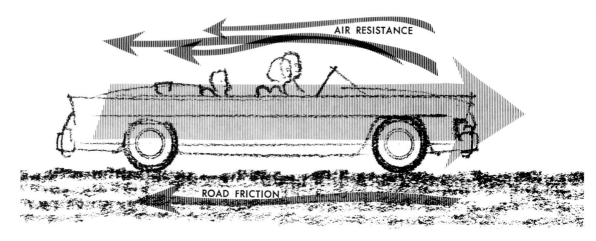

Fig. 6. As an automobile moves, it is held back by the forces of road friction and air resistance.

vehicle covers equal distances along the path in equal intervals of time, it has a constant speed, but its velocity cannot be constant since its direction constantly changes.

A familiar example of curvilinear motion is that of the bob, or weight, suspended at the end of a pendulum (Figure 5). In this case, neither the velocity nor the acceleration remain constant. The acceleration varies directly with the distance from the central position of the bob. This produces what is called *simple harmonic motion*, found in most vibrating objects. This concept is too complex to discuss further here.

Galileo seems to have been the first to observe that a pendulum takes the same time to complete each swing. The time is independent of the extent of the swing, provided the swing is not too large. The time required for a pendulum to make one complete vibration (that is, to swing back and forth once) is called the *period*. The extent of the swing is known as the *amplitude* of the pendulum.

The motion of the pendulum is expressed by the following laws:

1. *The period of vibration does not depend upon the weight of the bob.*

2. *The period of vibration does not depend upon the amplitude.*

3. *The period of vibration is proportional to the square root of the length of the pendulum.*

4. *The period of vibration is inversely proportional to the square root of the acceleration due to the force of gravity.*

The last law explains why a pendulum clock has to be adjusted when it is moved, say from Cambridge, Massachusetts, in the United States, to Rio de Janeiro, in Brazil. The pull of gravity is weaker in Rio de Janeiro and it will affect the rate at which the pendulum swings and thereby the length of time for a complete swing.

NEWTON'S THREE LAWS OF MOTION

So far we have discussed motion without considering the forces that cause it. We analyze these forces by applying three laws known as Newton's laws of motion because they were advanced by the great 17th-century English scientist Sir Isaac Newton. They are usually stated as follows:

1. *A body at rest stays at rest and a body in motion continues in motion at a constant speed in a straight line unless acted upon by an external force.*

2. *The rate of change of motion of a body is proportional to the applied force and takes place in the direction in which the force acts.*

3. *To every action, there is always an equal and opposite reaction.*

As we apply these three basic rules of motion in discussing various examples, their meaning will become clearer.

FIRST LAW OF MOTION

A body at rest stays at rest and a body in motion continues in motion at a constant speed in a straight line unless acted upon by an external force.

To understand the second part of this first law of motion, let us point out that when a body is in motion, two factors cause it to resist change. One is its velocity. The other is its mass. We know that it is hard to stop the motion of a bullet, because of its velocity. It is also hard for us to stop the motion of a slowly moving automobile, because of its mass. The two factors of mass and velocity of a body, when multiplied together, gives us what is called the body's *momentum.*

The mass of a body is always constant. Hence, to change the momentum, it is necessary to change the velocity. To change the velocity (either the rate of speed or the direction or both) and thus cause an acceleration requires a force. That is what Newton meant by saying that a body in motion will move in a straight line at constant speed unless acted on by an external force.

"But," you may say, "an automobile can travel at constant speed along a straight line only because we apply a force to it — the power supplied by the motor. Why shouldn't the automobile be able to travel at constant speed in a straight line without any force being applied to it, if Newton's first law is valid?"

The answer is that when an automobile moves along a road, it is continually held back by the resisting forces of road friction and air resistance (Figure 6). To counteract these resisting forces and to enable the automobile to maintain a constant speed in a straight line, there must be an external force, originating in the motor and applied at the wheels. We should, therefore, interpret the first law of motion to mean that when the forces acting upon a body are balanced, there is no acceleration. This condition would apply equally to bodies at rest and to those moving with constant velocity. For example, the automobile mentioned before, traveling at constant speed in a straight line, does so because balance exists between the driving forces developed at the wheels and the resistance of the air and the road. To change the velocity, we must upset this balance.

SECOND LAW OF MOTION

The rate of change of motion of a body is proportional to the applied force and takes place in the direction in which the force acts.

This second law of motion will apply when we upset the balance of forces by increasing the driving force. There will be a net accelerating force as long as the driving force is greater than the resisting force. However, air resistance increases with gain in speed. Hence the unbalance (lack of balance) between driving force and resisting force is gradually reduced, thereby decreasing the rate at which we gain speed. To continue accelerating, we must feed more power to the engine. Again the air resistance will increase, and again the unbalance between driving and resisting forces will be gradually reduced. It will become zero at the limit of engine power. When we have reached this condition, we shall then be traveling at our maximum possible speed.

Another way to upset the balance between driving and resisting forces is to ap-

Fig. 7. Whenever one fires a gun, the forward thrust of the projectile is matched by the backward thrust, which is called the recoil.

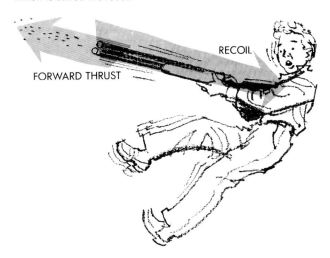

FORWARD THRUST

RECOIL

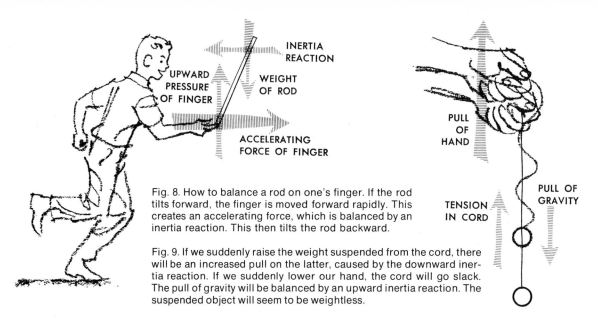

Fig. 8. How to balance a rod on one's finger. If the rod tilts forward, the finger is moved forward rapidly. This creates an accelerating force, which is balanced by an inertia reaction. This then tilts the rod backward.

Fig. 9. If we suddenly raise the weight suspended from the cord, there will be an increased pull on the latter, caused by the downward inertia reaction. If we suddenly lower our hand, the cord will go slack. The pull of gravity will be balanced by an upward inertia reaction. The suspended object will seem to be weightless.

ply the brakes. The large decelerating force created in this way brings the car rapidly to a stop. In a collision, a car is stopped almost instantaneously. The large deceleration that is developed corresponds to a large stopping force. It is this force that damages cars in collisions. The faster a car is moving at the instant of impact, the greater the decelerating force and the more damaging the effect upon the car will be.

Here is another example which shows the relation between force, acceleration, and stopping distance. If a milk bottle is dropped into a pile of sand, the bottle will not be damaged. But it will be shattered if it is dropped the same distance upon a concrete floor. This is because the sand is yielding and permits a lower deceleration than when the bottle drops against concrete. Since concrete is harder than sand, it will stop the bottle more quickly and will therefore exercise a larger force.

THIRD LAW OF MOTION

To every action, there is always an equal and opposite reaction.

The third law tells us that even when bodies are being accelerated, the forces that affect them remain in balance. The forces that bring about action and reaction always occur in pairs. We may not be aware of the reacting force, but it is always there. When we stand upon a floor, we exert a force. This is opposed by the force that the floor exerts upon us. If we kick a football, it reacts with equal force upon our foot. In the case of jet-propelled planes, the backward thrust of the gases issuing from the jet engine reacts against the engine and causes a forward thrust. Sometimes, we can feel the reacting force. If we fire a rifle, the forward thrust of the projectile is matched by the backward thrust, which we call the recoil or "kick" of the rifle (Figure 7).

The opposite and equal reaction that balances an accelerating force is called *inertia*. It is equal to the product of the acceleration and the mass of the body being accelerated. When we ride in a train, the starting forces that accelerate it and us forward create an inertia reaction which causes us to lurch backward. A sudden stop throws us forward because the deceleration is opposite to the forward motion of the car. The decelerating forces are balanced by the forward reaction effect of inertia.

We can demonstrate this effect very simply. Suppose that while we balance a rod on a finger, the rod tilts forward, as shown in Figure 8. The balance can be restored by a rapid forward movement of the finger. The movement of the finger creates an accelerating force which is balanced by an inertia reaction, acting at the center of gravity of the rod. If our forward motion is too rapid, the inertia reaction will tilt the rod backward. Then a quick backward movement of the finger will again restore balance by creating a forward inertia reaction. We can keep the rod in balance by alternate forward and backward movements.

The effect of inertia can be demon-

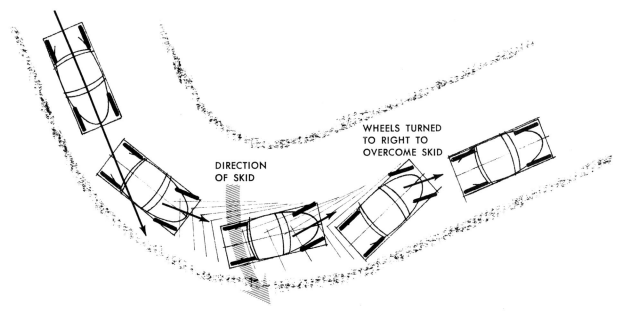

Fig. 10. If a driver makes a left turn too sharply or too quickly, centrifugal force will make the car skid to the right. In order to overcome the skid, the front wheels have to be turned in the direction of the skid.

DIRECTION OF SKID

WHEELS TURNED TO RIGHT TO OVERCOME SKID

strated by another simple experiment. Let us suspend a weight from a cord (Figure 9). If we hold the weight stationary, the tension in the cord just equals the pull of the weight. Suppose we raise the weight suddenly; that is, we accelerate it upwards. We will notice an increased pull of the cord, brought about by the inertia reaction to the accelerating force. If we lower the weight, not at a uniform velocity, but faster and faster, the pull becomes less when the weight is held still. This is because the weight acquires a downward acceleration, causing an upward inertia reaction. Since the inertia reaction counteracts the weight, the net pull on the cord is decreased. If we lower our hand suddenly, the cord will go entirely slack. The pull of gravity on the weight will cause it to fall with an acceleration of 9.8 meters per second. This gravity pull will be balanced by an upward inertia reaction, which will make the suspended object seem to be weightless.

CENTRIPETAL VS CENTRIFUGAL FORCE

The third law of Newton always applies to curvilinear motion. When we whirl a stone around by a string, our hand exerts an inner pull—a *centripetal force*— upon the stone to keep it moving around in a curve. We also become aware of an outer pull, which is called *centrifugal force*. If we

release the stone, so that there is no longer an inward pull, the outward pull, or centrifugal force, will cause the stone to fly off in a straight line. The centrifugal force is the inertia reaction to the inward pull our hand exerts on the stone. A centrifugal inertia force always acts on a body moving in a curve because such bodies are always being accelerated.

Centrifugal force, representing an inertia reaction, makes itself felt when a car begins to skid (Figure 10). Suppose we make a left turn too sharply or too quickly. The centrifugal inertia force of the turn becomes greater than the friction of the tires against the road, and the car will start to skid to the right. The only way to overcome the momentum of the skid is to create a force to overcome it. This we can do by immediately turning the front wheels to the right—that is, in the direction of the skid. The car will then tend to turn in a rightward curve. In this way it sets up a leftward centrifugal force which stops the skid. The car may overbalance and tend to skid leftward again. The driver, therefore, must be prepared to turn the front wheels to the left if this should happen. This maneuver is done expertly by racing car drivers, who deliberately skid their cars around turns in order to avoid losing speed.

One is likely to skid on an icy road,

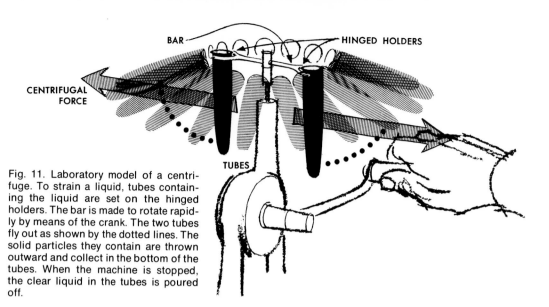

BAR · · · · HINGED HOLDERS

CENTRIFUGAL FORCE

TUBES

Fig. 11. Laboratory model of a centrifuge. To strain a liquid, tubes containing the liquid are set on the hinged holders. The bar is made to rotate rapidly by means of the crank. The two tubes fly out as shown by the dotted lines. The solid particles they contain are thrown outward and collect in the bottom of the tubes. When the machine is stopped, the clear liquid in the tubes is poured off.

because on such a road the road friction is reduced to a minimum. The turning of the front wheels will not bring us into an opposite turn as quickly as on a dry road, because the car will have traveled further in the direction of the skid.

A method commonly used to prevent skidding when rounding a curve is to construct the curved roadway so that the outer part of the road is higher than the inner. This construction is called banking the curve. It brings about the same effect as leaning inward when rounding a flat curve on a bicycle. The bicyclist leans inward to balance the effect of centrifugal inertia force. Of course the automobile cannot "lean inward," but the same effect is produced if the road is banked at the proper inclination.

A familiar application of inertia force is found in centrifuges, which are devices used to separate one material from another. A laboratory model of a centrifuge is shown in Figure 11. This device is often used to remove from a liquid the small solid substances it contains. A test tube of the liquid that is to be processed is put in a hinged holder at each end of a bar. As the bar is made to rotate by means of a hand crank or a motor, the two tubes in the holders fly out in an almost horizontal direction. The solids in the liquid, being heavier than the liquid itself, are thrown outward and collect in the bottom of the tube. The machine is then stopped and the clear liquid is poured off.

The same principle is used to separate cream from milk in the device known as the separator. In the separator, the milk is passed into a spinning bowl. The milk, being heavier than cream, is forced to the outward part of the spinning bowl and is drawn off. The cream, which is lighter, stays near the center and is drawn off from that point.

VECTORS AND SCALARS

Sometimes motion is brought about by combined forces, which can be analyzed by means of *vectors*. We can show what vectors are by a simple example. Suppose we walk east for four blocks and then north for three blocks of equal length. We diagram our walk east (Figure 12) by a line *AB* consisting of 4 equal units, each unit representing one block. The arrow on the line indicates the direction of the walk. Such a line is called a vector. Its length, drawn to some arbitrary scale, represents the magnitude, or size, of a quantity—whether it is a walk, a velocity, an acceleration, or a force. The direction of the quantity is denoted by the arrowhead drawn on the line or, as we can now call it, the vector.

The three-block walk north is represented by the vector *BC*. We now draw a line connecting *A* and *C*, and also provide it with an arrowhead. It constitutes another vector, *AC;* it represents the net change in our position and is called the *resultant*. It shows that we would have walked the equivalent of five blocks if we had walked straight from *A* to *C*, instead of using the

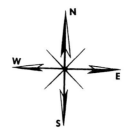

roundabout route from *A* to *B* and from *B* to *C*. In either case, we would be the equivalent of five blocks from our starting point, *A*, at the end of our walk.

The numbers 4 and 3 in this example are known as *vector quantities*. They answer the question "In what direction?" as well as the question "How much?" As a physicist would say, they possess direction as well as magnitude. They differ from what we call *scalar quantities*, which have only magnitude. We are all familiar with scalar quantities. When we speak of 3 apples, we are referring to scalar quantities, We can add 4 apples and 3 apples by the ordinary processes of arithmetic, giving the result 7 apples. We cannot add vector quantities arithmetically. We must add them by drawing vectors in a diagram such as the one shown in Figure 12.

When we add scalar quantities together, the results are always the same. The sum of 4 apples and 3 apples is always 7 apples. The sum of 4 bananas and 3 bananas is always 7 bananas. But the addition of vector quantities does not always give the same result, if their directions differ. To illustrate our point, suppose we had walked northeast for three blocks from *B* instead of north. We would now diagram our walk as in Figure 13. The resultant effect, *AC*, would be about 6½ blocks. If we had walked from *B* in a northwesterly direction (Figure 14), the net effect *AC* would be a little less than 3 blocks. A more extended walk along the path *A-B-C-D-E-F-G*, diagrammed in Figure 15, would produce an effect *AG* of 2.83 blocks. In fact, if after following the path *A-B-C-D-E-F-G*, we returned to *A*, there would be no net change in our position at all—the resultant would be zero.

Here are the rules for this new type of addition—the addition of vector quantities. We draw the vector quantities correctly to scale and indicate the direction. We carefully place the "tail" of each vector at the "tip" of the vector to which it is added. The resultant vector will be represented to scale and in direction by the vector drawn from the tail of the first vector to the tip of the last one.

Fig. 12. Suppose we walk east for four blocks and north for three blocks, and diagram our walk as indicated below. *AC* is the resultant. It is equivalent to five blocks.

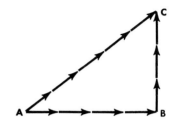

Fig. 13. If now we were to walk four blocks east and three blocks northeast, the resultant effect, *AC*, would be equivalent to something like six and one-half blocks.

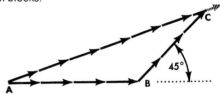

Fig. 14. We change our route again, walking four blocks east and three blocks northwest, as shown below. In this case, the resultant, *AC*, is a little less than three blocks.

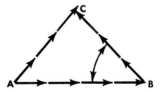

Fig. 15. If we adopted the roundabout route shown below, in walking from *A* to *G*, the resultant,, *AG*, would be equal to a little less than three blocks. If we continued our route from *G* to *A*, the resultant would be zero.

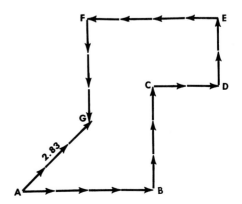

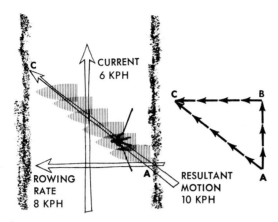

Fig. 16. Here we see the effect produced by rowing at 8 kilometers per hour across a stream which is flowing at 6 kilometers per hour. *AB*, in the right-hand diagram, indicates the rate of flow and the direction of the stream. *BC* indicates the rate of travel and direction of the boat; *AC* the resultant.

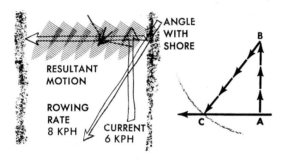

Fig. 17. As in the preceding illustration, a boat is traveling at 8 kilometers per hour across a stream flowing at 6 kilometers per hour. If the boat heads upstream in the direction BC (righthand drawing), it will reach the opposite shore at a point that is directly opposite the point of departure.

USING VECTOR ADDITION

Let us consider a few applications of vector addition. Suppose we row at 8 kilometers per hour straight across a river flowing at 6 kilometers per hour. Our net motion will be the vector combination of both movements. The resultant would be 10 kilometers per hour, as shown in Figure 16. The river would carry us downstream as we rowed across so that we would arrive at *C*.

To travel straight across the river, we would have to row in such a direction that the resultant movement would be at right angles to the flow. Figure 17 shows what to do. We draw the vector *AB* to represent the rate of flow of the river — 6 kilometers per hour. Then setting a compass to represent a value of 8 kilometers per hour, we draw an arc with *B* as a center to intersect *AC*, the direction we wish to travel, at point *C*. In order to reach the opposite shore, at a point directly opposite the point of departure, the boat would have to be pointed upstream in the direction *BC*. Once we know the angle *ABC*, we could set our course. To find the angle, we would use a protractor — an angle-measuring device.

Another example concerns a situation that often occurs in hockey. A player, *A* (Figure 18), moving in one direction, wishes to pass the puck to another player *B*, who is moving in another direction. Experience tells player *A* how to pass the puck so that it may reach *c* at approximately the same time as player *B*. If *A* could analyze the vectors involved at the time that he is making the shot, there would be no guesswork in the maneuver. Figure 18 analyzes the problem.

Fig. 18. Player *A*, skating toward point *a*, wishes to pass the puck to player *B*, traveling in the direction *c*, so that the puck will arrive at *c* at the same time as player *B*. If we know the velocity and direction of travel of players *A* and *B*, we can calculate how fast and in what direction the puck will have to travel to reach *c* in time.

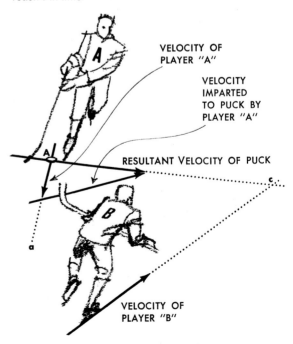

FORCES WITHIN LIQUIDS AND GASES

Every time you turn on the water in your kitchen or your bathroom, or put your foot down on the brake pedal of your car, or run the vacuum cleaner, or pump up a bicycle tire, certain powerful forces in liquids or gases are working for you. The safety of deep-sea divers and submarine crews depends upon a knowledge of how these forces behave. Engineers must calculate them with great accuracy when they build tunnels beneath rivers or construct airplanes.

Liquids and gases are made up of tiny, constantly moving particles called molecules. The forces in both liquids and gases are transmitted by these molecules as they strike one another and collide with the objects in their path. Both liquids and gases flow freely from a container and have no definite shape.

There are, however, some important differences between them. The molecules of a liquid are held together much more tightly than those of a gas. Again, a liquid will fill a container up to a certain level, and the upper surface of the liquid will be horizontal. A gas, however, will fill uniformly the whole of a container into which it is put. It diffuses, or spreads out, into the open air if released from the container.

DENSITY OF LIQUIDS

In order to measure accurately the forces within a liquid, we must know its density—that is, the mass, or amount of material, packed into a given volume. We express density in terms of mass per unit volume. We say of a given substance that it has a density of so many grams per cubic centimeter.

The density of pure water is about 1 gram per cubic centimeter when its temperature is 4° Celsius. Alcohol is less dense than water, and one cubic centimeter of it weighs only about 0.8 grams. Mercury,

however, is so dense that it weighs about 13.6 grams per cubic centimeter.

We can, therefore, find the density of any liquid—or of any solid—by finding out how much a cubic centimeter of it contains. If we compare the density of the liquid or solid with that of the same volume of water, we shall know what is called its *specific*

Liquids under pressure are used to perform many tasks. Top: paint under pressure is sprayed through a nozzle to produce an even coating on the car fender. Bottom: a hydraulic lift mechanism helps the truck dump its load.

White Motor Corp.   De Vilbiss

gravity. Since water is usually the basis of this type of comparison, we say that the specific gravity of water is 1. A substance half as dense as water would have the specific gravity of 0.5. Mercury, 13.6 times as dense, has a specific gravity of 13.6.

PRESSURES EXERTED BY LIQUIDS

A liquid exerts force on any surface with which it is in contact. The denser the liquid, the greater the force that it will exert. The water in a glass aquarium, for instance, presses against both the bottom and the sides of the aquarium, as well as against the fish and any objects that are submerged in the water. We measure a force exerted by a liquid in terms of so many newtons per square meter or per square centimeter. The newton and dyne are units expressing force or weight in the metric system. Weight or force per unit area is called *pressure*.

Imagine a cubical tank 10 meters long, 10 meters high, and 10 meters wide, and filled with water. Since water weighs 9,800 newtons per cubic meter, and since the volume of the tanks is 1,000 cubic meters, the weight of the water in the tank is 9,800,000 newtons. The thrust of the liquid against the 100-square meter bottom is equally distributed. The water will push against every square meter of the bottom with a pressure of 98,000 newtons per square meter—that is 9,800,000 newtons divided by 100. Since there are 10,000 square centimeters in a square meter, the pressure of the water in newtons per square centimeter will be 98,000 newtons divided by 10,000, or 9.8 newtons per square centimeter.

This figure represents the pressure due to the liquid only. To obtain the total or absolute pressure, you would have to add the pressure of the atmosphere upon the surface of the liquid. As we shall see, atmospheric pressure at sea level is about 10.3 newtons per square centimeter.

We have been considering the pressures within a tank that is a perfect cube. The pressure per square centimeter at the bottom would be the same in a tank of any size or shape, provided that the depth is the same (Figure 1). The reason is that pressure within a liquid depends only on the depth and density of the liquid. Suppose a piece of 2-centimeter pipe is placed in a vertical position and filled with liquid to a depth of ten meters. The pressure at its bottom is the same as it would be on a square centimeter at the bottom of a big tank filled to a depth of ten meters with that same liquid.

Liquids exert pressure not only on the bottom of a container, but in all directions and at all places throughout the liquid. The pressure downward or sideways increases with greater depth below the surface because of the added weight of the water above.

You can easily prove this by a very simple experiment. Punch holes in an empty tin can near the top and near the bottom. Then fill the can with water. The water will escape from the holes, thus proving that liquids exert pressure sideways as well as from above and below. Water will spurt from the holes near the bottom much more rapidly than from those near the top. This is

Fig. 1. There is the same amount of pressure per square centimeter at the bottom of these irregularly shaped water tanks because the depth of the water in each of the tanks is the same.

because the pressure of the water is greater near the bottom of the can, since the total weight of the liquid above that level is greater than the total weight above the hole near the top. Since pressure increases with depth, a dam is built so that the thickest and strongest part is at the bottom where the pressure of the water is greatest.

Because of the presence of salt and other impurities, seawater has a density of about 1.03 grams per cubic centimeter instead of 1. At a depth of only 180 meters, the pressure is about 180 newtons per square centimeter. A submarine designed to descend to this depth would have to be well built to withstand the weight of the salt water. At certain places off the coast of Asia, the ocean has been found to have a depth of more than 9.5 kilometers. At this depth the pressure is nearly 1,000 newtons per square centimeter.

BUOYANCY OF LIQUIDS

Everyone knows that some objects float in water and that others do not. An object that floats in water might sink in alcohol, and one that sinks in water might float in glycerin. All liquids possess in varying degrees the property that is called *buoyancy*—an upward push upon objects that are submerged within them.

If you throw a piece of dry wood into a pond or a pool, the wood floats. The up-ward push of the water makes this possible. An object that sinks in water—like a piece of iron—weighs less when under water than when in air. This also is because the water exerts an upward pressure upon it. When you are taking a bath in a well-filled tub, you can raise your whole body easily by a slight pressure of your hands on the bottom of the tub. You could not do this so easily if you tried to raise yourself from the living-room floor because, though air is buoyant, it is less buoyant than water.

In the third century B.C., the Greek mathematician and physicist Archimedes discovered the principle of buoyancy which bears his name. *Archimedes' principle* states that a body wholly or partly immersed in a fluid is buoyed upward with a force equal to the weight of the volume of liquid it displaces. Let us imagine that in a full pail of water we place an iron ball that weighs 45 newtons when weighed in air. We discover, however, that under water this ball weighs only 36 newtons—a loss of 9 newtons. The volume of water that spilled over when the ball was placed in the full pail weighs 9 newtons, which just equals the ball's loss of weight. In other words, our iron ball, though it sinks, is actually buoyed up by a force equal to the weight of the water it displaces.

A floating body always sinks to such a depth that the weight of the displaced liquid

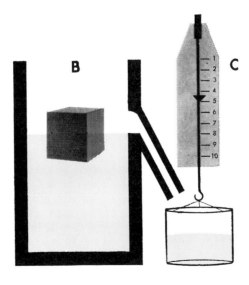

Fig. 2. A block of wood weighs 142 grams (A). It is put into an overflow can filled with water up to the spout (B). The displaced water flows into the container at the side of the overflow can (C). The weight of the displaced water equals the weight of the block.

Fig. 3. An application of Pascal's law: raising a car on a lift. The application of a small force on a small cylinder beside the lift exerts a large force on the large cylinder seen above.

is the same as that of the body itself (Figure 2). A body that weighs more than the liquid that it displaces will sink. A ship having a total weight of 5,000 metric tons sinks into the water until the weight of the water displaced is 5,000 metric tons. If a barge floating on the river is given a load of 2,000 metric tons of coal, the barge will sink farther down until an additional 2,000 metric tons of water is displaced. On the other hand, a block of iron will not float in water because it weighs more than the water it displaces. Since iron sinks in water, you may wonder why a ship made entirely of steel (an alloy of iron) can float. The reason is that a ship is not solid but contains much hollow space filled with air. Therefore, its density is less than that of solid iron.

A solid that is more dense than the liquid in which it is placed will sink. If it is less dense than the liquid it will float. Even a block of lead will float on mercury because mercury is denser than lead. If kerosene and water are poured into the same container, the kerosene will float on the water, since its density is less than that of water. For the same reason, cream rises in milk and floats on the top.

If an ocean-going vessel sails up a river, it sinks deeper in the river than it did in the sea. The reason is that seawater, because of the salts dissolved in it, is denser than fresh water. It is easy to show how objects are affected by the added buoyancy of salt water. Place an egg in a drinking glass full of water from your tap. The egg will sink to the bottom. Remove the egg and stir about four tablespoonfuls of salt into the glass. If you put the egg in this salt solution, it will float on the surface. Gently add fresh water until the glass is almost full, and then stir. This time the egg will sink about halfway to the bottom of the glass. This is obviously because the fresh water that you add to the drinking glass is less buoyant than salt water.

A submarine is built so that its total weight is a little less than the weight of the same volume of seawater. It has tanks into which seawater is allowed to enter when the craft is about to submerge. With the added weight of the seawater in its tanks, the submarine now weighs more than an equal volume of seawater, and so sinks. When it is to rise to rise to the surface, the water is forced out of the tanks by means of compressed air.

PASCAL'S LAW

Another property of liquids was discovered in the seventeenth century by the great French scientist Blaise Pascal. He showed that pressure applied to any part of a liquid in a closed vessel is transmitted with equal intensity to all parts of the liquid and acts in all directions. This principle is known as *Pascal's law*.

Let us see how Pascal's law works. Suppose we have two cylinders filled with some liquid — oil perhaps — and connected at their bottoms as those in Figure 3. Each cylinder is fitted with a movable piston. Suppose that a downward force of 5 newtons is exerted on the small piston, which has an area of one square centimeter. This force applies a downward pressure of 5 newtons per square centimeter at the top of the liquid in the small piston. Now, according to Pascal's law, this causes an increase in pressure of 5 newtons per square centimeter at every point throughout the liquid in both cylinders, since they are connected. Suppose the large piston has an area of 100 square centimeters. The upward force against this piston will be 500 newtons. Therefore, a downward force of 5 newtons against this piston will be 500 newtons. force of 500 newtons against the large piston.

As we look at Figure 3, it may seem strange that a small force on the small piston can exert such a large force on the large one. The fact is that the gain in the force or push upward is at the expense of the distance that the big piston is moved. When the small piston moves downward the entire length of the cylinder, the large piston will actually rise very little. Where the ratio of the areas is 100 to 1, the large piston will be forced upward only one hundredth of a centimeter when the small one is pushed downward one centimeter.

The city pumping station can pump water to your house only because of the forces explained by Pascal's law. If none of the faucets in the water system are open and if the faucets and the pumps are at the same level, the pressure at the pumping station and at your faucet is the same. In other words, the pressure exerted by the pumps at the pumping station, on the body of water there, has spread throughout the water in the system, passing through the pipes, which extend in many directions, perhaps many kilometers away, to your own faucets.

Pascal's law explains the working of the hydraulic press, shown in Figure 4. The small cylinder consists of a pump that forces liquid into the larger cylinder. As the small pump piston is pushed downward,

there is an upward force on the large piston.

The principle of the hydraulic press is used in many common appliances. In the hydraulic brake on automobiles, a small force applied to the brake pedal is so magnified on the brake drums as to stop the car. When your barber wishes to raise his barber chair, he works a lever that pumps oil into a cylinder containing the piston that supports the chair. A slight pressure on the lever is magnified so greatly that it raises the weight of the heavy chair. Hydraulic pressure operates the wing flaps of many planes. It also compresses cotton bales into comparatively small bundles so that they will take up less shipping space.

PROPERTIES OF LIQUIDS IN MOTION

Liquids that are in motion, such as water moving through a pipe, do not follow exactly the same rules as liquids at rest. For instance, although the pressure is the same for all points at the same depth below the surface of a liquid at rest, this is not true for liquids in motion. In a stream of water moving through a pipe that has everywhere the same diameter, the pressure decreases uniformly in the direction in which the wa-

Fred S. Carver Inc.

Fig. 4. The diagram below shows a hydraulic press compressing a cotton bale. At the right is a hydraulic press that is used in a laboratory.

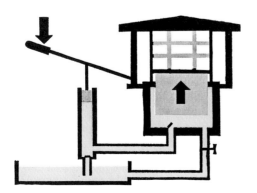

ter is moving. The pressure will be greatest nearest the point where the water enters the pipe. Where water is piped from a spring to a house, the pressure is greatest close to the spring. Where the pipe enters the house the pressure is much less. Pumps often have to be used in a case of this kind to step up, or increase, the pressure. One way of showing how the pressure becomes constantly less in the direction in which the water is flowing in a pipe is to place vertical glass tubes in the pipe at various points along it, as shown in Figure 5. The pressure at each point will be indicated by the height to which the water rises at that particular

point. The height of the liquid in the glass tubes decreases uniformly in the direction in which the liquid is flowing.

Now there is a different and very strange effect which we notice in liquids that are moving through horizontal pipes. You might imagine that the pressure in a liquid would be greatest where its speed is greatest, but this is not true. The pressure is least where the speed is greatest.

In a pipe whose inside passageway, or opening, is smaller at some places than at others, the liquid will flow more rapidly through the smaller sections than through the larger ones. In Figure 6, the liquid is

Fig. 5. The pressure on each of the vertical tubes in indicated by the height to which the water rises.

Fig. 6. Varying pressure in a horizontal tube. The liquid moves faster at B than at either A or C.

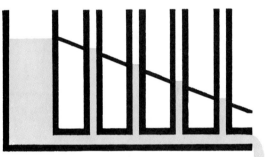

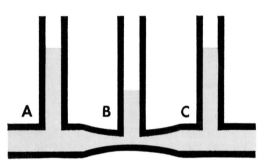

Fig. 7. The jet water pump operates on Bernoulli's principle. Water in a flooded basement is usually removed with this kind of pump.

Fig. 8. The two boats may collide because of the lessened pressure of the water moving between them — another phenomenon based on Bernoulii's principle.

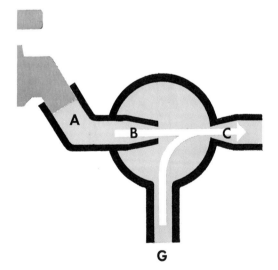

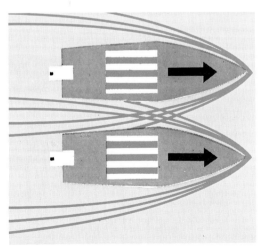

moving faster at B than at either A or C. But the pressure in the wide parts of the pipe at A and C is greater than the pressure in the small part of the pipe at B. When the pressure is greater in the pipe, the speed is less, and when the pressure is less the speed is greater. This law was discovered in the eighteenth century by Daniel Bernoulli. It is called *Bernoulli's principle*.

The jet water pump, which is often used to pump water out of basements, works on Bernoulli's principle (Figure 7). In this type of pump, the tube A is connected to a water faucet. As water from the faucet rushes through this tube, it travels fastest through the narrow part at B. The lessened pressure at B creates suction, which pulls water up from the basement through the tube G. The water from the faucet and the water from the basement flow out at C.

Bernoulli's principle also explains why two speedboats moving parallel and close to each other are likely to be pulled together and collide (Figure 8). As the boats move forward, water is funneled into the narrow region between them. The relative speed between the water and the boats is greater in this narrow region than if there were more space between the boats. As a result, there is a decrease in pressure of the water between the boats, and the greater pressure of the water upon the outer sides pushes the boats together.

GASES AND FORCES

Since we can see and feel liquids, it is comparatively easy to visualize the forces that they exert. The forces at work within gases are apt to seem far more mysterious, because most gases are invisible and many are odorless as well. Yet they surround us on every side. Air is a mixture of several gases—a great deal of nitrogen (about 78 per cent), a sizable amount of oxygen (about 21 per cent), and small quantities of carbon dioxide, argon, helium, krypton, neon, and xenon.

Many other kinds of gases are common in everyday life. Illuminating gas is piped to your kitchen range and perhaps to your furnace. Ginger ale fizzes because of a gas in it. Your electric refrigerator, the neon lights, the plumber's blowtorch, and hundreds of other familiar devices are operated by gases.

All gases have weight, just as solids and liquids have. A cubic meter of air at average temperature and pressure weighs about 12 newtons. The air in a room 10 meters long, 8 meters wide, and 6 meters high weighs about 55,200 newtons. Gases vary greatly in density. Hydrogen is the lightest gas. Helium, twice as heavy as hydrogen, is about $1/7$ as heavy as air.

As we pointed out, the atmosphere is made up of a number of gases. It is really an ocean of air, which surrounds the earth and which extends upward for hundreds of kilometers. The total weight of the atmosphere is about 5,810 million million metric tons. It presses down on every square centimeter of the earth's surface with a force of about 10.3 newtons per square centimeter at sea level.

ATMOSPHERIC PRESSURE

The existence of atmospheric pressure was demonstrated in a series of experiments in the seventeenth century by the Italian Evangelista Torricelli and the Frenchman Blaise Pascal. Torricelli took a long, slender tube, sealed at one end and containing mercury, and set it, open end down, in a dish containing mercury. He showed that the atmospheric pressure at sea level on the mercury in the dish would keep the column of mercury in the tube at a height of about 76 centimeters (Figure 9).

The modern mercury barometer, used to measure the pressure of the atmosphere, is based upon Torricelli's model. It consists of an inverted mercury-filled tube in a reservoir of the same liquid. There is another type of barometer that is more sturdy and that can be carried about much more easily—the aneroid barometer (Figure 10). It has no tube and no liquid. This type of barometer is an airtight box made of very thin and flexible corrugated metal fastened firmly to a base. Some of the air is removed from the box, and as a result the flexible top becomes very sensitive to changes in air pressure from the outside.

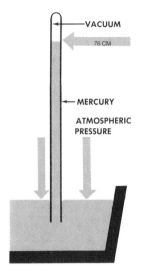

VACUUM

76 CM

MERCURY

ATMOSPHERIC
PRESSURE

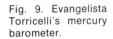

Fig. 9. Evangelista
Torricelli's mercury
barometer.

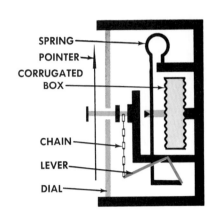

SPRING
POINTER
CORRUGATED
BOX

CHAIN
LEVER
DIAL

Fig. 10. The aneroid, or "liquidless," barometer—a
modern, portable, pressure-sensitive device.

When the pressure of the atmosphere increases, the top of the box is pushed down some distance. When the pressure decreases—pushes with less weight upon the top of the box—the top rises again. The motion of the box top is transferred by a system of springs, levers, and chains to a pointer that moves along a scale. The number on the scale at which the pointer comes to rest indicates the barometric, or atmospheric, pressure.

The ordinary lift pump that raises water from a well or spring works because of atmospheric pressure. The weight of the atmosphere pushes down against the water in the well and forces it up the pipe which connects the well and the pump.

The pump itself consists of a cylinder into which a sliding piston is tightly fitted (Figure 11). This piston is attached to the pump handle, and as the handle is moved up and down, the piston slides down and up

Fig. 11. How a lift pump works. In *A*, the cylinder valve is open as the water rises. *B* shows the piston descending. The cylinder valve is closed. In *C*, the piston valve is closed as the piston rises. The water above the valve is then forced out.

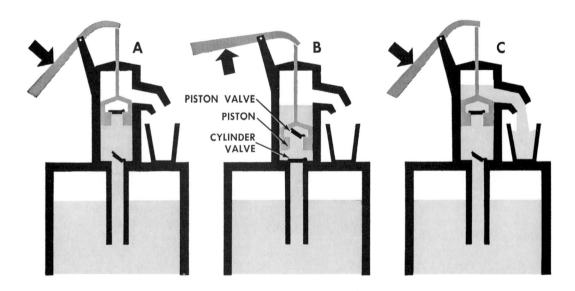

A

B

PISTON VALVE
PISTON
CYLINDER
VALVE

C

inside the pump cylinder. A valve opens upward in the piston, and another valve also opens upward in the bottom of the pump cylinder.

When the pump handle is pushed down, the piston in the cylinder rises, increasing the distance between it and the bottom of the cylinder. In this space a partial vacuum is created. Water from the well, forced upward by the pressure of the atmosphere, then passes through the inlet valve at the bottom of the cylinder and into the cylinder itself.

When the pump handle is pushed up, the piston descends in the cylinder with its valve open. The cylinder valve is closed. When the piston reaches the bottom of the cylinder, its valve closes because of the weight of the water above it. With the next stroke the piston is raised, bringing with it a load of water, which pours out through the spout of the pump. The cylinder valve remains open while the piston valve is closed. This process is repeated again and again.

Atmospheric pressure can hold up a column of water 10 meters high, equivalent to 76 centimeters of mercury. Usually, however, a pump will lift water only about 9 meters because even in the best pumps of this type there is some leakage.

The pressure of the atmosphere also makes it possible to move water or other

Fig. 12. The siphon is a bent tube having one arm longer than the other. Water flows from the higher to the lower level.

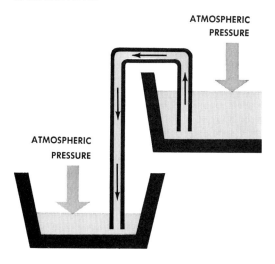

liquids by siphon from a higher level to a lower one when it is inconvenient to pour the liquid from one container into another. The siphon is a bent tube, one of whose arms is longer than the other (Figure 12). The tube is filled with liquid, and the shorter arm is immersed in the liquid in the higher vessel. The liquid immediately starts to flow into the container at the lower level.

The pressure of the atmosphere forces the liquid up the shorter arm rather than up the longer one because the weight of the liquid in the shorter arm is less. If the liquid being transferred is water, the bend in the siphon must not be higher than 10 meters above the water's surface at sea level.

With a siphon, one can draw off the upper part of the liquid in a container without stirring up the sediment that lies at the bottom. Laboratory workers use siphons to transfer liquids from containers that cannot be tipped. Certain enterprising thieves draw gasoline from the tanks of automobiles with siphons. To prevent such thefts, the cap of a gasoline tank is sometimes provided with a lock. The name "siphon" is often used for a bottle through which charged water may be drawn. This is not a true siphon, however, since the water is driven out by the pressure of the gas inside the bottle and not by the pressure of the atmosphere.

The ordinary atomizer works because of atmospheric pressure. You will remember that, according to Bernoulli's principle, the pressure in a moving liquid is least where the liquid is flowing most rapidly. The principle holds true not only of liquids but also of gases, including the collection of gases that we call the atmosphere. In the case of the atomizer, a rubber bulb is compressed suddenly to cause a jet of fast-moving air to pass over the top of a tube that extends down into the liquid in a bowl (Figure 13). The pressure above the liquid in the tube is greatly reduced by the air motion over the top of the tube. Atmospheric pressure upon the liquid in the bowl then forces liquid up and out at the end of the tube. The liquid is broken into a fine spray by the air jet and is carried along with it.

Bernoulli's principle reveals why a

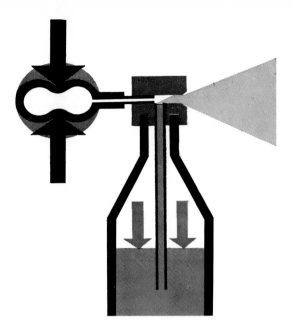

Fig. 13. An atomizer. Atmospheric pressure within the bowl of the atomizer, above, forces liquid up when the bulb is pressed. The liquid is emitted in a spray.

baseball or tennis ball can be made to curve. If a baseball is moving forward and spinning around as it moves, some air adheres to the ball and is carried around with it. Because of the ball's forward motion, there is a backward flow of air past the ball. At the top of the ball, the flow of air is opposite in direction to the motion of the air that is dragged around due to the ball's spinning motion. At the bottom of the ball, the two air motions have the same direction. There is, therefore, a greater relative speed past the ball on the underside than on the upper side. Because the pressure is less on the underside the ball will be forced to curve downward.

Anyone can make a ball curve and thus demonstrate Bernoulli's principle, with the help of a simple bit of apparatus. Roll up a sheet of sandpaper, making the diameter of the roll such that a Ping-Pong ball can pass through it easily. Wrap a string around the roll and tie the ends of the string together. Now put a Ping-Pong ball inside the roll and throw the ball away from you at high speed. Can you explain what happens?

Bernoulli's principle also explains the lift on an airplane wing. The shape of the wing causes air to rush over the top surface faster than it flows past the under surface (Figure 14). This causes a reduction in pressure above the wing, which usually accounts for about two-thirds of the lift. The rest of the lift on the wing is due to the upward push of air against the underside.

You can see how lift is brought about in accordance with Bernoulli's principle by making your own airplane "wing" as follows. Insert a sheet of paper between the pages of a book. Hold the latter in order to keep the "wing" in place. Blow across the top of the "wing" and see what happens to the sheet of paper.

AIR PUMPS

Under normal conditions, as we have seen, the pressure of the atmosphere at sea level is about 10.3 newtons on every square centimeter. There are times when we need higher air pressures than the pressure of *one atmosphere,* as 10.3 newtons per square centimeter is called. For instance, if the pressure within the tires of your automobile were just equal to the pressure of the atmosphere, you would have a flat tire. You must therefore force more air into the tire than it would have under normal atmospheric pressure.

The ordinary air pump is a very simple instrument. In one type (Figure 15), there is a cylinder with a sliding piston that is worked by a handle at the top. On opposite sides at the bottom of the cylinder are two valves, one for taking in air from the outside and the other for transmitting that air into the tube attached to the tire.

As you push the piston down (*A*, Figure 15), the inlet valve closes and the compressed air is forced through the outlet valve into the tire. As you raise the piston (*B*, Figure 15), a partial vacuum is formed in the cylinder and the atmospheric pressure forces more air through the inlet valve. With the next downstroke of the piston, pressure is again exerted on this new batch of air, and in its turn it is forced through the outlet valve into the tire. In this way the tire is filled with compressed air.

In another type of pump, the inlet valve is a flexible leather gasket connected to the piston. This gasket allows air to slip past it when the piston is raised, and then

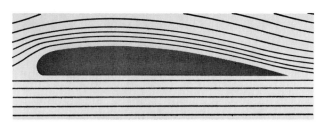

Fig. 14. Reduced pressure on the upper surface of an airplane wing causes most of the lift.

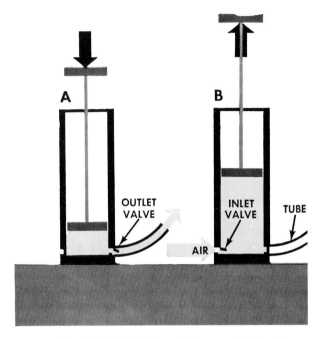

Fig. 15. A familiar type of air pump. Reduced pressure in the cylinder, together with the pressure of the atmosphere, forces air into the inlet.

expands tightly against the cylinder with the downstroke.

The use of gases under high pressure saves humans a great deal of labor. Much heavy work that once was done by hand and with backbreaking effort can now be done easily with the help of compressed air. It works air brakes on trains, riveting machines, various types of drills, and blasting machines for cutting stone and metal.

When new bridges are built across bodies of water, the foundations have to be laid underwater. For this purpose engineers often sink a caisson, which is a great tank open at the bottom. Air is then forced into the caisson at pressures high enough to hold back the surrounding water. Workers enter or leave the caisson, and materials are transferred to it, through air locks. The

people can work with comfort in the caisson once they have adjusted themselves to the very high pressure there.

Compressed air is also used to keep water from rushing in during the construction of tunnels, and in completed tunnels for ventilation. As we said before, submarines are built so that they submerge by admitting seawater into special compartments, in this way increasing their weight. To bring the submarine back to the surface, compressed air is used to force out this water and lighten the craft.

VACUUM PUMPS

If it is possible to pump air into a container, it should also be possible to pump air out of it and in this way to leave the vessel empty or, rather, almost empty. We can do this by means of a vacuum pump, which is really an air pump in reverse. A simple type is shown in Figure 16. As the piston P is raised, the pressure in the cylinder below the piston decreases. The valve at A is held firmly closed by the air outside, while the valve B is forced open by the air in C, which expands into the cylinder. At the downward stroke, the valve at B closes, and the air in the cylinder is forced out through A. Each stroke of the piston removes a fraction of the air from C. The pressure in C decreases greatly after many strokes, and, therefore, such a pump can never be very efficient. Eventually the pressure becomes too small to operate the valves.

A more efficient type of vacuum pump (Figure 17) uses a rotating inner cylinder that keeps turning in off-center fashion within an outer cylinder. As the inner cylinder rotates, it always makes contact with the inner wall of the outer cylinder and with a movable vane. The mechanism is immersed in oil, which serves as a lubricant and as a seal to prevent leakage. At each revolution of the inner cylinder, some air is drawn in through the intake, from the container to be evacuated, and is forced out through the outlet into the oil. Extremely low pressures can be achieved.

The vacuum plays a constant and practical role in our life today. If the air were

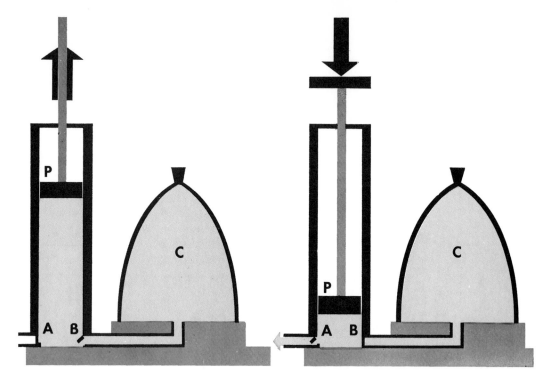

Fig. 16. The vacuum pump works on the same principle as the air pump, only the process is reversed. Vacuum pumps, however, cannot create a perfect vacuum.

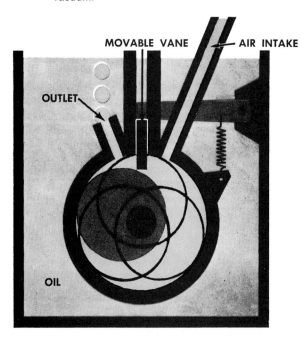

Fig. 17. How a rotating vacuum pump works. An inner cylinder rotates off-center within an outer (stationary) one, keeping in contact with the inner wall of the latter and with a movable vane. The turning inner cylinder moves air from the intake and toward the outlet.

not removed from the ordinary electric light bulb you use in your lamp, the oxygen in the air would cause the little filament or wire inside the bulb to burn out very soon. Many of the tubes in old radios are vacuum tubes. In some other types of bulbs and tubes, other gases are pumped in after the air has been pumped out. In neon advertising signs, the air is removed from the glass tubing, and neon gas at low pressure is admitted. As an electric current passes through the tubing, the neon gas glows with the familiar pinkish red color. If certain other gases are used, the tubes will glow with other colors. The gases must be under low pressure if they are to glow.

In modern dairies, cows are milked by machines that are nothing more than special kinds of vacuum pumps. The necessary mixture of fuel and air enters the cylinders of the automobile engine because a vacuum is created as the piston moves forward. In a vacuum cleaner, an electric motor causes a fan to whirl rapidly and this fan creates a vacuum inside the nozzle. As air rushes in to fill up the vacuum, it carries particles of dust, lint, and so on with it.

BUOYANT FORCE OF GASES

Gases exert an upward or buoyant force very much as liquids do. A body

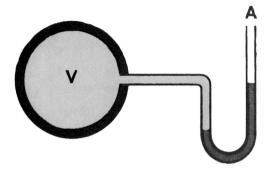

Fig. 18. Diagram of a manometer, an instrument used to measure the pressure of a gas occupying a closed vessel.

immersed in a gas is buoyed upward with a force equal to the weight of the gas displaced. This lifting power of air makes it possible for balloons to rise and to carry heavy loads. The buoyant effect becomes less the higher the balloon rises because of the decrease in density of the air. There is a limit to the height to which any balloon can travel. Sooner or later, it will reach an altitude where the buoyant force is not greater than the total weight. The balloon will then go no higher unless its load is reduced. A toy balloon made of rubber may expand until it bursts.

The earliest balloons were filled with hot air, which is not so dense as cool air. Later practically all balloons were inflated with hydrogen, the density of which is about one-fourteenth that of air. However, hydrogen is a highly inflammable gas, as many balloonists learned to their cost. In the case of dirigibles, too, the use of hydrogen sometimes led to disaster. Helium proved to be far more satisfactory for lighter-than-air craft because it will not burn. It is true that helium is twice as dense as hydrogen. However, since it is only approximately a seventh as dense as air, it still has abundant lifting power.

We can measure the pressure of gases in various ways. The mercury or aneroid barometer, as we have seen, indicates atmospheric pressure. The pressure of a gas in a closed vessel is measured differently. If the pressure of the gas is not too high or too low, a U-shaped glass tube—a manome-

ter—is used (Figure 18). This glass tube is partly filled with mercury and one of its ends is connected to the vessel (V) that contains the gas. The other end is left open to the atmosphere at A. If the pressure in V is greater than the atmospheric pressure, the mercury will be pushed up in the open side of the tube, as shown in the diagram. If the pressure in V is less than the atmospheric pressure, the mercury will stand lower in the open end of the tube. The difference between the height of the mercury in the two sides of the tube will show how the pressure in V compares with the atmospheric pressure.

The small pressures in closed vessels containing very thin gas cannot be measured by ordinary manometers. In such cases, a large volume of the low-pressure gas is tapped and then compressed into a small volume. The pressure in this small volume of gas and the relative change in its volume are then measured. According to Boyle's law, where there is no change in temperature the volume of a gas varies inversely as its pressure. By Boyle's law, the pressure in the low-pressure gas can then be computed.

High pressures, such as the pressure of steam in a boiler and in all cases where compressed air is used, are measured by the Bourdon spring gauge. The instrument consists mainly of a hollow tube bent into a circular arc. One end of this tube is closed and the other is connected to the container of gas. As the gas under pressure enters the curved tube, the tube tends to straighten out. As it does so, a pointer attached to it by a system of levers moves along a scale. The scale is usually so made that it reads zero at atmospheric pressure. Anything above zero, then, gives the pressure in excess of atmospheric pressure. A gauge pressure of 40 newtons per square centimeter would mean a total pressure of 40 newtons per square centimeter plus 10.3 newtons per square centimeter, or 50.3 newtons per square centimeter.

Today we have definitely harnessed the forces within liquids and gases. We can measure these forces, use them, and avoid dangers often inherent in them.

HEAT

by Albert J. Ruf

Heat plays an all-important part in our daily lives. Heating systems keep us comfortably warm in winter. In the summer, we prefer to keep indoor temperatures lower by means of air-conditioning systems. Heat engines supply power for automobiles, trucks, diesel-electric locomotives, tugboats, ocean liners, airplanes, and rockets. Certain heat engines run dynamos, which generate electricity. The electricity operates factory machines, television sets, telephone systems, and a host of other devices. We owe life itself to heat. Our bodies, in a sense, are heat engines: the food we eat is fuel that keeps us warm and supplies us with energy for our various activities.

THE SUN—A SOURCE OF HEAT

The sun is the source of most of the heat known to us. As a direct source of heat, the sun maintains life upon this planet, 150,000,000 kilometers distant. It is also an indirect source of heat. Since the earth is derived from the sun, it is to the latter that we must trace some of the earth's internal heat that reveals itself through volcanoes, geysers, and hot springs.

The sun is also the source of the heat stored in fuels such as coal and petroleum. Coal is derived from vegetation that flourished on the earth during the Carboniferous period, more than 200,000,000 years ago. At that time, huge trees and various other plants grew in the swamps that covered much of the earth's land surface. Through the process of photosynthesis, they manufactured food in the presence of sunlight and thus stored the sun's radiant energy. They died and their remains accumulated, only partly decayed, in the swamps. In later ages, this vegetation became part of the earth's crust and under the action of heat and pressure was transformed into coal. Like coal, petroleum is derived from organisms that lived in the remote past. The solar energy stored up in coal and petroleum is released when we burn these fuels.

Even the heat produced by electricity is ultimately derived from the sun. Much of it is generated in steam plants through the

ERDA's Sandia Laboratory

Solar energy is converted to thermal energy—heat—by this parabolic solar collector.

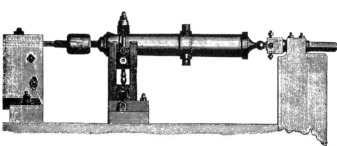

Apparatus used by Thompson in his experiments on heat. Upper engraving: water boiled due to friction when a filled test tube was rotated between wooden paddles. Lower engraving: a model of a cannon-boring device. The barrel became hot because friction produces heat.

Privat-Deschanel, A., *Traité Elementaire de Physique*

use of coal and of liquid fuels derived from petroleum. The electricity produced by the force of running water may also be traced to the sun. The reason is that our star plays the most important part in the water cycle, or hydrological cycle, of which running water is a phase.

NATURE OF HEAT

The scientists of earlier days had some rather curious ideas concerning the nature of heat. They regarded it as a fluid, which they called "caloric." All substances were supposed to contain caloric to some extent, depending on how hot they were. It was maintained that when one body was heated by another, it was because caloric was transferred from the hotter body. Supposedly this idea was confirmed by the fact that as fuels burn, they decrease in weight. But when scientists sought to determine the change in weight in heated substances, they found that the results were confused and often contradictory. This did not cause them to abandon the caloric theory. They simply revised it somewhat, by making caloric a weightless fluid.

It remained for the 18th-century physicist Count Rumford, born Benjamin Thompson, to contribute significantly to what is currently accepted as the correct theory concerning the nature of heat. While

Rumford was supervising the boring of a cannon in the late 1790s, he noted that the gun became very hot after being bored for even a short time. He put an insulated and watertight box containing about nine or ten liters of water in contact with the gun barrel. As the barrel was bored, the water in the box became hotter and hotter and in two

Benjamin Thompson, or Count Rumford.

and a half hours it began to boil. It continued to do so even when the tool became so dull that it was no longer cutting. Apparently the rotation of the drill in the bore of the cannon in some way transferred heat to the water.

It occurred to Rumford that perhaps the mechanical energy released by the rotation of the drill was transformed into heat. He sought to determine the relationship between the amount of work done and the quantity of water that boiled away. His measurements were so crude, however, that no definite relationship could be established. Sir James Prescott Joule, a 19th-century physicist, later proved definitely by refined experiments that a given amount of mechanical energy always produced the same amount of heat. This gave rise to a new concept of heat as a form of energy and to the development of the *kinetic-molecular theory.*

ENERGY OF MOTION

According to this theory, the heat that a body possesses is directly related to the kinetic energy, or energy of motion, of the molecules composing the body. The greater the kinetic energy involved, the hotter the body is.

This modular home uses a row of solar collectors to provide most of the home's heating requirements. In fact, solar-based heating systems could be used to cool buildings in summer.

LASL Photo by Johnnie Martinez

Heat, then, is energy of motion. How shall we define cold? Actually, there is no such thing from the scientific point of view. When we say that object A is colder than object B, we mean that it has less heat. The molecules of A are vibrating, if A is a solid, or moving about, if it is a liquid or gas, but they are vibrating or moving about less energetically than the molecules of B. Molecular motion stops altogether at the very low temperature called *absolute zero,* equivalent to $-273.16°$ Celsius.

Strictly speaking, then, all objects are hot, in the sense that they contain at least a certain amount of heat. To avoid confusion, however, we use such words as "cool," "cooler," and "cold" to indicate certain degrees of heat.

MEASURING HEAT INTENSITY

We must measure heat in order to acquire exact knowledge about it. Our senses cannot give us an accurate idea of how hot an object is, for they often mislead us.

In solids, the molecules attract one another strongly and their positions are fixed. Their motion, therefore, consists of vibration about a fixed point. The molecules of liquids can move about quite freely from one place to another, since they are not in fixed positions. They attract one another, but not quite so strongly as the molecules of solids. The molecules of liquids, because they can move, are able to assume the shape of the containing vessel, though the force of gravity keeps them from rising above a certain level in the container. In gases, the molecules are so far apart that the force of attraction between them is negligibly small. They move about even more freely than the molecules of liquids and occupy the entire volume of the container.

As solids or liquids or gases become hotter, they generally expand. As they become cooler, they generally contract. These effects provide an effective tool for measuring temperature—that is, the "degree of hotness" or "intensity of heat" of a body. The instruments designed to measure temperature are called *thermometers,* from the Greek *thermos,* meaning "heat," and *meter,* meaning "measure."

The famous early 17th century Italian scientist Galileo Galilei was among the first to devise such an instrument—a crude affair called a *thermoscope*. It consisted of a glass bulb provided with a long glass stem. The bulb was heated and the stem was dipped in water contained in another vessel. When the air in the bulb cooled, it contracted, and the water would rise up into the stem. The instrument could be used, then, to show changes in temperature. Galileo's thermoscope had two major disadvantages. It had no scale, and therefore could only indicate whether temperatures were rising or falling. Also it was sensitive to fluctuations in atmospheric pressure, which undoubtedly had something to do with the rise or fall of water in the stem. Somewhat later, water was used instead of air for the expansive medium, and the thermoscope was inverted. The water filled the bulb and a part of the stem.

Still later, alcohol was substituted for water, as it did not freeze so readily. The end of the stem was also closed to prevent the alcohol from evaporating. Another refinement was the use of mercury as the heat-measuring medium. Mercury is ideal for this purpose. It is opaque, has a low freezing point and high boiling point, and its volume changes uniformly with temperature. The ordinary mercury thermometer is made by fusing a glass bulb to the lower end of a narrow-bore glass tube. Mercury is then introduced into the tube through a funnel at the top end. The thermometer is boiled and annealed and the tube is sealed.

THERMOMETRIC SCALES

In determining any thermometric scale, so-called fixed points are selected. These are usually the freezing and boiling points of water at normal atmospheric pressure. The 18th-century Swedish astronomer Anders Celsius described a thermometer in which the value 100 was assigned to the boiling point of water and 0 to the freezing point. The range from 0 to 100 was divided into 100 parts, called degrees and indicated by the symbol °. It is not certain that Celsius developed this thermometer himself. Some authorities attribute it to the

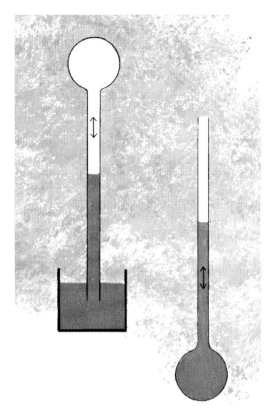

Left: the earliest form of Galileo's thermoscope. Air was the expanding and contracting medium. Right: a later version of the thermoscope, with water serving as the expanding and contracting medium.

18th-century biologist Carolus Linnaeus. It is called the centigrade ("hundred-degree") or Celsius thermometer, and is used extensively.

The German physicist Gabriel Daniel Fahrenheit, of the early eighteenth century, chose other fixed points. The lowest temperature he was able to attain with a mixture of ice and salt he called 0°. He gave the value 96° to the normal temperature of the human body. After his death the values of the freezing and boiling points were set at 32° and 212°, respectively. The Fahrenheit thermometer has been used widely in English-speaking countries. The simpler and more convenient Celsius thermometer has long been in common use in continental European nations and is now being adopted throughout the world.

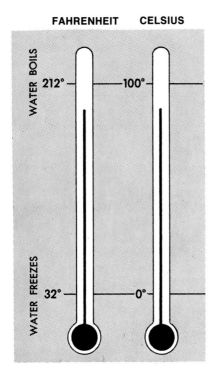

FAHRENHEIT CELSIUS

WATER BOILS — 212° — 100°

WATER FREEZES — 32° — 0°

A comparison of the Fahrenheit (left) and Celsius (right) thermometers.

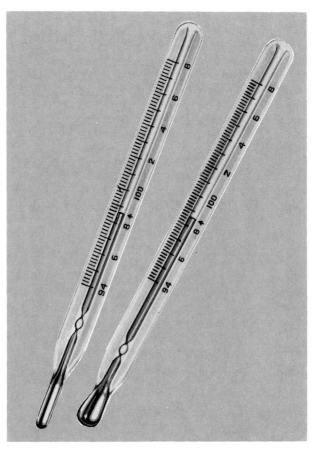

Clinical thermometers. In the clinical thermometer, a constriction in the tube near the bulb prevents the mercury above the constriction from returning to the bulb unless it is shaken down.

The measuring range of the common mercury thermometer is limited to the interval between the freezing point of mercury, −38.87° Celsius, and its boiling point, 356.9° Celsius. Higher temperatures may be read if the space above the mercury is filled with an inert gas such as nitrogen or helium. As the column of mercury rises, it compresses the gas and raises the boiling point of the mercury. The range of a thermometer may be extended downward by substituting for mercury a liquid with a lower freezing point.

SPECIAL THERMOMETERS

There are different kinds of thermometers. A widely known variety, the *clinical thermometer*, is used to determine body temperature. Normal body temperature is now known to be about 36.8° Celsius, or 98.2° Fahrenheit, for most people. One outstanding feature of the clinical thermometer is that the mercury column remains at the highest temperature reached. This is obviously most desirable, since otherwise the recorded temperature would change appreciably in the interval between removal from the body and reading.

The *differential thermometer* is used for very precise temperature measurement. This instrument may be adjusted by means of a small mercury reservoir so as to cover a given portion—say 5 degrees—of its total range, to an accuracy of 1/100 of a degree.

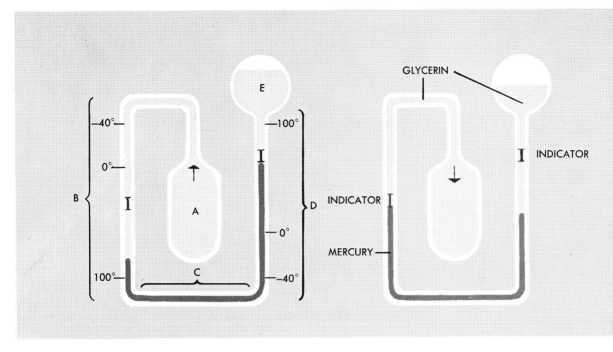

Diagram of a maximum-minimum thermometer.

Another interesting thermometer is the *maximum-minimum* type shown in the diagram on this page. Chamber A and part of section B of the tube are filled with glycerin or alcohol. The rest of section B, all of section C, and part of section D contain mercury. Above the mercury column in section D there is some more glycerin, which does not quite fill the top chamber, E. As the temperature rises, the glycerin in chamber A expands, pushing the mercury up section D of the tube. The rising mercury column pushes ahead of it a small steel indicator. Because it is provided with small springs, the indicator remains in place in the tube when the temperature becomes lower and the mercury column falls. With decreasing temperatures, the glycerin in chamber A contracts, and the mercury in section B of the tube rises. It pushes up a steel indicator set in section B, and thus indicates the minimum temperature. The indicators can be reset by means of a magnet.

To measure slight differences in temperature, a *thermocouple,* or *thermopile,* may be used. In this device, two different metals, such as iron and copper, are joined together in a circuit, forming two junction points. If one junction point is hotter than the other, an electromotive force is produced. This sends a current through the circuit. The electromotive force increases as the difference in temperature increases.

MEASURING THE QUANTITY OF HEAT

All of these devices measure the temperature, or the intensity of heat, but they are not a measure of the quantity of heat. Perhaps a simple example will illustrate the difference involved. Suppose you draw a ladleful from a vessel containing a quantity of hot water. The water in the vessel and in the ladle are at the same temperature but they contain different amounts of heat. If we immersed two identical specimens of iron, one in the ladle and the other in the vessel, the temperature of the iron in the vessel would be raised more than that of the iron in the ladle, because it would be subjected to more heat.

To measure the quantity of heat, in the

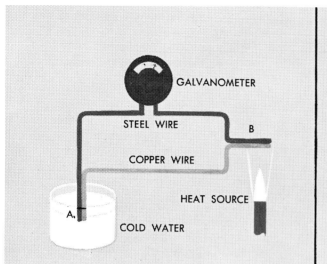

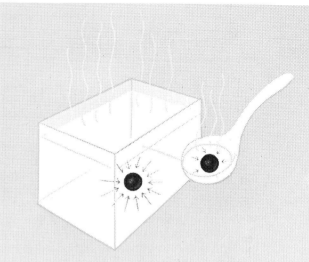

How a simple thermocouple works. Two different metals are joined together at junction points A and B. Since B is hotter than A, a current passes through the circuit, as is shown by the galvanometer. The greater the difference in temperature between A and B, the stronger the current will be.

A ladleful of hot water is drawn from the container at the left. Two identical specimens of iron are immersed, one in the container and the other in the ladle. The temperature of the iron in the container will be raised more than that of the iron in the ladle, since it is subjected to a greater quantity of heat.

metric system, we use the unit called the *gram-calorie*, or *calorie*. This represents the amount of heat required to raise the temperature of one gram of water 1° Celsius.

These definitions are accurate enough for most engineering work. It should be noted, however, that the amount of heat required to raise a unit mass of water through a given temperature change — say from 50° Celsius to 52° Celsius — will differ according to the initial temperature (50° Celsius in the above example). If a high degree of accuracy is required, the *mean calorie* is used. This is $1/100$ of the heat that is required to raise the temperature of a gram of water from 0° to 100° Celsius. The value in question is approximately equal to the amount of heat required to raise one gram from 15° Celsius to 16° Celsius.

Another heat unit — the *kilogram calorie* — is widely used in the field of nutrition. It represents the heat required to raise the temperature of one kilogram of water 1° Celsius. The calorie and the kilogram-calorie are sometimes referred to as the small and large calorie, respectively.

SPECIFIC HEAT

Not all materials absorb heat at the same rate when they are subjected to equal changes in temperature. For example, if equal weights of water, iron, mercury, and gold at the same initial temperature undergo an equal temperature rise, each will absorb a different quantity of heat. The unit called *specific heat* is used to indicate this heat-absorbing capacity. The specific heat of a substance is the number of calories needed to raise a gram of it through 1° Celsius. In the case of water, exactly 1 calorie is required. The specific heat of water, therefore, is 1. The specific heat of any substance other than water is less than 1. For pure iron it is 0.11; for mercury, 0.033; for gold, 0.0316.

To show how we use units such as these, let us see how much heat would be required to raise the temperature of 10 grams of copper through 50° Celsius. The specific heat of copper is 0.09; that is, 0.09 calories are required to raise the temperature of a gram of copper 1° Celsius. The amount of heat required to raise the tem-

perature of 10 grams of copper through 50°
Celsius, then would be $10 \times .09 \times 50$, or 45
calories.

The specific heat of a material is generally determined by heating a sample of the
material to a certain temperature and then
putting it in a quantity of water whose temperature is also known. We carefully note
the maximum temperature rise of the water.
Since the heat gained by the water is equivalent to the heat lost by the sample of material, we can calculate the specific heat of the
latter.

HEAT OF COMBUSTION

It is important to know the heat content of a fuel—that is, the heat made available when combustion, or burning, takes
place. This quantity is called the *heat of
combustion*. It is measured in heat units per
unit weight of the material—that is, in so
many calories per gram.

The heat of combustion of most solid
and liquid fuels is measured by means of a
combustion calorimeter, or *bomb calorimeter*. The substance that is to be tested is
put in a crucible mounted in a heavy steel
container called a bomb. This is lined with a
corrosion-resistant material such as nickel
or platinum. A small heating coil is connected to the substance that is being tested.
Oxygen under pressure is then admitted to
the bomb, which is then sealed and placed
in a container containing a known amount
of water. A current is passed through the
heat coil, igniting the sample. A thermometer measures the temperature rise of the
water in which the bomb is immersed.

The heat of combustion of gaseous
fuels is measured by means of a continuous-flow calorimeter. In this apparatus, gas
is burned at a constant rate in a chamber
surrounded by water which is circulating
uniformly. We can determine the heat of
combustion by noting the temperature of the
water as it enters and leaves the chamber.

EXPANSION AND CONTRACTION

We have already pointed out that most
solids expand when they are heated and
contract when they become cooler. When
railroad tracks are laid, a short space must

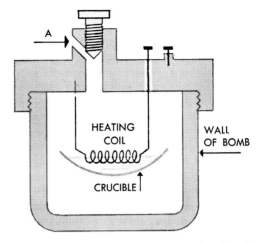

The bomb part of a combustion, or bomb, calorimeter. The substance to be tested is placed in the
crucible, through which a heating coil passes.
Some oxygen under pressure is admitted at A. The
bomb is then sealed off and placed in a container
holding a definite amount of water. When electricity is passed through the coil, the test material is
ignited. A thermometer measures temperature rise
in the outer container.

A continuous-flow calorimeter, which is used to
measure the quantity of heat that is produced by a
gas flame. Gas is burned at a constant rate in a
special chamber which is almost completely surrounded by water. This water is made to circulate at
a uniform rate. The heat of combustion can be calculated by recording the temperature of the water
as it enters the calorimeter and the temperature of
the water as it leaves the device.

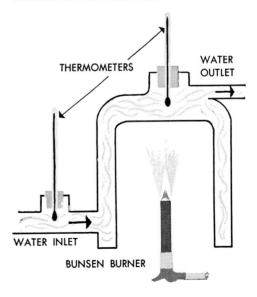

be left between adjoining rails to allow for expansion of the steel in summer. The rivets used to join structural steel members are a good example of the effects of cooling upon solids. The rivet is first heated and is then driven into place while still hot. When it becomes cooler, it contracts, drawing the two members together firmly.

The expansion of a solid may be expressed in terms of one, two, or all three of its dimensions. Thus we speak of linear (one-dimension), areal (two-dimension), and volume (three-dimension) expansion of solids. It is particularly important to know the linear expansion. Once we have this value, we can easily find out the areal and volume expansions.

We measure linear expansion in terms of the *coefficient of linear expansion*—the change in length of a unit length of a material for every unit change in temperature. This coefficient has been measured quite accurately for most materials. For example, if the temperature of a steel rod of a given length is raised 1° Celsius, the rod will increase in length by a factor of 0.000011. Likewise, the rod will decrease in length by that same factor if it is lowered 1° Celsius. Hence a steel rod 10 meters long, in a location where the temperature ranges from 0° to 35° Celsius during the year, will expand and contract over a range of $10 \times 0.000011 \times 35 = 0.00385$ meters, or about 4 millimeters, during the year. It is important to note that the amount of expansion depends on the length of the rod as well as on the temperature change.

Because different metals have different coefficients of expansion, one metal will expand more than another when it is subjected to the same degree of heat. An interesting illustration of this unequal expansion is the compound bar. It consists of two strips of different metals, such as copper and iron, riveted or spot-welded together so as to form a single bar. When such a bar is subjected to heat, one metal will expand more than the other, and the bar will bend or curl.

The compound bar finds numerous applications. The bimetallic thermometer commonly seen on kitchen ranges is made up of a compound bar wound into a spiral. The bar has copper on the inside and steel on the outside of the spiral. Copper has a much higher coefficient of expansion than steel. When the bar is heated, it tends to straighten out, and this causes an attached pointer to move over a scale. Compound bars are widely used as thermostats in temperature control.

Sometimes the expansion of solids as they are heated produces startling effects. It is well know that ordinary glassware will break when it is subjected rapidly to extreme temperature changes, as for example when one pours hot fat into a cold jar. This is because glass is a poor conductor of heat—another way of saying that heat is transmitted slowly through glass. The sides of the glass become unequally heated and consequently expand at a different rate. A strain is set up, and since ordinary glass is a brittle material, it will fracture. The glassware called Pyrex is subjected to a long cooling process called annealing, which results in a lower coefficient of expansion. This type of glass, therefore is less liable to break when there is a rapid drop or rise in temperature.

Liquids generally expand much more than solids when heated. They exert terrific pressures, therefore, if they are prevented

Heat is applied to a compound bar, made of copper and steel riveted together. The copper will expand more than the steel and the bar will bend.

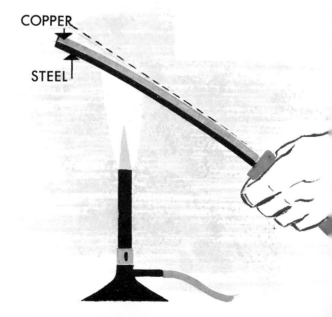

COPPER

STEEL

from expanding by being confined, say, in a container. In the case of liquids, we measure cubical rather than linear expansion. The coefficient of cubical expansion is the expansion per unit volume at 0° Celsius for a 1° Celsius rise in the temperature of a given liquid.

Water expands when heated and contracts when cooled like other liquids at temperatures above 4° Celsius. From 0° Celsius to 4° Celsius, however, it expands when it is cooled and contracts when it is heated. This means that water at 4° Celsius will expand and become lighter whether it is cooled or warmed.

IDEAL GAS LAWS

When gases are subjected to heat, they show greater and more regular expansion than solids and liquids. The French physicist Jacques Charles discovered in 1787 that all gases expand by the same amounts as they rise in temperature. He then worked out what is now called *Charles's law*. It states that a given volume of gas at 0° Celsius will expand approximately $1/273$ of its volume when heated 1° Celsius. It will contract an equal amount when cooled 1° Celsius. If we were to cool a gas to about −273° Celsius it would theoretically have no volume at all. −273° Celsius (more exactly, −273.16° Celsius) is, as you may remember, called absolute zero. Actually, a gas would turn into a liquid before this temperature would be reached. Charles's law holds only if the pressure of a gas remains constant during the temperature change.

What happens if the pressure varies while the temperature remains constant? This relationship was investigated by Robert Boyle and independently by Edmé Mariotte, two 17th-century scientists. The result of their researches was the formulation of the law that now bears Boyle's name. According to *Boyle's law*, the volume of a gas varies inversely as the pressure on it, provided the temperature remains constant.

If the volume of a gas remains constant, the pressure will increase $1/273$ of its original value for each degree rise in temperature. This relationship is some-times called *Gay-Lussac's law*, after the French chemist Joseph-Louis Gay-Lussac, who first formulated it.

The three laws described above apply exactly only in the case of what we call *ideal gases*. An ideal gas is one in which the space occupied by the molecules and the attractive forces between them are negligible. For most gases at comparatively low pressures and high temperatures, the three laws are accurate enough for ordinary purposes. There are cases, however, in which we have to take into account both the space occupied by molecules and the forces of attraction between them.

CHANGE OF STATE

An important effect of heat is *change of state*. This means that as a body is heated, it passes from one to another of the three states, or phases, in which matter can exist — solid, liquid, and gas.

What happens when a body passes from the solid to the liquid phase, as when ice is converted into water? We must recall that the molecules of a solid are closely packed together and exert considerable attractive force upon one another. That is why the original shape of the solid is maintained. Now, as heat is applied to the solid, the vibratory energy of the molecules is increased, and the individual molecules break loose from the bonds that formerly held them. What was formerly a solid has now become a liquid.

The cooling effect of ice is a consequence of the change of state from solid to liquid. The heat required for the melting of ice is absorbed from the surrounding objects, thus lowering their temperature. When ice cubes are added to a beverage, the cubes gradually melt as they absorb heat from the liquid. This causes the liquid to become cooler.

The temperature at which a solid changes to a liquid is called the *melting point* of the solid. The name *freezing point* is applied to the temperature at which a liquid solidifies. For a given crystalline material (one which forms crystals in the solid state), these two temperatures are the same. They are not identical in the case of certain substances, such as fats and glasses. Glass is not a true solid, but a supercooled liquid.

The melting points of different materials vary widely. Mercury, for example, melts at −38.87° Celsius; iron, at 1535° Celsius; tungsten, at 3370° Celsius. The low melting point of some alloys, such as Wood's metal, makes them valuable in fire-control systems. The sprinkler valves in such systems are held shut by plugs of the alloy. If the temperature in the room rises above 70° Celsius, the plug melts, causing the sprinklers to go into action. Wood's metal is often used in making trick spoons. A spoon of this sort will melt when it is used to stir a hot liquid.

The amount of heat required to melt a unit mass of a substance when it reaches the melting point is called the *heat of fusion*. This varies with different substances. Ice, for example, absorbs 79.71 calories for each gram of ice melted. Aluminum absorbs 94 calories; copper, 49 calories; lead, only 5.47. The heat energy absorbed at this stage does not show itself in a rise in temperature. Hence the heat of fusion is sometimes referred to as *latent* (hidden) *heat*.

If the temperature of a liquid is lowered, it will become a solid. Here too, the kinetic molecular theory furnishes an adequate explanation of what takes place. As the temperature of the liquid drops, the molecules of which it consists possess less energy. They move more sluggishly and they undergo, to a greater extent than before, the attraction of adjacent molecules. Finally, they are so strongly attracted by these molecules that they acquire fixed positions.

WATER—A SPECIAL CASE

Most liquids contract upon freezing. A notable exception is water, which expands considerably when freezing takes place. A cubic meter of liquid water will become 1.08 cubic meters of ice. Of course the ice will be less dense than liquid water. This phenomenon has certain fortunate effects from our viewpoint. Consider what happens when a body of water, such as a river or lake, is subjected to freezing tempera-

tures. When the surface freezes, the ice, being less dense than water, remains on top of the remaining liquid water. The body of water, then, becomes frozen from the top downward. Since the surface ice serves to insulate the water below, total freezing rarely takes place, except in the case of very shallow streams or pools.

If the ice were denser than water, it would sink to the bottom, leaving the top layer of water exposed to freezing temperature. Ultimately the whole body of water would be frozen solid. Most aquatic life in it would be destroyed. A long period of thaw would be required before the river or lake would be navigable.

Sometimes the expansive force produced by water as it freezes brings about certain undesirable effects. If the water in the pipes of our water-supply systems freezes, the expansion of the ice will cause the pipes to crack. This condition will first be revealed when thawing takes place. Water will then again flow through the pipes, which will leak badly. The freezing of water in the cooling system of cars may produce breaks in radiator tubes and may even crack the engine block. It is for this

When a liquid is heated, certain molecules (A,B) pass out of the liquid but are drawn back to it again because of the attraction of the molecules at the surface. Other high-energy molecules (C,D) escape from the attraction of the surface molecules, forming a gas.

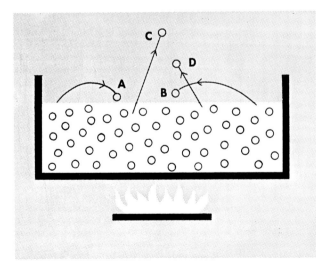

reason that antifreeze, with a lower freezing point than water, is used.

EVAPORATION

When a liquid is heated, a different sort of change takes place. At any specific temperature, a given liquid contains molecules possessing different amounts of energy. Some of the molecules will be energetic enough to pass through the boundary surface of the liquid, but the attraction of the surface molecules will draw the escaping molecules back again. Certain molecules, possessing even more energy, will pass to the outer air and will travel beyond the attractive influence of the molecules at the surface. The escaping molecules will constitute a gas, or vapor. The process of passing from the liquid to the gaseous state is called *evaporation*.

Evaporation goes on at all temperatures. It continues until the liquid disappears or until the space above the liquid becomes saturated. This condition occurs when as many molecules leave the liquid as return to it from the space above, thus bringing about a condition of equilibrium.

Since evaporation consists of the loss of high-energy molecules from the liquid, it is obvious that the average energy of the remaining molecules will decrease and the temperature will drop. This effect of evaporation has been known since ancient times and is still utilized in various tropical countries in order to cool drinking water.

Water evaporating from the skin helps maintain body temperature. When the weather is warm we perspire. As the perspiration evaporates, the skin is cooled. Evaporation is slowed in damp weather, since the concentration of water vapor in the air approaches the saturation point. That is why we feel cooler on dry days than on damp days, even though the temperature may be the same.

BOILING

As the temperature of a liquid is increased, it will reach a point at which the liquid will begin to boil. Boiling may be considered as evaporation taking place throughout the body of the liquid rather than just at the surface. Bubbles of vapor are formed in the interior of the liquid. They rise and then break through the surface.

Boiling can be brought about more readily either by increasing the temperature or lowering the air pressure. An increase in pressure will raise the *boiling point*. On the other hand, when the pressure is lessened, the boiling point is lowered. At high altitudes, the decrease in atmospheric pressure is such that certain cooking processes are slowed and others become practically impossible. Such difficulties can be eliminated through the use of a pressure cooker. This is a closed vessel in which pressure is built up by the steam that forms as water in the vessel is heated.

The reduced boiling point at low pressures finds considerable practical application in the field of vacuum evaporation (evaporation under low pressure). This process is of primary importance in the sugar industry. Boiling off the water from the syrup at normal atmospheric pressure would char the sugar. However, the pressure is kept so low in vacuum evaporation that the water may be removed at comparatively low temperatures.

The amount of heat required to change a unit mass of liquid to its vapor state at its boiling point is called the *heat of vaporization* of the substance. Each material has its own heat of vaporization. For example, it takes nearly 540 calories to vaporize a gram of water at 100° Celsius. Only 204 calories are required in the case of ethyl (grain) alcohol.

SUBLIMATION

In going from the solid to the gaseous state, most substances pass through the intermediate liquid state. Certain substances, however, go directly from solid to vapor form—a process known as *sublimation*. Naphthalene mothballs, iodine crystals, and the insecticide paradichlorobenzene are good examples of this phenomenon. Solid carbon dioxide also sublimes under ordinary conditions. It is often called "dry ice" because it is used to keep objects cold. It is excellent for this purpose, since it does not wet the substances it refrigerates.

HEAT TRANSMISSION

by John G. Albright

What we call *heat* is really the kinetic energy—energy of motion—of vibrating molecules. When a substance is cold, its molecules vibrate sluggishly. As the temperature is raised, the vibration becomes more and more rapid and the molecules possess more and more energy of motion. Greatly increased energy of motion—that is, heat—causes solids to become liquids and liquids to become gases.

A great deal of heat comes directly from the sun and accounts for the existence of life upon the earth. There are various other sources. Electrical energy may be changed into heat energy, as in the case of the electric toaster. Under ordinary conditions, gases give off heat when they are compressed and absorb it when they are expanded. Heat may be produced by friction. The mechanical energy used to overcome friction is transformed into heat energy. Chemical changes, such as the union of oxygen with carbon, also release heat energy.

One of the most striking things about heat is that it is constantly in motion. People spend much of their time trying to control the movement of heat so that they can get it into places where it is needed and keep it out of places where it is not wanted. They introduce heat into their dwellings in cold weather and seek to prevent its entry in hot weather. Winter clothes are designed to prevent the escape of heat from the body, summer clothes, to promote its flow from the body. Effective methods of transferring heat from one place to another in hundreds of industrial operations, such as the tempering of steel, the molding and curing of plastics and rubber, the processing of glass and pottery, and the preparation of chemicals, have been devised.

Heat may be transmitted in three different ways: by *conduction*, by *convection*, and by *radiation*.

CONDUCTION

When heat is applied to one part of a body, the temperature rises in that area. The molecules are made to vibrate more rapidly and, though they remain in their places, they bring about increased vibration in their neighbors. That is how heat is transmitted by conduction.

All substances conduct heat, some much more effectively than others. Copper, for example, is an excellent conductor, glass, a poor one. If a copper rod about the size of a pencil is held with one end in a flame, the other end will soon become too hot to handle. Heat has been transmitted from the heated end to the other one. If a glass rod of similar dimensions is held in a flame in the same way, the heat will pass

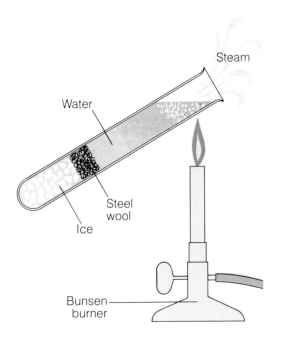

Experiment showing that water is a poor conductor of heat. The ice at the bottom of the tube has not melted though the water at the top has already begun boiling.

along the rod very slowly. It will be possible to hold the unheated end of the rod in the hand even though the other end is very hot. Very poor conductors, such as glass, are called *insulators*.

As a rule, solids conduct heat better than liquids, and liquids better than gases. Some solids, particularly metals, are fine conductors. Silver, copper, and aluminum are the best of all. On the other hand, many solids, such as wood, paper, and cork, are poor conductors because they have so many air spaces. Air is a good insulator.

We can show that liquids, such as water, are poor conductors of heat by a simple experiment. Place a small piece of ice at the bottom of a test tube and hold the ice down by means of a small lead weight or a wad of steel wool. Fill the tube about three quarters full of water, and hold it over a hot flame in such a way that only the upper portion of the water is heated. Because water is such a poor conductor, the ice will not be melted even after the water at the top of the tube is boiling. Water conducts heat only about $1/650$ as well as silver.

Air is a more effective insulator than water. In conducts heat only about $1/25$ as well. To insulate substances properly, however, the air must be kept motionless. If it is allowed to move, it carries heat by convection. The insulating properties of substances like sawdust, rock wool, felt, and cork are due largely to the fact that they are full of small spaces containing "dead" air. A woolen blanket or sweater provides warmth because of the air that is entrapped in its meshes. Air is also trapped among the long hairs of furs. That is why fur gives excellent protection against cold.

INSULATING HOUSES

There are many other instances of the insulating properties of air. Houses, for example, are often insulated by a double wall arrangement: an air space is left between the plastering and the outside wall or weather boarding. This sort of insulation is made more effective by placing some loose material, such as mineral wool, in the air space in order to stop any circulation of the air within the double wall.

Mineral wool is often placed between the joists in the attic floor or between the rafters of the roof to prevent the escape of heat. The effectiveness of such insulation is often strikingly shown. After a heavy snowfall, the roofs of some houses in a given neighborhood will remain covered with snow for weeks on end, while the snow on other roofs in the same neighborhood will melt rapidly. The houses on whose roofs

The solid red lines in the diagram indicate where insulation is placed in a typical frame house without an attic.

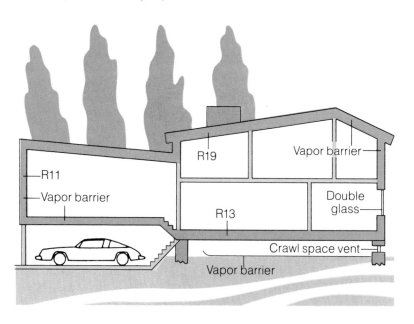

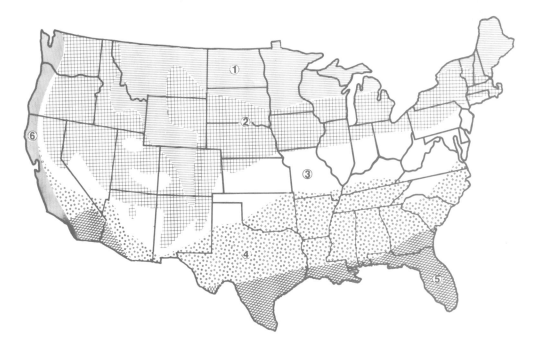

Specialized, climate-related standards for insulation have been developed. In the chart above, insulation needs progressively decrease from zone 1 to zone 6.

Insulating Values of Various Materials

Many kinds of materials are used as insulation. To give a clear understanding of the insulating quality of various materials, a rating system has been developed. The "R" value of a substance tells you about its resistance to heat flow. The higher the value of R, the less heat that can flow through the material, and the higher the material's insulating quality. The R values for several materials used together to provide insulation can be added to give the insulating quality of the combination.

Material	"R" per 2.5-cm thickness
Batt or Blanket	
Wood or cellulose fiber with paper backing and facing	4.00
Mineral wool (rock, slag, or glass)	3.80
Loose Fill	
Mineral wool (rock, slag, or glass)	3.80
Vermiculite, expanded	2.08
Board or Rigid	
Expanded urethane, foamed in place, sprayed or preformed	5.88
Polystyrene foam, extruded or expanded	4.50
Glass fiberboard	4.34
Construction Materials	
Wood fiberboard, laminated sheathing	2.90
Plywoods and softwoods	1.25
Plaster, stucco, brick	0.20
Doors	
Solid wood, 2.5 cm	1.56
Solid wood, 5 cm	2.33
Window (glass area only)	
Single glazing	0.88
Double glazing with 0.6 cm air space	1.64
Single glazing with storm window	1.89

the snow is melting are not properly insulated, at least in the upper attic area. The heat escaping through the roof melts the snow.

Storm windows set outside regular windows provide a nonconducting air space between two layers of glass. This cuts down loss of heat from the house in cold weather, and, as a result, the expense of heating the house is reduced. The windows of passenger trains often consist of double panes of glass with a thin layer of air between them to protect the passengers from extreme cold or extreme heat. Such glass is also manufactured for home use, as in sliding glass doors and picture windows.

The loss of heat from steam or hot-water pipes and boilers and water heaters is greatly reduced by covering them with a porous or fibrous material. The insulation is due in large part to the many air spaces in the material.

In some rural areas, where winters are very severe, ice from ponds or rivers is cut into large blocks and then packed in sawdust in a building or shed. The insulation provided by the sawdust preserves the ice for months. The ice will melt very slowly even in the hottest days of summer.

CONVECTION

What we call convection is really a combination of conduction through a fixed medium and convection proper, in which heat is transported by means of a moving medium. When flowing water, say, comes in contact with a hot surface, some of the fluid is heated by conduction and the heat is carried along with the moving fluid. When the warmed fluid later comes in contact with a cold surface, it will give up some of its heat to this surface, again through conduction.

Sometimes a stationary body of water or air can be set moving by being heated. Suppose there is a quantity of water in a rectangular tank. If heat is applied to the bottom of the tank at one end, the water at that end will become warm by conduction. It will expand in all directions and will become less dense than before. It will not be able to move along the bottom of the tank, since the cold water it encounters will be denser by comparison and will resist its passage. The cold water above the heated area will also be comparatively dense. But it will not offer the same resistance to the warm water, since the pressure of water decreases as the depth decreases. The warm water will rise, therefore, and will heat the area above it by convection. When the warm water comes to the surface, it will flow along the top until it reaches the other end of the tank. In the meantime, the cold water at the bottom of the tank has moved toward the warm end, where the water is warmer and less dense. In this way a complete circuit of moving water is set up. Circulatory movements of this kind are called *convection currents*.

Burning logs in a fireplace is a cozy sight, but it is an expensive way to heat even one room unless free firewood is available. Most of the heat goes up the chimney, and only a small amount warms the room (by radiation). Unless the house is very tight, convection currents set up by the fire pull cold air into the house.

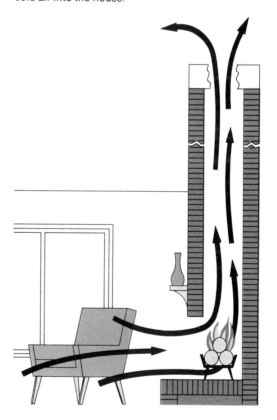

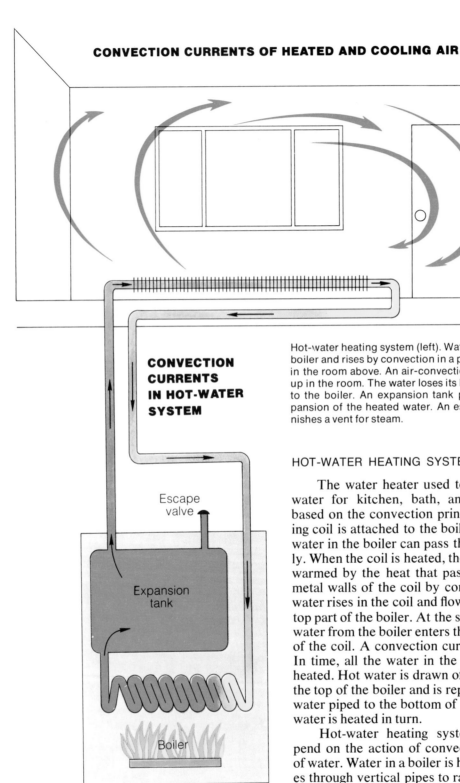

CONVECTION CURRENTS OF HEATED AND COOLING AIR

CONVECTION CURRENTS IN HOT-WATER SYSTEM

Escape valve

Expansion tank

Boiler

Hot-water heating system (left). Water is heated in a boiler and rises by convection in a pipe to a radiator in the room above. An air-convection current is set up in the room. The water loses its heat and returns to the boiler. An expansion tank provides for expansion of the heated water. An escape valve furnishes a vent for steam.

HOT-WATER HEATING SYSTEMS

The water heater used to provide hot water for kitchen, bath, and laundry is based on the convection principle. A heating coil is attached to the boiler so that the water in the boiler can pass through it freely. When the coil is heated, the water in it is warmed by the heat that passes along the metal walls of the coil by conduction. The water rises in the coil and flows out into the top part of the boiler. At the same time cold water from the boiler enters the bottom part of the coil. A convection current is set up. In time, all the water in the boiler will be heated. Hot water is drawn off for use from the top of the boiler and is replaced by cold water piped to the bottom of the tank. This water is heated in turn.

Hot-water heating systems also depend on the action of convection currents of water. Water in a boiler is heated and rises through vertical pipes to radiators in the different rooms of the house. Here it gives up its heat and then returns by other pipes to the base of the boiler. An expansion tank provides for the expansion of the heated water and also furnishes an escape vent for steam if the water boils.

HOT-AIR HEATING SYSTEMS

Convection currents are set up in air in much the same way as in water. However, the flow of air in such convection currents is more rapid than the flow in water. The reason for this difference between air and water is that air is much lighter and easier to move and expands more when it has been heated.

A convection current of air passes through the chimney of a house when logs are blazing away in the fireplace. Before the fire is started, the air inside the chimney has the same density as the air outside it. Once the fire is going, however, the air above it is warmed and expands. Made lighter as a consequence, it rises up the chimney. The cold air outside the fireplace makes its way into the chimney at the bottom, pushing the warmer air ahead of it. The rising hot air in a chimney is the ascending part of the convection current. The descending part of the current is in the air surrounding the chimney. The flow of the convection current can be reduced by partly closing the damper. In this way, the consumption of fuel can be regulated.

A tall chimney draws better than a short one—that is, a stronger convection current is set up within it. The taller the chimney, the greater the difference in weight between the column of gases in the chimney and the corresponding column of outside air, and the more effectively the cold air entering the chimney can force the hotter and lighter air upward.

When a room is heated by a steam radiator placed at one side of it, the air around the radiator is warmed first. It expands, rises, and spreads over the ceiling, while colder air from near the floor moves in around the radiator. This cold air is heated in turn and also rises. In this way convection currents are set up in the room, circulating and warming all the air.

In hot-air heating systems, the heat from the furnace is carried and distributed to the rooms of the house by convection currents of air. The air in the jacket surrounding the firebox of the furnace is heated. Rising through wide ducts, it enters the rooms through grilles, also called registers, set in the floor or at the base of the walls. Air is returned to the furnace for reheating by separate ducts. In modern homes heated by the hot-air method, electrically operated fans or blowers circulate the air through the rooms. The air is made to pass through filters made of fibrous material in order to remove dirt and dust.

Convection currents are common in nature. They account in large part for winds and ocean currents.

RADIATION

When you step out of a cabana at the beach and lie down upon a freshly spread blanket, fully exposed to the sun's rays, you

The radiometer is based on the radiation of heat. In a glass tube, from which almost all the air has been evacuated, there are four vanes of polished metal attached to a vertical pivot. One side of each vane has been blackened. When radiation falls on the vanes, the black sides of the vanes absorb more heat than the bright sides. The few molecules of air in contact with the black sides are heated and react violently. The blackened sides are repelled, and the vanes then spin rapidly about the pivot. The radiometer serves to indicate radiant heat. The more heat that falls on the device, the more rapidly do the vanes rotate.

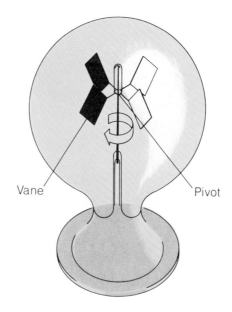

Vane   Pivot

feel an instant sensation of warmth. The heat that warms you is not transmitted by conduction, since the space between the sun and the earth's atmosphere is a more or less a complete vacuum. It is not transmitted by convection, since heated air would move upward and away from your body. In this instance, you are feeling the effects of *radiant heat*. Heat energy from the sun is being transmitted directly to your body without the aid of any intervening molecules. Radiant heat is emitted

Three types of radiant-heating units. In *A*, the unit is in the floor of the room; in *B*, it is in the wall; in *C*, in the ceiling. If the heating panels are in or near the floor, the air in the room will be gradually warmed by radiant heat and convection currents will then result. There will be practically no such currents if the heating panels are located in the ceiling.

A. Radiant heating unit in floor

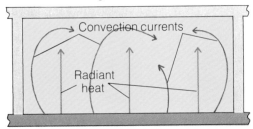

B. Radiant heating unit in wall

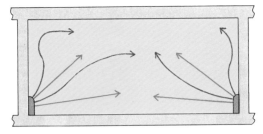

C. Radiant heating unit in ceiling

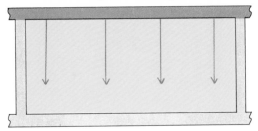

by all bodies. The higher the temperature of a body, the greater the amount of radiation that it will emit. Other kinds of radiation, such as light rays, may also be given off.

Radiant heat is closely akin to gamma rays, X rays, ultraviolet rays, visible light, infrared rays, and radio waves. These are all forms of electromagnetic radiation, traveling through space at the speed of light— 300,000 kilometers per second.

These electromagnetic radiations vary in wavelength. This variation is very great. X rays, for example, may be only about 0.00000001 centimeters in length. Radio waves may be several kilometers long. Some heat waves are longer than others. The higher the temperature of a radiating body, the shorter the waves.

Because of the differences in their wavelengths, electromagnetic waves behave quite differently when they encounter various substances in their path. Certain materials are transparent to light but are opaque to heat, and vice versa.

Glass, for example, is transparent to light, but opaque to heat. This fact brings about what is known as the *greenhouse effect*. The glass of a greenhouse, for example, transmits sunlight to the inside. The matter within absorbs the sunlight, becomes warmed, and gives off heat. Then by conduction and convection the air in the greenhouse becomes warmed. Because glass is opaque to heat rays, the glass does not transmit the heat, and the temperature of the interior builds up. The glass is warmed somewhat and loses some heat.

When radiant heat falls upon a body, some of it is reflected, some of it passes through, and some of it is absorbed. The body is warmed only by the part that is absorbed. If the body is smooth and highly polished, most of the heat will be reflected and little of it is absorbed. If the surface is dull and rough, a great deal of the heat will be absorbed and a little will be reflected. Color also plays an important part in the absorption or reflection of heat. Dark bodies absorb more heat than light-colored bodies and reflect less. Hence dark garments are best for cold weather and light garments for warm weather.

SOLAR-HEATING SYSTEMS

The panel heating of homes is an application of radiant heat. Hot-water pipes, or hot-air pipes, or electric wires are embedded in panels set in the floors or ceilings or walls of rooms. When the heating unit in the panel is operating, heat is radiated, warming all the objects that it strikes. If the panels are set in the floor, the air in the room will be gradually warmed and convection currents will arise. There will be no appreciable convection currents if the panels are in the ceiling.

Greenhouses and cold frames are based on the principle that the shorter waves of radiant heat penetrate substances more readily than do the longer waves. Since the sun is a very hot body, most of the radiant heat that comes from it to the earth is of very short wavelength. When this radiant heat falls upon the glass roof of the greenhouse, it easily passes through and is then absorbed by the soil and the plants in the greenhouse. The warmed interior of the structure then emits radiant heat. But since its temperature is so much lower than that of the sun, the waves of its radiant heat are too long to pass out through the glass roof, and so the heat is trapped within the greenhouse.

THERMOS BOTTLES

Another device that prevents radiant heat from escaping is the thermos bottle, or Dewar vessel. As a matter of fact, this ingenious device is designed to prevent loss of heat through conduction and convection as well as through radiation.

It is really a bottle within a bottle, the two being united only around the lip. The inner bottle contains the material whose temperature is to be kept constant. The loss of heat from the material by conduction is greatly reduced by providing no path along which the heat can flow to the outside except upward along the thin glass walls of the inner bottle and through the cork cover. Both the glass and the cork are poor conductors of heat.

Heat cannot be transferred from the inner bottle to the outer one by convection, since all the air has been removed from the space between the walls of the two bottles. This produces a near vacuum, and a vacuum is not able to transfer heat by convection as air does.

Finally, the transfer of heat from the inner bottle to the outer one is greatly reduced by silvering the outer surface of the inner bottle and the inner surface of the outer bottle. The bright, mirrorlike silver coating reduces the radiation of heat from the inner bottle. Any heat that penetrates the glass of the inner bottle is bound to be reflected from the silver-coated surface of the outer bottle.

Because of this threefold protection, the contents of a thermos bottle can be kept hot for twenty-four hours or longer. Thermos bottles also keep cold foods at a constant temperature. Heat is prevented from entering the inner bottle and raising the temperature of its contents.

The thermos bottle, or Dewar vessel. It is designed (1) to prevent loss of heat from liquids within its inner part and (2) to prevent heat from entering the inner part. The device keeps hot liquids hot and cold liquids cold for hours. It really consists of two glass bottles, one inside the other. The two are united around the lip. Air has been removed from the space between the bottles, leaving a vacuum. The outer surface of the inner bottle and the inner surface of the outer one have been silvered. The cork cap is a poor conductor of heat.

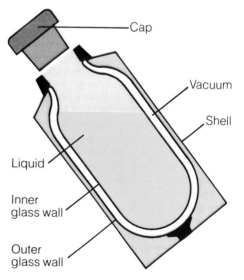

Crystals of the mineral quartz. In a solid such as this, the molecules are arranged in a definite order in relation to each other, occupying fixed positions in the crystal structure. The molecules have very limited movements. They are said to have a low degree of entropy.

Ward's Natural Science Establishment

ENTROPY

by Stanley W. Angrist

"This room is in a state of utter disorder!" How often have you heard your parents say this? Probably many times. And on such occasions your parents probably told you to do some work to help put things away in their proper places to make the room orderly. In working to create order in the room, you used up lots of energy.

A scientist is also interested in order and disorder. But, unlike a housekeeper, who may think of disorder in terms of an unmade bed or a table covered with dirty dishes, a scientist thinks of disorder in terms of the arrangement of atoms and molecules. The scientist calls this disorder *entropy*.

Entropy can be defined as the amount of disorder that occurs during a chemical or physical change, or the amount of disorder present in a system. The branch of science that deals with entropy and its changes is called *thermodynamics*.

THREE UNIFYING IDEAS

Thermodynamics is primarily concerned with the relationships between heat energy and work. Three fundamental laws form the basis for study in this area.

(1) Energy can be changed from one form to another, but it can never be created or destroyed.

(2) Natural events move in a predictable direction, toward a state of balance, or equilibrium.

(3) In a perfect crystalline solid at absolute-zero temperature, the entropy, or disorder, is zero.

In discussing entropy we are primarily interested in the second and third of these laws. Entropy is an abstract, or theoretical, quantity. It is used to predict the direction of natural events and to predict whether a given chemical or physical process is possible. If the process is found to be impossible, additional data may tell scientists what can be done to make the process possible. For example, is it possible to transform graphite into diamonds at a temperature of 25° Celsius and a pressure of 1 atmosphere, or approximately 1 kilogram per square centimeter? Entropy data, combined with certain other data, tell us that it is impossible to make diamonds from graphite under these conditions. Further calculations, however, indicate that such a transformation would theoretically be possible if the pressure were increased to 15,000 atmospheres.

Predictions of this sort are very valuable because they save us from trying to do something that is impossible. However, knowing that a process is theoretically possible does not mean we can carry out this process. Scientists endeavored for more than 100 years to make gem-quality diamonds from graphite. Only recently have they been successful—by using pressures approaching 70,500 kilograms per square centimeter and temperatures above 1,370° Celsius.

EQUILIBRIUM

Entropy data can also be used to calculate an *equilibrium constant*. This is the point at which there is no further change in the proportions of the various reactants in a process. For example, ammonia gas (NH_3) can be prepared from nitrogen gas (N_2) and hydrogen gas (H_2) by this reaction:

$$N_2 + 3H_2 \underset{P}{\overset{500°\,C}{\rightleftharpoons}} 2NH_3$$

(catalyst)

This reaction, as the double arrow indicates, does not go all the way to completion. No matter what the ratio of nitrogen to hydrogen is in the starting material, no matter how long the mixture is allowed to stand, not all of the hydrogen and nitrogen will be converted into ammonia. Some hydrogen and nitrogen will always remain. The equilibrium constant of this reaction tells us what the concentrations of nitrogen, hydrogen, and ammonia will be if the reaction is allowed to progress as far as it can— that is, until it reaches equilibrium.

Such information would be of great practical value if we wanted to manufacture ammonia by this process. The equilibrium constant would tell us, before we began, how much ammonia would be produced in each cycle of the operation and how much nitrogen and hydrogen would have to be recycled.

ORDER AND DISORDER

We mentioned earlier that a thermodynamicist thinks of disorder, or entropy, in terms of the arrangement of atoms and molecules within a substance. The way in which atoms and molecules are arranged with respect to one another will depend on whether the matter is in a solid, liquid, or gaseous state.

In crystalline solids, atoms and molecules are always arranged in an orderly manner. Look at the arrangement of the molecules in a crystal in Figure 1. Notice the regularity and order of the molecules in this crystal. An orderly arrangement such

Statue of Humpty-Dumpty. His fall, in the nursery rhyme, illustrates the irreversibility of many entropy processes. All the king's horses and men couldn't put him together again.

The Bettmann Archive

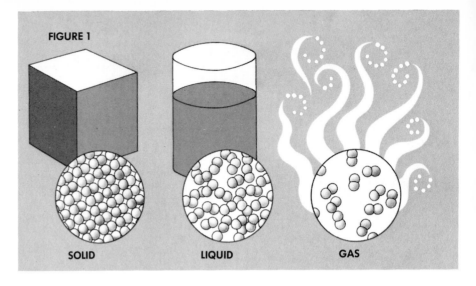

FIGURE 1

Fig. 1. The states of matter have different degrees of entropy. In the solid, entropy is least, because the molecules have a definite arrangement and little motion. Entropy is higher in the liquid, where molecules move about freely with no fixed order. The gas has the highest entropy, since its molecules move most freely, in a completely random manner.

SOLID   LIQUID   GAS

as this is typical of matter in the solid state. The molecules do have some freedom of motion, but, because they are so close together and are held in place by very strong forces, their freedom of motion is quite limited. The molecules can rotate about their own axes, and they can vibrate short distances about the midpoint of their tiny "cage" within the crystal lattice, but they cannot migrate throughout the bulk of the material. There is, in other words, a great deal of order. Or, to put it another way, there is little entropy in this system.

In liquids and gases, distances between molecules are greater and the forces between them are weaker than in solids. The molecules can migrate throughout the bulk of the material. There is more disorder in liquids and especially in gases than there is in solids. Hence, there is more entropy in these states of matter.

Figure 1 illustrates the molecular motion in solids, liquids, and gases. In all three of these states the atoms within each molecule may also move with respect to one another. This type of movement is illustrated in Figure 2. Such motion also produces disorder. In other words, it, too, increases the entropy of the material.

Different substances have different amounts of entropy. Hard substances, such as diamonds, have less entropy than do platinum and lead. This means the diamonds have a much more rigid, orderly structure than does platinum or lead. The diamond atoms and molecules are less able to move around. Of course, the entropy in all these—diamonds, platinum, and lead—is very little in comparison with the amount of entropy in carbon dioxide, water vapor, and other gases.

AN INCREASE IN ENTROPY

What do order and disorder have to do with the direction in which a natural event, such as a chemical reaction, will move? A chemical reaction tends to go in a direction in which the atoms, molecules, or ions involved become less ordered. That is, a reaction tends to go in a direction in which entropy increases. Why is this so?

Except at absolute zero, $-273.16°$ Celsius, atoms, molecules, and ions are always in motion. Because they are moving, they tend to become mixed up. Once mixed up, it is very unlikely that they will, by pure chance, fall back into an orderly arrangement rather than a disorderly arrangement.

Let us consider what happens, for

Fig. 2. Atoms in a molecule have entropy: they vibrate in their locations. Dark disks with white centers represent average positions of atoms. Lighter disks with arrows show their displaced positions and the directions of the vibrations.

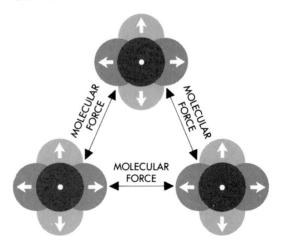

MOLECULAR FORCE

MOLECULAR FORCE

MOLECULAR FORCE

example, when a set of ten coins are tossed in sequence. If the original order, before tossing, was all heads or all tails, what is the chance that all heads or all tails will come up again? The chance is only 1 in 512. With 11 coins the chance is 1 in 1,024; with 12 coins it is only 1 in 2,048.

In the smallest piece of matter that can be seen through an optical microscope, there are about 1,000,000,000,000,000,000 atoms. If we were tossing this number of coins, what would be the chances for all heads or all tails?

Obviously, the chances of so many atoms forming an orderly arrangement are miniscule in comparison with the chances that they will form themselves into a disorderly arrangement.

Imagine an ice cube floating in a glass of water that is at room temperature. Imagine, too, that this is a *closed system*—that is, a system unaffected by anything around it (the table on which the glass sits, the air in the room, and so on). As the ice melts, it changes from an ordered, solid state to a disordered, liquid state. At the same time, its temperature increases, and the temperature of the water and the glass decreases. But calculations show that the entropy increase of the melting ice is greater than the entropy decrease of the water and the glass.

Does this mean that entropy is always increasing? Not necessarily. There is a tendency for entropy to increase, but there is also an opposing tendency: the tendency of matter to lose energy. For example, when a substance loses heat, its molecules move more slowly, and hence its entropy decreases. This occurs when supercooled water freezes. Supercooled water has a temperature of −1° Celsius. When it freezes, the water loses heat, and, because ice has a more ordered structure than does liquid water, the material undergoes a decrease in disorder, or, in other words, a decrease in entropy.

Events tend to approach a state of equilibrium. Equilibrium occurs in a material or a system when the tendency to an increase in entropy exactly counterbalances the tendency to an entropy decrease. At

The boiling of a liquid over a flame shows increase of entropy in the change of the liquid into a vapor. Heat increases the energy of the molecules until they rise out of the pot.

this point, no further change will take place in the material or system. This is what happens in the process for preparing ammonia gas that we discussed earlier.

TIME: A ONE-WAY STREET

An interesting implication of the concept of entropy and of the second law of thermodynamics is the indication that "time is a one-way street." That is, time moves in one direction only: toward the future.

This irreversibility is true of many natural events. You have seen an ice cube melt in a glass of water. Have you ever observed the opposite? Have you ever seen a glass and some of the water in it absorb energy from the rest of the water, thus producing a piece of ice in the glass? No, nor do you or anyone else ever expect to see this.

Similarly, if a steel bar is hot at one end and cold at the other, you expect the heat to flow from the hot end to the cold end until the temperature of the bar is uniform throughout.

Science is concerned with many variables. With two exceptions, all can either increase or decrease. Only entropy and time always increase. Entropy may decrease temporarily, but only in a localized area at the expense of an increase somewhere else. Like time, it is a one-way variable that marks the universe as being older today than it was yesterday. As the British astronomer Sir Arthur S. Eddington once said, entropy is "Time's Arrow."

Amano/Ziolo

A wonder of the modern world—electricity—powers another modern wonder—Tokyo's Ginza.

THE NATURE OF ELECTRICITY

Electricity is one of the great wonder-workers of our modern world—a force that is responsible for thousands of the inventions and appliances that make life in the twentieth century so pleasant. We may think of electricity as a fairly recent discovery. But people experimented with this force many centuries ago.

The word "electricity" comes from the Greek word *elektron,* which means "amber." Thales, a Greek philosopher, who lived about 600 B.C., noticed that a strange thing happened when he rubbed a piece of amber with a woolen cloth. The amber, as a result of the friction, acquired the property of attracting to itself lightweight objects, such as particles of dust, straws, feathers, and small pieces of lint. In other words, the amber became electrified. Thales, knowing nothing of electricity, thought that this property was peculiar to amber alone.

Today we know, however, that many kinds of substances can be electrified by rubbing. When electric charges are produced through the friction of one material upon another, these charges are called *static electricity*. The word "static" means stationary or at rest. The electricity that travels along wires is electricity in motion. It is called *current electricity*.

In our discussion of both static and current electricity, we shall, for convenience, often use the terms "production of electricity" and "creation of electricity." We must remember, however, that actually we cannot create a charge of static electricity or an electric current. We can only create conditions under which a charge of electricity can accumulate or under which a current of electricity can flow.

POSITIVE AND NEGATIVE CHARGES

While static electricity may be created by friction in various kinds of materials, the charges produced in this way are not always of the same kind. When you rub a glass rod with a piece of silk, you will have one kind of electrical charge. You will have a different kind of charge if you electrify a hard rubber rod by rubbing it with a piece of flannel or fur. We can prove this to be true by performing a simple experiment. If we hold a charged glass rod near a charged rubber rod, the two attract each other and move toward each other. In order to perform this experiment, we must suspend the rods so that they are free to move in space. If, on the other hand, we suspend two charged glass rods or two charged rubber rods near each other, they repel each other and move apart. It would seem, then, that there are two kinds of electricity, one kind on the charged glass rod and another kind on the charged rubber rod. The early students of electricity agreed to call the charge on the rubber rod *negative* and the one on the glass rod *positive*.

They knew that unlike electrical charges attract each other while like electrical charges repel each other. They knew, too, that most substances in their normal state are neutral—that is, that they carry neither a positive nor a negative charge. But they did not understand how a normally neutral body, such as a glass rod or a rubber rod, can become charged by friction. It was only when people came to have a more intimate knowledge of the structure of the atom that they began to understand the why and wherefore of such matters.

TINY PARTICLES

The atoms of which the molecules of

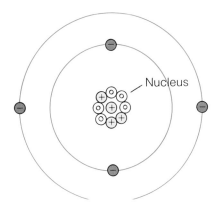

Simplified diagram of an electrically neutral beryllium atom, showing its negative and positive charges. In the nucleus there are four protons, each of which has a positive charge (+), and five neutrons, which have no electrical charge at all (O). Circling around the nucleus are four electrons, each of which has a negative charge (−).

matter are composed contain very tiny particles which are held together by electrical forces. The nucleus, which is the center of the atom, is composed mainly of particles called *protons* and *neutrons*. The proton carries a charge of positive electricity and the neutron is, as its name implies, without electrical charge of any kind. It is neutral. Revolving around the nucleus, somewhat as the earth and the other planets revolve around the sun, are minute particles called *electrons*. The electrons carry charges of negative electricity. The negative charge of the electrons in an atom ordinarily just equals or balances the positive charge of the atom's protons. In other words, the pos-

Each of these hard rubber combs possesses an excess of electrons. Therefore the two combs repel each other.

A hard rubber comb, which has an excess of electrons, is attracted to the glass rod, which has lost electrons.

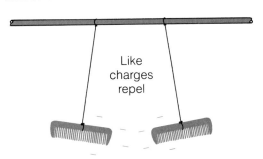

Like charges repel

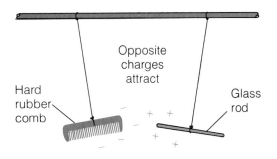

Opposite charges attract

Hard rubber comb

Glass rod

itive charge of the protons in the nucleus is neutralized by the negative charge of the electrons.

The outermost electrons are generally not firmly attached to the atom of which they are a part. The application of heat may in certain cases cause such violent motion of the atomic particles that some of the outer electrons are torn away from the body of the atom. In other cases the near presence of other bodies carrying large electric charges will tear some of the electrons loose. Thus an atom may lose electrons. It may also acquire electrons from another atom.

When an atom loses an electron, its electrical charges no longer balance or neutralize each other. It is left with an excess of positive charge. When an atom gains an extra electron, the atom has an excess of negative charge. When hard rubber is rubbed with a woolen cloth, electrons are actually removed from the cloth and added to the rubber. The electrons in the atoms of which the cloth is composed are attached less firmly to their atoms than are those in the rubber. As a result, the wool is left with an excess of positive charge, and the hard rubber has gained an excess of negative charge. Just the reverse happens in the case of the glass rod rubbed with silk.

Therefore, there are not two kinds of electricity as early experimenters believed. There is only one kind of electricity. The change in the charge upon an object is due to the excess or deficiency of electrons. When a body has an excess of electrons we say that it has a negative charge. When it has a deficiency of electrons, we say that it has a positive charge.

ELECTRON MOVEMENT

When electrons are deposited on certain materials, they can make their way freely in and out among the atoms. In certain other substances they cannot move freely. They remain more or less at the same point on the material at which they were deposited. Substances that allow electrons to pass freely through them are called *conductors*. They include such metals as silver and copper. We give the name of *insulator* to a substance that does not allow electrons to move through it freely. Hard rubber, mica, glass, and porcelain are all good insulators.

The electron theory explains the workings of static electricity. When you charge a rubber rod, you cause a great number of electrons to be piled upon it. Each electron repulses its neighbor, but since they are all on an insulator—the rubber rod—they cannot move apart. If you bring the rod close to your ear you will hear the crackling of electric sparks. What has happened is that the electrons on the rod nearest your ear

Static electricity has caused the strands of hair on this girl's head to repel one another. They have lost electrons and have been positively charged.

jumped across the space to your ear and rushed down from your body into the ground. Your body is a fairly good conductor of electrons. Once in the ground, electrons move far away from one another. This illustrates a very important point—namely, that electrons tend to move from where they are in excess to places where they can spread out. They flow particularly to places where electrons are lacking.

You yourself have probably had some experience with static electricity, whether you called it by that name or not. You may have noticed that when you were combing your hair with a hard rubber comb, the comb made a crackling sound. If the room was dark at the time, you may have seen the tiny sparks that accompanied the sound. Both the crackling and the sparks were caused by static electricity.

If you shuffle your feet on a thick rug on a cold day and then touch your finger to a friend's cheek, a spark will jump from your charged finger to his cheek and he will receive a slight electric shock. You can even give yourself a shock by touching a metal doorknob after rubbing your feet on the rug.

Drivers of gasoline trucks have to guard against static electricity. You may see metal-bearing straps hanging from the rear of such trucks and dragging on the ground. These straps are safety devices. They carry to the ground any sparks of static electricity that may be produced by the friction of the truck's wheels or by the flow of the gasoline into the tank. If the sparks could not escape into the ground along the strap, they might ignite the gasoline fumes and cause an explosion.

DETERMINING CHARGE

The kind and amount of electrical charge that is present on any body can be determined by means of an instrument known as an *electroscope*. One type of electroscope consists of a hollow cylinder of metal, with glass ends through which you can see what happens inside the cylinder. Through this cylinder, but separated from it by some insulating material, runs a vertical metal rod, the upper end of which is a knob

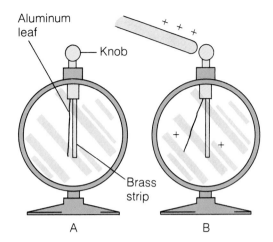

Two views of an electroscope, showing the effects of an electrical charge. In A, the instrument is uncharged and the aluminum leaf hangs limply beside the brass strip. In B, a positively charged glass rod touches the knob of the electroscope, positively charging the leaf and strip. They repel each other and the aluminum leaf swings out.

above the insulated top of the cylinder. From the bottom of the rod, inside the cylinder, are suspended a rigid brass strip and a light aluminum "leaf." The leaf is free to move. If the electroscope has no electrical charge, the leaf hangs down beside the brass strip.

If the knob of the electroscope is electrified by actual contact with a charged body, some of the charge is conducted to the brass strip and the aluminum leaf. When they receive such a charge, either positive or negative, the force of repulsion causes the leaf to swing out, since both the strip and the leaf have received the same kind of charge. The larger the charge, the wider the swing.

An electrical charge of one kind produces a charge of another kind on a nearby body when there is no direct contact. In this case, we say that *induction* has taken place. By using an electroscope and applying the principle of induction, we can find out whether the charge on a given body is positive or negative. First, we charge the electroscope by touching the knob with a body of a known charge—let us say a positive charge. For our experiment the aluminum leaf must swing out only slightly. We must therefore reduce the charge on the knob by grounding it. That is, we touch the knob with a finger. Some of the charge then flows through the finger and through the rest of the body to the ground and is dissipated. The brass strip and the aluminum leaf will remain slightly apart since they will continue to be repelled by the remaining charge on the knob. The charge will be positive, of course.

We are ready to use the electroscope to tell us the nature of the charge on the body that we are going to examine. If the body has a positive charge, a negative charge will be produced on the knob by induction. This will increase the positive charge on the brass strip and the aluminum leaf since more electrons will be attracted to the knob. Therefore the leaf will swing out more. If the body being examined has a negative charge, the positive charges on the electroscope will be attracted to the knob by induction. Less charge will remain on the brass strip and the aluminum leaf and the latter will swing closer to the strip.

STORING ELECTRICITY

The principle of induction is beautifully illustrated by the *condenser,* a device used to store up electricity. One of the earliest types of condensers was the Leyden jar, which was discovered by a Leyden University professor and, independently, by another scientist about the middle of the eighteenth century. The Leyden jar, which

Drawing showing the construction of a typical Leyden jar, a device that stores electricity.

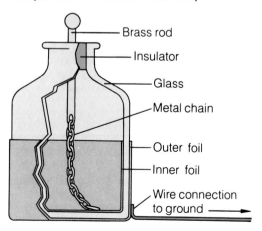

is now used mainly for experimental work and laboratory demonstration, is an extremely simple kind of condenser. It consists of a glass bottle partly coated inside and outside with aluminum foil. The glass of the bottle acts as an insulator between the two conductors—the plates of aluminum foil. The aluminum foil on the outside of the bottle is grounded by means of a wire connection. A metal rod passes through the insulated cover of the jar. There is a knob at the top of the rod. A metal chain hangs down from the bottom of the rod inside the jar and touches the inside aluminum foil.

To charge the jar, the inside aluminum foil is given a charge—let us say a negative charge—through the knob. Induction will now take place. The charge on the inside aluminum foil will bring about the opposite charge on the outside aluminum foil. The opposite charges continue to attract one another through the glass. Hence they will not escape from the jar. The Leyden jar is now charged. The glass in this case is called the *dielectric,* the name given to any insulating substance through which electrical attraction is maintained.

To discharge the jar, we touch its outside coating with the bare end of a piece of insulated wire and hold the other end of the wire near the knob of the jar. The electrons on the neutrally charged wire will be attracted to the outer tin foil coating, where there is a deficiency of electrons. Hence there will now be a deficiency of electrons in the wire. If we bring the other end of the wire near (but not touching) the knob of the jar, the electrons in the negatively charged inner foil will be attracted to the wire, where there is now a deficiency of electrons. They will jump the gap between the knob of the jar and the wire with such energy that a sizable spark will result.

LIGHTNING

Induction accounts for many natural phenomena. One of the best known of these is the thunderstorm. It generally occurs on days that have been hot and humid. It is caused by the rapid rising of the air. As the warm, damp air rushes upward, it is cooled

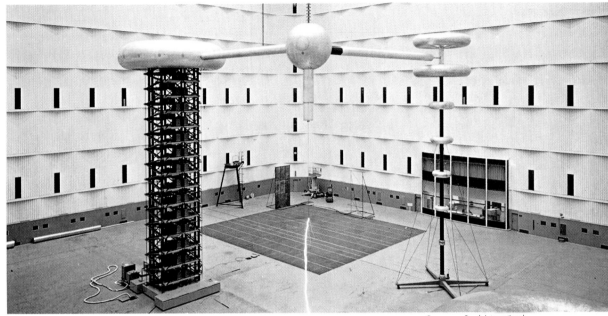

An electrostatic generator is used to study electricity. Here, an electric arc is created between a high tension wire and the ground.

by expansion and the water vapor in it begins to condense to droplets of liquid water. This condensation releases heat and warms the air further, giving added impetus to the rising air currents. The condensed moisture starts to fall.

As these drops of water fall through the mass of rapidly rising air, friction with the air tears some of the drops apart into spray or mist. In this process electrical charges are generated. The water drops falling to the lower levels of the cloud have a negative charge, the fine spray that rises to the upper levels, a positive one. A cloud may build up a positive charge or a negative one, or different sections of the same cloud may have different charges. Sometimes these clouds will accumulate charges until sparks jump from one cloud to another of opposite charge and the charges are neutralized. Such a spark or chain of sparks is called *lightning*. In other cases the charge on a cloud will induce charges of the opposite sign (positive $= +$ and negative $= -$) on objects below it on the earth. As the charge on the cloud increases, or as the distance from the cloud to the earth decreases, the

magnitude of the induced charges on an object below is increased until a spark jumps between the cloud and the earth to neutralize the charges. Such a spark discharge is also called lightning. The cloud and the earth are like the two conductors of a condenser and the air is like the dielectric — the insulator. When lightning occurs, the air proves to be not enough of an insulator to keep the two charges in question completely separated.

CHARGE ON THE SURFACE ONLY

A peculiar thing about the charge on a conducting body is that it resides on the outer surface only. A renowned English scientist, Michael Faraday, demonstrated this in an early experiment by placing a charge on a metallic pail. A small insulated metal ball touched to the outside of the pail was shown by the electroscope to become charged. But when it was lowered into the pail and touched to its inner surface, the ball received no charge. The principle has been demonstrated in a more spectacular way by placing a man in a metal cage and then charging the cage very strongly. To the man in the cage there was no evidence of such a charge, but it was necessary to discharge the cage before the man could safely leave it. Thus, the reason why the inside of an automobile is a safe place in an electrical storm is readily apparent.

So far we have been dealing with static electricity—charges at rest on a body. Almost all the very early experiments with electricity concerned static electricity. William Gilbert, who was physician to Queen Elizabeth of England, demonstrated that sulfur, resin, diamonds, glass, and various other materials could be charged by rubbing and that they would then attract lightweight objects. Otto von Guericke, who lived from 1602 to 1686, invented a very simple machine for producing static electricity. This device consisted of a large sulfur ball attached to a handle by means of which the ball could be made to rotate. As the ball rotated, von Guericke charged it with electricity by holding his hand against it. With his sulfur ball von Guericke discovered that certain kinds of objects were not attracted by the electrically charged sphere but were instead repelled by it. He also found that other objects could be charged

An oscilloscope. This instrument produces a visual display of an alternating voltage current on a fluorescent screen.

by induction when they were brought near the big ball. Sir Isaac Newton later used this same type of device but with a glass ball instead of a sulfur one. Many other experimenters produced static machines of one kind or another.

ELECTRICITY IN MOTION

In the eighteenth century scientists began to investigate the possibilities of current electricity—electricity in motion. Stephen Gray, an Englishman, and Charles du Fay, a Frenchman, showed that an electric current could be conducted along metal wires or lengths of hemp string. In 1752 Benjamin Franklin performed his famous kite experiment. He prepared a silk kite, with a pointed wire attached to the top of it. A long linen string, by which the kite was flown, led from the wire, and a short silk ribbon was tied to the end of the string. The linen string served as a conductor, and the silk ribbon as an insulator. A key was fastened to the string at the place where it was attached to the silk ribbon. When Franklin flew the kite during a thunderstorm, holding the silk ribbon in his hand, he drew sparks from the key by bringing his knuckle near it. Thus he proved that he could draw electricity from the clouds down the wet string to the key.

Franklin knew enough to stand under a shelter while he was making this experiment. If he had allowed the ribbon that he was holding to become wet, it would have become a conductor and he would probably have been electrocuted. It was still a quite dangerous experiment, however. A Russian scientist was killed attempting the same experiment shortly afterward.

It had become clear by that time that an electric current could be conveyed by means of a conductor from a source of electricity to some other place. Experimenters now began to investigate sources of supply for current electricity. Luigi Galvani, a professor of anatomy at Bologna University, in Italy, took what was probably the first step in this direction in the course of certain experiments on frogs. He was trying to find out how the nerves make the muscles of the body contract. He discovered that an elec-

These electric generators use strong magnets to produce electrical currents in large coils of wire. They are responsible for the power used in electrical appliances, heating, and lighting in our homes.

tric spark or a flash of lightning would make the legs of a freshly killed frog move violently. In another experiment Galvani found that when a brass hook attached to the frog's leg came in contact with another kind of metal, the leg would jump in just the same way. He came to the conclusion that the cause of this jumping was electricity in the nerve itself—"animal electricity," as he called it.

Alessandro Volta, professor of physics at the University of Pavia, Italy, was convinced that Galvani's conclusions were wrong, that the frog in his experiments had nothing to do with the production of electricity. "It is clear," he wrote in 1792, "that the cause of this electricity is in the metals themselves." To prove his point he set up an experiment in which he piled up silver and zinc disks with cloths between them that had been wet in salt water. He succeeded in obtaining electricity from this pile of metal disks, which came to be known as the voltaic pile. Obviously what was involved was not "animal electricity" but chemical action between the two metals.

Volta's discovery brought the knowledge of electricity to a distinctly new phase. Previously it had been possible to charge Leyden jars, from static-producing machines, and to discharge them, but it had been practically impossible to produce a current that could be made to flow for any appreciable length of time. In the voltaic pile, however, current continues to flow at a relatively steady rate for a considerable period. Since the current flow is dependent upon chemical action, it will continue to flow as long as this chemical action is forthcoming.

WHAT IS AN ELECTRIC CURRENT?

Just what is this mysterious electric current? To put it as simply as possible, it is

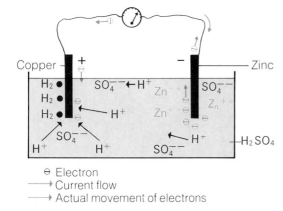

- ⊖ Electron
- → Current flow
- → Actual movement of electrons

A diagram of a voltaic cell. The two electrodes are set in an acid solution which acts upon the metals causing the copper electrode to lose electrons, becoming positively charged and the zinc to gain electrons, becoming negatively charged. Thus, an electrical current is produced in the wire connecting the two plates.

the flow of electrons along a conductor. In one type of modern voltaic cell there are two metal plates of different kinds of metal—zinc and copper, for instance—set in a solution, generally of hydrochloric acid. The two plates are known as *electrodes*. The solution is called the *electrolyte*. The hydrochloric acid in the voltaic cell acts upon the metals. The zinc plate becomes negatively charged, because it acquires an excess of electrons. The copper plate acquires a positive charge. It now has a deficiency of electrons.

We pointed out before that electrons stream from places where they are in excess to places where they are lacking. Therefore, when we connect the two plates by means of a wire, electrons flow in a steady stream from the zinc plate through the wire to the copper plate. An electric current has been produced. The path along which the current travels is called a *circuit*.

If one material, such as the zinc plate, has a greater concentration of electrons than another material, such as the copper plate, it is said to have a *higher potential*. In order to obtain an electric current—a stream of electrons—we must have two metals with different potentials. Since the difference in potential sends electrons moving along a circuit, it is called the *electromotive* (electron-moving) *force*.

PRODUCING AN ELECTRIC CURRENT

An electric current can be produced by chemical action or by a generator. In all cases, we are not really producing the current, but rather creating the conditions under which electrons can flow.

CHEMICAL ACTION

A common source of electric current by chemical action is the dry cell, which is used in devices such as the flashlight. The lead storage cell, or storage battery, also produces a current through chemical action.

Dry cell. The dry cell consists of a zinc, cuplike container, tightly sealed at the top. The zinc is the negative electrode. The

How a flashlight works. When the circuit is closed by means of the switch, current flows through the tiny lamp and it produces light. At the bottom is a simplified diagram of a dry cell, showing the various parts.

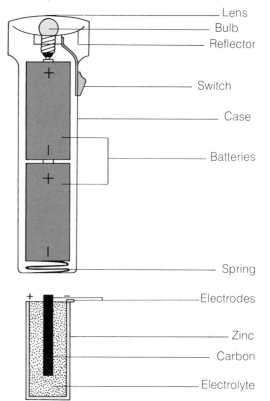

- Lens
- Bulb
- Reflector
- Switch
- Case
- Batteries
- Spring
- Electrodes
- Zinc
- Carbon
- Electrolyte

positive electrode is a vertical carbon rod set in the middle of the cup and extending beyond the top. The inside of the container is lined with blotting paper soaked in sal ammoniac (ammonium chloride). Between the carbon rod and the blotting paper is a pasty mixture of sal ammoniac, pulverized carbon, and manganese dioxide. The so-called dry cell is not really dry, but only comparatively dry. It is made up of two-fifths by weight of water.

Let us see what happens when we use a dry cell in a flashlight. Chemical action goes on in the cell whenever we close the circuit by pressing the switch. The current then flows through the circuit, which includes the fine wire filament in the small electric lamp. The flow of the electricity through the filament heats it to such an extent that visible light is produced. When the circuit is broken, the current ceases to flow. The chemical action is greatly reduced, though not entirely stopped. You have probably learned from experience that if you keep the batteries of your flashlight from one summer to the next, you cannot use them the second summer in most cases. Some slight chemical action has been going on and the chemicals have been used up.

Storage battery. Chemical action is also used to produce electricity in the *lead storage cell* or *storage battery*—the kind that is used to produce current in automobiles and portable radios. In this kind of electrical cell, there are a number of positive electrodes and an equal number of negative electrodes, separated from each other by partitions of wood or hard rubber. The electrolyte is sulfuric acid. When the plates are charged and ready to supply current, the positive electrodes are composed of lead peroxide, the negative electrodes of spongy lead. In some lead storage batteries, the lead peroxide or spongy lead is contained in holes that have been punched in the plates. In a lead storage cell, electrons are liberated much as they are in a dry cell. In the course of time both the lead peroxide and the spongy lead are converted into lead sulfate. Chemical action stops and there is no further flow of current. But it is not necessary to discard

the cell. It can be restored to its former condition by recharging it from some outer source of electric current. As current flows into the cell, lead peroxide forms on half the electrodes and spongy lead on the other half. In an automobile, the storage battery is being constantly recharged by a special generator attached to the motor as long as the motor itself is running. But this is only a special case.

The electric current produced by chemical action would not do for household circuits, where current may be required for many hours at a time day after day. It would not be strong enough for the powerful electric motors used by industry, nor could it produce the intense heat required by such a device as the electric furnace. To produce strong current or current that must flow constantly and over long periods of time, we use the device known as the electric generator.

ELECTRIC GENERATORS

It is almost impossible to exaggerate the importance of electric generators in our

Diagram of an alternate-current generator. The armature turns between the poles of the magnet. The ends of the armature are connected to the slip rings. Two brushes, connected to the external circuit, rub constantly against the slip rings. As the armature turns, its wire cuts the lines of magnetic force of the magnet. Therefore current is produced. The wire moves first up and then down between the poles. Electric current flows first in one direction in the external circuit and then in the other one. This shifting flow is called alternating current.

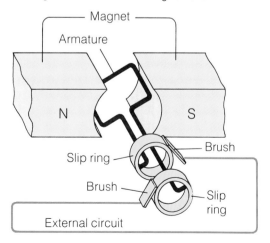

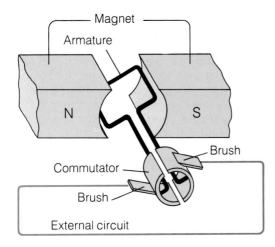

Magnet
Armature
N
S
Brush
Commutator
Brush
External circuit

A direct-current generator. The slip rings of the alternate-current generator have been replaced by a commutator—a ring split into two insulated segments. A brush rests on each segment. The armature generates alternating current when it turns between the poles of the magnet, as it does in the alternating-current generator. But the brushes are arranged so that each shifts from one segment of the ring to the other, as the current is reversed in the armature. As a result, the current keeps flowing in the same direction as in the external circuit.

modern civilization. Without these machines in our power plants, industry would come practically to a standstill. The lights would go out in our homes. Our subways and street cars would stop running. We could not operate our electric toasters, ironers, washing machines, and refrigerators. There would be no radio, no television, no moving pictures. Doctors would no longer be able to use the X ray.

The electric generator is based upon the property known as *electromagnetism.* It was Hans Christian Oersted, a Danish physicist, who first discovered that there is a relationship between electricity and magnetism. During one of his experiments he found that the needle of a compass, that happened to be lying near wires carrying an electric current, was deflected from the north. Oersted came to the conclusion that an electric current acts upon a compass much as the magnet does. In other words, it sets up an *electromagnetic field.*

Michael Faraday became interested in Oersted's work. He argued that if electricity could produce a magnetic effect, a mag-

net could produce an electrical effect. He proved that this is true and thus laid the foundation of the development of the modern electric generator.

We know now that if we pass a conductor, say a wire, through which no current is flowing, between the poles of a strong magnet, an electric current is induced in the conductor. This conductor must form part of a closed circuit. The wire must be moved in such a way that it cuts across the lines of magnetic force. When the wire is moved upward between the poles of the magnet, the electrons forming the current flow in one direction. When it is moved downward, the electrons flow in the other direction. A current that flows first one way along a conductor and then the other way is known as an *alternating current.* This differs from *direct current,* which flows in one direction only—the sort that we obtain from dry cells and storage batteries. Some electric generators produce alternating current; others produce direct current.

In a generator that produces alternating current there is at least one magnet (sometimes more) and there is an *armature*—that is, a wire conductor coiled around a core of iron. The armature is so mounted that it is free to turn between the poles of the magnet. The ends of the armature coil are connected to slip rings—metal rings that move with the armature. Two brushes of metal or carbon are so mounted that they rub constantly against the slip rings. The brushes are connected with the electric circuit through which the electricity produced in the generator flows.

To produce electricity, the armature of the generator rotates constantly between the magnetic fields. As it does so, its wire coils cut the lines of magnetic force of the magnet and current is produced. Since the wire in the coils first moves down between the magnets and then up, an alternating current is set up. The current will flow first one way and then the other.

POWER FOR ELECTRIC GENERATORS

Sometimes it is more efficient to move the magnetic fields rather than the armature, but the result is the same. Of course there must be some source of power to turn the armature or the magnets. This source may be a steam engine or gasoline engine. It may be the energy of flowing water, or it may be the wind that turns the blades of a windmill on a farm.

Alternating current is used widely for electrical heating and lightning purposes as well as for radios, television sets, motors, and a great many different kinds of electrical devices, such as toasters, clocks, irons, and washing machines. Alternating current is preferred for the long-range transmission of electric current.

In a direct-current generator a *commutator* replaces the two slip rings used in the alternating current generator. The commutator consists of a ring that has been split in two insulated halves, called segments. These segments move with the armature. One of the brushes rests on one of the segments, the other brush, on the other segment. As the armature turns, an alternating current is induced in the wire coil. The brushes are so arranged that each shifts from one segment of the ring to the other just as an electric current is reversed in the coil. As a result, the current keeps flowing in the same direction in the external circuit.

The direct current can be used for lighting, heating, and for running motors and electric appliances in general – in short, for most of the purposes for which we can also employ alternating current. Only direct current can be used for electroplating and for charging storage batteries. However, we can change alternating current into direct current by means of a device called a *rectifier*.

THE ELECTROMAGNET

Oersted's discovery that an electric current produces a magnetic effect led to the development of one of our most useful electrical devices – the *electromagnet*. In this, a coil of wire is wound around a core of soft iron. When an electric current is sent through the coil, the iron becomes magnetized. When the current is turned off, the iron loses practically all its magnetism. The electromagnet is used in electric motors, electric bells, telegraphs, telephones, magnetic hoists, and many other devices.

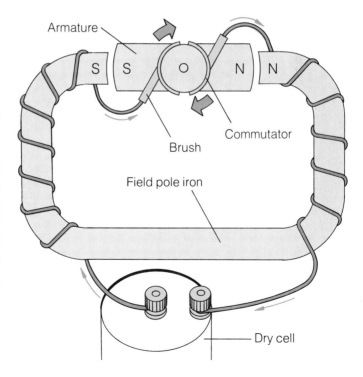

A simple electric motor. When current from a dry cell passes through the coils of the field pole iron, the latter becomes a magnet. As current passes through the brushes, the armature also becomes a magnet, whose south pole will be opposite the south pole of the field-pole iron. Repulsion between like poles will cause the armature to make a half turn. Each brush now makes contact with another segment of the commutator and, as a result, the current is reversed. So are the poles in the armature. The north pole of the armature is now opposite the north pole of the field pole iron. It will be repelled, and the armature will continue to turn.

Armature

S S O N N

Brush

Commutator

Field pole iron

Dry cell

It is very important to be able to measure the electric current that passes through a circuit or the amount of electric charge on a given body. Otherwise you may use too much or too little current. The companies that supply electricity must be able to measure it so that they will know how much to charge their customers. Now, just as there are units of measurements for liquids, solids, and distances, there are units of measurement for electricity. The most important of these are the coulomb, the ampere, the ohm, the volt, and the watt.

When a grocer sells bananas to a customer, he measures their quantity by counting them one by one. Obviously we could not apply this type of calculation to electrons, since we reckon them in the thousands of millions. There is a measure of quantity for electrons nevertheless. It is the *coulomb,* a unit equal to about 6,-280,000,000,000,000,000 electrons, or, as a physicist would put it, 6.28×10^{18}. The coulomb was named after the French physicist Charles Augustin de Coulomb.

Suppose 10 coulombs passed by a given spot on the conductor in just 5 seconds. We could then say that the rate of flow of the electrons in this case was 2 coulombs per second. Actually, instead of coulombs per second we use the word *ampere,* derived from the name of the French physicist André Marie Ampère. The rate of flow of the current described above then would be 2 amperes. The larger the current, the greater the amperage—the number of amperes—will be. Electrical appliances use varying amounts of current. An ordinary electric bulb may use only about half an ampere; a toaster, 5 amperes, a big motor, 75 amperes; an electric furnace, 10,000 amperes or more.

THE UNIT OF RESISTANCE—THE OHM

A conductor of electricity offers a certain amount of resistance to the passage of the electrons in an electric current. Some conductors are more resistant than others. The amount of resistance is measured by a unit known as the *ohm,* named after the

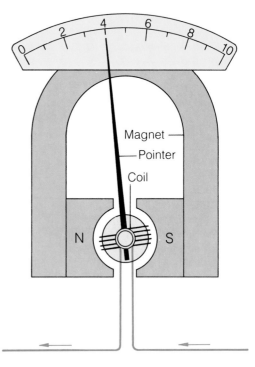

A galvanometer. Current passes through a coil, setting up a magnetic field, which becomes stronger as more current is provided. As the strength of the field increases, the coil will be affected by the magnetic field's permanent magnet, and the coil will be twisted around. The amount of twist is recorded on the dial by a pointer.

German physicist Georg Simon Ohm. By international agreement scientists have defined the ohm as the electrical resistance of a column of mercury 106.3 centimeters in length and 1 square millimeter in cross section at a temperature of 0° Celsius. Electric motors have very small resistances—from 2 to 5 ohms. On the other hand, the resistance of electric lamps is large. A lamp giving a fairly strong light may have a resistance of as much as 500 ohms. It is because of this large resistance that the temperature of the filament in the lamp becomes so high, and it is because the temperature is so high that the filament begins to glow.

Materials that offer no resistance to the flow of electrons in a conductor are called superconductors. Until recently these materials have become superconduc-

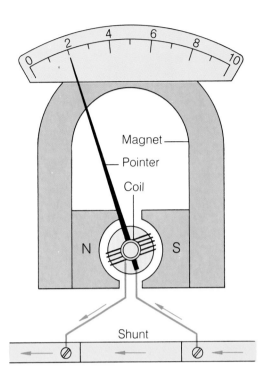

An ammeter. This instrument is really a galvanometer, in which the greater part of the current is passed through a shunt. The rest passes through the coil and is measured in terms of amperes by means of a pointer.

tors only at the temperature of liquid helium. In the mid 1980's new ceramic materials were discovered that become superconductors at the much higher temperature of liquid nitrogen. This breakthrough may lead to new uses of electricity. (See "Superconductors—Hot and Cold.")

FORCE AND POWER

As we have seen, we must have an electromotive force—a difference of potential—to drive electrons along a conductor to produce an electric current. We measure this force by the unit called the *volt* (from Alessandro Volta, the creator of the voltaic pile). One volt is the amount of electromotive force that is needed to drive an electric current of 1 ampere through a resistance of 1 ohm. The ordinary house current is usually either 110 or 220 volts. The voltage of the ordinary dry cell is 1.5 volts.

The *watt* measures the power of the electric current. One watt is the power produced by a current of 1 ampere under a pressure of 1 volt. When you buy an electric light bulb, you may notice that it is marked with a voltage number and the number of watts—indicated as W100 or W60 or W25 as the case may be. A 100-watt bulb will give you more light than a 60-watt bulb. It will also use up more electrical power.

MEASURING DEVICES

There are various devices, called *electric meters,* that measure currents in terms of the above units. These devices are based on the magnetic, chemical, and heat effects produced by electrons as they flow along a conductor. Most electric meters utilize the magnetic effects of the current.

A *galvanometer* is an instrument for measuring small amounts of electric current. In the galvanometer the electric current to be measured is passed through a small coil of wire. As we have seen, an electric current passing through a conductor sets up a magnetic field of its own. If the current is small, the magnetic field will not be very strong. If the current is larger, the magnetic field will be stronger.

The *ammeter* is a galvanometer that is used to measure larger currents. The pointer of the ammeter measures the current in terms of amperes. Because the passage of a large current through coils might damage the instrument, the larger part of the current—a known proportion—is directed through a wire called the *shunt* and thus bypasses the coil. The small percentage of the current that passes directly through the coils of the ammeter can be measured on the dial. Since this is known to be a certain proportion of the entire current, the total current can be obtained easily. Some ammeters have several shunts of different resistances, allowing different amounts of current to pass through them.

Certain types of ammeters measure, instead of the magnetic effect of the current, the heat produced by the electrons as they pass through. Still others make use of the chemical effects of the electric current, though this type is less common.

The measuring devices that we have spoken of so far measure current strength, usually in a wire. To measure difference of potential, or voltage, an instrument known as a *voltmeter* is used. The voltmeter is really a form of galvanometer, but with a very high resistance. Because of this high resistance only a small amount of current is able to flow through it. Voltage can be read directly from the dial of the voltmeter. The coil used in the voltmeter is much like that used in a galvanometer. A shunt is never used in this kind of instrument.

A *wattmeter,* as its name suggests, is an instrument that measures in watts the power passing through a circuit. It is the watt-hour meter that measures the amount of electrical power that we use in our homes. The watt-hour has a special kind of motor with an armature, or moving part, composed of rotating coils. The armature turns with a speed relative to the amount of electric energy flowing through the meter. The amount of electric power consumed is indicated on the meter's dial. Readings of the meter taken at the beginning and end of each month indicate the amount of electricity that has been used in that period. The readings are in *kilowatt-hours*. One kilo-watt-hour is the same as 1,000 watt-hours. A *watt-hour* is the electrical energy consumed in one hour's time at the constant rate of one watt.

TYPES OF CIRCUITS

As we have seen, current electricity flows through a circuit. A circuit is the current's path from the source of the electricity through the conductor, into the lamp or radio or special appliance and back again to the generator or battery that produced the electric power in the first place. A circuit is a completely closed conducting path, through which or along which the electrons of the current travel. There can be no break anywhere in the circuit or the electric current will cease to flow. When you turn off an electric lamp, you have pressed a switch that opens the circuit of which the lamp is a part, and the electricity ceases to flow. When you press the switch again, closing the circuit, the electrical energy once more lights up the lamp. The same things happens when you press the button that rings an electric bell. The pressing of the button closes the circuit and establishes the path the electrons must travel.

The three lamps shown below are set in series. If one of these bulbs burns out, the current flow is stopped.

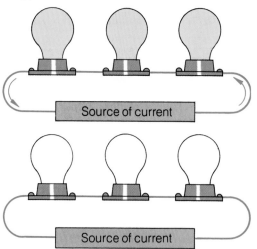

These lamps, set in parallel, are typical of a house circuit. If one burns out, the current is not interrupted.

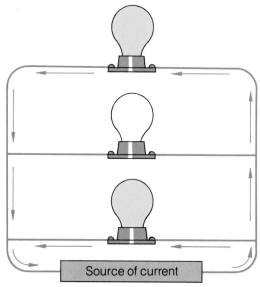

No matter how efficient a conductor may be, there are no conductors that do not offer some resistance to the electron stream flowing through them. For this reason we are likely to speak of conductors in a circuit as resistances. The conductors or resistances in a circuit may sometimes be connected in series and sometimes in parallel. The electric lights used as decoration — say, on a Christmas tree — are perhaps in series. A *series circuit* is one in which the current passes through the different resistances, one after the other, and then returns to the main circuit. In a series circuit, if the electron stream is halted or a break occurs at any place along the circuit, the current is stopped and all the lights go out. This, as you have probably noticed, is also true if one of the lights burns out, because the defective bulb provides a break in the complete circuit path.

A *parallel circuit* is one in which all the lamps or electrical appliances are supplied by separate resistances from the same circuit, side by side. By the use of such a circuit, the various electrical appliances in a home can be turned off or on one at a time.

CIRCUIT BREAKS OR OVERLOADS

When a wire breaks in a circuit or a connection is faulty, a leak of electricity is likely to occur. Since an electric current produces heat, such a leak may be very dangerous and may cause a fire. A break of this kind is known as a *short circuit*.

When electrical circuits become overloaded, too much heat is produced and fire may result. To avoid overloading, a device called a *fuse* is inserted in series with the circuit, usually in what is known as a *fuse box,* in which there are fuses for other circuits. The fuse is a piece of wire made of a bismuth alloy and is generally encased in a porecelain cover that screws into the fuse box. The fuse offers a great deal of resistance to electric current and melts at a fairly low temperature. When the circuit becomes overloaded, as it may if you use a number of appliances at the same time, the heat on the conductor melts the fuse, the circuit is broken, and the electric current ceases to flow. We then say that the fuse has blown out.

In powerhouses, *circuit breakers* take the place of fuses in circuits that carry heavy loads. A circuit breaker is a switch that opens automatically when too much current passes through the wires. Current flows through a wire coil wound about a movable iron core. If there is not too much current, the strength of the magnetic field around the coil is too weak to move the core. But when the current is too strong, the core moves and the circuit opens.

STEPPING CURRENT UP OR DOWN

Electric current can be stepped up or down in an electrical device known as a *transformer.* This yields stronger current

In the circuit breaker below, current flows through a wire coil that is wound around a movable iron core, making the core an electromagnet. If the current is not too strong, the magnetism of the coil is too weak to move the core. However, if there is an excessive flow of current in the circuit, the increased magnetism of the coil moves the core up against the catch. The spring pulls the switch back; as a result the circuit is broken.

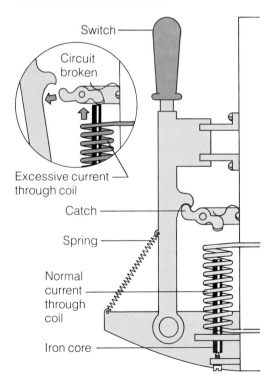

Switch

Circuit broken

Excessive current through coil

Catch

Spring

Normal current through coil

Iron core

and lower voltage, or weaker current and higher voltage. Sometimes it is necessary to have either one combination or the other. It is more convenient for an electric power company to send power at a high voltage and low current. However, it is safer and cheaper to use a low-voltage and high-current combination in electrical equipment that we use daily in the home.

Let us see how the transformer works. We know that the electric current passing through a conductor is surrounded by its own magnetic field. The strength of this field depends upon the power of the current. It can also be strengthened by coiling the conducting wire; the more coils there are, the stronger the magnetic field will be. When a second coil of wire, not connected directly with the first, is placed near the first one, a strange thing happens. When the electric current in the first coil moves in only one direction, nothing happens in the second coil even though it is exposed to the magnetic field of the first. When the electric current in the first coil changes direction, however, as it does in alternating current, its magnetic field undergoes a change also. The change in magnetic field induces an

electric current in the second coil. This current will continue to flow just as long as the current in the first coil constantly continues to change direction. The first coil, in which electric current flows from a source of alternating current, is called the primary. The second coil, in which current is induced, is known as the secondary. It is interesting to note that induced current in the secondary coil always flows in the direction opposite to that in the primary. The magnetic field in the secondary is also opposite to that in the primary.

Now in a transformer, which can be used only with alternating current, both the primary and the secondary coils of wire, which have been thoroughly insulated, are wound around a solid core of iron. The iron core carries the magnetic field back and forth from the primary to the secondary, and in this way it induces an alternating electric current in the secondary coil. The amount of current in the latter depends upon the strength of the current in the primary coil and also on the number of turns in the coils of wire. If there are more turns on the secondary coil than on the primary coil, the voltage is stepped up (that is, increased), and the current is decreased proportionately. On the other hand, if there are more turns on the primary coil than on the secondary coil, the voltage is stepped down (decreased), and the current is increased proportionately.

Without the transformer, we should not be able to use many of our electrical devices. An electric doorbell, for instance, uses only a fraction of the full 110 volts that are usually supplied to homes. A transformer is employed to step down the voltage from 110 volts to between 5 and 15 volts. Electricity is generated at very high voltages in powerhouses. This means that the electric current can be sent along the wires to distant locations at high voltage (sometimes exceeding 200,000 volts) and low current. When it reaches its destination, the voltage is stepped down by a transformer. To send electricity at high current and low voltage would mean that the wires would be overheated and much of the electrical energy would be lost.

Below is a diagram of a simple transformer. Alternating current, in the transformer, is conducted through the primary coil. Current is induced in the secondary coil.

Core

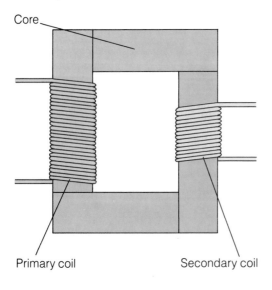

Primary coil   Secondary coil

MAGNETISM

by John A. Fleming

Everybody is familiar with the toy magnet, that mysterious little U-shaped device that picks up needles or pins and that holds them indefinitely, through what seems to be sheer magic. But the magnet is far from being a mere toy. It is an essential part of a great many machines and tools and measuring devices, without which the world's work could not be done. A magnetized needle set within a compass helps the navigator to keep to his course at sea. When you hold a phone receiver to your ear, a magnet records the vibrations set up by the voice of the person talking into the mouthpiece. The electric motor, the electric generator, and the ammeter could not possibly work without their built-in magnets.

KNOWN SINCE ANTIQUITY

Magnetism, the natural force that causes magnets to function as they do, revealed itself to men many centuries ago. A number of persons in antiquity knew that the black metallic iron ore called magnetite, or loadstone, had the property of drawing particles of iron to it. The Greek philosopher Thales, who lived during the sixth century B.C., is said to have been the first to call attention to this property, but it may have been known long before. After Thales' time the loadstone was often mentioned in ancient writings. It was sometimes given the name of "magnet," from Magnesia, a district in Asia Minor where large magnetic deposits were to be found.

Socrates remarked, "The stone that Euripides calls a magnet . . . not only attracts iron rings but also imparts to them similar power of attracting other rings, suspended from one another, so as to form quite a long chain, and all of these derive their powers from the original stone." The Roman Lucretius, who lived in the first century B.C., tried to explain magnetism in terms of his atomic theory.

There are many legendary accounts of the properties of magnets. The *Arabian Nights* contains the story of a ship that approached an island made of magnetic rock. The ship fell completely to pieces because all the iron nails were pulled out of it through the attraction of the rock. Other ships, according to legend, avoided a similar fate by substituting wooden pegs for iron nails. There are stories, too, of huge statues of iron and bronze held in mid-air through the force exercised by magnetic domes.

Loadstones attracting iron objects, as shown in a French treatise on the loadstone, published in Amsterdam in 1687. The properties of the loadstone were known long before that time.

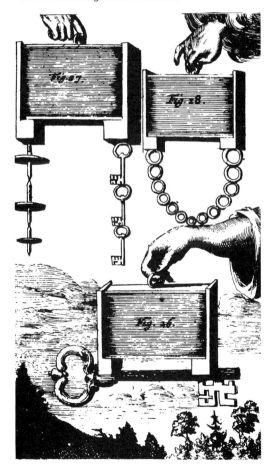

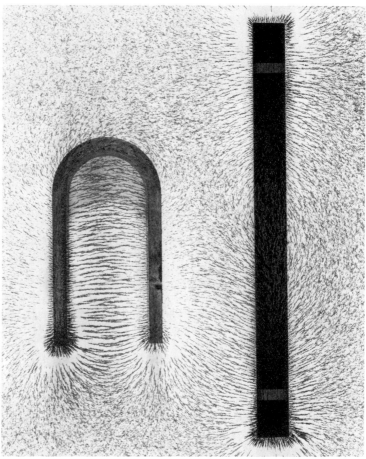

Fig. 1. A bar magnet and a horseshoe magnet, which is really a bar magnet bent into a U-shape, arranged near each other. Fine iron filings sprinkled near them have aligned with the lines of magnetic force in each magnet's field. In each the magnetism seems to be concentrated in two regions, or poles, near the ends—the regions where the filings have clustered.

Another tale gives a fanciful account of the origin of the word "magnet." It seems, according to this tale, that one day a shepherd called Magnes was tending his flock on the slopes of Mount Ida, in Asia Minor. Suddenly he noticed that the iron tip of his staff was being pulled toward the ground. He dug up the earth in the vicinity and found that his staff was being attracted by the loadstone in which the place abounded. Thereafter the loadstone was called a magnet in honor of the shepherd who had discovered it. Scholars have pointed out that this story originated long after the word "magnet" was in common use.

NO LONGER LEGEND

In the course of the centuries much of the mystery that once surrounded magnetism has been dispelled. In the place of out-and-out legends and pseudoscientific speculation, we now have at our disposal a respectable amount of scientifically proved fact. Furthermore, we have been able to put the force of magnetism to work for us in a great many different ways.

Today that natural magnet, the loadstone, no longer figures prominently in the study of magnetism or in its many applications, for practically all magnets nowadays are artificial. It is easy to make such a magnet out of a steel object, such as a needle, if you have a permanent magnet. Simply draw one end of the magnet along the needle, stroking in one direction only. The needle will become an artificial magnet and will have the property of drawing various particles to it. The original magnet has lost none of its strength. It will be capable of magnetizing any number of other steel needles.

Suppose, now, that we break our magnetized needle in two: each of the fragments will be a magnet.

Until about 1820 artificial magnets were made by stroking bars of steel with a loadstone or with an artificial magnet. But then a Danish scientist, Hans Christian Oersted, revealed that a magnetic field can be produced by sending an electric current through a coil of wire, called a *solenoid*. If this wire is wound around a core of steel, the core will be permanently magnetized when current is passed through the coil. Nowadays the manufacture of artificial magnets is based on this principle.

MAGNETIC FIELD

Every magnet, natural or artificial, produces a magnetic field in the space around it. We can map out this field in various ways. If we place a magnet under a piece of paper whose surface is covered with iron filings, the filings will arrange themselves in lines of force like those shown in Figure 1. We can also use a small compass to mark out the magnetic field about a magnet. If we put the compass in different places in the vicinity of the magnet, it will assume the direction of the lines of force at any given point (Figure 2).

In any magnet there are regions near the ends of the long axis where the attractive forces are greatest. These regions are called the *poles* of the magnet. The line joining the two poles is the magnetic axis. If a straight, or bar, magnet is suspended so as to move freely in a horizontal plane, the magnetic axis will assume, roughly, a north-south direction.

The pole of a magnet that is directed northward is known as the north-seeking pole, or *north pole,* or *positive* (+) *pole.* The pole that is directed southward is called the south-seeking pole, or the *south pole* or the *negative* (−) *pole.* Similar, or like, poles repel each other. Dissimilar, or unlike, poles attract each other.

When a bar magnet is bent double, it forms what is known as a horseshoe magnet. It will attract a given substance more powerfully than a bar magnet of corresponding size. The reason is that since the

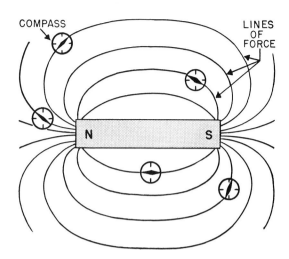

Fig. 2. The magnetic field about a magnet may be marked out by a compass, which will assume the direction of the lines of force at any point.

two poles are close together, the lines of force are crowded in a comparatively small space and exert a greater effect upon their surroundings.

Magnets lose their strength when heated, but they regain it when they become cool again. However, if they reach a tem-

Fig. 3. The diamagnetic substance takes a position across the field of force.

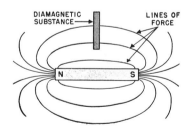

Fig. 4. The paramagnetic substance aligns itself along the field of force.

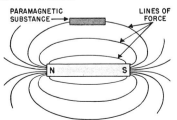

perature that is known as the *Curie point,* they become entirely demagnetized. The Curie point for iron is about 750° Celsius. It is different for other magnetic materials. In general, magnets tend to lose their strength as they grow older. To provide longer life, they are aged. That is, they undergo various treatments by heat, shock, and repeated magnetizations and demagnetizations in weak alternating fields.

VARYING PENETRATING ABILITY

Magnetic lines of force cannot penetrate certain materials as easily as they can penetrate air or a vacuum. Such materials are called *diamagnetic* ("magnetic across," in Greek) because rods made of these substances tend to take up a position across the field of force of a strong magnet (Figure 3). Bismuth, antimony, and most of the other chemical elements and their compounds are diamagnetic.

Magnetic lines of force can penetrate other materials more easily than they can penetrate air or a vacuum. Such materials are called *paramagnetic* ("magnetic alongside," in Greek) because rods made of them tend to line up alongside the field of force of a strong magnet (Figure 4). Iron and liquid oxygen are paramagnetic substances.

A few paramagnetic materials are so much more easily magnetized than the rest that they are generally put into a separate class, called *ferromagnetic.* The ferromagnetic materials derive their name from the fact that one of them is iron (*ferrum,* in Latin). Nickel and cobalt are also ferromagnetic. So are various alloys of these metals and a few other materials. Sometimes ferromagnetic substances are called magnetic materials, while all the rest are lumped together as non-magnetic materials.

PERMANENT MAGNETS

Magnets that retain their strength for long periods of time are called *permanent magnets.* They have a wide range of uses in industry and in research. They serve in various measuring devices, such as ammeters, voltmeters, galvanometers, cardiograph recorders, seismographs, magnetic compasses, magnetometers, and so on. They form an essential part of many kinds of scientific equipment. Everyday devices such as traffic signals are also controlled by permanent magnets. They are used in certain kinds of lathes, chucks, conveyors, hand tools, and separators, and also in various toys and novelties.

Improvements in permanent magnets are constantly being made. New alloys with superior magnetic properties have been developed. Some of these alloys are called *quench-hardening* steels. They are first heated to a very high temperature and then hardened by rapid cooling, after being quenched in water or oil. Quench-hardening steels include alloys of steel with carbon-manganese, tungsten, chromium, cobalt, and cobalt-chromium. They all contain carbon, since steel is basically an alloy of iron and carbon.

Other alloys are carbon-free. They include alloys of aluminum-nickel-iron, aluminum-nickel-cobalt-iron, aluminum-nickel-titanium-iron, and cobalt-molybdenum. All these substances require special casting, grinding, magnetizing, and aging.

ELECTROMAGNETS

Not all magnets are permanent. If electric current is passed through wire that is wound around a soft-iron core, the core will be magnetized, but only as long as the current is turned on. If the current is shut off, the iron will lose practically all its magnetism. If the current is increased, the iron will have greater magnetic strength. This temporary sort of magnet—a core of soft iron set within a coil of wire through which an electric current is passed—is called an *electromagnet.*

The electromagnet is used in a great variety of modern electrical devices. It is the heart of the electric motor, and of such devices as the telegraph, the telephone, and the electric bell. It serves to separate iron from its ores. Big electromagnets are frequently employed to load and unload iron and steel materials in railroad yards, steel plants, and junk yards. Doctors sometimes use electromagnets to remove bits of iron or steel that have become imbedded in the eye.

Powerful industrial magnets can lift extremely heavy materials—including some three metric tons of scrap motors.

THE EARTH IS A MAGNET

We pointed out before that a magnet that is free to swing in a horizontal plane adopts, approximately, a north-south position. The reason is that it is attracted by magnetic forces arising within the earth itself. For the earth is a great natural magnet, with a North and South Magnetic Pole, a magnetic axis, and a field of force that extends far out into space. Even 6,500 kilometers above the surface of the earth, the magnetic intensity of this field is still one-eighth as great as it is at the surface. The lines of force of the earth's field are parallel to the surface near the equator, but as they approach the two magnetic poles, they bend and converge.

Certain substances within the earth—its magnetic deposits, for example—are highly magnetic, but that is certainly not true of most of the materials that make up our planet. As a matter of fact, a magnet of steel would be something like 10,000 times as powerful as one just as large made of typical earth-stuff. But in view of the earth's size, its total magnetism is awe-inspiring.

Naturally, if the earth is a huge magnet, other magnets will be influenced by it. The north-seeking pole of a bar magnet, moving freely in a vertical axis, will evidently be attracted to the earth's North Magnetic Pole. It is true that the North Magnetic Pole does not correspond exactly to the geographical North Pole, as we shall see, but if we make certain calculations, we can determine in what direction true north lies. Of course we can then find the other points of the compass. This is the principle of that supremely useful instrument, the magnetic compass.

THE SOUTH-POINTING CHARIOTS OF THE CHINESE

The Chinese have been credited with the invention of the compass. It is said that the Emperor Hwang-ti, who lived twenty-five centuries before the birth of Christ, built a chariot on which a dummy, mounted on a pivot, always indicated south. By using this chariot at a time when a thick fog had closed in upon his army during a battle, Hwang-ti was able to defeat the foe. This account is almost certainly mythical. Unfortunately it is difficult to see how much truth there is in later Chinese accounts of south-pointing chariots.

A Chinese document of the third century A.D. contains a statement that a needle enables a ship to follow a southward course. But it is not until the twelfth century that we find in any Chinese work a detailed description of the manner in which a needle is made to point to the south. The invention of the compass has also been attributed to other peoples, including the Arabs, the Greeks, and the Etruscans, but there has been no authentic proof of such claims.

The earliest definite mention in Euro-

Fig. 5. The compass needle that is shown below points in a direction that is 18° west of true north. Hence the declination of the compass is 18° west.

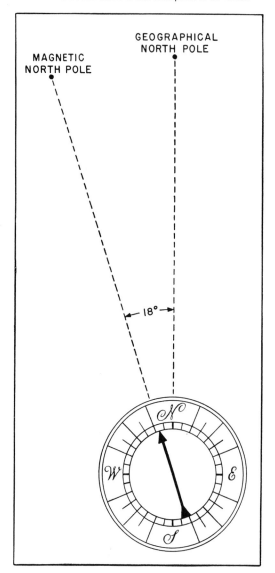

pean literature of the directive property of the magnet and its use in navigation appears in two Latin treatises by the Englishman Alexander Neckam. These treatises, entitled *Of Instruments* and *Of the Nature of Things,* were written toward the end of the twelfth century A.D. The first of these works describes the use of the magnetic needle to indicate the north. Neckam points out that sailors use the needle to find their direction when the sky is overcast and the stars cannot be seen. The second treatise gives a description of a magnetic needle mounted on a pivot.

AN EARLY AUTHORITY—PETRUS PEREGRINUS

In the following century, Petrus Peregrinus de Maricourt, a soldier monk, discussed the directive property of magnets in his famout *Epistle Concerning the Magnet.* "Take a loadstone," he says, "and put it in a wooden cup or plate and set it afloat, like a sailor in a boat, upon water in a larger vessel, where it will have room to turn. Then the stone so placed in the boat will turn until the north pole of the stone will come to rest in the direction of the north pole of the heavens, and its south pole toward the south pole of the heavens. And if you move the stone away from that position a thousand times, a thousand times will it return by the will of God." Peregrinus suggested a number of improvements in the nautical compass.

In the years that followed, the magnetic compass was gradually perfected. Even the most learned, however, had no idea why the needle always pointed in a more or less northerly direction. It was not until the year 1600 that the reason was definitely revealed.

In that year the learned English physician Sir William Gilbert published a treatise called *Of the Magnet, Magnetic Bodies and That Great Magnet the Earth.* This was one of the greatest scientific works of all time. In it Gilbert, who has been called the Galileo of magnetism, set forth his theory that the earth is a magnet, and that a magnetic compass needle points north because the north-seeking pole of the needle is attracted by the North Pole of the earth. Gilbert also pointed out that the Magnetic North and South Poles do not correspond to true north and south. He observed, too, that a magnetic needle free to move up and down dips toward the earth at many places. Gilbert's work laid the foundations for our present-day knowledge of the earth's magnetism.

We realize now that, as a result of the magnetic field surrounding it, the earth and its atmosphere constitute a great magnetic laboratory. In this, nature continually performs experiments, utilizing as apparatus not only the earth, but also the sun, the moon, and radiations from outer space.

The North and South Magnetic Poles are at quite a distance from true north and south—more than 1,500 kilometers. They are constantly shifting their positions. The Carnegie Institution of Washington has determined their average positions for various periods, as shown in the table. This table indicates that both magnetic poles have shifted in a direction that is generally north by northwest in recent years.

THE DECLINATION OF THE COMPASS

Since the magnetic poles of the earth do not correspond to true north and south, it is very important to know how much they diverge. Otherwise a navigation officer would have only an approximate idea of his direction. We can find out how great this divergence is by examining the *declination* of the compass. The declination represents the angle between the magnetic needle, as it points to the North Magnetic Pole, and the geographic meridian—that is, the line passing through a given point on the earth's surface and connecting the North and South Poles of the earth. If the declination is 18 degrees west at a given point, it means that here the compass needle points in a direction that is 18 degrees west of true north (Figure 5).

Another factor that is also important in many calculations is the *inclination*, or dip, of the compass. To determine this, a magnetic needle, called a dipping needle, is mounted on a horizontal axis and is allowed to swing in a vertical plane. The needle will follow the direction of the earth's lines of force. The dip of the needle represents the angle between the plane of the horizon and the magnetic needle (Figure 6). The dipping needle is vertical at the North and South Magnetic Poles. It is horizontal at the magnetic equator. It occupies intermediate positions between horizontal and vertical at other places on the earth's surface.

The degree of declination or inclination at different points on the earth's surface may be shown on charts (maps) showing the entire world or a single region. In

Fig. 6. Sketch of the dipping needle with which Robert Norman measured dip at London, England, in 1576. The needle, mounted on a horizontal axis, followed the direction of the earth's lines of force.

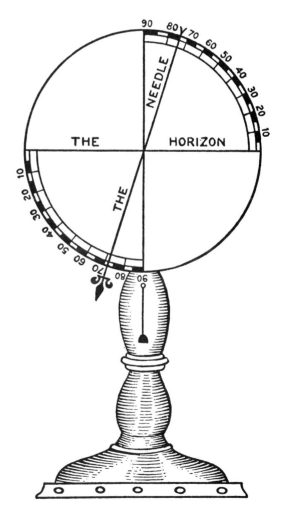

An isogonic chart of the entire world, showing the North and South Magnetic Poles and the lines of equal magnetic declination. The magnetic declination of the compass is the same for all points on a given line in the chart. Thus the declination is 10° E. in all places through which the line marked 10° E. passes. Navigators find charts like this invaluable. Without them, the compass reading would give only a rough idea of true north.

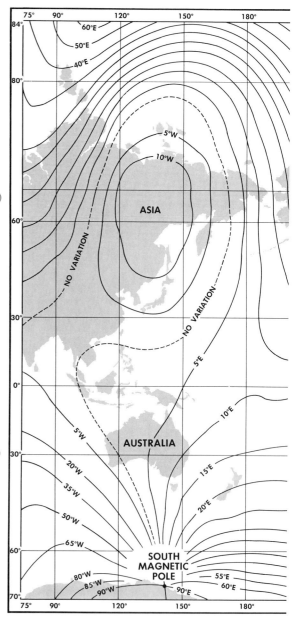

POSITION OF NORTH MAGNETIC POLE

Year	Latitude, N	Longitude, W	Authority
1904.5	70.5	96.6	Amundsen
1912.5	70.9	96.8	Carnegie Inst. of Wash. (CIW)
1922.5	71.4	97.2	CIW
1932.5	71.9	97.6	CIW
1935.0	70.0	97.0	United Kingdom
1940.0	70.7	96.1	U.S.N.O.O.
1942.5	72.6	97.9	CIW
1943.0	70.5	95.5	U.S.S.R.
1945.0	72.8	98.0	CIW
1947.0	73.1	94.5	United Kingdom
1948.0	73.9	100.9	Canada
1950.0	73.0	100.0	National Ocean Survey
1955.0	73.8	101.0	National Ocean Survey
1960.0	74.9	101.0	National Ocean Survey
1962.5	75.1	100.8	Canada
1966.0	75.8	100.8	Canada
1970.0	76.2	101.0	National Ocean Survey
1975.0	76.1	100.0	U.S. Geological Survey

POSITION OF SOUTH MAGNETIC POLE

Year	Latitude, S	Longitude, E	Authority
1903.0	72.8	156.4	Chetwynd:B.N.A.E. 1901-4
1912.0	71.2	150.8	Mawson:A.A.E. 1911-14
1922.0	70.1	149.3	Carnegie Inst. of Wash. (CIW)
1932.0	69.0	148.0	United Kingdom
1935.0	72.5	156.0	CIW
1939.0	70.3	149.0	Farr
1942.0	68.2	146.3	CIW
1945.0	68.2	145.4	CIW
1950.0	68.0	144.0	National Ocean Survey
1952.0	68.7	143.0	Mayaud
1955.0	68.0	144.0	National Ocean Survey
1957.0	69.0	141.0	Nagata
1959.0	67.7	141.0	France
1960.0	67.1	142.7	National Ocean Survey
1962.0	67.5	140.0	Burrows and Hanley
1965.0	66.5	139.9	National Ocean Survey
1970.0	66.0	139.1	National Ocean Survey
1975.0	65.8	139.4	U.S. Geological Survey

preparing a chart showing magnetic declination, the degree of declination is determined at a number of different places. A line is then drawn through all the points where the degree of declination is 0°, another, say, through the places where the degree of declination is 5°, and so on. If the places where the declination is determined were spaced closely enough, it would be found that the lines would form an intricate pattern of complex bends and closed loops. It is customary to smooth out the lines somewhat and to disregard irregular values.

The same method is used in preparing charts showing magnetic inclination. The charts that show lines of equal magnetic

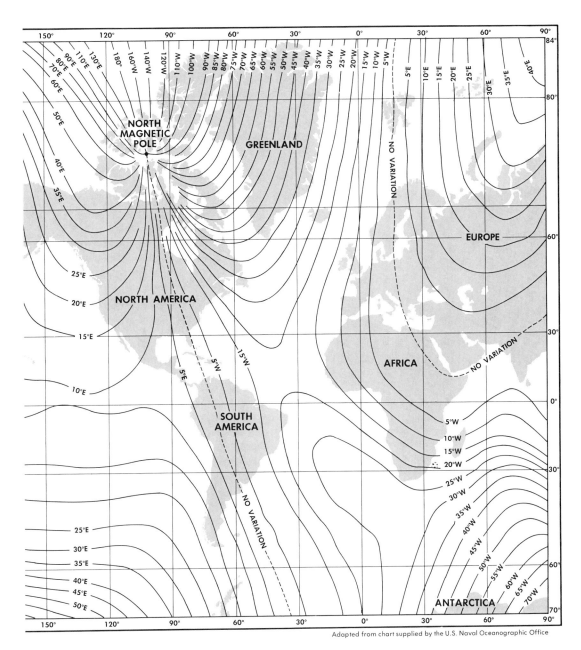

| 150° | 120° | 90° | 60° | 30° | 0° | 30° | 60° | 90° |

NORTH MAGNETIC POLE

GREENLAND

NO VARIATION

EUROPE

NORTH AMERICA

AFRICA

NO VARIATION

SOUTH AMERICA

5°W
10°W
15°W
20°W
25°W
30°W
35°W
40°W
45°W
50°W
55°W
60°W
65°W
70°W

NO VARIATION

ANTARCTICA

Adapted from chart supplied by the U.S. Naval Oceanographic Office

declination are called *isogonic* (having equal angles). Those that show lines of equal magnetic inclination are called *isoclinic* (dipping equally). Both isogonic and isoclinic charts are called isomagnetic. The above is a world isogonic chart.

The usefulness of the magnetic compass has been enhanced with the development of the accurate isogonic charts that

we have just described. It is true that on large vessels the gyrocompass is now the chief directional instrument. In all such ships, however, magnetic compasses are held in reserve. For, after all, the gyrocompass is dependent on a source of motive power for its operation and it is also subject to mechanical failure. The magnetic compass, on the other hand, practically never goes out of order and the only "power" that is required to run it is the attraction of the North Magnetic Pole.

REGULAR CHANGES IN THE FIELD

There are certain more or less regular changes in the earth's magnetic field, known as *secular variations*. There are secular variations from place to place, from season to season, from Northern to Southern Hemisphere and, for the same place, from year to year. For example, observations made at London indicate that the magnetic needle pointed 11° east of north in 1580 and 24° west of north in 1812. Since that time the needle point has shifted eastward and now points to about 10° west of north. All this indicates that this particular secular variation will complete a cycle in about five hundred years. The cycle of secular change varies from one place to another. Efforts have been made to predict such cycles on the basis of theoretical studies, just as astronomers predict eclipses of the sun and moon. It is now recognized, however, that because of the many unknown factors involved, there is as yet no basis for predictions of secular change. The fact is that secular variations are still an unsolved mystery.

It has been found that as the earth revolves around the sun during the year, there are corresponding fluctuations in the earth's magnetic field. The maximum variations, or crests, occur during the equinoctial months of March and September, when the sun crosses the equator in its apparent yearly journey through the heavens. The minimum variations, or troughs, come in the solstitial months of June and December, when the sun is farthest from the equator in that same yearly journey.

Eriez Magnetics

Magnetism has many industrial uses. Here, magnets attached to a lifting beam provide sure lifting power for raising this steel plate.

OTHER IRREGULAR FLUCTUATIONS

The presence of magnetic-ore deposits brings about anomalies, or irregularities, in the magnetic field of the earth. Such irregularities do not greatly alter the overall picture of the whole field. They make it possible, however, to locate deposits of magnetic ores in a given area. Aerial surveys of the area are made with a magnetometer, an instrument for measuring the intensity and direction of magnetic forces. By examining the findings of such surveys, prospectors can determine the extent and depth of magnetic-ore deposits. In some cases, soil and rocks are only slightly magnetic, and there are only minor anomalies in the earth's magnetic field.

Other variations in the earth's magnetism are caused by disturbances known as *magnetic storms*. These frequently occur simultaneously over the whole globe. They are much more violent in the polar regions because of the nearness of the earth's magnetic poles. There is not a regular cycle of quiet and disturbed days. An old disturbance may die out and a new one may occur at any time. Generally any marked disturbance in the magnetic field reappears in several successive months before it permanently disappears.

One of the most important factors in bringing about these magnetic storms is the existence of numberless electrified particles that stream out from the sun. The earth's lines of magnetic force, extending far out into space, entrap these charged particles, which spiral around the lines of magnetic force. Since these lines are steepest in the polar regions, the electrified particles penetrate most deeply into the earth's atmosphere in these areas and cause the natural displays called polar lights, or auroras. These result from the resistance offered to the electrified particles by the earth's atmosphere—a resistance that makes the particles glow. Polar lights are known as the aurora borealis (northern lights) when they occur in the Northern Hemisphere and the aurora australis (southern lights) when they occur in the Southern Hemisphere.

It has been discovered that the earth's magnetic field is responsible for the formation of a vast zone of radiation surrounding the earth. This zone is called the Van Allen radiation belt. It is made up of charged particles (from the sun and beyond) captured by the magnetic field of the earth.

The space near the earth is filled with a stream of charged particles from the sun known as the solar wind. Satellite explorations have shown that the solar wind distorts the earth's magnetic field, which thus forms a long "tail" on the side of the earth away from the sun.

Observations in the United States and in various other countries have shown that there is a very close connection between sunspot activity and disturbances in the earth's magnetism. When the sunspots are the most active, there are the greatest variations in the earth's magnetic field and the most brilliant displays of polar lights.

Cosmic rays from outer space also contribute to magnetic variations.

Thus far we have dealt with the effects of magnetism and with its numerous applications. The question now arises: What is the ultimate cause of magnetic fields of force and of polarity? Thus far the scientists have not been able to give a definite answer. It is quite generally agreed that magnetism results from the orderly arrangement or interactions of particles of matter within magnetic materials.

IS THERE A MAGNETIC MONOPOLE?

In August 1975 scientists were startled to hear a report that the long-sought fundamental unit of magnetism—a *magnetic monopole*, having only one magnetic pole—might have been detected. Scientists were able to identify all but one of the tracks left by fundamental particles in their cosmic-ray detector. They reported that this one track was left by a heavy fundamental particle that they called the magnetic monopole. Many scientists questioned the interpretation of the detector track, but in February 1982 a physicist at Stanford University in California reported observing a single event consistent with the properties predicted for a monopole. Researchers at the Imperial College in England claimed to have detected a magnetic monopole in 1986.

ELECTROMAGNETIC RADIATION

by James Stokley

If you stand in front of a campfire, even though you do not actually touch the flames, something that is emitted from the blazing wood makes itself felt on your face or hands. Out in the noonday sun, on a summer's day, you get the same sensation of warmth. You know that if you stay out in the sun too long you will get sunburned, possibly painfully.

You listen to the radio or watch television. So do thousands of others, all hearing or watching the same program. Something is sent out from the transmitting station. The receiver picks up this something and converts it into the particular program that you hear or see.

These are all examples of the process called *electromagnetic radiation,* whereby energy is carried across space. Ordinarily the energy travels outward from its source, spreading equally in all directions like the radii of a circle.

VISIBLE LIGHT

The most familiar kind of radiation is *visible light,* which makes it possible for us to see the world about us. Long ago men were puzzled as to just what light was, and even today we do not have the final answer. Some believed in the *corpuscular theory.* They held that the sun, the flame of a candle, and all other luminous bodies give off tiny particles, or corpuscles, which fall upon an object in their path, bounce off, and are detected when they hit the eyes. Others thought that light is something like the waves that are formed in a pond when you drop in a stone. These travel outward on the surface of the water. If the waves hit some solid obstruction, they are reflected, producing a new series of waves.

One of the basic discoveries about light came in the year 1666 when the celebrated English scientist Sir Isaac Newton

held a glass prism over a small hole in a window shutter, through which a beam of sunlight entered the room. The result, instead of a spot of white light, was a band of colors—a *spectrum*—red at one end, violet at the other. Thus he showed that what we ordinarily call white light is a mixture of colors. Six are now generally accepted as the principal colors—red, orange, yellow, green, blue, and violet. Newton held up a second prism, which was turned in the opposite direction, and let the band of colors fall upon it. The light that finally emerged was white once more. He had taken white light apart and put it back together again.

Although Sir Isaac never expressed a very positive opinion as to the nature of light, he leaned toward the corpuscular theory. His high position in science did much to maintain it, even though in 1690 a Dutch physicist, Christian Huygens, published a well thought out wave theory, which later became widely accepted. According to this theory in its modern form, light is made up of waves, and their length determines the color. The longest—about 16,000 to the centimeter—produce the light we call red. The shortest, which are about $1/28,000$ of a centimeter in length, produce violet. In between are the wavelengths of the other colors of the visible spectrum.

INVISIBLE RAYS

About the beginning of the nineteenth century, scientists found that this spectrum represents only the range of radiation that the eye can see, and that beyond both ends there are other rays, which are invisible. The first discovery of this sort was made by the English astronomer Sir William Herschel, already famous as the discoverer of the planet Uranus. In 1800, just as Newton had done, he put a prism in front of an opening through which sunlight was shining

ELECTROMAGNETIC SPECTRUM

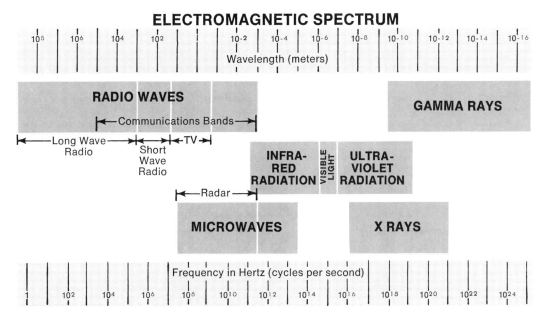

Diagram showing the wavelength ranges of various electromagnetic radiations. From the diagram it is easily seen that the spectrum of visible light is quite narrow relative to other types of electromagnetic radiation. Also note that the ranges assigned to various radiations often overlap.

and obtained a spectrum. Then he placed several thermometers at the various colors, to see which would produce the greatest heating effect. Of all the visible parts, the red light made the temperature go up the most. Then Herschel placed one of the thermometers beyond the red end of the spectrum. The mercury went up even higher, clearly showing that invisible heat rays were present.

In this way Herschel demonstrated the existence of what we now call *infrared rays*. "Infra" means below. The infrared rays have a lower frequency than the red rays. He performed a whole series of experiments with these heat rays and discovered that he could reflect them with mirrors and focus them with lenses, just as visible light is reflected and focused. He also found that these rays are emitted from a bed of hot coals in a fireplace, as well as from the sun.

A year later, a German physicist, Johann Wilhelm Ritter, discovered that there was also something at the other end of the spectrum. This did not cause heating, but was capable of producing chemical effects.

Ritter knew that the white compound called silver chloride turns black when exposed to light. He put some of this substance in line with a solar spectrum, out beyond the violet part, in a place where no visible light was shining. The silver chloride blackened even more rapidly than it did in the region where the various colors could be seen.

Thus *ultraviolet rays* were discovered. "Ultra" means beyond. These rays are located beyond the shortest wavelength of the visible spectrum. These rays could also be focused and reflected but they did not penetrate glass very well. By 1852, the English physicist Sir George Stokes found that the ultraviolet rays could pass through quartz quite easily. So he made a spectrum-producing device out of quartz lenses and prisms. With the light from an electric spark, he could produce an ultraviolet spectrum ranging to a wavelength about half that which produces the deepest violet light. To detect these invisible rays, Stokes made use of fluorescence. This is an effect shown when ultraviolet rays fall on certain chemicals, causing them to shine visibly.

Since silver chloride and similar substances darken when exposed to either visible or invisible light, plates coated with these substances can be used to photograph the invisible spectrum. With refinements in this method, it was possible to extend the study of the spectrum far into the ultraviolet region. Special improved plates were used in a spectrograph from which air had been evacuated, so as to avoid the absorbing effect of the air upon ultraviolet waves.

STUDY OF THE INFRARED REGION

While some photographic materials are sensitive to infrared, they do not record rays far beyond the visible spectrum. Other methods must be used, therefore, in studying these rays. Some of the most fundamental work on the infrared region was done by the U.S. scientist, Samuel P. Langley. To study infrared rays, Langley invented an electrical device called the *bolometer*. With the bolometer, which detects and measures small quantities of radiant heat, he studied

A fluorescent sunlamp, suspended from the ceiling of a pigpen, gives the hogs a dose of ultraviolet radiation. Livestock thus treated reach maturity faster.

Westinghouse

the infrared spectrum of the sun well beyond the wavelengths of visible light.

Later, with the bolometer and other even more efficient detectors, scientists found out that still longer waves are given off from earthly sources that do not even have to be what we commonly call "hot." Infrared rays are emitted, though rather feebly, from a glass of ice water, for example, in a long range of wavelengths, with the maximum at about $1/1,000$ of a centimeter. If the water is boiling, the emission is stronger, with the maximum wave length now slightly shorter. For still hotter objects, the total amount of radiation given off increases, and the maximum wavelength becomes shorter and shorter. This is in accordance with the so-called *displacement law* discovered by the German physicist Wilhelm Wien in 1893. According to this law, the product of the wavelength at which a radiating body gives off the greatest energy and the absolute temperature of the body is a constant.

DISCOVERY OF RADIO WAVES

The first clue that there might be additional waves, still longer than those of the infrared, came from the Scottish physicist James Clerk Maxwell. In 1864 he published a series of papers dealing with the nature of light, and showed, theoretically, that it is a movement of electrical and magnetic waves. He reached the conclusion that these waves could extend far beyond the infrared. One of his suggestions was that a vibrating electrical charge could set up such waves, and that they would travel through space at the speed of light — more than 300,000 kilometers per second.

Perhaps the first person to produce such waves experimentally was Elihu Thomson, a brilliant young science teacher at the Philadelphia Central High School. In 1871, while experimenting with high-voltage electrical sparks which could jump across a gap of several centimeters, he connected one of the terminals, through which the electrical discharge passed, to a water pipe, the other terminal to a metal table top. While the sparks were jumping, he found that he could go to distant parts of the

X rays are used extensively in physical and chemical research. Structural analysis of crystalline substances is usually done by means of X-ray diffraction patterns.

The Franklin Institute Research Laboratories

building, hold a knife blade near a metal object, and draw sparks from it. After some other experiments, a few years later, he realized that he had confirmed Maxwell's prediction. The energy that produced the sparks at the knife point was transmitted from the original spark by waves that traveled through space.

As sometimes happens in scientific history, Thomson's experiments were almost entirely ignored. In 1887 Heinrich Hertz in Germany obtained similar effects with what was essentially the same equipment. The world acclaimed the German scientist, and the radiation that was produced came to be known as *hertzian waves*. They are now generally called *radio waves*.

USING RADIO WAVES

Various people realized that these waves might make it possible to send messages between distant points, without any wires to carry them. The first to succeed was a young Italian engineer, Guglielmo Marconi. After some preliminary experiments in which he sent signals across a vegetable patch near his home, he improved his equipment and finally, on December 12, 1901, sent the letter "S" in the Morse telegraphic code across the Atlantic Ocean from England to Newfoundland. From this humble beginning came such developments as television, radio broadcasting, and radar.

The waves used in broadcasting are about 300 meters in length. Still longer ones, more than a kilometer in length, are sometimes used in transmitting radio messages across the oceans. Certain waves, many kilometers long, are emitted from the power lines over which alternating electrical currents are transmitted. This equipment is designed, of course, to deliver electrical power—the waves are a sort of by-product.

The range of wavelengths of radiation has been extended, therefore, from those of the shortest ultraviolet to others a million million times as long. What of the other end of the spectrum? Are there any rays shorter than the ultraviolet?

SHORTER WAVES—X RAYS

In November of 1895, Professor Wilhelm Konrad Roentgen, at the University of Wuerzburg, discovered that some new kind of ray was given off from a Crookes tube, an evacuated glass bulb in which an electrical discharge was produced. This ray caused some materials to glow by flourescence. It was possible to record its effect on a photographic plate. The most striking characteristic of the new rays was that they could pass through many solid materials— wood, paper, and even human flesh. In fact, with the rays one could see and photograph the bones in one's body. Roentgen gave the name of *X rays* to the newly discovered and mysterious emanations.

Finally it turned out that X rays—or *roentgen rays*, as they are also called—are radiations with wavelengths even shorter than those of ultraviolet. Their wavelength depends on the voltage of the electricity used to create them: the higher the voltage, the shorter the length. For 100,000-volt X rays, most of the radiation has a wavelength of about $\frac{1}{540,000,000}$ of a centimeter, or about $\frac{1}{19,000}$ of that of the shortest visible light.

With some of the large particle accelerators that are used by physicists, X rays of hundreds of millions of volts energy are produced, and their wavelengths are still shorter. Beyond these are certain radiations that occur in nature. These are associated with the cosmic rays that rain upon us from outer space.

All the radiations that we have been discussing — light, ultraviolet rays, infrared rays, X rays, and the rest — are known as electromagnetic radiations. They have received this name because they are sent out into space as the result of periodic variations — variations occurring at regular intervals — in electric and magnetic fields. They are all transmitted at the same speed — 300,000 kilometers per second. The complete range of wavelengths of these radiations is known as the *electromagnetic spectrum*. The known electromagnetic spectrum extends from the cosmic rays, with wavelengths so tiny that some 1,250,000,000,000 would have to be lined up to equal a centimeter, all the way up to alternating-current power waves, more than many million million times as long.

QUANTUM THEORY OF LIGHT

So far we have been discussing the wavelike properties of light and similar radiation. The wave theory of light, which, as we have seen, had been originally proposed by Huygens, had won wide acceptance in the nineteenth century. However, about 1900 the work of Max Planck, in Germany. and later that of Niels Bohr, in Denmark, showed that there were some effects that the wave theory could not explain. Apparently, at least in certain respects, light acts as if it were a stream of tiny corpuscles of energy called *photons*. A photon is the smallest amount of light energy possible. Many other radiations, especially those that come from excited atoms, seem to show the same property: radiant energy is emitted and absorbed in separate, particle-like units, or *quanta*. These ideas of Planck and Bohr were eventually united in a revolutionary theory of atomic structure. However, the overwhelming weight of experiment still seemed to uphold the wave nature of light. Indeed, it was said that one had to believe the wave theory on Monday, Wednesday, and Friday, and the quantum theory on Tuesday, Thursday, and Saturday.

WAVE MECHANICS

A compromise was introduced about 1925, as a result of the work of a French physicist, Louis de Broglie, and an Austrian, Erwin Schroedinger. They originated the *theory of wave mechanics*. According to this theory, all forms of matter and energy have both wavelike and particlelike qualities, but the two aspects never appear together under the same conditions. All particles have waves associated with them, whose wavelengths depend on the mass and velocity of the particle. The bodies of our familiar world, such as a bullet fired from a gun, have relatively huge masses and low velocities, and their associated waves are therefore too small to be detected. But an electron, which is one of the particles of which atoms are made, has an infinitesimal mass and moves with a velocity close to the speed of light. The wavelength associated with the electron is therefore great enough to measure.

Beams of electrons, and many other elementary particles, have been diffracted to produce interference patterns just like those of a conventional wave. From these patterns, the wavelength of the electron is measured and is found to agree with the theoretical prediction of wave mechanics. Therefore, the seemingly paradoxical ideas of Schroedinger and De Broglie have been generally accepted by science.

WAVE GENERATION

For an atom to generate waves of light, changes must take place in its structure. An atom consists of a positively charged nucleus, around which are moving negative electrons — from one in the case of hydrogen to

Courtesy of the Metropolitan Museum of Art

Sometimes infrared rays are used to detect alterations in paintings. When a painting by Tintoretto— *The Doge Alvise Mocenigo Presented to the Redeemer* (above)—was photographed by infrared rays, the figure of an angel hovering over the kneeling Doge was revealed. Tintoretto had apparently changed his mind about the composition of his painting, as the photo on the left clearly shows.

shifted to other orbits, farther away from the nucleus. Scientifically, they are said to be shifted to a higher energy level. Sometimes electrons may be removed completely. Then the electrically unbalanced atom is said to be ionized.

After an electron has been thus shifted or removed, it tends to return to its former position. When it does, energy is given off in the form of radiation. Visible light, as well as the nearby ultraviolet and infrared, is a result of transitions of the outer electrons in atoms and molecules, while those of the inner electrons in atoms give rise to very short ultraviolet rays and to X rays. *Gamma rays,* still shorter in length, which come from radioactive materials, are a result of changes of energy state in the nucleus itself. In the case of the longer infrared rays, the atoms are not affected. Vibrations and rotations of molecules, consisting of a number of atoms, are responsible.

more than a hundred in the case of the heaviest elements known. Normally, these electrons move in certain regions, or orbits, but when energy is absorbed they can be

USES OF X RAYS

Let us examine some of the applications of radiation, starting at the shortwave end. So far cosmic rays have not been used by man. Indeed, it is only comparatively recently, with particle accelerators operating at energies of thousands of millions of volts, that some of their effects have been duplicated for the first time in the laboratory. But X rays have found wide application. The enormous penetrating power of fifteen- to twenty-million-volt rays, which are produced by several commercial devices, have been utilized in the X-ray examination of very thick metal parts. Steel materials from 20 to 25 centimeters thick can be examined with one- and two-million-volt radiation. X rays of this same energy range are also used in treating cancer. Properly controlled, they do more damage to the cancerous tissue than they do to the healthy tissue.

For more general medical use, as in finding the place where a bone has been fractured, and in industry, rays of about 100,000 volts may be used. X rays have been employed to analyze the structure of matter. A beam of X rays, passed through a crystal in which the atoms are arranged in regular layers, is scattered by the atoms and forms a pattern of spots, characteristic of the particular kind of crystal, on a photographic film. This is called *X-ray diffraction.*

The longest-wave X rays overlap into the shortest of the ultraviolet waves. One of the chief uses that man has found for the ultraviolet is in fluorescent lighting, which has now become widespread. Inside these familiar tubular lamps, an electrical discharge produces ultraviolet rays, which shine on the special material lining the tube. The invisible rays are thus turned into visible light by the process of fluorescence. It is the ultraviolet part of sunlight that produces sunburn, in addition to forming vitamin D in the body and preventing the disease called rickets. Ultraviolet rays can also be used to kill germs. Lamps producing these rays are employed in medical treatment.

USES OF VISIBLE AND INFRARED RADIATION

The uses of rays of visible light are obvious enough. One can readily imagine what the world would be like without light.

Next above the light waves in the electromagnetic spectrum are those of the infrared—the heat rays. Any incandescent lamp gives off these rays in large amounts. There are special heat lamps in which the proportion of visible light is reduced, while that of infrared is increased. Such lamps are employed for drying purposes. For example, after an automobile body is painted, it is often run through a tunnel in which it is exposed to the rays from a battery of heat lamps. As a result it dries in minutes. The drying process would take much longer with other methods.

RADIOWAVES AND RESEARCH

The shortest radio waves, which overlap the region of the longest infrared, are often known as *microwaves.* One use is in chemical research—in the study of the structure of molecules. Some gaseous molecules are able to absorb these tiny waves, and the energy the molecules acquire will start them spinning, much as the blades of a windmill turn in a breeze. The windmill can turn rapidly or slowly, or at any speed between. The molecule, on the other hand, can absorb only certain well-defined amounts of energy, and none that are intermediate between these. The amounts in question depend on the weights and arrangements of the atoms that make up the molecule. When the rotational energy that a molecule absorbs from microwaves is determined, it is possible to calculate the relative locations of the atoms in the molecule.

Radio waves of various lengths, down to the microwaves, are used in radar. This development makes it possible to detect ships and airplanes at night and through fog, and to determine their exact direction and distance. With radar, navigators of ships and airplanes can "view" clearly the objects and geographical features around them under adverse conditions of visibility.

Microwave ovens have become more and more common in home kitchens. Microwaves cook food—even frozen food—in a very short time. There have, however, been some reports of possible health hazards from microwaves.

General Electric

All this is done by sending out a short pulse of radiation. When the pulse hits the distant target it is reflected and the echo is picked up. Even though radio waves travel nearly 300 meters in a millionth of a second, electronic circuits determine accurately the time it takes for the echo to return, and thus give the distance. Radar pulses have even been bounced off the moon, at a distance of some 385,000 kilometers, and the echo has been detected and measured.

Radio waves emitted from various celestial objects have opened up a new field of research into the heavenly bodies—including many that cannot be detected by other means—and have given rise to the science known as radio astronomy. By studying the radio waves coming to us from outer space, radio astronomers have demonstrated, for example, the presence of clouds of hydrogen gas between the stars. The hydrogen atoms in such clouds can have two different formations. They switch back and forth between one and the other, emitting waves about 20 centimeters in length as they do so.

Of course, the most familiar uses of radio waves are in the fields of communication: broadcasting, telecasting, and radio telegraphy and telephony between ships and across continents or oceans.

Radio waves of microwave length, in particular, are finding more and more uses. For example, microwaves are used in the manufacture of materials as diverse as rubber tires and potato chips. In homes, microwave ovens are popular for their ability to cook food quickly. Microwaves are also used extensively for communication between the earth and spacecraft, as well as for tracking the orbits of such objects.

HEALTH THREATS

As we've discussed, electromagnetic radiation that is strong enough to alter atomic structure—like that emitted by radioactive material—poses a health threat to humans. In recent years, much attention has focused on nonionizing radiation—the type emitted by power lines, video display terminals, microwave ovens, radio and TV transmitters, and countless other modern devices. Although no definitive link to cancer has been proven, nonionizing radiation at extremely high levels can heat living tissue and cause burns, cataracts, and sterility. Even though such high doses are rarely encountered in everyday life, public pressure for concrete answers has stimulated much vital research on the potential danger of nonionizing radiation.

Sound plays a vital role in the communication of ideas. The transmission of sound from the record to the child's ears involves the vibrations of molecules.

SOUND

by Harold P. Knauss

The sounds that reach our ears provide us with an immense amount of information. They make us aware of the voices of friends, the rustling of leaves, the crash of thunder, and the sounds of animals and machines—to name a few. Yet the source of these various sounds is simply a series of pressure changes.

A tree crashes to the ground. As it topples, it produces disturbances in the air, and these disturbances give rise to pressure waves that travel out in all directions. Some of these waves strike our eardrums and set up vibrations that are recorded in the brain. And it is then that we hear the crash of the tree. The same set of happenings is involved when we listen to the strings, woodwinds, brasses, and percussion instruments of an orchestra. The instruments produce different kinds of disturbances which strike the ear and are translated, in the brain, into recognizable characteristic sounds.

The word "sound" may apply to the whole set of happenings that we have just mentioned, or it may refer to the pressure waves sent out by a particular disturbance, such as the crashing of a tree or the blowing of a bugle. Finally, it may apply only to the sensation experienced by the listener. To avoid confusion, it is best to use the phrase "unheard sound" for the pressure waves resulting from a given disturbance, and the phrase "heard sound" for the sensation that is produced. In a forest, say, there may be plenty of unheard sound, as disturbances set up pressure waves. But there will not be any heard sound unless the ears of some person or animal are affected by the waves.

PRESSURE WAVES ARISE

Let us see just what happens when pressure waves arise. We shall suppose that the sound source is a tuning fork that is set to vibrating. The prongs are in the midst of a sea of air, made up of molecules. As the prongs swing toward each other in their vibration, they bring about a *compression*—that is, they crowd together the molecules of air between them. These molecules will push against those of the next layer of air, and a wave of compression will spread out in all directions. As the prongs of the tuning fork swing apart, they leave a more or less empty space behind, called a *rarefaction*, or thinning out, of molecules. The molecules nearby rush to fill the gap, and as a result they will leave another gap directly beyond—another rarefaction—and so on, so that a wave of rarefaction travels away from the fork.

If we could see the molecules of air in an area where a tuning fork has just been struck, we would find a regularly repeated alternation of regions where the molecules are crowded together (compressions) and of regions where they are few in number (rarefactions). Between each crowded area and low-density area we would see a "neutral zone," where the concentration of molecules is about what it was before the tuning fork was struck. The alternating compressions and neutral zones and rarefactions make up a series of concentric spheres, with the source of sound—the tuning fork—at the center. This is clearly shown in Figure 1. It should be pointed out here that the diagram represents an instantaneous picture of sound waves traveling away from the source. One wave is thought of as an expanding sphere of both compression and rarefaction.

UPI
A sonic "boom" is produced as this SST breaks the sound barrier. The loud blasts are a great concern to people living below the plane's flight path.

Canadian Government Travel Bureau Photo
Electric trains and modernized tracks and platforms help control the noise produced by mass transit systems.

UPI
Traffic noise is common and often reaches harmful levels. Some noise can be controlled: the blasting of auto horns is illegal in many towns.

We may compare it with the water wave that is created when we toss a stone into a pond. The stone pushes aside the water as it enters the surface, producing a more or less circular ridge. This in turn disturbs the water beyond it, and so the crest of the wave travels outward. Meanwhile, the stone continues to sink and it leaves a depression in the surface. As water rushes into this region from the surrounding region, the trough, or depression, starts outward. Enough water rushes in to more than fill the original depression, so that the level rises above the original surface and oscillates up and down before gradually subsiding. Troughs and crests alternate, as ever-widening circles spread out over the surface of the pond. These become less and less noticeable until finally they disappear entirely. In much the same way, the pressure waves set up by the vibrations of a tuning fork, or any other source of sound, travel outward in space, becoming increasingly weak as they do so, until finally they cease to press upon the molecules of air. Of course, they may strike our eardrums if we happen to be in the vicinity.

Like other kinds of waves, sound waves have a definite *wavelength* and *amplitude*. In the case of a sound wave, the wavelength represents the distance from one area where the concentration of molecules is greatest or least to the next similar area. The amplitude represents the pressure difference between the compression or the rarefaction and the normal undisturbed pressure. We can show wavelength and amplitude most clearly by representing the sound wave as a curved line alternately passing above and below a straight line, as shown in Figure 2. The straight line represents normal surrounding air pressure, and the curved line the pressure above and below normal caused by the sound wave. The distance between the crest, or highest point, of one curve and the next, or be-

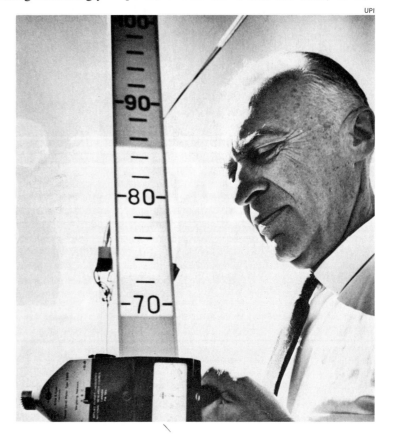

Using a noise meter to measure the intensity of sound in a given area.

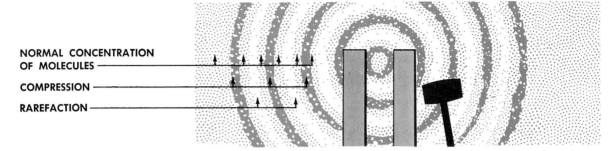

Fig. 1. Whenever a tuning fork is struck, it vibrates, bringing about alternating compressions and rarefactions. These make up a series of concentric spheres with the source of sound—that is, the space between the two tips or the tuning-fork prongs—at the center. The complete series is called a sound wave. It may be compared to a water wave that is created when a stone is tossed into a pond.

tween the trough (lowest point) of one curve and the next, is the wavelength. The amplitude is the perpendicular distance from the crest or trough to the straight line.

A sound wave passes a given point in space at the rate of so many compression-rarefaction pairs, or cycles, per second. This is called the *frequency* of that particular sound. For example, a tuning fork that produces the musical tone of A above middle C vibrates at the rate of 440 times a second, producing 440 compression-rarefaction pairs. It is said to have a frequency of 440 cycles a second, or 400 *hertz*. The higher the frequency, the shorter the wavelength and vice versa. Each body has its natural frequency—the rate at which it vibrates when it is made to send out a sound wave.

FREE, MAINTAINED, AND FORCED VIBRATIONS

If a tuning fork is not interfered with in any way after it has been set vibrating, the vibration will die out in time—generally pretty quickly—and the radiation of sound will stop. This is called *free vibration*. In many cases, however, we find it advisable to keep up the vibrations for a time before they die out. Thus, in the case of a sustained note of a bugle, as long as the player keeps blowing, the bugle continues to send out sound waves of the same frequency. When vibrations like these are produced, we speak of *maintained vibration*.

A violin bow produces a somewhat different type of maintained vibration. The bow pulls the string against which it is

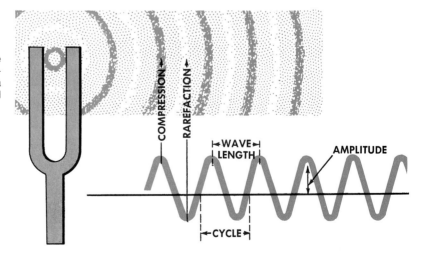

Fig. 2. Wavelength and amplitude may be indicated clearly by representing the sound wave as a curved line passing above and beneath a straight line.

pressed until the string slips back underneath it. Then the string starts moving with the bow until it slips back again and so on. The kind of maintained vibration is known as *stick-and-slip*. It accounts for, among other things, the screeching sound that is sometimes produced when a piece of chalk is drawn across a blackboard.

In some cases an object is made to vibrate at a frequency different from its natural frequency. The vibrations that result are called *forced vibrations*. The wooden body of a violin, for example, has its own rate of vibration. When a violinist draws his bow across one of the strings, the vibrations of the string are communicated, by way of the bridge, to the body. The wooden surfaces now begin to vibrate, not at their natural frequency, but at the frequency of the string. As the vibration is transmitted to the air around the wooden surfaces, a large volume of sound is produced. Were it not for such forced vibrations, the tone of the violin would be thin and altogether undistinguished.

Like the wooden surfaces of the violin,

Not just noisy — but harmful. Jackhammer operators must use ear plugs to protect their hearing.

the eardrum has its own natural frequency. But it, too, is made to vibrate at a different rate as it transmits the frequencies of incoming sound waves. The diaphragms of telephones and loudspeakers are also applications of the principle of forced vibrations.

RESONANCE

If the frequency of a sound wave arising from another source coincides with the natural frequency of a body, this body is set into strong vibrations. This is called *resonance*. Suppose, for example, that you push down the right-hand pedal on a piano, thus lifting all the dampers on the strings inside. If you sing a loud tone into the piano and bring your ear close to it, you will hear strings whose frequency corresponds to the frequency of the tone. Sometimes destructive vibrations may be built up because of resonance. If a company of soldiers starts marching across a bridge in parade step, the frequency of their steps may correspond to the natural frequency of the structure. If that is the case, the bridge may be set to vibrating so violently that it will come crashing down. That is why soldiers are required to break step when they march across a bridge.

POWER PROVIDED BY SOUND WAVES

Each sound wave releases a certain amount of energy with its alternate compressions and rarefactions. It produces a certain amount of power. We use the word "power" here in the same sense as when we speak of the power produced by a dynamo or by the motor of an automobile. The amount of power made available by sound waves that produce heard sound is extremely small compared to that produced by a dynamo or motor. It is even smaller than that which is required for an electric light bulb. One hundred watts of power is converted from electric energy into heat and light in a fairly bright bulb. When an orchestra of seventy-five performers is playing its loudest, it produces only seventy watts of acoustic, or sound, energy. The power output of an average speaking voice is roughly only one three-millionths as great. Imagine the infinitesimal power out-

A properly designed bandshell should reflect and concentrate sound to a point outside the shell.

put of a voice when it sinks to a whisper.

Fortunately, it is not necessary for the power of a sound wave to rival that of a dynamo in order to act upon our eardrums. Under laboratory conditions the ear can detect tones with a power output of a millionth of a millionth of a watt per square meter of area, at frequencies to which the ear is most sensitive.

The power produced by a sound wave as it passes through an area at right angles to the direction of the wave is called the *intensity* of the sound at that area. The intensity decreases rapidly as a sound wave moves from its source. The reason is that the energy of an expanding wave has to be spread over a larger and larger area, and therefore less energy travels through a unit area far from the source. The intensity drops as the square of the increasing distance. Suppose that a sound wave strikes your ear first at point *A*, 2 meters away from the source, and then at *B*, 6 meters away (see Figure 3). The distance from the source at *B* would be three times greater than at *A*. The square of three is nine. Therefore the intensity of the sound wave

at *B* would be only one-ninth as great as it would be at *A*.

The intensity of sound can be accurately measured by the electronic device known as a *noise meter*, or *sound meter*. It measures intensity in *decibels*.

TRANSMISSION THROUGH DIFFERENT MEDIA

Sound is always transmitted through a definite medium. This may be either a gas, such as air, or a liquid, such as water, or a solid, such as steel. It cannot be transmitted at all where there is no medium—that is, in a vacuum. We can show that this is so by putting a ringing electric bell in a glass jar, sealing the jar and then gradually pumping the air out of the sealed jar. The ringing of the bell becomes increasingly feeble as more and more air is withdrawn from the jar. Finally, when there is practically no air left, the bell will no longer be heard. Because outer space, beyond our atmosphere, is a good vacuum, no sounds arising there could be transmitted to us.

The speed of sound depends upon the medium through which it travels. It passes

RINGING BELL **A** **SOUND WAVES OF RINGING BELL** **B**

2 meters

6 meters

Fig. 3. At a distance of 6 meters (B), our eardrums are three times as far from the source of sound—a ringing bell—as at 2 meters (A). The intensity of sound at a distance of 6 meters is only ¹/₉ as great as at 2 meters.

through dry air at 0° Celsius at a speed of 330 meters per second. Its speed increases as the temperature rises, reaching 700 meters per second at a temperature of 1,000° Celsius. In water, the rate is about 1,400 meters per second. The exact rate depends on the composition and temperature of the water. The speed of sound in copper is 3,560 meters per second. It is 5,100 meters per second in aluminum.

Such variations in the rate of speed are due to differences in elasticity and density of the medium through which the sound wave passes. The speed increases as the density decreases. It also increases with increasing elasticity. You may ask why sound is transmitted through water more rapidly than through air, since water is much denser than air. The answer is that

while water is about eight hundred times as dense as air, it is much more than eight hundred times as elastic as air.

The speed of sound in air, as we have seen, is about one-third of a kilometer per second. The speed of light is about 300,000 kilometers per second. These sharply contrasting rates account for many interesting phenomena. Puffs of steam from a distant locomotive make their appearance before the sound arrives. You see a distant woodsman's ax striking a tree before you can hear the blow. During a thunderstorm, the crash follows the flash with a delay that decreases as the storm approaches. The distance of the lightning may be estimated by allowing one kilometer for each three seconds of delay. (Figure 4).

We make use of the known speed of

Fig. 4. Because of the sharply contrasting rates of speed (indicated in the drawing) of sound and light, the workers at the left see the flash of the explosion before the sound of the explosion reaches their ears.

Sound wave (explosion)
Speed: Approximately 330 meters per second

Light ray (flash of explosion)
Speed: 300,000 kilometers per second

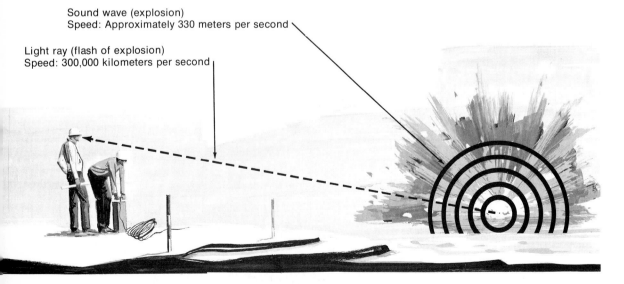

sound in water in the fathometer, a device that sends a short pulse of sound downward from a ship's hull and picks up the echo from the bottom. The time required for the echo to return indicates the depth of the water at that place. The same principle is employed in the device known as sonar.

All sounds transmitted through the same medium have the same rate of speed, even though their frequencies are different. It is fortunate that this is so. Imagine the chaos that would result if the high-frequency sounds of the flute or the piccolo kept reaching our ears before the low-pitched sounds of the bassoon or the tuba.

The medium in which a sound wave is traveling may undergo certain changes. If the medium is air, it may become denser or more rarefied. Or else the sound wave may strike an entirely different medium. A sound traveling through air may reach the surface of a pond. When the medium changes in any way, various things may happen to the sound. Some of it may be reflected at the boundary between one medium and the next, and may start traveling back toward the source of the sound. Some of the sound may penetrate the second medium. Some of the energy may be converted into heat.

REFLECTION OF SOUNDS

Perhaps the most familiar reflected sounds are natural echoes. They bounce back from a surface far enough away so that the reflection is distinct from the original sound (Figure 5). Reflection of sound may occur within a room. Usually, however, no distinct echo is heard even in a very large room because so many reflections return, as the sounds bounce from one wall and then from another, that the sound simply appears to be somewhat prolonged. The clatter that follows a single handclap illustrates this effect, which is known as *reverberation*. We shall see later how architects deal with reverberation in designing auditoriums.

The rolling of thunder is caused by reflection, not from solid objects necessarily, but from layers of air that have different temperatures. The sound wave in this case is produced by the sudden expansion caused when the air is heated by the passage of lightning. As this sound wave meets a colder or warmer layer of air, a part of it is reflected from the boundary.

Blind persons can sometimes be trained to become sensitive to faint reflections of sound. They can then detect obstacles in front of them. A bat is guided even in darkness by the echoes that result when its cries are reflected from objects in its path.

REFRACTION OF SOUNDS

When sound waves penetrate a different medium, they are *refracted*—that is, they change their direction. Such refraction accounts for the fact that sound can sometimes be heard for surprising distances over the surface of a lake. Suppose that there is a layer of cool air above the water and, above that, a layer of considerably warmer air. The upper part of a sound wave begins to penetrate the warm layer of air. It travels faster in this layer and gains on the adjoining part of the sound wave in the lower, colder layer. As a result, it is redirected toward the surface of the water. In other words, instead of going off into the upper atmosphere, it comes back within hearing of people near the water (see Figure 6).

Fig. 5. When sounds are reflected from surfaces, they produce echoes. If sounds bounce from a number of surfaces, as shown below, there are several echoes.

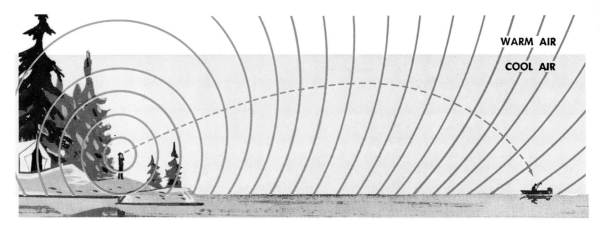

Fig. 6. A familiar sound phenomenon. The sound wave produced by the person's voice travels faster in the warm upper layer of air. It gains on the part of the wave in the cooler lower layer. The wave is then redirected towards the water.

Just the opposite effect is produced when there is a layer of cold air above a layer of warmer air. If a cannon is fired where such a condition exists, the upper end of the wave will be slowed up as it reaches a cooler region of air. The part in the lower, warmer layer will travel ahead at the same rate as before. The difference in speed of the two ends of the wave front will cause the wave to be lifted up, so that a silent zone is created on the earth's surface. A person standing in this silent zone will not hear the gun, even though he may see the puff of smoke as it is fired. Now as the explosion wave continues to ascend, it may reach a height at which the temperature starts to rise. It will then be bent downward and may reach the ground many kilometers

Fig. 7. A cannon is discharged. The upper part of the sound wave that results is slowed down as it enters a layer of cool air above the warmer one. As a consequence, the lower part of the wave, gaining on the upper part, is lifted upward, leaving a "silent zone." Should the sound wave continue to rise, it may ultimately reach a layer of warmer air. If this happens, the wave will be bent downward and may reach the ground many kilometers from the place where the cannon is. There will be a "zone of audibility" farther from the gun than the silent zone.

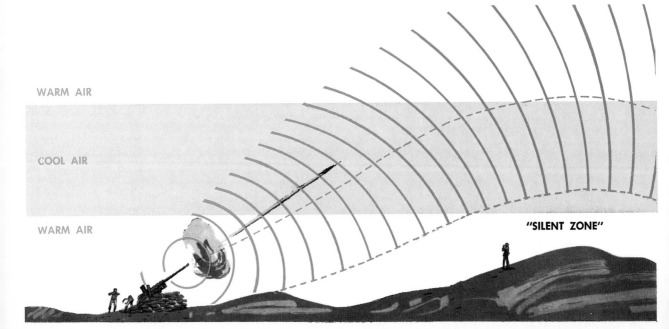

from the place where the cannon was fired (see Figure 7). There will be a zone of audibility at a far greater distance from the cannon than the zone of silence.

ABSORPTION OF SOUNDS

When sound energy is transformed into heat, the sound is absorbed. This phenomenon is a common one. We observe it, for example, when the ground is covered with a layer of soft snow. The snow acts in much the same way as the acoustical tiles that are specially designed to absorb sound. In both the tiles and the snow there are a great many small holes. In the snow these holes are to be found in the intervals between the snow crystals. The sound enters the tiles or the snow and fails to find its way out. Or, to express the matter more accurately, the energy of the sound waves is converted into heat by friction with the air contained in each one of the innumerable small holes. The quantity of energy in sound is very small, as we have seen, and therefore the amount of heat produced in the absorption process is small, too. Otherwise we would be able to melt snow by simply shouting at it.

The notion that sound waves in air produce heat may seem to be pretty farfetched, but it has been proven by actual experiments. We know that these waves consist of a series of compressions and rarefactions. We know, too, that when a gas is compressed its temperature rises and that when the gas expands its temperature falls. By using the temperature-measuring devices known as thermocouples, investigators have been able to detect the small temperature changes that are brought about by the alternate compression and expansion of sound waves.

WHEN SOUND WAVES ACT UPON THE EAR

It is when the pressure waves of sound act upon the ear that they become a superb means of communication. The human ear is an elaborate and complex instrument. The sound goes into a hole called the *external auditory canal* and then strikes the *eardrum*. Three bones in the middle ear act as a system of levers to transmit the motion of the drum to the inner ear. These bones are called the *hammer,* the *anvil,* and the *stirrup*. The stirrup acts upon a membrane known as the *oval window*. On the other side of the oval window is one end of the *cochlea,* a tube coiled like a snail shell of 2½ turns. Within the cochlea is the tough *basilar membrane*. On this, we find the *organ of Corti,* containing the endings of the *auditory nerve*. A clear fluid fills the cochlea.

The basilar membrane is narrowest at the end near the oval window and it becomes gradually wider. The fibers of the basilar membrane have been compared with piano strings—long, loose, thick strings for low frequencies and short, tight, thin strings for high ones. It is generally assumed that the membrane is under tension, but no direct evidence of such tension has ever been obtained.

This is probably what happens when a sound wave strikes the ear. The pressure variations in the outer canal set the eardrum into vibration. This vibration is transmitted through the hammer, the anvil, and the stirrup to the oval window. Waves set up there travel in the fluid along the basilar membrane. A particular spot in the basilar membrane responds to the vibration, whereupon impulses that correspond to a definite frequency are set up in the organ of Corti. The nerve endings in the immediate

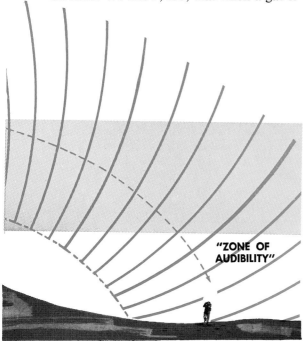

"ZONE OF AUDIBILITY"

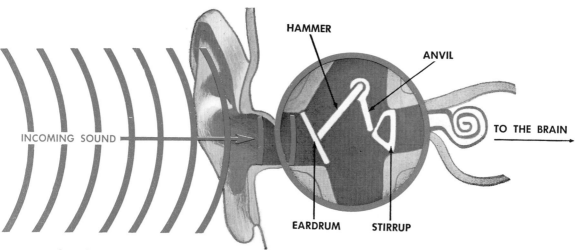

Sound waves strike the eardrum. Then three bones in the middle ear—the hammer, anvil, and stirrup—act as levers to transmit the motion of the drum through the middle ear to the inner ear. Impulses are generated in appropriate nerve fibers, and they are then transmitted to the brain. The diagram does not show the exact shapes or relative sizes of the bones.

vicinity are stimulated, and nerve impulses travel to the brain. Finally, these impulses are duly interpreted in the brain. All this may seem to be extremely complicated, yet it is a highly simplified version of what actually takes place.

PITCH, LOUDNESS, AND QUALITY OF SOUNDS

To interpret incoming sounds adequately, the brain analyzes (1) their pitch, (2) their loudness, and (3) their quality. The *pitch* of a tone—its place in the musical scale—depends mainly on the frequency. (It is affected somewhat by the intensity level of the tone.) If the vibrations that strike the eardrum are of low frequency—say 30 to 40 hertz—we hear a low tone. If the frequency is in the neighborhood of 15,000 hertz, we hear a high tone. The human ear can detect sounds with frequencies ranging from 20 hertz to about 20,000 hertz. Sounds with a higher frequency than that are called *ultrasonic*.

The use of the word "pitch" is not restricted to the sounds that we recognize as music. The speaking voice follows a definite pitch pattern. So does the rat-a-tat of a riveter and the meowing of a cat.

Pitch depends upon the mass and tension of the sound source. Take as an example the next to the highest string of a violin as it vibrates under the bow. If no finger is placed upon the string, it will vibrate from the nut to the bridge. The string has a frequency of 440 hertz and its pitch corresponds to the tone of A above middle C in the musical scale. If we press a finger against the string about eight centimeters from the end of the fingerboard, we decrease the length and therefore the mass of the part of the string that vibrates. The frequency is now raised to 594 hertz and the pitch is now the second D above middle C. Theoretically at least, we could also raise the pitch to B above middle C, even if we did not touch the string with a finger, by turning in a clockwise direction the peg to which the string is attached. In that way, we would increase the tension on the string. The chances would be good, however, that the string would break because of this increased tension before we could reach the desired pitch.

The *loudness* of a sound represents the intensity of the sound-wave vibrations as they strike our eardrums—that is, the quantity of sound energy that actually gets into our ears. Loudness depends on a number of factors—among others, the distance of the

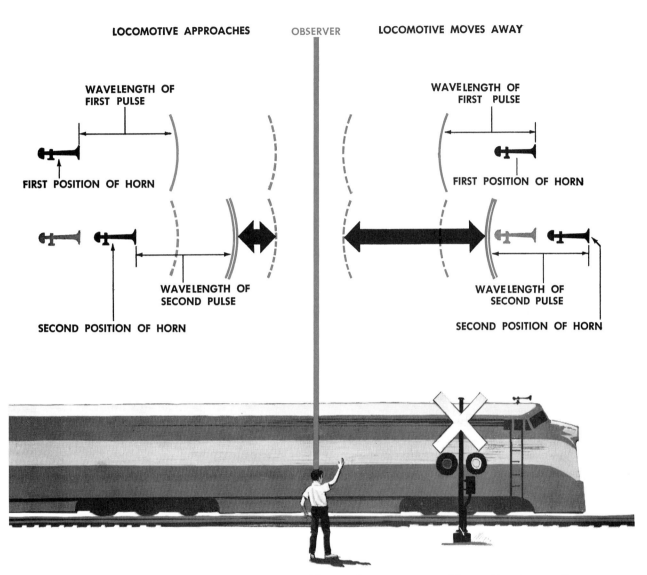

WAVELENGTH OF FIRST PULSE

FIRST POSITION OF HORN

WAVELENGTH OF SECOND PULSE

SECOND POSITION OF HORN

WAVELENGTH OF FIRST PULSE

FIRST POSITION OF HORN

WAVELENGTH OF SECOND PULSE

SECOND POSITION OF HORN

When a train is approaching an observer, waves sent forward by a horn on the forward-moving train tend to "catch up" with prior waves (left), thereby shortening the perceived wavelength for the listener. In other words, the sound will have a higher perceived pitch. When the train moves away from the listener (right), the opposite occurs: extended wavelengths and lower perceived pitch. These phenomena are known as the Doppler effect.

source of sound from our ears, the frequency of the sound wave, and the duration of the sound. A small change in the intensity level of a low-frequency pure tone produces a very large change in its loudness. You can note this effect when you listen to a marching band that is approaching you. The bass tones are inaudible until the band is quite near. Then they rapidly become prominent and, if you are very close, quite deafening.

If we were aware only of the pitch and loudness of sounds, it would be difficult to identify them. All sounds of the same frequency would sound more or less alike. We would not be able to tell the difference between the middle C produced by a singer's voice and the middle C that we hear when we strike certain tuning forks. We can distinguish between such sounds largely because of their differences in *quality,* or *timbre.* Quality is chiefly a matter of overtones. If a violin string is tuned so that it plays middle C, corresponding to a frequency of 264 hertz, it will vibrate as a whole. In doing so it will produce middle C, which is its fundamental tone. But smaller sections of the string will also vibrate at the same time, and each of these will produce a higher frequency. These additional frequencies are the overtones. In the case of a violin, they are fairly numerous and strong. They account for the rich and full sound of the violin as compared with a tuning fork, whose overtones are few and weak.

DEFECTS IN HEARING

Some persons are unable to derive as much information as they should from incoming sounds because they detect such sounds imperfectly or not at all. There are various kinds of hearing defects. In *transmission deafness,* the inner ear is normal, but difficulties in the outer or middle ear reduce the loudness. Sometimes this condition is temporary. It may occur if the Eustachian tube, connecting the middle ear with the pharynx, is clogged up. Or it may result from rapid changes of pressure, as when an airplane swoops downward from a great height. Another type of defective hearing is *nerve deafness,* arising from conditions in the inner ear. In this, the loss may be limited to certain frequencies, or it may be total. In *central deafness,* difficulties in the higher nervous centers prevent the interpretation of sounds, even though the ear is functioning normally.

Persons with defective hearing may often be helped by hearing aids. In the electronic hearing aid, sounds are picked up by a microphone, transformed into electric signals, amplified electronically, and then transformed again, in an earphone, into sound waves of the same frequency as the original sounds. In severe cases of middle-ear deafness, bone conduction is used to transfer vibrations to the inner ear, and a special vibrator is used instead of an earphone.

The early electronic hearing aids were heavy and bulky and were a definite burden to the user. The development of extremely small transistors, and small batteries, microphones, and earphones has reduced to a minimum the inconvenience of wearing hearing aids.

By measuring loudness, pitch, and quality, we can distinguish between various sounds even if they arrive at our ears at the same time. Thus, we can identify the individual tones that make up a simple chord. Or, if we listen to an orchestra, we can make out some of the instruments playing at the same time. Each sound wave generally affects our ears as if the other sound waves were not there.

There are certain exceptions, however. Suppose that there are two similar sources of sound, each sending out pressure waves at the same number of hertz, one stationed a little farther away from our ears than the other. At any given moment, while the wave from one source tries to push all the air near our ear to one side, the wave from the other source pulls the same layer of air to the other side. Since the waves from the two sources will constantly oppose each other as they arrive at our ears, they will produce no vibrations at all upon the eardrums. This is called *destructive interference* of sound. If the two sources of sound in question were placed differently, the two waves would reinforce each other. The resulting sound would appear to us to be twice as loud as each of the sounds by itself.

HOW BEATS ARE PRODUCED

If you listen to two sounds having a slightly different pitch, one wave will reinforce the other one or possibly interfere destructively. There is a gradual change from one condition to the other, say in half a second. Once every second, therefore, the sound will fade. Once every second it will become louder. In other words, our two sources of sound have brought about a succession of regular beats. In this case, one beat per second will be produced. The number of beats per second is equal to the difference of the two frequencies.

THE DOPPLER EFFECT

Sometimes the pitch that we hear does not correspond to the real frequency of the sound wave. For example, the sound of a locomotive whistle has a higher pitch as the train approaches you, but there is a definite lowering of pitch as the train moves away again. This is known as the *Doppler effect* from Christian Doppler, the 19th-century Austrian physicist who first discovered it. The explanation is this. If the frequency of the whistle is 2,000 hertz throughout the time that it is blowing, the wavelength will be about 15 centimeters. That means that each pulse will travel about 15 centimeters before the next one is emitted. If the whistle is mounted on a locomotive that is approaching you, each pulse will be a little nearer than 15 centimeters from the preceding pulse, as both approach your ear.

Since the distance between the pulses coming toward your ear is now less than 15 centimeters, the wavelength has become smaller. As we have pointed out, the smaller the wavelength, the higher the frequency. Since more than 2,000 pulses per second will reach your ear, the pitch will seem higher to you. Of course, after the locomotive passes you, just the opposite will be true. Doppler's effect applies to the wavelengths of light as well as to those of sound.

THE VISIBLE SPEECH TRANSLATOR

Pitch and other aspects of heard sound can be studied with an ingenious device called the Visible Speech Translator. The translator makes it possible to "read" the highly complex variety of sounds that make up human speech. It analyzes the speech sounds by a set of twelve filters, each 300 cycles wide, covering the range from zero to 3,600 cycles. The intensity in each of the 300-cycle bands is represented by a band of light on a moving screen. The pattern gradually fades as the screen moves to the left, and new patterns appear as the sounds succeed each other.

Operators of the translators have been trained to recognize the patterns and to carry on conversations with nothing more to go by than what they actually see on the screen. An engineer who was born deaf mastered this particular art. He was able to improve his speech very noticeably by

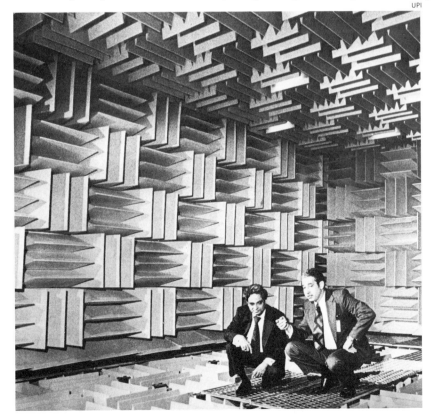

UPI

Acoustics engineers check out an anechoic (no-echo) chamber. This chamber will be used to investigate the causes of noise pollution.

comparing his own speech patterns with those of persons with normal hearing.

PROTECTING THE EAR FROM LOUD SOUNDS

It is sometimes desirable to protect the ear from the effects of unduly loud sounds, produced by explosions, powerful sirens at close range, and the like. This may be done by wetting a wad of cotton or daubing it with Vaseline and then putting it in the ear. Wax pellets have also been employed for this purpose. Specially designed ear defenders furnish quite satisfactory protection without reducing ordinary sounds too much.

In some cases, it is important to ward off the effects of continuously noisy surroundings, which make hearing difficult and cause fatigue. If the offending noise is concentrated in a room or building, it can be reduced at its source by substituting quiet machinery for noisy machinery. Noise may be absorbed, too, by installing special surfaces on walls, ceiling, or floor.

The general sound level in cities may be too high for the peace and comfort of the people living in them. Automobiles, horns, grinding brakes, radios going full blast, and the banging of ashcans may contribute to the general din. Various cities have tried to reduce the noise level.

THE IMPROVEMENT OF ACOUSTICAL QUALITY

The acoustical engineer is also concerned with improving the acoustical qualities of auditoriums, broadcasting studios, and the like. The pioneer in such studies was the 19th-century U. S. physicist Wallace C. Sabine, of Harvard University's Department of Physics. He became interested in this particular problem when a newly completed auditorium in the Fogg Museum at Harvard proved to be unbearably bad acoustically. The university authorities were mystified at first, because this auditorium was almost an exact replica of another that had proved to be very satisfactory. The only difference was that fireproof materials had been substituted for wooden surfaces.

The reason for the unfortunate situation soon became clear. The hard, reflecting surfaces of the new auditorium kept sounds bouncing back and forth, so that each spoken syllable had to compete with the slowly dying reverberations of preceding syllables. Sabine found that the reverberation time was five seconds against one or two seconds in auditoriums where the acoustic conditions were good. He borrowed a number of seat cushions from nearby Sanders Theater and installed them in the new auditorium; the reverberation time was cut down radically. He then worked out a reverberation-time formula that took into account the absorbing effects of the various surfaces in the room exposed to the sounds in question.

Nowadays the acoustical engineer tries to reduce all possible factors to definite formulas. Among other things, he has to take into consideration the sound absorbed by members of the audience. It may interest you to know that, from this viewpoint, you are equivalent to four-tenths of a square meter of open window.

In general, large concave surfaces are avoided, since they tend to concentrate the reflected sound at definite points within the auditorium. In a newly built concert hall, an orchestra conductor was appalled to find that sounds reflected from the dome over the room seemed to be concentrated on his podium, so that they sometimes drowned out the sounds coming from the orchestra. It was necessary to build up ridges on the dome; the construction was then hidden by a false ceiling that was transparent to sound waves.

In certain broadcasting studios, the walls and ceilings are provided with convex surfaces, forming portions of cylinders; the floors are left flat. In studios of this sort, sound is so perfectly diffused that microphones can be set almost anywhere.

Despite modern advances in acoustical engineering, some concert halls have revealed serious deficiencies. Apparently acoustics experts have still much to learn about the important factors involved in the transmission of sound and in the analysis of reverberation time.

Bruce Coleman

A Peruvian plays a rondador, or panpipe, used since ancient times.

MUSICAL SOUNDS

by Harold P. Knauss

Scientific knowledge has had little or nothing to do with the invention and improvement of most of our musical instruments or with the techniques involved in playing them. Modern laboratories devoted to the analysis of sound have failed to improve on the violins built in the early eighteenth century by Antonius Stradivarius. The amazing techniques and interpretative skills of past and present master performers have existed independently of science.

Yet musicians and scientists alike realize the importance of the scientific analysis of musical sounds. Such analysis helps the musician to recognize and solve many problems of musical composition. It also lays the foundation for more effective teaching and criticism. The scientist finds the study of musical sounds most rewarding. Since he is interested in the why and wherefore of phenomena, he wants to find the answers to many questions concerning tone production. Why does a violin not sound like a piano or a flute? Why does the tonal quality of a Stradivarius violin differ from that of a brand new fiddle that sells for much less? What are the mathematical relationships between the sounds that make up a musical scale?

Research on the scientific basis of mu-

Nogues – Gamma

Percussion instruments consist of rods, plates, and membranes that are designed to produce certain tones when they are struck.

sic has been beset with difficulties. Scientists are trained to measure and analyze, but few of them have enough musical training to understand the artistic problems of the musician. Musicians have an intimate knowledge of such problems, but are seldom familiar with *acoustics* – the science of sound – or with the mathematical language in which its findings are often expressed. Despite these difficulties, great progress has been made in the analysis of musical sounds. In the pages that follow, we shall give you some idea of what has been learned.

CHARACTERISTICS OF MUSICAL TONES

When we listen to a passage from a symphony or an aria from a familiar opera, we hear many tones blended so as to form a rich musical pattern. To analyze this pattern adequately, we must know something about the individual tones of which the pattern consists. Each tone has a certain pitch, from which the musician often deliberately strays in order to produce pleasing effects. It also has a certain degree of loudness, a definite tone quality, and longer or shorter duration.

Pitch. Pitch is the relative highness or lowness of a tone, and depends primarily upon the frequency of atmospheric vibrations that convey the sound. The more vibrations there are per second – that is, the greater the frequency – the higher the pitch; the fewer there are, the lower the pitch. In

what is called *standard,* or *philharmonic, pitch,* the frequency of 440 vibrations, or cycles, per second is assigned to the A above middle C. The pitch of a tone is indicated in musical notation by the place of the note on the staff.

Certain musical instruments, such as the piano, the organ, and the saxophone, are of the fixed-pitch variety. Once such an instrument has been tuned, it will automatically respond with a tone of a definite pitch when a certain key is pressed down. In the case of the human voice and such instruments as the violin and the trombone, the performer is entirely responsible for pitch.

Vibrato. You might suppose that one of the distinguishing marks of a good vocalist is the ability to stay on pitch. Actually, extensive studies at the University of Iowa showed that such is not the case. A trained singing voice wanders at the rate of about six times a second above and below the nominal pitch of the note that is being sung. This variation of pitch, accompanied by pulsations of loudness and tone color, is called *vibrato.* It gives a pleasing flexibility, tenderness, and richness to the tone.

Similar effects are produced on musical instruments. The tremulant stop of the organ and the shake of the left hand of the violinist are examples of this. Probably, the use of the vibrato in such instruments was derived from its natural presence in the human voice, which was used long before any man-made instrument. The ear responds more agreeably to varying stimulation, such as is given by the vibrato, than to a monotonous steady tone. Experiments

The pitch of a tone is indicated by the place occupied by the corresponding note on the staff. Here we show middle C and A above middle C. In standard pitch, A above middle C has a frequency of 440 cycles a second.

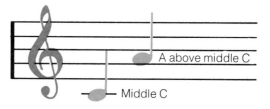

A above middle C

Middle C

have shown that the sensation produced by a steady tone quickly builds up to a maximum, and then drops off to a steady level, as if the stimulated sense organs had tired. The sensation produced by a vibrato maintains itself at or near its peak.

Loudness. Variations in loudness, or the ear's interpretation of the intensity of sound, are obtained in various ways, as we shall see when we come to examine the different musical instruments. The composer can indicate the degree of loudness he wishes in a particular passage by means of various symbols. Among these are *pp* (*pianissimo:* "very soft"), *p* (*piano:* "soft"), *mf* (*mezzoforte:* "moderately loud"), *f* (*forte:* "loud"), and *ff* (*fortissimo:* "very loud"). Much, however, depends on the musical taste of the performer.

Degrees of intensity of sound can be measured in units called *decibels*. The decibel is based on the ratio between the sound intensity to be measured and a standard intensity—the zero of the decibel scale. This standard represents one ten thousand millionth of a microwatt of sound pressure passing through one square centimeter of space. A microwatt is a millionth of a watt. A decibel may be described as the smallest degree of difference in loudness that a normal human ear can distinguish. A symphony orchestra has a seventy-decibel range, from a barely audible solo passage to an extremely loud one for full orchestra.

Some musical instruments of the past

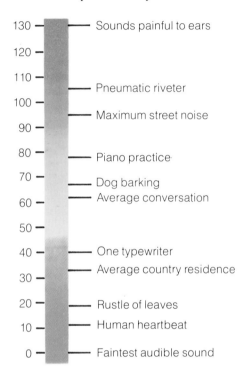

130 —	Sounds painful to ears
120 —	
110 —	Pneumatic riveter
100 —	Maximum street noise
90 —	
80 —	Piano practice
70 —	Dog barking
60 —	Average conversation
50 —	
40 —	One typewriter
30 —	Average country residence
20 —	Rustle of leaves
10 —	Human heartbeat
0 —	Faintest audible sound

Above: some familiar sounds are placed on the scale according to their approximate decibel levels. Below: a simple classification of sounds.

Classification of Sounds

Relatively simple . . . .
("Hooty," smooth, dark, dull at low pitches)

Moderate complexities . . . .
(Richer, normally most pleasant)

Greater complexities . . . .
(Rich, bright, brilliant, cutting, blaring or even strident)

Tuning forks
Falsetto "o"
Blown bottles
Wide, stopped, organ flue pipes
Female and boy soprano voices
Flutes
Female mezzo or alto voices
Open organ flue pipes
Soft horns
Soft male voices
Pianos
Strings
Wood winds, not including flutes
Organ reed pipes
Loud male voices
Loud brass
Chimes and bells
Ensembles
Noise

produced lovely sounds, but could not be played loudly enough to compete with other instruments in an orchestra or band. They have now been all but forgotten. With modern amplifying techniques, these instruments could be put to good use.

Tone quality. A flute may play a note corresponding in pitch and loudness to one played by a clarinet. Yet trained musicians and many others can easily distinguish between the instruments. That is because each has a distinctive tone quality, due to the presence of *overtones*, also called *upper partials*. Overtones have a higher pitch than the *fundamental* or *principal* tone and sound together with it. According to one theory—the *harmonic theory*—the differences between instruments are due to the frequencies of the overtones as compared to the fundamental frequency. According to the *formant theory*, the characteristic quality of an instrument is due to the strengthening of certain overtones because of resonance with some parts of the instrument.

The tuning fork produces a pure tone—that is, one that has only one frequency. A tone of this kind is quite uninteresting musically. The addition of overtones yields richer and more pleasant effects.

Duration and decay. The effect produced by tones depends to a certain extent upon their duration. Piano tones, for example, build up rapidly and decay, or die down, gradually, if the damper pedal is held down. This causes the dampers to leave the strings and allows the vibrations to continue unchecked. A bass drum has a rapid rise and fairly rapid decay.

MELODY AND HARMONY

The relationships between tones of different pitch are of the utmost importance. This holds true for the tones that follow each other in a melody, as well as for those which are sounded together in a chord. When two combined tones have a whole-number ratio, as explained later, they are called *consonant*. If they do not, they are called *dissonant*. A consonant tone is produced, for example, when you strike C and E at the same time on the piano. If you strike two adjacent keys on this instrument, you produce a dissonant note. Its different character is due to the clashing beats produced by the fundamentals and overtones of the two notes.

Not all dissonances are barred in music. The composer may introduce them because he wishes to arouse some certain emotion in the hearer. The extent to which dissonance is used in a musical composition depends on the taste of the composer and also on that of his public. The rules of harmony once prohibited many dissonant

TRUE DIATONIC, OR JUST, MAJOR

Letter	Syllable	Relative Frequency	Interval	Interval above C	Ratio to C
C	do	48		perfect octave	$2/1$
			semi $16/15$		
B	ti	45		major seventh	$15/8$
			major $9/8$		
A	la	40		major sixth	$5/3$
			minor $10/9$		
G	sol	36		perfect fifth	$3/2$
			major $9/8$		
F	fa	32		perfect fourth	$4/3$
			semi $16/15$		
E	mi	30		major third	$5/4$
			minor $10/9$		
D	re	27		major whole tone	$9/8$
			major $9/8$		
C	do	24		union	1

combinations that are used freely today.

The basis of harmony is found in the structure of the musical *scales* and *intervals*. A scale is simply a pattern or succession of tones, ascending or descending in pitch. The relative position of two tones on the scale is called the musical interval between the tones.

It was learned long ago that the interval between two tones depends on the ratio of their frequencies. The *octave*, for example, is the name of the interval between two tones, of which the higher has a frequency double that of the lower. The octave is found in all types of scales. Many ways of dividing it into smaller intervals have been used at various times and in various regions. Experience has shown that an interval between two tones is consonant if the ratio of frequencies between the tones can be expressed by small whole numbers. We may think of the scale as a solution to the problem of getting as many consonant intervals as possible into an octave.

The direct ancestor of the scale with which we are most familiar was the *just*, or *true, diatonic* scale. In the table on page 274, note how the scale is built up of three types of intervals, called a *major whole tone* ($9/8$), a *minor whole tone* ($10/9$), and a *semitone*, or *semi* ($16/15$). The figures in parentheses represent the ratio of frequencies between one tone and the next lower tone on the scale.

The tones of the just scale can be combined to produce consonant intervals in a great many ways. It has been shown mathematically that this way of dividing an octave produces more consonant intervals than any other. However, the just scale is not practical for instruments with intervals fixed in advance, such as the piano or organ. The reason is that the frequencies called for would be correct only in the key of C. To modulate, or change over from one key to another, on an instrument such as the piano, an excessive number of different tones would be required. The keyboard would be far more complicated than the one we now use. Even in the case of the unaccompanied voice, or of instruments such as the violin, in which the performer is free to

Victoria Beller-Smith/Photo Trends

The sounds of a piano are made by felt-tipped hammers striking strings of various lengths.

adjust the pitch of each tone, the just scale could not be followed exactly. The human ear would not be sensitive enough to perceive the slight changes in pitch required for just intonation.

A slight modification of the just scale leads to a scale that permits modulation from key to key. This is the *tempered scale*, the one that we actually use. In this, the octave is divided into twelve equal semitones. The ratio between a semitone and the one directly below it in the scale is approximately 1.0595. For example, in the key of A, based on A-440 (440 cycles per second) we would have the following table of frequencies: A-440; G♯ -415.3; G-392.0; F♯ 370.0; F-349-2; E-329.6; D♯ 311.1; D-293.7; C♯ -277.2; C-261.6; B-246.9; A♯ -233.1; A-220. If we multiplied any one of these frequencies by 1.0595, we would get the frequency of the note next above it.

As we have pointed out, consonance is associated with ratios between frequencies that can be expressed in small numbers. The more consonant intervals, arranged according to the degree of consonance, are unison (1/1), the octave (2/1), the fifth (3/2), the fourth (4/3), the major third (5/4), and the major sixth (5/3).

SURVEY OF MUSICAL INSTRUMENTS

The piano, the organ, and the instruments used in a band or symphony orchestra represent the survivors of a tremendous number of music-making devices that have served at one time or another. From the scientific point of view, all musical instruments fall into a few groups, depending on the mechanism used for generating the sound. The vibrations that result in music may be produced by bowing, plucking, or striking strings (stringed instruments); by blowing air across a sharp edge (flutes and organ); by setting a vibratory reed into motion (reed instruments); by blowing through the tensed lips (brass instruments); by striking various surfaces (drums, bars, bells, triangles, cymbals, gongs, and so on). The human voice has a unique generating apparatus, as we shall see. There are also some new claimants for a place in the musical world—the electronic instruments.

Stringed instruments. The laws of vibrating strings were discovered long before the days of modern laboratory equipment. Experiments were performed with long pieces of rope, which moved slowly enough

so that the vibrations could be counted directly. From the way in which these vibrations responded to changes in the tensile force, the length, and the weight of the rope, it was possible to deduce the frequencies of strings moving rapidly enough to produce tones. It was found that long, loose, and heavy strings vibrate more slowly than short, tight, and light ones.

The frequency of a vibrating string is inversely proportional to the length, other conditions being constant. For example, when the vibrating part of a string is cut in half, the fundamental frequency is doubled. Another factor controlling frequency is the force with which the string is stretched. Four times the force on a given string will double the frequency: in other words, the frequency is proportional to the square root of the force. A third factor is the weight of the string. If you increase the weight and therefore the inertia, you slow down the vibration rate, other conditions being equal. The frequency of a vibrating string varies inversely as the square root of its density — that is, its weight per unit length.

In a piano all three factors come into play, in a way dictated by the taste of the manufacturer and the length of the sounding board. Strings for the low notes are weighted by wrapping copper wire around them to obtain the desired rate of vibration

The slightly sticky (rosined) hairs of a bow rub the violin strings and make them vibrate.

without excessive length. The extra length of a concert grand piano makes possible the use of longer strings throughout and a larger sounding board. In this way more sound energy is released.

When a piano key is struck, lever action causes a felt-tipped hammer to strike a string or strings, corresponding to a given frequency, and sets the string or strings vibrating. It has been pointed out that in producing a single tone, the most accomplished piano virtuoso can do no better than a novice pushing down a piano key with the tip of an umbrella. Of course, music does not consist of isolated tones. The umbrella-wielding novice could hardly compete with a Vladimir Horowitz in the playing of a Liszt rhapsody.

The action of a violin bow is less obvious. As the bow, whose hair has been rubbed with rosin, begins to move in a given direction, the string sticks to the bow. It is pulled along with it for a moment and then slides back. As the bow continues to move in the same direction, the string is again pulled along and again slides back. The frequency of this stick-and-slip action is very nearly that of the free vibration of the string when it is plucked.

Variations of pitch in stringed instruments of the violin family are produced by pressing the strings with the fingers at different places on the fingerboard. Quality and loudness are controlled by varying the pressure of the bow on the string, and also by a device called a *mute,* which is set on the bridge of the instrument. The mute not only reduces the amplitude of vibration but also smothers the higher overtones more effectively than it does the lower ones. Different effects are obtained by plucking the strings of bowed instruments. This is known as *pizzicato* playing. Plucking is the normal method of playing the harp, mandolin, guitar, banjo, and similar stringed instruments.

Wind instruments. Air may seem to be a rather insubstantial vibrating body. Yet the thrilling chords produced by a big organ are eloquent evidence that vibrating air can produce tones of great variety, power and beauty. The elasticity of air was first studied by the seventeenth-century English physicist and chemist Robert Boyle. He described this elasticity as the "spring" of the air. Inertia is another quality of air. The combination of elasticity and inertia produces vibrations. The musical instruments in which a column of air is made to vibrate, producing musical sounds, are called wind instruments.

To study the vibrations of an air column, a tube of glass is fitted with a movable plunger, or piston, to change the effective

A chamber orchestra of violins, violas, and a cello.

length of the air column. A tuning fork at the mouth of the tube serves to set the air into vibration. As the piston is moved back from the mouth, a position is found for which a loud tone is observed. The air column of this length is said to resonate to the fork. Now if you put aside the tuning fork and blow across the end of the tube, you will hear a tone that is obviously of the same pitch as the tuning fork.

Instruments in which the vibrations of an air column produce musical sounds include the organ, the woodwinds, and several other kinds of instruments which we shall now take up in turn.

Edge-tone instruments. Did you ever blow against the edge of a blade of grass held between your hands, so as to produce a piercing tone? This is known as an *edge tone,* generated by the rapid eddying of air flowing past a sharp obstacle. The process is used to excite vibrations in certain organ pipes, known as flue pipes. Each of these pipes has a lip across which air is blown to make the pipe sound. Most of the pipes are open to the outer air at the ends. Some, however, are closed by a cap or stopper. A stopped pipe has a note that is an octave lower than an open pipe of the same length. The flue pipes make up the basic stops of an organ. The organ's reed pipes are based on a different principle. In them, a curved tongue of brass vibrates over the surface of a reed. We discuss the method of tone production by reeds elsewhere in this article.

Variety of tone color in organs is obtained not only by the method of setting the air into vibration and stopping or opening the end, but also by varying the wind pressure and adding additional resonating chambers. Tone color is affected, too, by the nature and thickness of the wall material of the pipes and by the ratio between their diameter and length.

The flute and the piccolo are excited by edge tones. In these instruments, the pitch is controlled by opening and closing properly placed holes in the side of the tube, and also by changing the blowing so as to produce the desired frequency.

Reed instruments. The air column of a reed instrument, such as the clarinet or saxophone, is excited by a reed made of bamboo or other elastic material and set in motion by the breath of the performer. The column of air vibrates at a constant rate, determined by its length. By opening or closing side holes in the tube, the effective length of the column is changed. The corresponding required change in the rate of vibration of the reed is accomplished by altering the lip pressure.

The oboe and bassoon are equipped with a pair of reeds. These alternately come together and move apart as the performer's breath strikes them, causing a series of puffs of air to enter the air column in the instrument. The oboe and bassoon yield a pleasing, penetrating tone color.

The reed pipes of the organ are designed to imitate various orchestral instruments, such as the trumpet, tuba, oboe, and clarinet. When an organ reed pipe sounds, it is possible to recognize the instrument that is being imitated. However, the analysis of wave forms shows that the imitation is not exact.

Small areas on steel drums are heated by a blow torch to various temperatures, giving the area a specific note. They were first made in Trinidad.

The sounds made by the brasses are produced when a column of air in a metal tube is made to vibrate by the performer's lips. The pitch is determined by the number of lip vibrations and the length of the tube.

Percussion instruments. The percussion section of the orchestra includes a variety of instruments that are sounded by striking. They contribute to rhythms and colors of a musical performance. Some of the percussion instruments, such as snare drums, cymbals, and tambourines, do not produce musical tones, but rather sounds more correctly described as noise. At the other extreme are tuning forks and triangles, which produce a pure tone. The vibrating elements of percussion instruments — used with or without resonating air columns — are membranes in the case of drums. They are rods in the xylophone; bent rods in the triangle or tuning fork; plates or modified plates in the celesta, glockenspiel, cymbals, gong, and bells.

The membranes of drums are stretched, flexible sheets, usually circular in form. Bass and snare drums have no definite pitch and are used simply to emphasize the rhythm. In the kettledrum, a hemispheric cavity forming the body of the instrument resonates to the fundamental frequency of the membrane. The membrane is tuned by adjusting the tension in order to produce a range of pitches. The name timpani is applied to a set of kettledrums.

The tuning fork consists of two prongs vibrating on a common base. It is important mainly as a standard source of single-frequency tones, rather than as an instrument of musical expression. Consequently, it is far more valuable in the laboratory than in the musical studio. Its nearest relative in the orchestra is the triangle. The xylophone is a collection of straight bars with resonators, excited by striking with hammers.

Vibrating plates are capable of various modes of vibration. These are readily studied by dusting sand on the plate. The sand dances as the plate vibrates and it forms definite patterns. These are called *Chladni figures* in honor of the 18th-century German physicist Ernst F. F. Chladni, who first called attention to them.

A bell is a sort of curved plate. While it is possible to study the vibrations of bells with great accuracy, there is no very direct connection between the vibration frequencies and the pitch as judged by the ear. One of the interesting characteristics of a bell tone is the throbbing or pulsating of the tone as it dies down. This is undoubtedly caused by the interference of neighboring but distinctly different overtones.

Brass instruments. The simplest of the brass instruments is the bugle. It consists of a tapering air column in a brass tube, which is curved for the sake of compactness. The lips of the performer act as a double reed, through which the breath is blown to excite the air column. The tones that can be produced on the bugle by the control of lips and breath are simply the harmonic overtones of the fundamental frequency of the air column.

The trombone is provided with a sliding device by means of which the effective length of the air column can be varied. In this way the performer can obtain all twelve notes of the scale in each of three octaves. The performer has full control of the pitch, just as a violinist does. Other brass instruments, including the French horn, the trumpet, the cornet, and the tuba, have valves. These bring into play one or more side tubes to change the length of the column.

The voice. The voice is at once the oldest and the most versatile of musical instruments. It excels in expressing emotion and in variety of tone color. The singer never has full conscious control of all the elements influencing the tone, nor is the mechanism known or understood in all of its details. Different teachers of singing have obtained outstanding results by using methods based on contradictory theories.

The basis of the voice is a pair of cushionlike vocal cords, set in vibration by air from the lungs. These vocal cords are like a double reed, allowing puffs of air to get through as they alternately come together and move apart. The quality and tone color depend on the size and shape of various resonating cavities, including the windpipe, the back of the throat, the mouth, and various other air-filled parts of the head. Apparently the higher-pitched tones of sopranos are generated as almost pure tones. In this case, there are no strong overtones.

An outstanding characteristic of good voices, as we have noted, is the vibrato, or rhythmic variation in pitch of sustained tones. Smoothness and regularity of the vibrato seem to be invariably present in voices with the most beautiful tone quality. Singers speak of placing the voice. This means simply using the vocal cords and the resonating cavities for a desired effect.

Electronic musical instruments. Some of the most striking new musical instruments are based, in part at least, on electronic principles. In some of these instruments, as in the electric guitar, electric current and electronic circuits are used to amplify tones that have been produced mechanically. In other cases, electric current and electronic circuits produce the tone, control it, and amplify it.

In the electronic organ, there are a number of metal disks, which are made to rotate by an electric motor. Each rotating disk is set in front of a magnet on which a coil of wire is wound. There is one disk for each fundamental tone represented on the keyboard. The rim of a given disk is made up of alternating teeth and slots. As teeth and slots move past the tip of the magnet, they bring about a regular series of changes in the magnetic field. As a result, an alternating electromotive force is generated in the coil of wire. This is amplified in an electronic circuit, and it operates a loudspeaker. The frequencies obtained in this way can be mixed so as to provide a great variety of tones. The instrument is equipped with two manuals (keyboards for the fingers) and a pedal keyboard.

An experimental electronic music synthesizer, developed by the engineers of RCA, was publicly demonstrated in 1955 and caused a sensation. It imitated individual instruments, such as the piano, with astonishing fidelity. It was not so effective in combining the sounds of many instruments, as in symphonic music. Today, improved synthesizers can accurately imitate the sounds produced by a large orchestra.

In an electronic music synthesizer, the properties of desired musical sounds are built up into a short musical phrase by utilizing a great number of electronic circuits. A series of such phrases is then combined to form an entire musical composition. The composition is in the form of electrical impulses, which are transformed into audible sound in a loudspeaker.

Besides the obvious possibilities of electronic imitation or duplication of familiar musical sounds, electronics makes possible the exploration of hitherto unattainable varieties of sounds. Tape recorders can be used to transpose the frequencies of conventional instruments upward or downward. One simply runs the tape faster or slower during reproduction than during recording. Sound tracks for sound films can be made by hand, or drawn on a large scale and copied photographically.

Electronics has enriched music by radio, television, records, stereophonic and quadraphonic systems, magnetic tapes, and synthesizers. And now, completely electronic music can be created. What will result from the new technical tools available to composers remains to be seen.

OPTICS

by J. H. Rush

Scientists have realized for years that light is a form of radiant energy. Light, X rays, radio waves, and gamma rays are all electromagnetic radiations. They are so called because they are sent out into space as a result of variations in electric and magnetic fields.

Not so many years ago, scientists were quite sure that light and other radiant energy consisted of electromagnetic waves rippling through space. They distinguished sharply between the wave character of radiation and the particle character of matter. Today the picture is not so clear. In some respects, at least, light and other radiations behave like streams of particles. What is even stranger, perhaps, electrons and other very small particles sometimes behave like waves.

The peculiar dual nature of radiation and of material particles is the subject matter of wave mechanics. It would take a great deal of space to explain this theory, even nonmathematically. It is perhaps enough to say here that neither radiation nor material particles ever exhibit both wave and particle characteristics at the same time. In transit—that is, as it moves from place to place—light behaves like a system of waves. It has a fixed speed in empty space. Its wavelengths can be measured by a variety of methods. In many ways its behavior is entirely consistent with the theory that it is an electromagnetic wave disturbance. But in its emission from electrons in an atom and in its absorption or other reactions with atoms in its path, light behaves more like a stream of very small particles of energy—*photons*.

The old argument over the nature of light seems to have ended in a draw. Instead of saying that light, or any other kind of radiation, consists of waves or particles, physicists recognize that it behaves like waves in certain respects and like particles in others.

The speed of light plays an important part in astronomical and physical research. It is extremely important, therefore, to measure it as accurately as possible. For nearly three centuries scientists have gradually refined and improved their techniques, until now the speed of light is known to within about one part in a hundred thousand. It is nearly 300,000 kilometers per second.

Strictly speaking, *light* means any radiation whose wavelengths excite a sensation of brightness, or illumination, in the retina of the eye. These wavelengths range from about 0.00004 centimeters to 0.000076 centimeters. Certain types of radiation—ultraviolet light and infrared light—are similar to visible light, though their wavelengths are somewhat shorter or longer than those of the visible range. In the following pages we will use the word "light" in the sense of all wavelenths that behave similarly to those that excite vision.

WHEN LIGHT ENCOUNTERS MATTER

When light encounters some form or other of matter, it behaves in different ways. For example, when it strikes an opaque object (one that does not transmit light), the object casts a shadow. Much of the early argument as to whether light is made up of waves or particles hinged on observations of shadows.

If light consists of a shower of flying particles, those striking an opaque barrier

Fig. 1. Shadow formation. In A we assume that light consists of particles, in B, that it has a wavelike nature.

 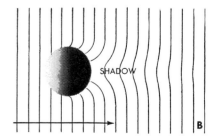

in their path would be stopped. Other particles would miss the edge of the barrier, as shown in Figure 1A. If the latter struck a screen set beyond the obstacle, a geometrically sharp shadow would appear on the screen. That is apparently just what happens. In this respect light seems to act unlike waves of water or sound, which bend around obstacles. That is why Sir Isaac Newton and other thinkers came to the conclusion that light does not consist of waves, but of tiny flying particles.

Careful observation, however, shows that light does bend around obstacles (Figure 1B). Short waves of any kind bend less than long ones, and light waves are very short indeed. Yet under special conditions this bending of light at edges—a phenomenon called *diffraction*—produces striking effects. A certain amount of light makes its way into the shadow area. Ordinarily it is not noticeable. However, in the case of a small ball or disk, all the light diffracted around the edge crosses at the center of the shadow, and a bright spot appears. Immediately outside the shadow cast by the sharp edge, the light is broken up into a series of bands, or fringes, that alternate in intensity.

When light falls on a material barrier, some of it is *reflected*. It bounces back from the surface, like a ball from a floor. If the material is not quite opaque, some of the light is transmitted through it at reduced speed. In its course it is *refracted*, or bent. Usually a large portion of the light is absorbed in the material. It disappears as light, and its energy is changed into other forms.

The most common effect of absorption is the change of light energy into heat. The rapidly oscillating electric and magnetic fields involved in the light wave set the electrons in the material to vibrating. As a result, all the molecules of the material vibrate more energetically and the material becomes hotter. To understand why this is so, one must bear in mind that heat is simply the kinetic energy, or energy of motion, of the molecules of a substance. The more vigorously its molecules vibrate, the hotter the substance is.

Light energy is not always transformed directly into heat. If it strikes a clean metal surface in a vacuum, the light energizes some electrons so violently that they jump out of the metal and fly through the adjacent space. This so-called *photoelectric effect* is used in television cameras and many other electric-eye devices. Light also causes chemical changes in many substances. It can bleach various dyes, cause intense irritation of the skin, or produce subtle changes in the silver chloride of a photographic film. From the human standpoint, the most important photochemical effect of light takes place in a substance in the retina of the eye, called visual purple. This pigment is partially bleached by light, and the chemical change that takes place is an essential part of the visual process.

Light affects matter chiefly by acting upon the outer electrons in atoms. When light energy is absorbed by electrons, each electron acts like a tiny radio antenna, radiating back part of the energy it has received. Some of this reradiated light is reflected back in the direction from which the light originally came.

In describing what happens when light strikes a surface, we shall have occasion to speak of rays and beams of light. Rays will be represented by arrows in our diagrams. We must think of a *ray* as representing the direction in which a wave front is advancing. A bundle of rays is called a *beam*.

When light is transmitted through space, the waves spread out at equal speed in all directions from the source, as shown in Figure 2. By the time a wave of light from a point on the sun reaches the earth, some 150,000,000 kilometers away, the part of the wave that strikes a surface such as a mirror or a pavement is only very slightly curved. For all intents and purposes the wave front is a straight line and the rays of light are parallel to each other (Figure 3).

REFLECTION OF LIGHT RAYS

The light rays that fall upon a surface are called *incident rays*, from the Latin *incidere*, meaning "to fall upon." If the rays strike a surface perpendicularly, they are

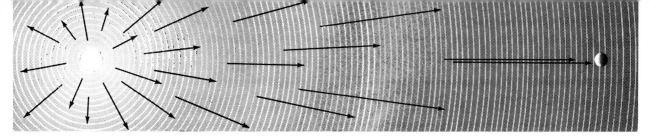

Fig. 2. In the diagram above, we are considering light as being wavelike in nature. From the light source at the left the light waves spread out equally in every direction.

Fig. 3. By the time light waves from the sun strike the earth, each wave front is so slightly curved that it is almost a straight line. Light rays are practically parallel.

reflected straight back. Figure 4 shows what happens when they fall obliquely upon the surface. AB is the incident ray, which strikes the surface at B. BC represents the ray after it has bounced back from the surface, as a reflected ray. Let us draw a line, BD, at right angles to the surface at the point B. This perpendicular is called the *normal*. The angle ABD (or I), formed by the incident ray and the normal, is the *angle of incidence*. The reflected ray and the normal form another angle, CBD (or R), which is known as the *angle of reflection*. The angle of incidence is always equal to the angle of reflection. The incident ray, reflected ray, and normal, all lie in the same plane. This statement is the *law of reflection*.

If the entire reflecting surface is very smooth, regular reflection takes place — that is, light is reflected in a regular, orderly way. The surface need not be flat. If it is curved, the curve must be smooth and gradual. If such a smooth surface, flat or curved, is capable of reflecting most of the light that falls on it, it is called a *mirror* (Figure 5).

Mirrors are scarce in nature. Most objects would absorb too much light to be good mirrors, even if they were smooth, and comparatively few are smooth. Dirt, stones, leaves, wood, hair, skin — nearly all common materials have surfaces that are very rough and irregular, from the atom's-eye view of a light wave. We may think of

such a surface as made up of a great number of very small smooth surfaces (Figure 6). Light is reflected from each small surface element according to the law of reflection, but the rays from the various elements go off in many different directions. The reflected light is diffused, or scattered. This process is called *diffuse reflection*.

Our visual judgment of the textures of different surfaces depends mainly upon the degree of diffuseness of the light they reflect. Consider, for example, the difference in appearance between a piece of black velvet and one of polished black marble or obsidian, or between a good mirror and a slab of plaster of Paris.

Good mirrors are relatively hard to make. Not only must the surface be smooth to within a few millionths of a centimeter. It must also absorb no more than a small amount of the incident light. With a few special exceptions, metals are by far the best mirror surfaces. Why? Metals conduct electricity, and that means that their electrons are not bound to their respective atoms so strongly as in the nonmetals. Consequently, the electrons in metals have more freedom to vibrate in unison with the incoming light waves, and to reradiate the energy they absorb. In nonmetals, more of the absorbed energy is trapped as heat. Except for gold, silver, and copper, metals are rarely found uncombined with other elements. That is why natural mirrors are scarce.

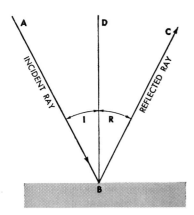

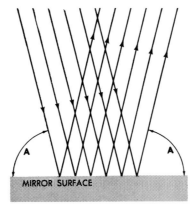

In Fig. 4, extreme left, we see light striking an ordinary surface and reflected from it. Angle I is the angle of incidence, angle R, the angle of reflection. The other drawing (Fig 5) shows light being reflected from a mirror surface. Angles A of incoming and reflected waves with the surface are equal.

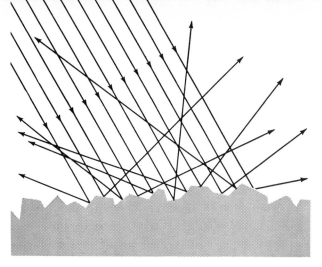

Fig. 6. When light strikes a rough surface, made up of many tiny smooth surfaces, it is reflected in many directions.

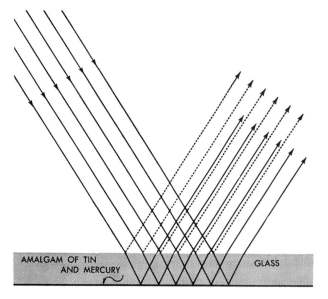

AMALGAM OF TIN AND MERCURY   GLASS

Fig. 7. Light that strikes an ordinary mirror of plate glass is reflected from the coating on the back of the glass.

Fig. 8. Light reflected from a precision mirror is reflected from a metal film on the front surface of the mirror.

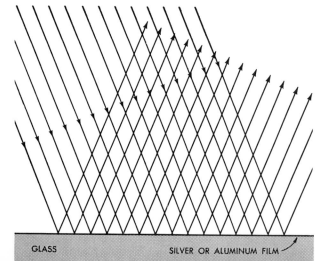

GLASS   SILVER OR ALUMINUM FILM

Ordinary household mirrors are made of plate glass coated on the back with an amalgam of tin foil and mercury, which is painted black to protect it. The metal film itself is the mirror. The glass serves only to shape the film and protect it (Figure 7). Such a mirror is very durable, but it is not satisfactory for a telescope or any other precision optical instrument. For such exacting purposes, a piece of plate glass is so irregular that it may be compared to a plowed field. The front surface of the glass in an ordinary mirror reflects enough light itself to produce an undesirable secondary image.

Glass for a precision mirror must be ground and polished to near-perfect smoothness. A thin layer of metal—usually aluminum—must then be evaporated in a vacuum onto the front surface of the glass. The glass itself merely supports the metal film (Figure 8). Such a mirror reflects up to 95 per cent of the light that falls upon it.

ABSORPTION OF LIGHT

Most substances absorb practically all the light they receive before it penetrates to any appreciable distance. Such substances are said to be opaque. A few natural substances, besides gases, and many more artificial ones are transparent. They permit substantial fractions of light to penetrate through moderate thicknesses of material. All substances absorb some light energy. Transparency is only relative. Good optical glass will transmit more than 99 per cent of the light entering a plate 2.5 centimeters thick, but such transparency is uncommon.

The light that strikes transparent substances usually is a mixture of all the different wavelengths that we see as white light. Transparent materials absorb some wavelengths more than others. Since different wavelengths cause different sensations of color to arise in the brain, very few transparent materials appear colorless in transmitted light. Ordinary glass favors the green wavelengths slightly. It is practically opaque in the ultraviolet region. A thin film of silver, such as is used on some mirrors, is quite opaque to visible light, but is transparent in the ultraviolet range.

Some materials allow light to pass through, but they scatter and diffuse it. Such a material is called *translucent*. Its effect can be compared to that of diffuse reflection. Opal glass and certain kinds of paper are very familiar examples of translucent substances.

REFRACTION OF LIGHT RAYS

When an incident ray of light enters a transparent substance at some angle other than the perpendicular, or normal, to the surface, its direction is changed. This effect is called *refraction* (bending). The ray whose course has been changed in this way is known as a *refracted ray*. Refraction takes place because the speed of light in a vacuum (or in air, which in this case amounts to nearly the same thing) is greater than in water, say, or glass.

To illustrate the principle involved in refraction, let us suppose that a column of marching men crosses the boundary between a smooth parade ground and a plowed field at an angle, as in Figure 9. They cannot march as rapidly on the rough ground as they can on smooth terrain. In each rank the men who first reach the rough ground slow down while those to their left gain on them. As a result of this difference in marching speed, the direction of march of the column is changed.

In Figure 10 we apply this principle to the refraction of light. The rays AA′ indicate the direction of a train of light waves advancing in air and approaching the surface of a transparent material SS. The speed of light in this material is less than the speed of light in air. As the advancing wave front reaches the position CD, the part near C enters the material and is slowed down. The portions of the front nearer D continue to advance in air at their former speed. As a result, the advancing wave front swings toward the normal, or perpendicular.

By the time the wave front at D has reached the surface at E, traveling the distance *a*, the light that entered at C has traveled a lesser distance, *b*, to the point F. Other points of the wave front will have traveled intermediate distances, so that the new wave front in the transparent sub-

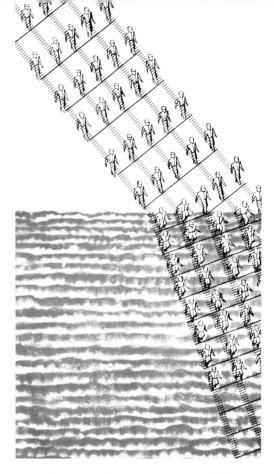

Fig. 9. A column of marching men would change direction as it passed from a smooth parade ground to a rough field. On the plowed field the men could not march so rapidly.

stance will be EF. Note that the angle between the refracted ray and the normal NN (angle R), called the *angle of refraction,* is smaller than the angle of incidence (angle I). A ray will be turned toward the normal when its velocity is reduced. It will be turned away from the normal when the velocity is increased. This happens, for example, when light passes from water into air.

INDEX OF REFRACTION

The ratio of the speed of light in air to the speed of light in a given substance is called the *index of refraction* of that substance. The index of refraction can be defined more exactly as the ratio of the sine of the angle of incidence to the sine of the angle of refraction. Let us see what this means. In Figure 11, a ray of light strikes the surface of a transparent substance — say, water — so as to form the angle of incidence I. It is then refracted; the angle of

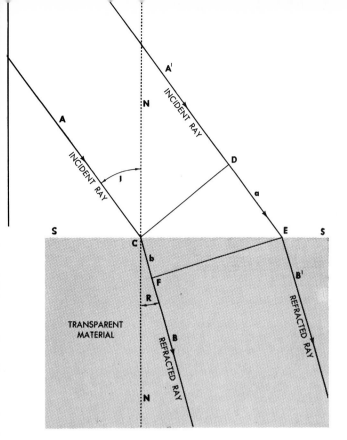

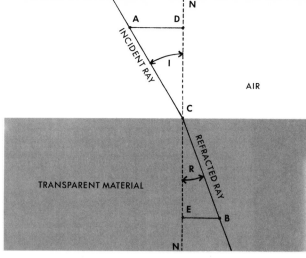

Fig. 11. How to find the index of refraction of a transparent substance. Details are given in the text.

Fig. 10. This diagram shows how a wave of light is refracted. Details about refraction are given in the text.

refraction is R. We mark off equal distances AC and BC on the incident ray and the refracted ray. Then we draw perpendiculars AD and EB to the normal NN. ADC and ECB are two right-angled triangles, with equal hypotenuses.

The sine of any angle of a right-angled triangle is the ratio of the side opposite that angle to the hypotenuse. The sine of angle I, therefore, is AD/AC. The sine of angle R is EB/CB. Since the index of refraction is the ratio of the sine of the angle of incidence (I) to the sine of the angle of refraction (R), it is

$$\frac{AD/AC}{EB/CB}$$

The hypotenuse AC is equal to the hypotenuse CB. Therefore, the index of refraction is equal to AD/EB.

The sines for all angles up to 90° are given in trigonometric tables. Once we know the angle of incidence and the angle of refraction as a ray of light strikes and then penetrates a given substance, it is a

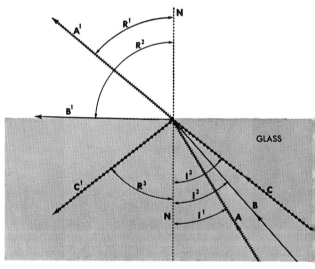

Fig. 12. Total internal reflection: ray C is reflected from the surface of the glass as C'.

simple matter to find the sines of the two angles and to determine the index of refraction. The index of refraction of water is about 4/3, or 1.3. That of common glass is about 3/2, or 1.5.

We saw that when light passes obliquely from a transparent substance into air, the rays bend away from the normal. This may result in the effect called *total internal reflection*. In Figure 12, the ray of light A passes through glass and is bent away from the normal as it moves through air (as A'). In this case, the angle of incidence is I[1]. The angle of refraction is R[1]. If the angle of incidence is increased, the angle of refraction will be increased even more, proportionately, and the refracted ray will be closer to the surface of the glass.

When incident ray B passes through the glass, the angle of refraction (R^2) as it penetrates the air (as B') is 90°. The ray will graze the surface of the glass, as shown. Angle I^2, formed by the incident ray B and the normal, is called the *critical angle*. If the critical angle is exceeded, the ray does not leave the glass at all but is totally reflected from the surface. This happens, for example, when ray C passes through the glass, forming the angle of incidence I^3, which is greater than the critical angle I^2. The ray is reflected from the surface (as C'), following the law of reflection.

The critical angle depends on the index of refraction on either side of the boundary surface. It is about 41° for a glass-air boundary, and about 49° for water-air. Total reflection can occur only in the medium that has the higher index of refraction.

The effects we have been describing apply to light of one particular wavelength. Light is usually made up of many wavelengths, which are not refracted uniformly. Nearly all transparent materials refract short waves more than long ones. This effect is the basis of the most common type of spectroscope, in which a refracting prism is used to separate the wavelengths that make up light.

Refraction effects may be used to identify small samples of mineral crystals. This is done by immersing them in a succession of liquids, each with a different refractive index. A transparent crystal or other material is visible in air mainly because it refracts light passing through it. It also reflects some of the incident light. It is visible for the same reasons in liquids that have a different index of refraction from its own. Suppose, however, that the crystal is placed in a transparent liquid whose index of refraction is identical with that of the crystal. The light rays will then pass through both liquid and crystal without reflection or refraction and the crystal practically disappears. An imitation diamond of ordinary flint glass vanishes in carbon disulfide, while a real diamond continues to sparkle in the liquid.

FORMING IMAGES WITH MIRRORS

The simplest way to form an optical image is to allow light from a glowing or illuminated object to pass through a pinhole and to let it fall on a white screen. Some light will be diffracted from the edge of the pinhole. Most of it, however, will pass through the hole and will form an inverted image on the screen. We show how such an image is produced in Figure 13. Light from the bottom of the head passes through the pinhole and forms the top of the inverted image on the screen. Rays from the top are directed by the pinhole to the corresponding part of the image. Of course light is spreading in all directions from the illuminated object, but the pinhole allows only the light traveling in certain definite directions to reach the screen.

An important factor in any image-forming system is its magnification. This term means the ratio of any dimension — height or width, for example — of the image to the corresponding dimension of the object. In a pinhole system, such as is shown in Figure 13, the magnification is the ratio of the image distance q to the object distance p. These distances can be made whatever one likes.

To obtain a brighter image than the pinhole permits, we need a device that will collect a substantial quantity of the light radiating from an object. The mirror is just such a device.

Plane mirrors. The most familiar type of mirror is the plane mirror, which has a flat, or plane, surface. Figure 14 shows a typical mirror of this kind. Light coming from a series of points on the object is reflected by the mirror to the eye as shown, creating the illusion of coming from a corresponding series of points behind the mirror. As a result, a *virtual,* or apparent, reversed *image* is formed. It is the same size as the object itself and is at the same distance behind the mirror as the object is in front of it. The image is reversed from left to right. The plane mirror is the only optical device that is capable of forming a completely accurate

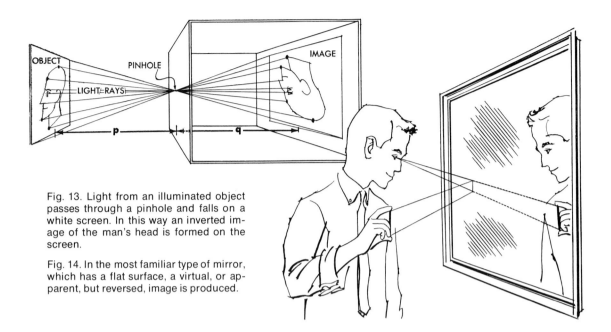

Fig. 13. Light from an illuminated object passes through a pinhole and falls on a white screen. In this way an inverted image of the man's head is formed on the screen.

Fig. 14. In the most familiar type of mirror, which has a flat surface, a virtual, or apparent, but reversed, image is produced.

image (in reverse, of course) of an object of appreciable size.

Spherical Mirrors. Some mirrors are curved. Such mirrors generally form part of the surface of a sphere. In Figure 15 we show a concave spherical mirror. The center of the sphere of which the mirror forms a part is called the *center of curvature.* The line passing through the middle of the mirror and the center of curvature is the *principal axis.*

Most spherical mirrors are relatively flat. This means that the diameter of the mirror is small compared with the radius of curvature — any straight line drawn from the center of curvature to the surface of the mirror. In concave mirrors of this type the rays of light parallel to the principal axis will pass through, or very near, a common point after they have been reflected from the mirror surface. This point (F in Figure 16) is called the *principal focus.* It is on the principal axis, half way between the center of curvature and the surface of the mirror. It is known as a *real focus* because the rays of light actually pass through it. We can show that this is so by placing a piece of paper at the focus of a concave mirror upon which sunlight is falling. The rays converg-

ing at the focus will soon cause the paper to burst into flames.

Not all the rays parallel to the axis of a concave spherical mirror meet exactly at the principal focus. This failure of the rays to converge at a single point is called *spherical aberration* (see Figure 17). Spherical aberration is comparatively small when the mirror is almost flat. It is larger when the mirror is more rounded. This effect can be remedied by making the curve of the mirror a paraboloid instead of a sphere (Figure 18). This figure is commonly used in searchlights, automobile headlights, and various other devices. Large mirrors used in telescopes are usually paraboloids.

Some spherical mirrors use the convex side of a spherical surface (Figure 19). In this case, parallel rays diverge from the surface when they are reflected. However, they all seem to come from a single point behind the mirror (F in the diagram). This is called a *virtual* (apparent) *focus,* because the rays do not actually converge at this point but only appear to do so. If one exposed a convex mirror to sunlight and placed a piece of paper at the virtual focus, nothing would happen to the paper.

Real and virtual images. If all the rays

that struck the surface of a spherical mirror were parallel to the principal axis, they would not produce an image of appreciable size. The reason is that they would all really meet (concave mirror) or apparently meet (convex mirror) at about the same place. Actually, however, many of the rays are not parallel to the principal axis. As a result, they produce an appreciable image.

The images produced by concave spherical mirrors may be smaller or larger than the objects. They may be virtual or real. As in the case of the plane mirror, an image is virtual if the light rays only seem to form it after they are reflected from the mirror surface. The image is real if it is actually produced by the rays. Both types of images look the same. However, a real image can be thrown on a screen, while a virtual one cannot. A real image is an actual convergence, or concentration, of light. A virtual image is an optical illusion.

If an object is placed at the center of curvature of a concave mirror, the image is real, inverted, and the same size as the object (Figure 20). The image is also real and inverted if the object is placed farther away from the mirror than the center of the curvature. In this case, however, the image will be smaller than the object. It will dwindle to a point as the object is moved to a very great distance from the center of curvature.

Suppose we place the object between the center of curvature and the principal focus. The image will still be real and inverted but it will be larger than the object. We could also put the object between the principal focus and the mirror. In that case a virtual, erect image, larger than the object, would appear apparently behind the mirror (Figure 21). This is what happens when one uses a concave shaving mirror. One sees an erect image of one's face larger than life.

Convex spherical mirrors can produce only virtual images, which seem to be located behind the mirror. These images are erect (not inverted) and smaller than the object (Figure 22). Small convex mirrors are sometimes fastened outside a car near the windshield to bring into the driver's

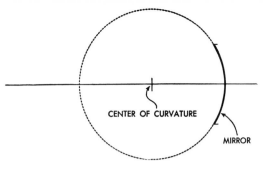

Fig. 15. A concave spherical mirror, part of the surface of a sphere whose center is called the center of curvature.

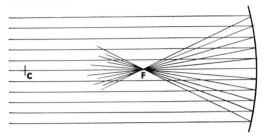

Fig. 16. The light rays that are reflected from the surface of a concave mirror pass through the principal focus, F.

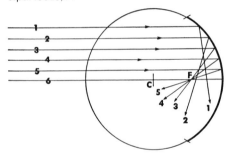

Fig. 17. Spherical aberration, which is explained in the text.

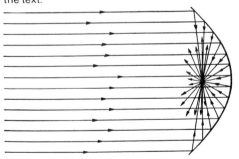

Fig. 18. Spherical aberration can be remedied by making the curve of a mirror a paraboloid rather than a sphere.

Fig. 19. Some spherical mirrors use the convex side of a spherical surface. The virtual (apparent) focus is F.

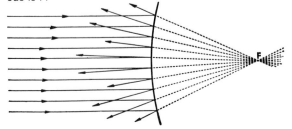

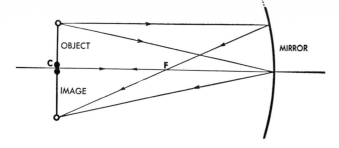

Fig. 20. If an object is placed at the center of curvature of a mirror that has a concave surface, the image is real, inverted and also of the same size as the object.

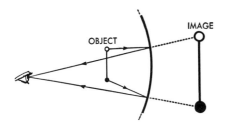

Fig. 21. When an object is placed between the principal focus and the mirror, a virtual, erect image, larger than the object, will appear — and apparently behind the mirror.

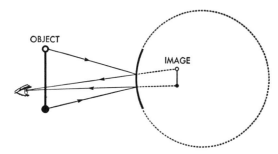

Fig. 22. The image in a convex spherical mirror is virtual, erect instead of inverted, and smaller than the object.

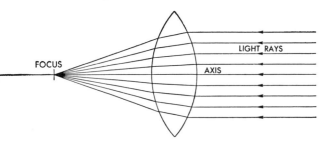

Fig. 23. A converging lens. Rays parallel to the principal axis meet at the principal focus.

Fig. 24. A diverging lens. Rays parallel to the principal axis seem to meet at the virtual principal focus, F.

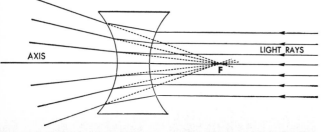

view cars that are approaching him from behind.

FORMING IMAGES WITH LENSES

So far we have been dealing with images formed by mirrors. Images can also be produced by lenses. A lens is a piece of transparent material, such as glass, shaped in such a way that it intercepts rays of light from an object and bends them by refraction so that they meet at a common focus. A simple lens has two opposite regular surfaces, which are generally spherical. In some cases, one surface may be a plane.

There are two chief kinds of lenses — *converging* and *diverging*. A converging lens is thicker at the center than at the edges. Light rays parallel to the axis converge, or meet, at a point beyond the lens (Figure 23). This point is the principal focus. A diverging lens is thicker at the edges than at the center. When light rays parallel to the principal axis pass through the lens, they are spread apart. In this case the rays seem to come from a virtual focus on the same side of the lens as the light source (Figure 24). This point, F, is the principal focus. In both converging and diverging lenses, the distance from the principal focus to the lens is called the *focal length*. The focal length depends on both the index of refraction of the material and the curvatures of the two surfaces of the lens.

In Figure 25 we show some common types of converging and diverging lenses. Note that in each case the principal axis is a straight line drawn through the centers of curvature of the two spherical surfaces. If one surface is plane, the principal axis passes from the center of curvature to the center of the lens.

The relationship between the object and the image formed by a converging lens is much the same as in a concave mirror of the same focal length. The only difference is that since light passes through the lens, real images are formed on the side opposite the incoming light, while virtual images are formed on the side from which the light comes. As in the mirror, the nature of the image will depend upon the distance of the object from the lens (Figure 26). Diverging

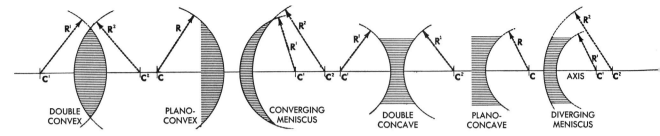

Fig. 25. Six important types of lenses are illustrated above. The first three—the double convex, the plano-convex, and the converging meniscus—are converging lenses. The last three—the double concave, the plano-concave, and the diverging meniscus—are diverging lenses. C stands for "center of curvature"; R stands for "radius."

lenses produce much the same type of image as convex mirrors. The image is virtual, erect, and smaller than the object. The only difference is that the virtual image is behind the mirror, while it is in front of the lens (Figure 27).

DISTORTIONS IN IMAGES

Optical images, except those formed by plane mirrors, are distorted to a greater or lesser extent as a result of various defects. We have already mentioned spherical aberration in spherical mirrors. This defect occurs also in lenses. Another defect is *astigmatism*. In this case rays that come to a lens or mirror from a point far from the principal axis⁄do not meet at a common point of the image. A blurred effect is produced. Astigmatism in lenses can be corrected by using several lens elements. This arrangement is found in the expensive camera lenses called *anastigmats*.

Chromatic aberration is a familiar type of lens defect. The index of refraction depends upon wavelength. Hence any simple single-element lens will focus short-wave light nearer to the lens than it will longer-wave light. This chromatic aberration causes serious blurring and confusion in the image in an instrument of any power. It is usually reduced to a tolerable level by a two- or three-element combination called

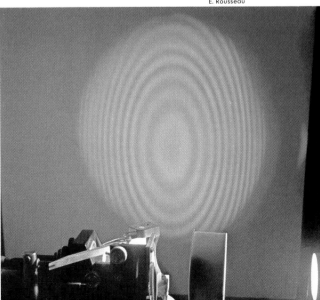

Diffraction can produce striking effects, such as that in upper right photo in which light is passed through a trellis. Lower photo shows an interferometer, a device used to measure the effects that occur when light waves interfere with one another.

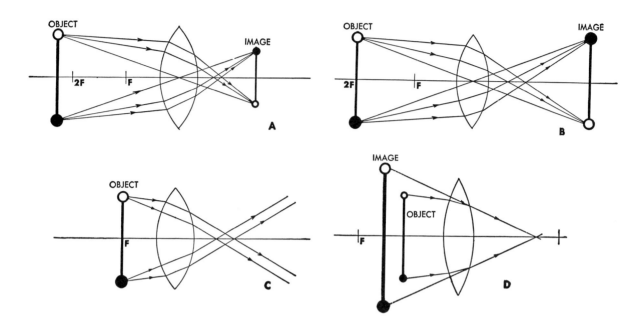

Fig. 26. Converging lenses. A, object beyond twice focal distance—image inverted, real, and smaller than the object. B, object at twice focal distance—image real, inverted, and same size as object. C, object at focus—no image. D, object between focus and lens—the image is erect, virtual, and larger than the object.

an *achromatic* lens. The principle of this correction is somewhat complicated. It depends upon the use of two or three different kinds of glass to bring all wavelengths approximately to the same focus

Fig. 27. Diverging lenses produce very much the same type of images as those produced by convex mirrors. The images are virtual, erect, and smaller than the objects. The only difference is that the virtual images are behind the mirror while they are in front of the diverging lens.

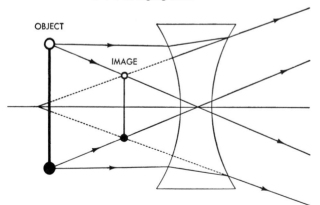

without lessening the over-all power of the lens combination.

SOME FAMILIAR OPTICAL INSTRUMENTS

A number of devices, called *optical instruments,* are based on the refraction or reflection of light. Here is a brief survey of a few of the better known ones.

A prism is used to study light polarization, or the effects produced when light is made to vibrate in a certain pattern.

Cameras. One of the most familiar optical instruments of any complexity is the camera. Basically this familiar device is a converging lens arranged to form a real image on a piece of sensitive photographic film. Usually the image is smaller than the object, but this is not always the case.

A projector is simply a camera in reverse. If a transparent photograph or drawing is placed in the film position in a camera and strongly illuminated, an image will be projected on a screen placed at the normal object position in front of the camera. This is the principle of the photographic enlarger, the slide and motion-picture projector, and other such devices. Opaque objects can be projected if suitably illuminated.

Magnifiers and microscopes. The simple magnifier, commonly called a magnifying glass, is a converging lens used to view an object placed nearer to the lens than its principal focus. Under these conditions the eye sees a magnified virtual image, as in Figure 26D. This is a very convenient arrangement because the image is upright. For low-power work, up to magnifications of three or four times, a single lens element works acceptably as a magnifier. For more power, however, it is necessary to use a more complicated lens to correct the chromatic and other aberrations that cannot be avoided in the simpler instrument. The eyepiece of a telescope or microscope is a high-quality simple magnifier, used to view a real image formed by other optical elements—lenses or mirrors or both.

The compound microscope is used to obtain high magnifications—up to several thousand times—of very small objects. Condensing lenses concentrate light on the object, which is placed just outside the principal focus of a small converging lens, called the *objective*. This objective forms a magnified real image of the object a few centimeters away. Since this image, which is inverted, is still quite small, a magnifier (consisting of another converging lens) is used as an eyepiece to produce a further enlarged virtual image of the first image.

Telescopes. The oldest optical instrument of any complexity is the astronomical refracting telescope. As originally devised by Galileo, the instrument used a diverging lens as an eyepiece. However, this system has serious defects. (See the article "Telescopes" in Volume 1 of *The New Book of Popular Science*.)

THE EYE—A CAMERA

The eyes of animals are complex image-forming devices. The eyes of human beings and many other species are similar in principle to the ordinary camera. They use a small lens that forms an inverted real image on the sensitive film of the retina. The retina then relays informa-

E. Rousseau

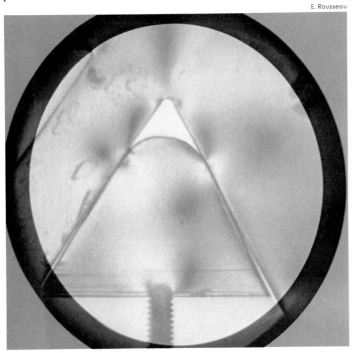

Polarized light, here used to study the structure of Plexiglas.

tion on light intensities to the brain. The eye has an iris diaphragm to control the amount of light entering the organ. Instead of varying the lens-to-film distance in order to focus for various ranges, the eye makes this adjustment by varying the curvature, and consequently the focal length, of the lens itself. Thus it performs a feat that would be quite impractical with artificial optical materials.

It is mainly the higher animals—the vertebrates—that have this complex eye structure which permits the eye to adjust in various ways. However, a few invertebrates—the squid and the octopus, for example—also possess similarly complex image-forming devices. The evolution of their eye development was quite different from that of the vertebrate eye, but the end result is much the same.

Some lower animals have long since appropriated the simplest image-forming device—the pinhole. The eyes of such animals are merely hollow, enclosed cups with a small hole leading in from the exterior. On the inside, facing the hole, is a primitive retina.

The compound eyes that insects possess are rather complicated versions of the same principle. That is, they consist of a great number of narrow tubes pointing in different directions. As a result, each tube "sees" one tiny portion of the environment, different from the portions seen by the tubes around it. All of the tubes, together, build up a picture. The greater the number of tubes making up the compound eye, the more refined the composite image is. People have taken photographs through the eyes of insects. The result has been a number of tiny, crude images of the object being viewed through the eye. However, in the insect itself, this does not happen. Special screening mechanisms keep the image seen by each tube from being more than a tiny portion of the whole image, and there is only one image of the whole object.

Mirrors are described by their surface. The still water here is a plane mirror. Spherical mirrors are either concave or convex, depending on whether they cause light rays to converge or diverge.

The refraction of white light by a glass prism. Shown are the manuscript, prism, and compass of Isaac Newton.

COLOR

by J. H. Rush

The perception of color adds greatly to our knowledge of the world around us, and is also a never ending source of pleasure. It causes us to appreciate fully the beauty of a sunset, the striking hues of autumn leaves, or the many colors of the rainbow. Men have been keenly aware of color since very early times, as countless writings attest. Yet its nature remained a complete mystery until the seventeenth century. Sir Isaac Newton's classic experiments in 1666 yielded the first clues to the physical basis of color.

Newton admitted a narrow beam of sunlight into a darkened room through a very small hole in a window shutter. When he placed a triangular glass prism in the beam, the light falling on a normally white screen appeared as a series of colors. This is because light is refracted in passing through a prism. In other words, its path is bent away from the original direction. This effect had long been known, but Newton's experiments showed that different colors are bent through different angles. Red is bent least; then, in order, orange, yellow, green, blue, indigo, and violet. To Newton it seemed evident that these colors must be derived from the white sunlight. To prove this point, he used a second prism to bend the colored rays back to their original path. When the rays were recombined in this way, they produced white light.

Much later, when Thomas Young and other investigators established that light is a wave disturbance, they found that color is related to the wavelength of light. Waves of light are not directly visible, as are the crests and troughs of waves on water, but they can be demonstrated by indirect methods.

LIGHT — ELECTROMAGNETIC RADIATION

Light is one of many different kinds of electromagnetic radiations — disturbances that travel through the universe in the form of waves. These radiations all travel at a fixed speed of about 300,000 kilometers per second in free space. They may have any of a very wide range of wavelengths. Lightning generates waves that may be several kilometers long. The waves of gamma rays emitted from radioactive materials may be less than $5/100,000,000,000$ of a centimeter in length.

Scientists express very small distances

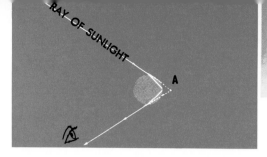

Fig. 1. How light is refracted and reflected (once) when it strikes a water droplet floating in air. The path of only one wavelength is shown.

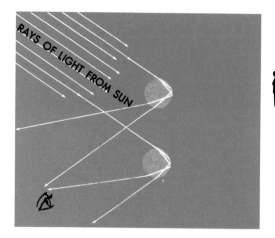

Fig. 2. How two wavelengths of light, corresponding to two different colors, are refracted and reflected (once) upon striking a raindrop. Note the effect on the eye of an observer at ground level.

Fig. 3. The light that produces a rainbow's outer bow is reflected twice inside water droplets. In this drawing we show the path of a single wavelength.

Fig. 4. How a two-bow rainbow appears to an observer.

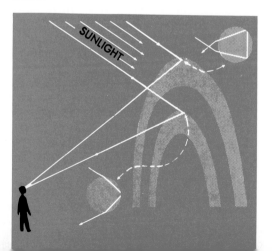

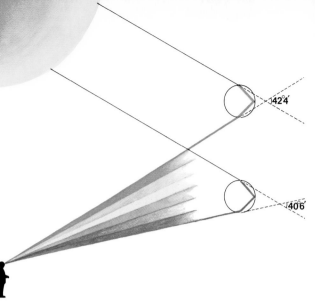

Fig. 5. Here we see how the precise angle at which a raindrop refracts and reflects the white light coming from the sun determines the wavelength of the light and the color that appears to the observer.

in units called *angstroms*. An angstrom is equal to $1/100,000,000$ of a centimeter. The wavelengths of visible light range from about 7,500 angstroms in the red end to 4,000 angstroms in the violent end. Electromagnetic waves in this range excite certain nerve endings in the retina of the eye. The impulses transmitted to the brain by these nerves give rise to sensations of light and color. We shall consider this process later in more detail.

A wavelength of light is not a color. Rather, it is related to a sensation of color. Different wavelengths excite different color sensations. The sensations can be produced without the agency of light. If you have reasonably good visual imagination, you can "see" colors at will by closing your eyes and imagining them. You can generate the sensation of color by pressing gently on your closed eyes. A blow on the forehead may cause you to see colors.

ISOLATING COLORS

We pointed out that a ray of light is refracted, or bent, as it passes through a prism. It is always refracted whenever it crosses the boundary between air or a vacuum and a transparent substance, except when it strikes the substance at right angles. The paths of the different wavelengths of light will diverge within the substance, because each wavelength is bent at a different angle. This divergence is the basis of many natural phenomena and is applied in various ways.

A familiar example of color dispersion by refraction (and also by reflection) is the rainbow. This phenomenon is produced as sunlight strikes water droplets floating freely in the air. The droplets are nearly spherical. The sunlight that strikes them is refracted and reflected in such a way that the observer sees a huge arc extending above the horizon and showing the full range of color. A rainbow generally shows a bright primary, or inner, bow and a fainter secondary, or outer, bow.

The rays of light that produce the inner bow are refracted in water droplets, reflected once at the internal surface, and then refracted again on emerging (Figure 1). Since the different wavelengths of sunlight are refracted differently, they follow different paths within a given drop and also as they emerge from it. Figure 2 shows how two wavelengths of light—those corresponding to violet and red—are refracted from a drop.

An observer can see color only from certain raindrops. In each case a given wavelength of light, refracted and reflected from a drop, must strike the observer's eye. This happens only if the ray striking the drop and the emerging wavelength that strikes the eye form the appropriate angle (A in Figure 1). The angle in question ranges from about 42° for red to 40° for violet. The observer will obtain the sensation of color from a vast number of drops, which from his viewpoint form an arc in the sky. The different colors of the inner bow follow one another in order from the outer part of the arc to the inner part, red being outermost.

The secondary, or outer, bow of the rainbow is formed by light that has been reflected twice inside each droplet (Figure 3). The order of colors is reversed from that of the primary bow and the colors are fainter. The two bows will appear to an observer as in Figure 4. Theoretically a third bow can be produced by three internal reflections. However the third bow is rarely, if ever, intense enough to be visible. (See also Figure 5.)

The effects called *halos* and *sundogs* are produced quite similarly by the refraction of sunlight or moonlight in ice crystals high in the atmosphere. These colored rings or spots occur at definite angular distances from the sun or moon. They result from somewhat more complex optical processes than those that produce the rainbow.

SPECTROSCOPES

The dispersion of wavelengths by a prism of glass or other material is the basis of one of the most useful instruments of science—the *spectroscope*. Basically it is simply a prism of the type used by Newton, plus various accessories. The first section of the instrument is the *collimator*. This consists of a tube with a slit at one end and a lens at the other. The slit limits the width of the entering beam, so that the final images in different wavelengths will be narrow

A beautiful rainbow, intense enough to show most of the colors of the spectrum.

Maury—Galliphot

Sundogs are produced by the refraction of sunlight on ice crystals found high in the atmosphere.

and will not overlap. The lens of the collimator bends the rays of light so that the rays from any point in the slit are all parallel as they enter the prism.

As the prism disperses the light, wavelengths are sent off in different directions. The lens on the other side of the prism brings each beam of parallel rays to focus again, forming a sharp image of the slit in each wavelength. These slit images are viewed through a magnifying eyepiece. The second lens and the eyepiece form a small telescope. The telescope, which can be swung around on a vertical axis, can be focused on the different wavelengths of light that appear.

There are various other kinds of spectroscopes. In the instrument called the *spectrograph,* the spectral images are focused on film and are photographed. The photographic records they produce are known as *spectrograms.* The prism is not the only device used to disperse the different wavelengths of light. The *diffraction grating,* described later, serves the same purpose. It is superior in certain respects to the prism spectroscope.

CHARACTERISTICS OF LIGHT SOURCES

The band of colors, or wavelengths, produced by a spectroscope is called a *spectrum.* All wavelengths are present in white light. Its spectrum forms a continuous, unbroken spread of color from the longest visible waves of red light to the shortest visible waves of violet. This is called a *continuous emission spectrum.* Incandescent solid or liquid substances, or even gases under unusual conditions, radiate white light and show a spectrum of this kind. Sunlight is essentially white, but its spectrum is interrupted by thousands of narrow dark lines, representing missing wavelengths. As we shall see, these wavelengths have been absorbed by atoms in the sun's atmosphere. The sun's spectrum is called an *absorption spectrum.*

When a gas at low pressure is heated or excited by electrical discharge, it shows a different kind of spectrum. The molecules of the gas will always radiate certain sharply defined wavelengths. This kind of radiation appears in the spectroscope as a *bright-line spectrum*—a series of isolated sharp images of the slit. Each of the images represents a different wavelength.

Even iron and other metals, normally solids, emit bright-line spectra in the vapor state. The vapors of the chemical elements and of many compounds can be identified by the characteristic wavelengths they radiate. Sodium vapor, for example, emits only two wavelengths in the visible spectrum: these are in the yellow-orange region. They are only six angstroms apart, and are easily mistaken for a single line in a small instrument.

The wavelengths of the bright sodium lines match those of two strong dark lines in the spectrum of the sun. This is why. Any relatively cool vapor absorbs exactly the same wavelengths of light that it radiates when it is hotter. The region outside the brilliant white photosphere of the sun is cool, compared to the photosphere itself. Therefore the sodium vapor in this outer region absorbs its characteristic wavelengths out of the continuous spectrum radiated from the sun's dense interior. In-

Spectrum of incandescent light produced by diffraction grating.

stead of bright lines of sodium the sun's spectrum shows dark lines of this element. Other dark lines in the solar spectrum similarly reveal the presence of other elements in the sun.

Many substances radiate visible light even when cool and in the solid or liquid state if they are excited by other radiation. Such substances are said to be *fluorescent*. Materials that store the exciting energy and

When light from a given light source is broken up by the instrument called the spectroscope into different wavelengths, the resulting color combination is called a spectrum. Below are some spectra.

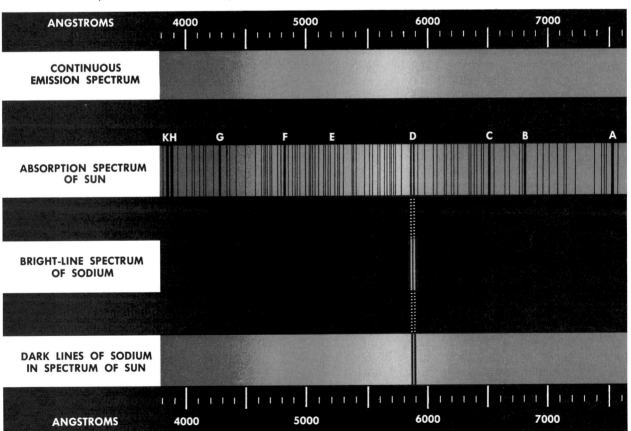

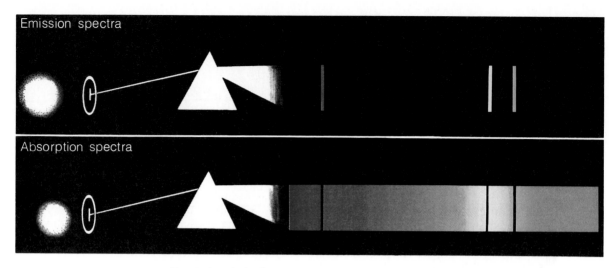

Spectral analysis. On top, an emission spectrum of an incandescent gas. Lower: absorption spectrum, in which the brillant-colored rays have been replaced by dark bands situated in the same places in the spectral band.

reradiate it visibly later are said to be *phosphorescent.* Zinc sulfide glows yellow under blue or ultraviolet light or X rays. Barium platinocyanide glows blue. Mix these two materials, and you get a fluorescence that approaches white light.

This principle is used in fluorescent lamps. A glass tube is coated inside with a mixture of substances known as *phosphors* that emit a great number of wavelengths when they are excited. The total effect of these wavelengths on the eye is similar to that of white light. The tube is filled with mercury vapor at low pressure. When the vapor is excited by a high-voltage discharge, it radiates strongly in the violet and ultraviolet regions. These radiations bring about visible fluorescence in the coating.

INTERFERENCE

Bubbles of nearly colorless soap solution develop brilliant colors before they break. Similar colors appear in cracked glass or ice, or on the surface of very old glass, or in insect wings, or in oil films on water, or in light viewed through a window screen or through one's eyelashes. All of these color effects are caused by the *interference* of light waves and by a related effect called *diffraction.*

Interference can be illustrated by a familiar example, illustrated in Figure 6. When a motor boat cuts through calm water, it throws out a series of waves that run off at an angle from the bow. These waves move across the water at a definite speed. If two boats cruise on parallel courses some distance apart, their bow waves meet and

Fig. 6. Waves produced by the boats show an interference pattern. The water is choppy where crests or troughs intersect each other; quiet where crests intersect crests or troughs intersect troughs, quiet where crests intersect troughs.

CREST
TROUGH

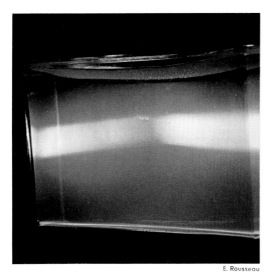

Light waves are affected by passing through certain types of solutions, as shown above.

cross each other. As they do so, they interfere. If crests on both waves pass a certain point at the same time, the water at this point rises to twice the height of one wave. This effect is called *constructive interference,* or reinforcement. At an adjacent point, a crest of one set of waves arrives simultaneously with the trough of another set, and the two cancel each other. This is *destructive interference.* If the two wave motions interfere with each other in this

way, the water at the meeting point will hardly be disturbed. As a result of reinforcement and destructive interference, the water between the two boats will show a checkerboard pattern — smooth at some points and with up-and-down motion at others.

Such interference effects are commonplace in the case of light waves. Let us see what happens when light is reflected from a thin film, such as the wall of an ordinary soap bubble. When light encounters a surface of transparent material, such as this, most of it goes on through, but a small fraction of its energy is reflected. We see a soap bubble, therefore, partly by light that is reflected back to us from the outside surface of the watery film. But some of the light is reflected from the second, inside surface also, and most of this light also comes back to our eyes. What we then see depends on the degree of interference that takes place between these two reflected beams of light. It will depend on the wavelength of the light and the thickness of the bubble film.

In certain instances the two reflected wave trains for a given wavelength will be in phase: that is, their crests will coincide and so will the troughs of the waves. The film will then appear bright in the color corresponding to that wavelength because of what is known as constructive interference. The reflected wave trains for another wave-

Soap bubbles are transparent, but they reflect a small fraction of light. Thus we see them partly by light reflected from their inside surfaces.

RAY OF SUNLIGHT

VIOLET

RED

AIR

SOAP BUBBLE FILM

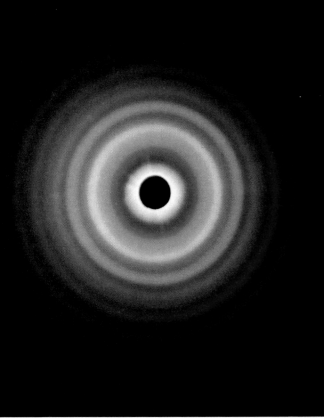

Interference fringes produced by two identical and parallel light diffusers.

length will be out of step, and will cancel one another as a result of destructive interference. The film will then appear dark as far as this particular wavelength of light is concerned.

Usually we see a soap bubble film in white light, which presents the entire visible range from red to violet. Since the wavelengths differ, the film cannot cancel or reinforce all colors at once. If the thickness of the film is such that one wavelength will be canceled, another wavelength will be partially reinforced. The colors that are reinforced and that are therefore visible will change continually as the film grows thinner by evaporation.

An air space between two surfaces of glass, ice, mica, or other transparent material behaves much the same as a water film in air. If two such surfaces meet at an angle, the thickness of the air gap between them will vary along the wedge that is formed. If light passes through, one color and then another is reinforced. The colors of the

spectrum are repeated along the wedge like a series of rainbows. Extremely small distances can be measured by instruments that depend on slight changes in the width of the air gap and the consequent shifts of the colored fringes.

DIFFRACTION

The phenomenon called *diffraction* is basically an interference effect. A wide, uninterrupted wave front moves forward without appreciable change of form. If the wave is interrupted, however, by an opaque edge or a narrow obstacle, the light bends around the obstacle into the shadow zone. Light and dark fringes then develop outside the edge of the shadow. When white light is diffracted in this way, differences in wavelengths lead to color effects as in other cases of interference.

Diffraction effects ordinarily become noticeable only when light passes through very small openings or around very narrow obstacles. A very common and beautiful demonstration is the appearance of the moon through a door screen or window screen. Four bright streaks of light appear to radiate from the round globe, parallel to the horizontal and vertical wires of the screen. These streaks appear nearly white when seen with the naked eye. But a binocular or a small telescope will reveal that the streaks consist of a series of white-light spectra, in which the sequence from violet to red is repeated over and over again. This principle is applied in the diffraction-grating spectroscope. The diffraction grating used in this instrument consists of a surface of glass on which many thousands of fine grooves have been ruled. There are generally several thousand evenly spaced grooves to the centimeter. When white light strikes the diffraction grating, a series of continuous spectra is produced.

POLARIZATION

Striking color effects can be produced when certain objects are viewed in *plane-polarized light*. To understand what this kind of light is, it must be pointed out that light does not vibrate in the direction in which it travels, but at right angles to that direction. It vibrates in all possible directions perpendicular to the ray—up and down, sideways, obliquely. (Figure 7).

Suppose we let a beam of white light pass through a thin plate cut from a crystal of the mineral called tourmaline. If the beam is directed to a screen after it passes through the tourmaline plate, the light will appear practically unaltered. Yet a definite change in its character has taken place. The tourmaline plate has permitted only the vibrations in a single plane to pass through (Figure 8). Let us now put a second tourmaline plate behind the first. This second plate can be set in such a way that light vibrating in one plane will pass through both plates and will appear on the screen (Figure 9A). If we now turn the second plate 90°, the light vibrations in the single plane can no longer pass through and no light at all will fall on the screen (Figure 9B).

When vibrations of light can vibrate only in a single plane as in Figures 8 and 9A, we say that the light has been plane-polarized. The plate that polarizes the light is called the *polarizer*. The second plate, which serves to analyze the light, is called the *analyzer*.

If we put a flake of mica between the plates, a series of beautiful spectral colors will appear on the screen. This is due to variations in the thickness of the crystal and in the wavelengths of light that pass through it. Various other transparent crystalline substances will show similar effects in plane-polarized light. If glass or transparent plastic is stretched or compressed, the stresses set up in this way alter its structure. When plane-polarized light is passed through it, colored fringes will appear.

COLOR EFFECTS IN NATURE

Why is the sky blue? Why is the setting sun red? Why is deep water blue or green? The cause in each case is the scattering of light—a diffraction effect. Dust particles and even the molecules of atmospheric gases interrupt the wave fronts of sunlight advancing into the atmosphere. The light that was advancing in a straight line from the sun to the earth is dispersed. Tiny new wave fronts develop in all directions. In a word, the light is scattered.

Like all diffraction and interference effects, the resulting color patterns depend on wavelength. A given obstacle "looks

A color wheel for the analysis of color. When two primary colors are mixed, they give a secondary color, which lies half way between them on the color wheel. At right, illustration showing how yellow and magenta produce red.

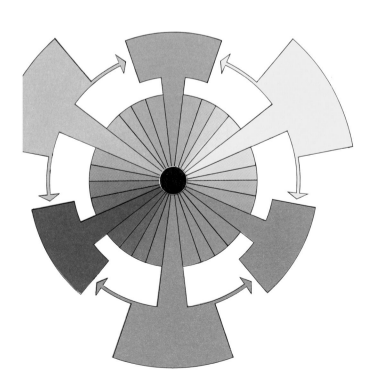

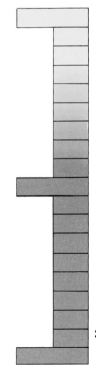

bigger" to a short wave of violet light than it does to the longer red. Therefore, the effect of the obstacle on the violet wave will be more drastic. Consequently, violet light is scattered most, blue somewhat less, and green, yellow, and orange still less, in that order. Red is affected least of all.

The sky generally looks blue because the short blue waves are scattered more than the longer waves of red light. It is true

Fig. 7. Ray of unpolarized light. The light vibrates in all possible directions in a plane at right angles to the ray.

Fig. 8. The plate permits only light vibrations in a single plane to pass through; the light has been plane-polarized.

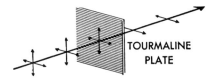

Fig. 9. In A, below, plane-polarized light passes through both plates and appears on the screen. In B, plane-polarized light passes through the first plate but is stopped by the second; it does not appear on the screen.

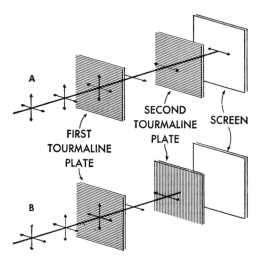

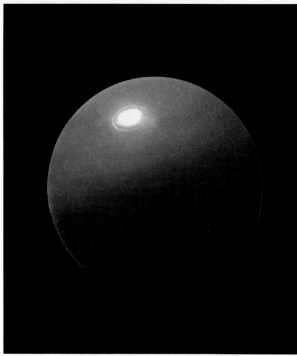

Fraudreau

that the violet waves are dispersed even more than the blue. However, the sky does not appear violet because the sun is relatively weak in violet light. Besides, the eye is less sensitive to violet light than to blue. Deep water appears blue for the same reason. However, impurities in the water often absorb the blue, and green becomes the predominant tint. The sky looks red near the horizon because, at that angle, the path through the atmosphere is long and traverses much low-flying dust. As a result the bluer light is effectively scattered out of the direct beam before it can be observed by the viewer.

Twilight is also the result of the light scattered by particles and molecules in the atmosphere. Were it not for this scattering, we would be in darkness as soon as the sun went below the horizon.

SELECTIVE ABSORPTION

A white dress, red roses, green leaves, brown paint, blue enamel, a black diamond—each of these terms implies that a definite color is associated with a particular kind of material. How do such colors arise?

First, let us see what we mean by "black" and "white." We have already pointed out that white light is a mixture of all visible wavelengths of light. Obviously, when an opaque object reflects all wave-

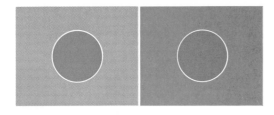

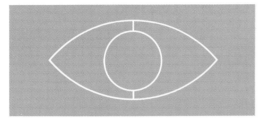

The intensity of a color appears different to us depending on background and lighting. Left: the red of the ball appears paler in the region of white light, darker in areas adjacent to the black background. Above: the blue circle appears brighter on the red background than on the green background. A related effect causes the thin white lines on the red background to appear slightly bluish to us.

lengths equally well, it will appear white in white light. Surfaces that reflect all wavelengths weakly will look gray. Extreme gray merges into black. A truly black surface would be one that absorbed all the light falling upon it, reflecting none whatsoever. Actually there is no absolutely black surface.

If an opaque object absorbs certain wavelengths and reflects others, we see a blend of the colors that it reflects. A red rose absorbs violet, blue, green, and yellow wavelengths much more strongly than it does red. Blue enamel absorbs red and other longer waves and reflects blue and perhaps some green. If an object absorbs wavelengths at both ends of the spectrum — violet and red — and reflects the middle lengths, we see yellow-green.

The apparent color of an opaque object depends upon the quality of the light that falls upon it. Suppose the molecular structure of a substance is such that it can reflect only blue light. If it is illuminated by white light, it will absorb all wavelengths except that of blue. It will reflect blue light and will appear to be blue. If a blue beam of light falls upon the object, the beam will be reflected: the object will still look blue. Suppose, however, that we direct a red light upon the object. The red light will be absorbed. The object will reflect no color at

all. Therefore, it will appear to be black.

VISUAL JUDGMENTS OF COLOR

Visual judgments of color are very rough. They tell only the predominant behavior at the surface of an opaque object. Only the spectroscope can reveal just what the object really does with the light that falls upon it. Frequently the actual color pattern is quite complex.

What happens when light passes through a colored transparent or translucent substance? Some wavelengths are absorbed as they pass through the material. Those that emerge in perceptible strength determine the color. The absorption pattern is complex. It varies according to the material. For example, if cellophane in a certain shade of green is folded and refolded and viewed against a strong white light, the green color gradually darkens as the number of folds increases. After the cellophane has been folded fifteen or twenty times, the light that comes through will no longer be green but deep red.

The spectroscope would show why this mystifying color change occurs. The dye in the cellophane is only relatively transparent to green. It is even more transparent to deep red than to green — that is, it permits a greater percentage of the red light to pass through. But the eye is not very sensitive to deep red, and so it responds to the green instead of to the red. When enough thicknesses of cellophane have been provided, the absorption of the green light becomes so much greater than the absorption of the red that the eye at last perceives red as the predominant wavelength.

COLOR MIXING

We know that when wavelengths of roughly equal intensities from all regions of the spectrum are combined, they give the color sensation of white light. Curiously enough, however, by mixing certain color pairs, we can deceive the eye into believing that the entire spectrum is present. Blue and yellow are one such pair. Red and cyan (bluish green) are another. The colors in a pair of this kind may consist of light from very narrow portions of the white-light

Human

Frog

Bird

Reptile
Fish

Butterfly
Beetle

Bee
Mosquito

spectrum. Yet, if the wavelengths and intensities are properly balanced, the total impression they give is the same that we would get from viewing the whole spectrum of light.

Obviously there is more to the perception of color than the simple matter of wavelengths in the spectrum. As a matter of fact, this very complexity makes the task of the artist or dyer easier than it would be if every color had to correspond to a definite wavelength. Two colors can match perfectly, as far as the eye can see, and yet appear altogether different in the spectroscope. Because of the flexibility of color sensation, it has been difficult to develop a science of color, as distinguished from a purely physical science of wavelength.

The most familiar approach to the analysis of color is based on a set of three primary colors. Many different three-color combinations can produce the sensation of white light when they are superimposed. Red, green, and blue are often chosen. They can be used as the basis of a color wheel, which will show the relationship between the primary and secondary colors. When two primary colors, such as red and green, are combined, they give a secondary color. Red and green light produce yellow; blue and green give cyan; blue and red yield magenta, a shade of reddish purple. Each secondary color produced by the mixing of two primaries lies half way between them on the color wheel. For example, we saw that red and green produce yellow. Therefore, yellow lies half way between red and green on the wheel. There is an infinite variety of intermediate hues between adjacent primary and secondary colors.

It is interesting to note that the secondary color magenta is not a spectral color. That is, it does not correspond to any wavelength in the white-light spectrum. It is produced when we bend the spectrum back on itself, so to speak, allowing the red and violet ends to overlap.

Each primary color on the wheel is opposite the secondary formed by the mix-

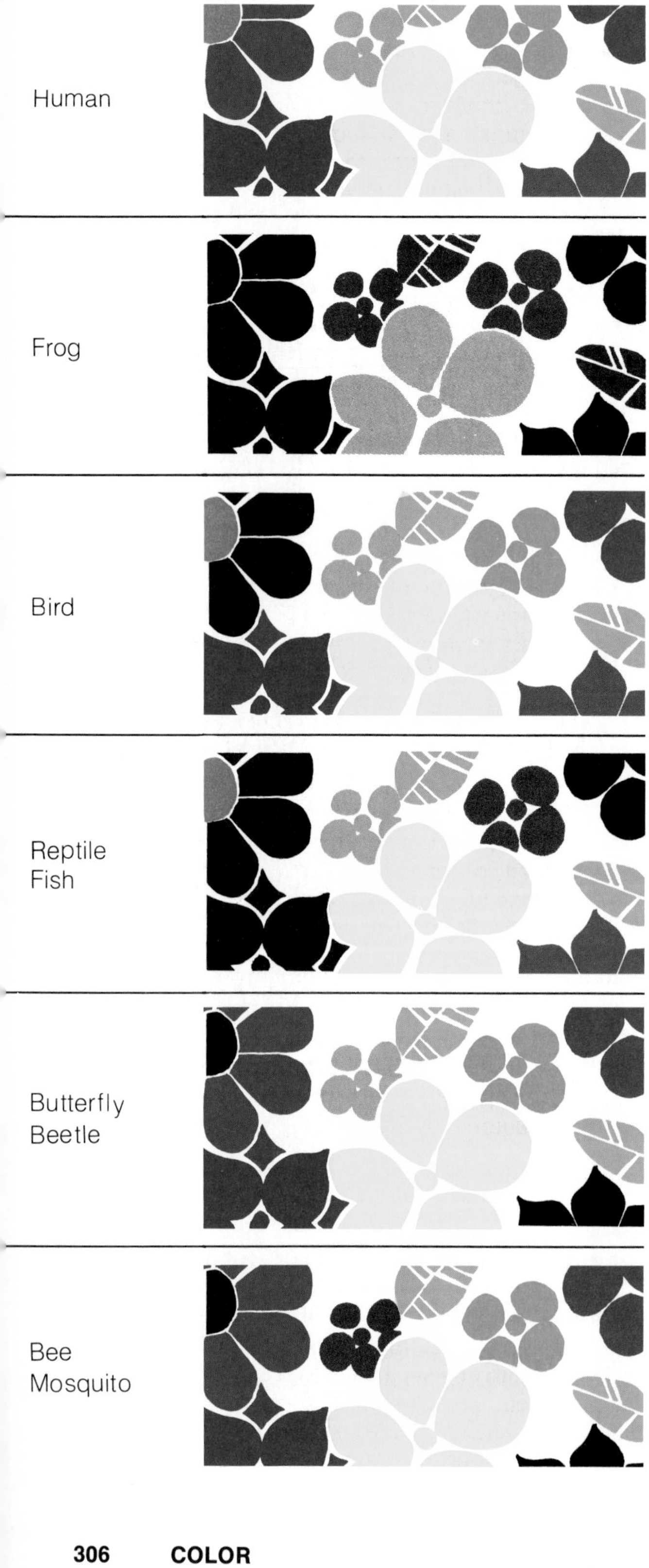

Differences in the perception of color in different animals.

ture of the other two primaries. The colors in such a primary-secondary pair are called *complementary*. When they are combined, they produce white. Green and magenta are complementary; so are red and cyan; so are blue and yellow.

ADDITIVE vs SUBTRACTIVE MIXTURES

The color relations we have been discussing apply only to *additive* mixtures of colored lights, in which one color is *added* to another. This kind of color mixing must be distinguished from the process of *subtractive* combination. In this process, blended colors *subtract* from white light the colors that they cannot reflect.

Let us examine the difference between the two processes. To illustrate the additive process we project yellow and blue light onto the same area of a white screen. We add one color (yellow) to the other (blue). Both colors are reflected and the eye registers the combination as white light. When we mix yellow and blue paints, we are using the subtractive process. The resulting mixture, as any child who has dabbled in water colors knows, is green, not white.

To understand why subtractive combination produces this result, we must point out that paint contains coloring ingredients called *pigments*. These are finely divided particles of a substance that absorbs certain wavelengths and reflects others. Pigments do not absorb or reflect sharply defined regions of the spectrum. Yellow pigment absorbs violet and blue light almost completely. It reflects some green and red, together with the yellow that it reflects most strongly. Similarly, blue absorbs practically all the red and yellow but reflects some green and violet, as well as blue. When yellow pigment and blue pigment are mixed and exposed to white light, each is free to absorb any light not absorbed by the other. The only region of the spectrum that is not fully absorbed by either pigment in this case is green. Therefore the mixture of yellow and blue pigments gives green.

Subtractive color mixing is the basis of color printing. The color printer applies his colors one at a time. Usually he uses three colors—blue, yellow, and red—plus black for shading and emphasis. Each of four printing plates carries those details of the picture that appear in one of these colors. The plate is rolled with ink of that color and printed. When the four colors are printed on the paper one after the other, in correct position, the colors of the pigments combine subtractively and reproduce approximately the colors of the original picture being used.

CHARACTERISTICS OF COLOR

Everyone recognizes that a word such as "yellow," or "green," or "blue" is not enough by itself to describe a color impression adequately. We commonly use the terms "strong," "rich," "pale," "weak," "dark," and "light" in our efforts to describe variations of colors. It is possible to analyze color impressions more accurately by giving three qualities or characteristics: *hue, saturation,* and *brightness*.

Hue. Hue is a color's spectral characteristic. In other words, it corresponds to a definite wavelength of the spectrum. What of colors, such as magenta, that do not appear in the spectrum? In such cases, the hue represents the value obtained by the additive mixing of two spectral colors, representing two definite wavelengths of light. When we say that something is red, or green, or yellow, we refer to its hue. No amount of shading or thinning can turn red into green. Hue is really the fundamental quality of color sensation.

Saturation, or intensity. Saturation expresses the quality that we commonly call richness or strength of color. Suppose we project a beam of red light from the spectrum onto a white screen. The color we see is saturated. All of the light that comes to the eye from the screen is capable of exciting the sensation of red. Now, suppose we project a beam of white light onto the same spot as the red. The color is now diluted, or paler. It is a combination of red sensation and the general sense of illumination that we call white. By varying the intensities of the white and red beams, we can get any degree of saturation. In handling pigments, adding white or gray to a hue is equivalent to adding white light. This results in a decrease in saturation.

Brightness. Brightness, lightness, brilliance, or value is a measure of the effectiveness with which an object reflects light. A brightly colored object is one that reflects or transmits a large fraction of the light falling on it, so that it appears brilliant, or luminous. We can demonstrate brightness by projecting a spectral color onto a series of screens of various shades of gray. The brightness of the resulting color will vary according to the reflecting qualities of these gray shades. On a white screen, which reflects the greatest possible amount of light, the brightness of the color would be at its maximum. A black screen would not reflect any light; the brightness would be zero.

CLASSIFYING COLOR

To provide a basis for comparing and cataloguing color samples, several systems of classification have been developed. All depend upon the fact that a color can be uniquely classified, provided its hue, saturation, and brightness are specified. One of the systems is the *Munsell color atlas,* giving colored samples of different hues and various degrees of saturation and of brightness for each hue. The intervals between the successive hues, saturations, and brightnesses are small enough so that they form a practically continuous series.

The Munsell atlas and other catalogue systems depend upon the matching and identification of colors in terms of the color sensation they arouse, and not in terms of their wavelengths. For many purposes, a more accurate standard is desirable. Instruments called *colorimeters* have been developed to measure color properties according to wavelengths of light. One type of colorimeter uses an ingenious approach, based on the hue, saturation, and brightness of the sample. A spectroscope selects the wavelength that matches the sample hue. An auxiliary beam of white light is then added to the hue and its intensity is adjusted to match the saturation of the sample. Finally the brightness of the combination is adjusted to match the brightness of the sample. In this way the three factors of color are related to the white-light spectrum.

Of course, no measurement means anything unless the standard to which it relates is accurate and consistent. It is useless to refer a color sample to a white-light spectrum if the reference light is not really white, according to an agreed standard. Much of the work of *colorimetry,* or color measurement, has gone into specifying and developing stable, reproducible illumination sources.

COLOR VISION

Many attempts have been made to explain the perception of colors. A number of puzzling details must be accounted for. That is why, even today, no theory of color vision is completely satisfactory. Most scientists agree, however, upon one basic approach suggested by Thomas Young and developed by the distinguished German physicist and physiologist Hermann von Helmholtz a century ago.

According to the Young-Helmholtz theory, the cones in the retina of the eye are of three types. One type is most sensitive to the red region of the spectrum—that is, the wavelengths corresponding to the red region. Another is particularly sensitive to the green region, a third to the blue-violet. Each type of cone is sensitive to a wide range of other wavelengths. The red-sensitive cones respond not only to red, but also, though less strongly, to orange, yellow, and green. The green-sensitive elements respond also in some degree throughout the range from red to blue. The blue-sensitive cones respond also to violet and green.

Light of almost any single wavelength must excite at least two sets of cone-elements to some extent, because of their overlapping ranges of sensitivity. We pointed out that light from the red and green regions of the spectrum, received as a single impression, is perceived as yellow. If we accept the three-color theory, yellow is a composite sensation that can be produced by stimulation of both the red-sensitive and green-sensitive cones by wavelengths near their points of maximum sensitivity. We also know that a single wavelength in the spectrum between the red and green also produces the sensation of yellow.

If all three sets of retinal cones are equally stimulated, the resulting sensation is that of white light.

EVIDENCE FOR THE THREE-COLOR
  THEORY

There is nothing in the structure of the retina to support or refute the three-color theory of color vision. All the color-sensitive cones look alike under the microscope. The theory has not successfully explained some details of color sensation. However, it is supported by many bits of indirect evidence. One of the most interesting of these is *retinal fatigue.*

If you stare fixedly for at least fifteen seconds at a colored pattern, so that the colored image is held on a definite region of the retina, and then close your eyes or look steadily at a white surface, you will see the pattern in its complementary color. This seems to be due to the fact that exposure to a red pattern, for example, produces an effect similar to fatigue in the retina or the brain and reduces the sensitivity to red. Then, when you expose the retina to white light, green- and blue-green-sensitive cones in the fatigued area respond more strongly than the recently exposed red-sensitive cones. The resulting color sensation is that of the white-light spectrum with the red region removed—that is, cyan, the color that is complementary to red.

Color blindness affords further evidence for the three-color theory. A person with green color blindness, for example, responds to various light stimuli as if his eyes were not green-sensitive. Other types of color blindness are just as consistent.

PSYCHOLOGY OF COLOR PERCEPTION

The visual effects we have been discussing depend upon stimuli affecting the retina and other structures of the eye. Many subtle peculiarities of color vision involve the complex processes by which visual sensations are interpreted in the brain. Among them are harmony or discord of colors, the effects of adjacent colored areas on each other, and the influence of colors upon emotional moods.

Most people are aware that certain combinations of colors are pleasing, while others are distinctly displeasing or even painful to one of artistic sensibilities. The harmonious blending of colors is a fine art, which is based on certain definite principles and practices.

Colors influence each other strongly by contrast, especially when they are adjacent to one another. Such contrasts usually follow the principle of black-and-white contrast effects. For instance, a small gray circle on a white background appears distinctly darker than the same circle on a background of black. In much the same way, a color generally appears lighter in contrast to a dark shade than to a light tint, regardless of the contrasting hue. Such effects are obviously important in clothing design and interior decoration, as well as in painting.

Color can deceive the eyes in many ways. For example, red walls seem nearer than blue walls. This effect seems to depend upon the way in which the eye accommodates to radically different wavelengths.

Certain more or less definite emotional effects are produced by colors. Green is restful, while red is exciting and disturbing. These two colors have long served for traffic signals. Green indicates that an automobile driver can safely proceed; red, that it would be dangerous to go forward. Blue is a "cool" color; do we think of blue water in this connection? Blue is favorable to sleep, but it can also be depressing.

These psychological effects of different colors are obviously significant when it comes to designing interiors, whether homes or work spaces or public buildings. Different colors are usually chosen for different rooms of a house, such as a restful blue for the bedroom, a cheerful yellow for the kitchen, and vibrant or light colors for the family's recreation areas. Industrial designers have found that color choices for factory or office interiors can have a very important influence on the work performance and the psychological well-being of employees.

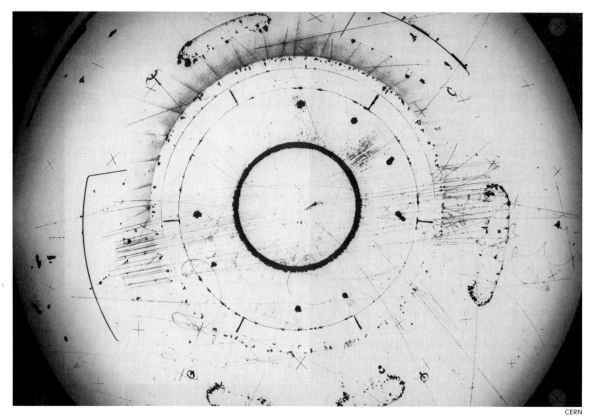

CERN

When elementary particles are made to interact in a special bubble chamber, they leave characteristic tracks seen here as dark lines against the bright background of reflected light in the chamber.

ELEMENTARY PARTICLES

by Gerald Feinberg

When a rock is crushed, it breaks up into tiny particles of sand. When a cube of sugar or a peanut butter cookie is crumbled, it breaks up into tiny grains. Everything that we see is made up of tiny bits of matter held together by forces. Even air has weight and is composed of particles of matter.

Since ancient times man has tried to discover the basic units of matter. What, he has asked, are the smallest particles from which all the objects in the universe are made?

Many people in ancient Greece thought that all matter was made of various combinations of four basic "elements": earth, fire, air, and water. But one Greek philosopher, Democritus, had a different theory. He suggested that matter was composed of tiny particles called *atoms*. The word "atom" comes from a Greek word meaning "unable to be cut," or "indivisible."

Democritus' theory was largely ignored for some two thousand years. Then, in 1802, an English chemist and physicist named John Dalton revived the atomic theory. He was the first scientist to define the atom as it is understood today—the smallest particle of an element that behaves chemically like that element.

BASIC ATOMIC STRUCTURE

Atoms are incredibly small. A tiny speck of dust contains many million millions of atoms. Some molecules, such as certain of the protein molecules, contain hundreds of thousands of atoms. Yet, a protein molecule is so small compared with things we can see with the unaided eye that a powerful electron microscope is needed to view it. Even then, the individual atoms cannot usually be seen. But in 1976 two physicists, A. V. Crewe and M. Isaacson, used a special scanning electron microscope to view what they believe are uranium atoms interacting with carbon atoms. They produced a 30-second film of atoms in action.

Small as the atom is, however, it is not the smallest component of matter. Beginning in the late 1890's, scientists discovered that the atom is composed of still smaller particles. The first of these *subatomic particles* to be discovered was the *electron*. Then it was found that the atom has a central core, which was named the *nucleus*. Surrounding the nucleus are the electrons.

The nucleus of the simplest atom, that of ordinary hydrogen, consists of a single particle, the *proton*. A single electron moves around the nucleus. A heavier nucleus, such as that of an oxygen or an iron atom, contains two kinds of particles: protons and *neutrons*. Neutrons are about the same size as protons. Electrons are much smaller than protons and neutrons.

Electrons, protons, and neutrons are the chief components of an atom. Since the 1930's, however, several hundred other particles have been discovered. Some of the most important of these are listed in the table on page 314, along with some of their properties. In some cases the existence of these particles was predicted on the basis of complicated theories. The particles themselves were later detected during experiments. In other cases the particles were discovered first.

PROPERTIES OF PARTICLES

Mass. Perhaps the most important property of a particle is its *mass*. The mass of an object is the quantity of matter it contains. It is closely related to weight. The weight of an object tells how much it is pulled down by gravity.

The masses of subatomic particles are incredibly small compared with the masses of familiar objects. The water that fills a thimble is more than 1,000,000,000,000,000,000,000,000 times as heavy as a proton or a neutron. And a proton is almost 2,000 times as heavy as an electron.

View of the 6.9-kilometer-long underground ring of the colliding beam particle accelerator of the European Laboratory for Particle Physics (CERN) near Geneva, Switzerland. The machine was used in 1983 to find the W- and Z-bosons.

Physicist Samuel C. C. Ting (foreground) and the team of scientists of Brookhaven National Laboratory in New York that discovered a massive subnuclear particle, named J, in late 1974.

Sometimes the mass of a particle is given instead in terms of its equivalent energy. According to the *theory of relativity*, proposed by Albert Einstein near the beginning of the 20th century, mass and energy are equivalent. The theory also says that the mass of an object is not constant. Instead, the mass gradually increases as the object moves faster and faster.

The minimum mass that an object has when it is not moving at all is its *rest mass*. Notice in the table on page 314 that *photons* and *neutrinos* have zero rest mass, or at most very small rest mass. If these particles were brought to rest, they would have no mass at all. But these particles do not come to rest. In fact, they move only at the speed of light—about 300,000 kilometers per second.

Charge. Another property of subatomic particles is electric *charge*. Electrons have a negative charge, while protons have an equal but opposite positive charge. The attraction between these opposite charges keeps the electrons around the nucleus, which contains the protons. The neutron has no charge at all. Its name comes from the fact that it is electrically neutral.

Neutral particles have 0 charge. A negatively charged particle, such as an electron, has a charge of -1. A positively charged particle, such as a proton, has a charge of $+1$. Theoretically, particles could have charges of $+2$, -2, $+3$, -3, and so on, and such particles have been found.

Because the charge of a subatomic particle is always a multiple of a certain value, it is said to be "quantized." The energy of a light beam of a definite frequency, such as red light, is also quantized. A particle of light of a definite frequency is called a *photon*, or a *light quantum*.

Spin. Photons are important in our discussion of still another important property of particles: *spin angular momentum*, or *spin* for short. Physicists have found that most particles have an intrinsic spin, one that does not depend on their relation to other particles. These particles turn on their axes, just as the earth rotates once each day on its own axis.

Particle spin, like electrical charge, is quantized. It can have only a certain value. The spin of a particle is expressed in relation to the photon, which has a spin of 1. Certain types of particles are allowed to have only spins that are integral multiples of 1, such as 0, 1, 2, 3, and so on. The other types of particles can have only half-integral spins, such as 1/2, 3/2, 5/2, 7/2, and so on. For example, electrons, protons, and neutrons all have spin 1/2.

Magnetic moment and parity. In addition to mass, charge, and spin, several other properties of subatomic particles are described by physicists. For example, the concept of the *magnetic moment* of a particle describes the tiny magnetic field associated with a spinning particle. *Parity* is a term dealing with similarities between an event or an object and its mirror image.

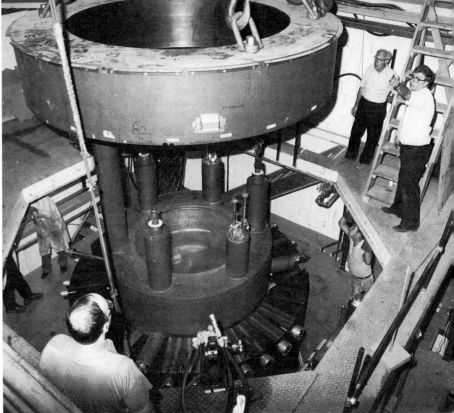

The first superconducting cyclotron, being assembled in 1982 at Michigan State University. The 500 MeV machine is able to accelerate nuclei of certain elements to half the speed of light, using intense magnetic fields generated in magnets that are cooled by liquid helium to temperatures close to absolute zero. Such supercooled magnets lose their resistance to electricity, reducing the energy required to accelerate particles.

TYPES OF PARTICLES

Physicists use the different properties of subatomic particles to group them into various types and classes. The most basic classes are grouped according to interaction strength:

(1) One class of particles, the *hadrons*, interacts a great deal when its members are near one another, through a force called the *strong force*. We say that hadrons take part in *strong nuclear interactions*. The strong force is responsible for keeping protons and neutrons together inside the nucleus. When strongly interacting particles collide, they easily convert into one another. Hadrons include two groups of particles: the *baryons*, such as the proton and neutron, which have half-integral spin; and the *mesons*, which have integral spin.

(2) Another class of particles, the *leptons*, includes the electron, muon, and tau, and their associated neutrinos. All leptons have spin 1/2. Leptons do not undergo strong interactions. However, they do take part in what are called *weak interactions*, as do the hadrons. The weak interactions are forces responsible for some particle decays—the gradual disintegration of one particle into other particles—and for a form of

radioactive decay in which electrons are emitted from the nucleus (nuclear beta decay).

Another classification of particles is based on their spin. As we mentioned earlier, some types of particles have integral spins while others have half-integral spins. This difference has a significant effect on the behavior of the particles when many of them are together.

All particles of one type, such as electrons or photons, are identical and indistinguishable. The mathematics that describes particles with integral spins is called *Bose-Einstein statistics*, and these particles are known as *bosons*. Mesons and photons are bosons. Since many of the properties of particles are quantized, the state of a particle is often described by a series of numbers called *quantum numbers*. The members of a collection of bosons of one type can each have specific quantum numbers independent of those of the others.

Some bosons, such as photons, have the value zero for the conserved quantum numbers. Because of this, they can be created in large numbers during collisions of particles in a giant particle accelerator or during cosmic-ray showers.

ELEMENTARY PARTICLES   313

SUBATOMIC PARTICLES

MASS CLASS	PARTICLE	SYMBOL	MASS (MeV)	STATISTICAL CLASS	SPIN	ELECTRIC CHARGE	STRANGE-NESS	AVERAGE LIFETIME (seconds)	ANTIPARTICLE
zero-mass	photon	γ	0	boson	1	neutral	0	stable	same as photon
boson	graviton		0	boson	2	neutral		stable	same
lepton	neutrino, electronic	ν_e	0	fermion	½	neutral	0	stable	electronic antineutrino
	neutrino, muonic	ν_μ	0	fermion	½	neutral	0	stable	muonic antineutrino
	neutrino, tau	ν_τ	0	fermion	½	neutral	0	stable	tau antineutrino
	electron	e^-	1	fermion	½	negative	0	stable	positron (antielectron)
	muon	μ^-	207	fermion	½	negative	0	2.2×10^{-6}	antimuon
	tau	τ	1	fermion	½	negative	0	1×10^{-13}	anti–tau
meson	pions	π^+	273	boson	0	positive	0	2.61×10^{-8}	same
		π^-	273	boson	0	negative	0	2.61×10^{-8}	same
		π^0	264	boson	0	neutral	0	0.9×10^{-16}	same
	kaons	K^+	966	boson	0	positive	+1	1.23×10^{-8}	
		K^0	974	boson	0	neutral	+1	0.87×10^{-10} or 5.7×10^{-8}	antikaons
	eta	η	1074	boson	0	neutral	0	more than 1×10^{-22}	same
	j/psi	ψ	3105	boson	1	neutral	0	1×10^{-13}	same
	psi-prime	ψ'	3695	boson	1	neutral	0	unknown	same
baryon	proton	p	1836	fermion	½	positive	0	stable	antiproton
	neutron	n	1838	fermion	½	neutral	0	1,010	antineutron
	lambda hyperon	Λ^0	2183	fermion	½	neutral	−1	2.51×10^{-10}	anti–lambda hyperon
	sigma hyperons	Σ^+	2328	fermion	½	positive	−1	8.1×10^{-11}	
		Σ^-	2343	fermion	½	negative	−1	1.6×10^{-10}	anti–sigma hyperons
		Σ^0	2334	fermion	½	neutral	−1	1×10^{-20}	
	xi hyperons	Ξ^-	2585	fermion	½	negative	−2	1.75×10^{-10}	anti–xi hyperons
		Ξ^0	2573	fermion	½	neutral	−2	3×10^{-10}	
	omega hyperon	Ω^-	3276	fermion	3/2	negative	−3	1.5×10^{-10}	anti–omega hyperon

Particles with half-integral spins are described by different mathematics. They follow the *Fermi-Dirac statistics* and are known as *fermions*. These particles have one very important characteristic: no two fermions can have all their quantum numbers alike. This makes possible the ordering of electrons around the nucleus of an atom and the ordering of protons and neutrons within the nucleus itself. Protons and neutrons are strongly interacting fermions. The electron, and other leptons, are weakly interacting fermions.

QUANTUM THEORY PREDICTIONS

Many things happen on the subatomic level that have no counterpart on the level of things we are familiar with. Particles

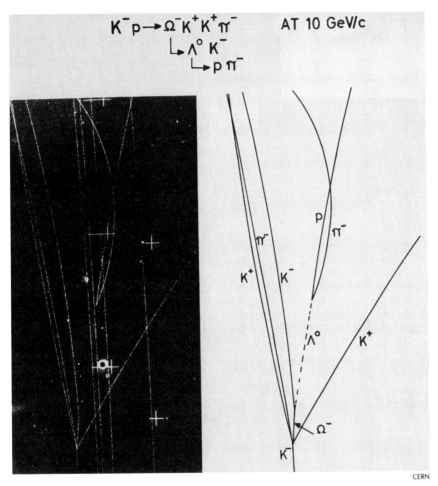

$$K^-p \rightarrow \Omega^-K^+K^+\pi^-$$ AT 10 GeV/c
$$\llcorner \Lambda^\circ K^-$$
$$\llcorner p \pi^-$$

CERN

Collision of a negative kaon against a proton produces several secondary particles, including a negative omega that decays into a neutral lambda. The full process is shown in the drawing.

appear out of nowhere and vanish again almost immediately. Other particles pass through electrical barriers that should have been able to stop them—as if an automobile should drive into a solid brick wall and appear on the other side unmarked, leaving the wall still in one piece.

This kind of behavior is due to the "split personality" of particles and waves, and is a logical outcome of the quantum concept in physics. Particles and waves, according to the quantum theory, are not really separate, distinct entities. Light waves and other types of electromagnetic radiation sometimes behave like particles. And beams of moving particles sometimes act like waves; they can be reflected and refracted and can cause interference pat-

terns just as light waves can. It is as if the moving particle were no longer a small, compact bit of matter moving in a precise way but were instead spread out into a moving bundle of energy having both particlelike and wavelike properties.

This dual personality has very important consequences. It is not possible, for example, to measure the position of a particle exactly and at the same time determine its speed and the direction in which it is moving. You ean do only one or the other at any given instant. The results of this are often strange and unexpected.

ANTIPARTICLES PREDICTED

Another result predicted by quantum theory was the existence of antiparticles for each "ordinary" subatomic particle. The antiparticle, said quantum physicists, would have the same mass and spin as the ordinary particle, but opposite electrical charge and magnetic moment.

The first antiparticle to be detected was the antielectron, or *positron* (its name comes from "positive electron"). Since then, antiparticles have been discovered for all the known particles. Even some neutral particles have distinct antiparticles, whose magnetic properties are opposite to those of the ordinary particles. In other cases, the particle is its own antiparticle.

When a particle collides with its antiparticle, both particles may be destroyed, and energy, in the form of lighter particles, appears in their place. The reverse process also happens. When two high-energy photons meet, they are converted into an electron and a positron.

Just as protons, neutrons, and electrons make up the atoms of ordinary matter, so do antiprotons, antineutrons, and positrons combine to form *antimatter*. An antiatom of antihydrogen, for example, would contain one antiproton and one positron.

While ordinary particles are very common on earth, antiparticles are very rare. As soon as they meet their ordinary counterparts, they are destroyed. In fact, no naturally occurring antimatter has ever been observed, although physicists have produced a few simple antiatoms in the lab.

It is believed that the surplus of matter over antimatter is true everywhere in the observable universe, and is the result of processes that occurred very early in the evolution of the universe.

LAWS OF CONSERVATION

Thus far, we have discussed some of the strange behavior and properties of subatomic particles without saying much about how physicists detect, keep track of, and describe all these particles. In spite of the apparent chaos of the subatomic world, a kind of order does exist. This order is embodied in laws drawn from traditional, or classical, physics and from the modern developments of relativity and quantum theory. These laws make possible a kind of "bookkeeping" of particle activity, just as certain business laws enable an accountant to keep track of the business activities of a company.

The appearance of things on the subatomic level may seem much more chaotic than the orderliness of a modern, efficient business. But the laws that govern particle behavior are actually much more exact than those describing the operation of a business. These basic laws of particle physics are called *conservation laws*.

Matter and energy. One of the most important conservation laws is the law of conservation of matter and energy. This law says that, in any particle interaction, the total mass plus energy after the interaction must equal the total before the interaction. Some mass may be changed to energy, and some energy to mass. But the combined total must remain the same. One important consequence of this law is that a particle cannot decay and become heavier.

Momentum and charge. Another conservation law is the law of conservation of momentum. Momentum is a measure of the amount of motion of an object. An object moving in a straight line has *linear momentum*. Linear momentum is equal to the mass of the object times its velocity. An object traveling in a curved path or rotating on its axis will have *angular momentum*. Angular momentum is a very important idea in the understanding of particles. The total momentum of a set of objects acting only on one another remains constant.

A third general conservation law is that of electrical charge. If the total charge of several particles before an interaction is +3, then it must be +3 after the interaction. What matters is not how this charge is distributed among the particles, but only that the total remains unchanged. Thus, when a neutral particle decays into charged particles, equal numbers of positive and negative particles must be produced.

Baryon and lepton number. The laws we have just described relate not only to subatomic particles but also to larger-scale, more familiar things. In addition, there are conservation laws that have meaning only

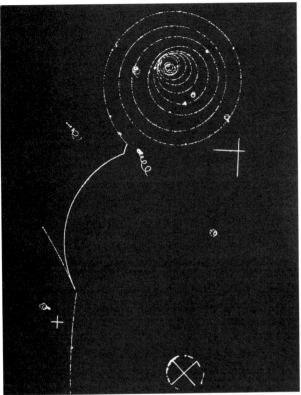

Aesthetically pleasing—yes—but also informative to physicists, who can determine the characteristics of a particle from analysis of its tracks.

in the subatomic world. An important one determines the appearance and disappearance of baryons and leptons. Each particle in each group is assigned the number 1, which is its *baryon number* or *lepton number*, as the case may be. The antiparticle of a baryon or a lepton is given the number −1.

In all known particle interactions, baryon and lepton numbers are conserved; that is, they cannot just disappear. To understand this better, let's see what happens when a neutron decays into a proton and an electron. When this occurs, a third particle is also produced—an antineutrino.

$$n^0 \rightarrow p^+ + e^- + \bar{\nu}_e^0$$
[neutron→proton + electron + antineutrino]

When we apply the conservation laws, we see that energy conservation is not vio-

lated: all the particles on the right are lighter than the neutron. The charge on the left is zero and the charges on the right add up to zero, so charge is conserved. The baryon family number of the neutron is +1. This is balanced by +1 of the proton (right).

Next, we come to the leptons. There are no leptons on the left side of the equation. On the right, the electron has a lepton number of +1; the antineutrino has a lepton number of −1. The total of +1 and −1 is zero, which equals the lepton number on the left.

INVARIANCE PRINCIPLES

A conservation law, as we have seen, says that some quantity or property remains unchanged during all types of physical events, such as collisons, particle decay, and so on. Closely related to conservation laws are *invariance principles*. These say that all physical laws should appear the same when viewed by different observers at different times or places and when involving either matter or antimatter. For example, when a neutron decays, the same particles always emerge. It doesn't matter if the neutron decays today in Rome, Italy, or 10,000 years from now on the planet Jupiter. These invariance principles are an example of symmetry of physical phenomena.

Each invariance principle is related to a conservation law. The uniformity of space, for example, results in the law of conservation of momentum. The uniformity of time results in the conservation of energy. In some cases, physicists do not yet understand which conservation law corresponds to which symmetry principle or invariance principle. Usually, it is the conservation laws that are studied. However, there are three invariance principles that have been of great importance in physics.

Time reversal. The first of these invariance principles is that of time reversal. This says that all physical laws are valid no matter whether they occur in "normal," or forward, order or in reverse, or backward, order. For example, all the physical laws that apply when a movie film is being run forward would also apply when the film is run backward, from end to beginning.

In our everyday world, for time to go backward—that is, for events to reverse themselves—is very unlikely. But there is no scientific law that says that this could never happen. It is only the great number of atoms and molecules involved in large-scale events that makes one time direction more probable than another.

In the subatomic world, this difference disappears. There is nothing, theoretically, to prevent the reversal of events. Thus, the production of a positron-electron pair from two photons is easily reversed by the annihilation of the positron and electron and the appearance of two photons:

$$\gamma^0 + \gamma^0 \rightarrow e^- + e^+$$
$$e^- + e^+ \rightarrow \gamma^0 + \gamma^0$$

You may be wondering about the law of conservation of mass and energy, which requires particles to decay "downhill," into particles of smaller mass. Can the reverse occur? Can lighter particles come together to form a heavier particle?

Yes, this is possible. First, however, the lighter particles must be accelerated. That is, they must gain energy. The energy is then used to create the mass of the heavier particle. This process occurs naturally in cosmic rays. Scientists use giant accelerators, which generate tremendous energies, to create heavy particles. In fact, most of the known ones were first discovered in this way.

Charge conjugation. The second invariance principle is that of *charge conjugation.* This states that particles and their corresponding antiparticles obey the same laws of physics. Thus, a world composed of antimatter should behave exactly the same as our own world, which is composed of ordinary matter. For this reason, it is impossible to tell from the light and other energy reaching us from a very distant galaxy whether that galaxy is composed of ordinary matter or antimatter. The physical processes that release energy in both types of worlds follow the same laws.

Parity. The third invariance principle is that of *parity*, or space inversion. This says that a mirror-image world, in which left and right are switched, should have the same laws as the ordinary world. In other words, nature should not prefer a "left-handed" world to a "right-handed" world.

EXCEPTIONS TO THE RULES

One surprising discovery is that nature does occasionally distinguish between left and right. Physicists have found that certain weak interactions do not look the same through a mirror. Nor is charge conjugation valid in weak interactions.

However, the physical laws for weak, as well as strong, interactions appear valid if we use the *TCP theorem.* This general principle is a combination of time reversal (T), charge conjugation (C), and parity (P). According to the TCP theorem, the combined result of (1) reversing the direction of time, (2) switching particles and their antiparticles, and (3) switching left and right still leaves all physical laws valid.

STRANGENESS

Before leaving the subject of conservation laws, we must mention one such law that is valid only for some situations. It was discovered in trying to explain why certain mesons and baryons lasted much longer than expected.

Particle stability is measured by the "lifetime" of each particle. For example, massless particles do not decay. Their infinite lifetimes are due to the fact that there are no lighter particles into which they can decay. And the law of conservation of energy prohibits them from decaying "uphill" into a heavier particle.

Similarly, the electron is stable because it is the lightest particle with nonzero rest mass. Even though the proton is quite heavy, it is the lightest baryon. It, therefore, cannot decay. For a proton to decay into a lighter particle, such as a meson, would be a violation of conservation of baryon number. Conservation of energy keeps the proton from decaying into a heavier baryon.

A partial conservation law applies to the K-meson and certain heavy baryons called *hyperons.* These particles are created suddenly in accelerators through the strong force. But they do not disappear just as suddenly through the strong force, as would be expected. Instead, they decay slowly through the weak force, a seeming violation of conservation laws. Physicists referred to them as "strange" particles.

Physicists have now accepted this *strangeness*. Strangeness is said to be conserved for certain particles that undergo strong interactions. But in weak interactions, it is not conserved. Thus, although strange particles decay through weak interactions, they do so much more slowly than they could through strong interactions.

OTHER PARTICLES

In spite of their many unusual properties, the particles we have discussed so far are real enough. In most cases, their presence and movements can be directly detected by instruments.

Still other particles are known to exist from their indirect effects on the behavior of other particles. The *W-bosons* (W^+ and W^-) and the *Z-boson* are the carriers of the weak force. Both the W- and Z-bosons have been observed as the products of high-energy collisions between protons and antiprotons. Their properties confirm a theory that gives a unified description of weak and electromagnetic interactions.

The *graviton* is the theoretical carrier of the gravitational force, analogous to the photon, the carrier of the electromagnetic force. Another proposed particle is the *tachyon*. Unlike ordinary particles, tachyons would always travel faster than light. This would result in some strange properties. For example, a tachyon would speed up when it lost energy! Experimenters have not yet been successful in efforts to detect tachyons.

The *resonances* are an interesting group of particles. They exist for such tiny fractions of a second that they cannot be detected directly. Instead, scientists find them indirectly, by observing sharp changes in the rates at which other particles collide with one another.

Resonant particles are created in high-energy accelerators through the strong force. Hundreds of them have been observed. All of them are related to low-energy baryons and mesons.

It is possible that resonances, as well as more easily identified particles such as nucleons, may represent a combination of other particles. This opens up the possibility that many of the particles now known may be made up of more basic ones.

Spurred by questions such as this, physicists have presented new theories to try to make sense out of the jungle known as particle physics. One theory suggests that the hadrons now known are not elementary at all but are themselves made up of smaller units. The suggested particle is called a *quark*. According to the theory, three quarks form a baryon, while a quark and an antiquark make up a meson.

Unlike the known particles, a quark would have a fractional electrical charge of 1/3 or 2/3 the basic unit of charge. It is believed that quarks can exist only when bound together into hadrons. The force responsible for this binding is thought to be carried by hypothetical particles called *gluons*, which are analogous to photons. Several different types of quarks are known to exist. Thus far, the quark theory has been found to give an accurate description of the properties of hadrons.

UNRESOLVED QUESTIONS

Some physicists have proposed that the law of conservation of baryon number is not an empirical rule, and that violations of it allow protons to decay into leptons and mesons. The basis of this prediction is a theory that allows for the conversion of quarks into leptons and vice versa.

It has been suggested by other physicists that quarks and leptons, presently thought to be the most fundamental particles, are themselves made up of still simpler particles. Whether this is true, or whether there is some other explanation for the properties of the particles that we now know, remains a question for future physicists to answer. Helping them in this quest will be the giant new particle accelerators and the superconducting supercollider now being built in Texas.

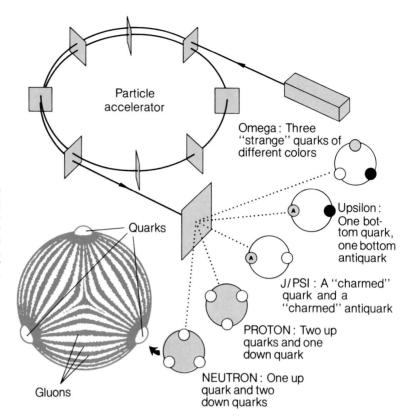

A conceptual diagram of an accelerator creating several kinds of particles. The enlarged particle represents a neutron made of three quarks, which are glued together by hypothetical force-particles called gluons.

Particle accelerator

Omega: Three "strange" quarks of different colors

Quarks

Upsilon: One bottom quark, one bottom antiquark

J/PSI: A "charmed" quark and a "charmed" antiquark

PROTON: Two up quarks and one down quark

NEUTRON: One up quark and two down quarks

Gluons

Illustration by Frank Senyk

QUARKS

By James Trefil

One of the long-cherished goals of science has been to discover a simple "basic stuff" of the universe — something, perhaps a few particles, from which all matter is made. Another goal has been to find one or two basic laws that can explain all forces, motions, and changes in the universe.

It has been a long quest. It began about 2,500 years ago when a Greek philosopher named Empedocles stated that all matter in the universe is a mixture of only four elements: earth, air, water, and fire. Only two basic forces exist, he said: love and strife.

Other theories followed. Then, in the 17th century, Isaac Newton discovered the three laws of motion, and the law of gravitation. The whole universe could be explained by these easy-to-understand laws. The world seemed so very simple. Especially when, in the 19th century, John Dalton's atomic theory was announced. Dalton said, in effect, that all matter is composed of atoms — and that an atom is the smallest particle of an element that behaves like that element.

In the early part of the 20th century, it was discovered that each atom's nucleus consisted of just two elementary particles — the proton and the neutron. This discovery introduced an important simplicity into our view of nature. Instead of saying that there are hundreds of different nuclei, each equally elementary, we now say that there are two elementary particles which, when put together in different combinations, make up all the known nuclei. The discovery replaces a complex view of nature with one that is relatively simple.

As the century progressed, however, more and more particles were discovered. Soon there were hundreds. It then became clear that there are too many elementary

Caltech

Murray Gell-Mann theorized that with three more particles, all matter could be explained.

particles for them to be the "basic stuff" of the universe.

THE THREE QUARKS

In 1964 Murray Gell-Mann, of the California Institute of Technology, and George Zweig, of the European Center for Nuclear Research in Geneva, independently developed a new theory. They suggested that the hundreds of elementary particles that had been discovered up to that point could be understood if we assumed that the particles weren't really elementary at all. The particles, they said, were made up of other, still more elementary things, which they named "quarks." In the original hypothesis, there were three different kinds of quarks. The name itself comes from a line in James Joyce's novel *Finnegan's Wake*—a line that runs "Three quarks for Muster Mark."

The quark model replaces a view of nature in which every particle is supposed to be elementary, with one in which the apparent complexity of the particles is explained in terms of a small number of constituents. In this analogy, the quarks play the same role in the study of elementary particles as the proton and neutron do in the study of the nucleus.

The three quarks that were suggested in 1964 explained all the elementary particles that were known at the time. These

particles are grouped into two classes, according to their masses. (See "Elementary Particles," page 312 in this volume.) The groups are hadrons (baryons and mesons) and leptons. Hadrons are believed to be made up of quarks. Leptons are not.

UP, DOWN, AND STRANGE

The quark model holds that every baryon is made up of three quarks, while every meson is made up of a quark and an antiquark. In the table below, the properties of the original quarks are given.

quark name	symbol	charge	strangeness
up	u	$\frac{2}{3}$	0
down	d	$-\frac{1}{3}$	0
strange	s	$-\frac{1}{3}$	-1

The terms "up" and "down" are not chosen at random. They refer to mathematical properties of the quarks. The term "strange" refers to the property of elementary particles that causes some of them (the so-called "strange particles") to decay more slowly than might be expected. All quarks have the same spin as the proton and neutron. (Spin is a property of the elementary particles. See page 312 of this volume.)

Using the quark properties in the table, it is easy to figure out how the elementary particles might be made. For example, the proton has an electric charge of one positive unit. In order to get $+1$ from the quark charges in the table, we take two u quarks and a d quark, giving a total charge of $\frac{2}{3} + \frac{2}{3} - \frac{1}{3} = 1$. Similarly, the neutron would be made of two d quarks and one u quark.

Following rules similar to these, it is possible to put together different combinations of quarks whose properties match every one of the hundreds of known particles. In fact, up until 1974, it was possible to make two statements about the quark model: every known particle could be explained as being made up of quarks, and every particle whose existence was predicted by the model was actually found. Our claim that the model introduced a special simplicity into the world of elementary par-

Accelerators speed up beams of particles to extremely high levels and smash them into other particles in an attempt to discover what the bits and pieces are.

Brookhaven National Laboratory

ticles, then, was well borne out.

What happened in 1974 was that two experiments, one by Samuel C. C. Ting at Brookhaven National Laboratory in New York and the other by Burton Richter at Stanford University in California, uncovered an unexpected new particle which did not fit into the quark scheme. It is called the J/ψ (each group gave the particle a different name before they were aware of the other's findings), and it is now accepted that it is a particle that cannot be made from the u, d, and s quarks. This means that we must either abandon the quark model or accept the idea that there might be a fourth kind of quark.

MORE QUARKS

In fact, a number of theoretical physicists had already suggested that there might well be a fourth quark. A short time after the discovery of the J/ψ, conclusive evidence that it was, indeed, made of the new kind of quark had accumulated. In our nuclear analogy, it would be as if someone had discovered a new kind of nucleus—one that wasn't made up of protons and neutrons, but of some other kind of particle.

Before too long, scientists at the Fermi National Accelerator Laboratory near Chicago found yet another kind of particle, made of yet another kind of quark. The number of quarks had grown to five. Finally, in 1984, scientists detected the long-sought (and predicted, since quarks seem to come in pairs) sixth quark, thereby further establishing the quark theory.

The three new quarks are called the "charm," or c-quark, the "bottom," or b-quark, and the "top," or t-quark. The J/ψ is made from the charm quark, and the Fermilab particle from the b-quark. The c-quark has a charge of $+\frac{2}{3}$; the b-quark has a charge of $-\frac{1}{3}$; and the top quark has a charge of $+\frac{2}{3}$.

When we refer to the kind of quark being discussed (that is, when we specify whether we are dealing with a u, d, s, c, b, or t), we say that we are specifying the *flavor* of the quark. In other words, there are six flavors of quarks now known to exist.

Actually, the term "known" is a little misleading. What are actually seen in the experiments we've talked about are particles, not the quarks themselves. The nature of the quarks inside of a particle is deduced from the way the particle behaves, not because we can actually see the quarks. In fact, one of the main problems with the quark model is the fact that at present there is no experimental result that the scientific community is generally willing to accept as a definite proof of the finding of a quark. It would be as if we knew about nuclei, but could never see protons or neutrons in our laboratory apparatus.

It's not that we haven't looked. Any number of searches have been made for quarks, either synthesized or natural. Only one such experiment, by William Fairbank of Stanford University, has come up with a measurement that could be the first sighting of a quark, about which there is still some controversy. Because the quarks have charges that are either ⅓ or ⅔ as large as the charges on ordinary particles, they would be very easy to identify if one were around. Consequently, physicists have started to ask themselves the question: "What if a quark is never seen in the laboratory?" Does it make sense to talk about the particles if they are never seen?

Actually, the continued failure of quark searches has led theorists to suggest that it might be possible that elementary particles are made of quarks *and* that quarks can never be seen by themselves in the laboratory.

One scheme that would accomplish this sort of thing would be the following: suppose the elementary particles we "see" can be thought of as rubber bands, with the quarks corresponding to the "ends" of the rubber bands. A meson, then, would be a simple band, one end of which is the quark and the other end the antiquark. If you think about it for a while, you will realize that no matter what you do to the rubber band, it is impossible to isolate one end. For example, if you cut through the rubber band, all you have are two shorter bands, each of which has two ends. This would be seen in the laboratory as the creation of an ordinary particle, and not the isolation of a quark.

QUARK CONFINEMENT

A scheme like this is one example of something called "quark confinement". There are now many different theories in which quarks are confined to the interiors of elementary particles, never being seen in isolation. So we have progressed to the point that even if a quark is never seen in the laboratory, the quark model can still be used.

If the failure to find a quark is the major experimental difficulty with the model, the major theoretical difficulty is somewhat more subtle. It turns out that there are some particles whose description in terms of quarks seems to violate one of the basic laws of quantum mechanics. For example, one of the elementary particles, called the Δ, is supposed to be made of three d quarks sitting and spinning in the same direction. According to the Pauli Principle (after Wolfgang Pauli, a Swiss physicist), three identical particles like quarks simply can't arrange themselves in this way.

The principle says, in essence, that once one particle is there and spinning in a given direction, it has taken up all the space allotted for that type. No more particles can be put in unless they are spinning in a different direction or are different from the first particle in some way. In the case of the Δ, this rule seems to be violated because all three particles are identical.

THREE COLORS

The way around this apparent dilemma is to find some way to think of the quarks in which the quarks of the same flavor are not identical. For example, suppose the three quarks in the Δ are different colors? Then there would be no conflict with the Pauli principle, because it applies only to identical particles, and the quarks would no longer be identical.

A word of warning. No physicist actually imagines that the quarks have different colors in the everyday meaning of the word. No one went around the Δ with a paintbrush and three buckets of paint, any more than when we use the term "flavor" we imagine that a quark can be tasted like a candy bar. The word "color" as it is used

Properties and Quark Compositions of Some Particles

Greek Symbol	Common Name	Mass (Proton = 1)	Quark Composition	Charm	Strangeness
π^+	pi-plus	0.149	ud	0	0
K^+	K-plus	0.526	us	0	1
K^0	K-zero-bar	0.530	sd	0	−1
η	eta	0.585	ss	0	0
p	proton	1	uud	0	0
n	neutron	1.001	udd	0	0
Ω^-	omega-minus	1.783	sss	0	−3
D^0	D-zero	1.986	cu	1	0
D^+	D-plus	1.991	cd	1	0
F^+	F-plus	2.164	cs	1	1
ψ	psi	3.299	cc	0	0

Burton Richter created J (psi) particles by colliding electrons with positrons.

here has a more restricted meaning. The term refers to whatever property it is which makes the three quarks distinguishable. In order to explain particles like the Δ, there ought to be three different kinds of quarks—three different colors. So we say that quarks come in six flavors, and each flavor comes in three colors.

Although the idea of "colored" quarks seems a bit fanciful, it turns out that there are a few experiments that can be done which would be different if there were 6 kinds of quarks in the world (one "color" for each flavor) or 18 (three colors for each flavor). The experiments strongly support the idea that there are 18 different kinds of quarks. Although the idea started out as a purely theoretical abstraction, it does have some important consequences in terms of things that can be measured.

The net effect of the theoretical and experimental investigations of the quark model since it was first proposed is simply that the number of quarks has grown from 3 to 18. Everything else has pretty much stayed the same. In particular, the essential statement that all elementary particles are made up of a (relatively) small number of quarks is still valid.

There has been one more important outgrowth of the quark model that we can understand best if we go back to the analogy between the nucleus and the elementary particle that we discussed earlier. We know that the nucleus is made up of protons and

neutrons. If we look at the nucleus closely, though, we see that its structure is far from a static thing. The protons and neutrons aren't just stacked together like oranges in a crate. Instead, other particles (called mesons) are zipping around between the protons and neutrons, providing the force that holds the whole thing together. In many ways, these particles, constantly being exchanged between the constituents of the nucleus, provide the "glue" that keeps the nucleus from flying apart because of the electrical repulsion between the protons.

GLUONS

The elementary particles are similar. They are not static collections of quarks, but instead are dynamic systems like the nuclei of atoms. Instead of mesons, however, the things that hold the particles together are called "gluons", a name which arises from their role as the "glue" that keeps the structure together. There are eight kinds of gluons in all and, like the quarks, they have color. In 1988, physicists found evidence of a new, theoretical fourth state of matter—a quark gluon plasma.

The important result of the quark picture referred to above is the development of a theory that describes the interactions of quarks and gluons. The theory is called "quantum chromodynamics"—the "chromo" referring to color as being important in the behavior of quarks. For the first time since the discovery of the nucleus itself, it appears that we have a theory that is capable of describing the so-called strong force—the force that holds the nucleus together. For if we understand how the particles are built of quarks and how the quarks interact with each other, then we can understand what happens when two separate particles, each with its complement of quarks, come together, since the interaction of the particles is just the sum of the interactions among all the quarks.

So that in addition to bringing a great deal of simplicity into the otherwise very complex world of elementary particles, the quark model has also led us to the brink of understanding the strongest (and until now the most mysterious) force in nature.

RADIOISOTOPES

by Eugene Fowler

The forms of nuclear energy known as *radioisotopes* have come to play an increasingly important part in everyday life. They are used in the manufacture of many products, such as plastics, food, clothing, and medical supplies. They are used in scientific and medical research and in medical treatment. They provide power for unmanned buoys, weather stations, and space satellites.

WHAT IS A RADIOISOTOPE?

To answer this question, we must first ask: "What is an isotope?" Matter is made up of elements. Carbon, oxygen, iron, and copper are examples of elements. All elements are made up of invisible building blocks, called atoms. Atoms, in turn, consist of protons (with a positive electric charge), electrons (with a negative electric charge), and neutrons (with no electric charge). The neutrons and protons occur together in what we call the nucleus, or core, of the atom. The electrons revolve about the nucleus.

All atoms of any one element have the same number of protons and electrons. All carbon atoms have 6 protons and 6 electrons. All oxygen atoms have 8 protons and 8 electrons. However, atoms of a particular element may differ from one another in the number of neutrons. For example, some naturally-occurring carbon atoms have 6 neutrons, while others have 7. Some oxygen atoms have 8 neutrons, some have 9, and some have 10. Atoms of the same element that differ only in the number of their neutrons are called *isotopes*.

When the ratio of neutrons to protons falls within certain numerical limits, the isotopes are stable. However, sometimes atoms will have too many neutrons for the number of protons or vice versa. In such cases, the nucleus will undergo a rearrangement, or readjustment, called *decay*, which results in the release of energy in the form of radiation.

The isotopes emitting radiation are said to be *radioactive*. Radioactive isotopes are commonly referred to as *radioisotopes* or *radionuclides*. They are usually indicated by the name of the element plus the mass number, or the number of protons plus the number of neutrons, with a hyphen in between. Thus cobalt-60 stands for the radioisotope cobalt with mass number 60. The mass number is often written at the upper left of the chemical symbol standing for the element. Thus ^{60}Co stands for cobalt-60.

THREE TYPES OF RADIATIONS

The radiations that are usually given off from radioisotopes are alpha particles, beta particles, or gamma rays, depending upon what kind of decay occurs in the nucleus. The *alpha particles* are helium nuclei, each consisting of 2 protons and 2 neutrons. Usually the *beta particles* are negative electrons. Occasionally these electrons are positive and in this case are called *positrons*. *Gamma rays* are a form of electromagnetic radiation.

WHAT RADIATIONS DO

The alpha, beta, and gamma radiations are responsible in one way or another for the uses of radioisotopes. Generally speaking, they act or are acted on in one of four different ways. (1) *Radiations affect materials*. This permits such applications as the sterilization of medical supplies, the treatment of cancer, the mutation of plants, and the pasteurization of foods. (2) *Radiations are affected by materials*. This makes possible radiography (taking pictures "through" objects), determining the thickness of objects, and protecting ourselves from radiation by shielding. (3) *Radioactivity traces materials*. Radioisotopes mingle and combine readily with stable isotopes. Their chemical and physical behavior is similar, and we can trace them by the signals they give off. Hence we can trace the chemical mixtures or combinations of which they form a part. The substance to which a radioisotope is attached is said to be *labeled* or *tagged*. (4) *Radiations produce heat.* This makes it possible to use radioisotopes in the production of heat and power.

MEASURING RADIOISOTOPES

The decay of a radioisotope is spread out, occurring at a predictable rate that is characteristic of the particular radioisotope. The *half-life* of a radioisotope is the time it takes for one half of any starting amount of radioactive substance to decay. Half-lives vary from fractions of a second to many millions of years.

The relative concentrations of the radiations from radioisotopes are measured by a unit called the *curie*. One curie is the quantity of a substance in which there are 3 $\times$ 10^{10} disintegrations, or individual decay events, per second (dps). One-half of a curie of a radioisotope is undergoing only 1.5 $\times$ 10^{10} dps.

PRODUCTION AND PROCESSING

A few radioisotopes occur in nature, including uranium-235, radium-226, potassium-40, rubidium-87, and carbon-14. Most of the radioisotopes used in agriculture, medicine, industry, and research are not natural. They are referred to as *artificial radioisotopes*. They are prepared in the following ways.

(1) *Bombardment of targets with neutrons in nuclear reactors.* Irradiation may take a few minutes or several years. In most cases a neutron is simply "captured," thereby increasing the neutron-to-proton ratio beyond the stability ratio. In other cases—usually with the capture of more energetic neutrons—a proton may be ejected simultaneously. These isotopes are referred to as *neutron excess* isotopes and usually decay by ejecting a beta particle.

(2) *Bombardment of targets in accelerators.* Accelerators bombard targets with charged particles, such as protons and alpha particles. The charged particle is accepted by the nucleus. Many times a neutron is ejected. This results in what we call *neutron-deficient* isotopes. They decay either by ejecting a positron or by capturing an electron, thereby converting a positively charged proton to a neutron.

(3) *Fission, or splitting, of the molecules of reactor fuels.* In all fission-type nuclear reactors, the fissioning of uranium or plutonium results in a large number of fission-product isotopes that can be separated and used individually. Some of these are stable, while others are radioactive.

There are several important differ-

Fig. 1. This diagram illustrates the comparative penetrating powers of alpha, beta, and gamma radiations.

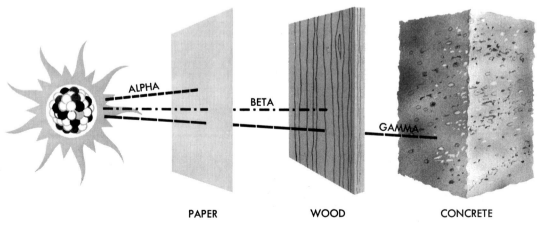

PAPER   WOOD   CONCRETE

ences between radioisotopes that result from the fission of atomic fuels and those produced in other ways. For one thing, far larger quantities are made available—sometimes in the hundreds of thousands of curies. Likewise, the isotopes are produced whether they are needed or not.

The chemical processing of isotopes after they have been removed from nuclear reactors or accelerators may be as simple as dissolving a soluble salt in water. In other cases, it may require evaporation, precipitation, extraction, distillation, and ion exchange.

USE IN SCIENTIFIC RESEARCH

Radioisotopes are used a great deal in scientific research—in chemistry, the life sciences, oceanography, and water studies, to name just a few areas. Radioisotopes have, for example, revealed much about the nature of chemical reactions, their speed, and the role catalysts play in the reaction process.

Researchers in the life sciences have made particular use of isotope tracer techniques. They have studied diffusion in living cells, the buildup of protein molecules, the synthesis of deoxyribonucleic acid (DNA), and the steps involved in photosynthesis. They have also used radioisotopes to determine the lifetimes of various cells. Human red blood cells tagged with iron-59 were, for example, found to have an average life span of 120 days.

In oceanography radioisotopes have been used to analyze oceanic currents and coastal drift. A radioisotope tracer—usually iodine 131—is injected into sea water in the center of a circular array of sensitive detectors. The various detectors respond to the amount of radioisotope reaching them. It is thus possible to determine quickly and accurately both the direction and rate of ocean currents. The movement of coastal sands along the ocean floor can be traced in a similar way if typical sands are tagged with radioisotopes—in this case, usually xenon-133, barium-lanthanum-140, or iodine-131.

The flow of river sediments can also be traced and measured using radioisotopes.

And radioisotopes have been used to measure the thickness and density of mountain snow layers. This allows the quantity and rate of meltwater formation to be predicted.

USE IN MEDICAL DIAGNOSIS AND TREATMENT

Medical diagnosis and treatment is probably the best known use of radioisotopes. X-ray machines have been important diagnostic tools for many years. They are bulky devices and require electric current for their operation. It is now possible to use the gamma rays from radioisotopes to obtain the same results. Since the devices using gamma rays are portable and require no electricity, they can serve in the field or where patients cannot be conveniently moved to an X-ray room. The isotopes thulium-170, iridium-192, cobalt-60, cesium-137, and iodine-131 have provided radiation for this purpose.

Radioisotope tracers are used to locate tumors and to trace blood movement through the body. Certain elements tend to concentrate in particular parts of the body. The amounts involved differ for healthy and abnormal tissues and for skeletal structures. Calcium and strontium are "bone-seekers"; iodine concentrates in the thyroid gland; arsenic and copper in the brain.

If some atoms of a given element are radioactive, they will act chemically and physiologically just like the normal atoms of the element. They will, however, indicate their paths and ultimate locations by the radiations they give off. Special detecting devices—one type is known as a scanner—locate and identify concentrations of abnormal tissues because more concentrated radiation signals are given off at these sites than anywhere else in the body.

Radioisotopes attached to blood cells allow the tracing of blood flow through the body, the determination of heart rate, and the identification of sites of blood loss.

Medical treatment with radioisotopes depends generally on the fact that malignant cells are more sensitive to radiation than are healthy cells. Malignant cells can be killed by controlled doses of radiation while healthy cells in surrounding tissues are not affected to nearly the same extent. Therapy can be carried out externally with beams of radiation—usually in the form of cobalt-60—concentrated on the cancerous area. Or small amounts of certain radioisotopes can be given internally. Iodine-131 has, for example, been given internally to treat thyroid cancer.

USE IN INDUSTRY

The applications of radioisotopes in industry are quite numerous and involve a wide variety of techniques and products. We will discuss a few examples. In the chemical reaction called polymerization, a number of small molecules called monomers join together to form a long chain of molecules known as a polymer. That is the way in which many plastics are built up. The polymerization of various monomers within the structure of wood has only now been done through the use of radioisotopes that are commercially available. Wood is first placed under vacuum. The pores are then filled with a monomer, and the wood-monomer system is exposed to radiation. Not only is the monomer polymerized, but a small amount of it appears to be "grafted" to the cellulose of the wood, thereby enhancing some of the wood properties. These "wood-plastics" appear to be particularly suitable for use as veneers, for furniture, flooring, and for musical instruments.

Radioisotopes can also be used to analyze coal, ores, liquids, grains, slurries, and other substances while they are being processed. A neutron generator bombards the stream of materials being processed as they pass by. This produces short-lived radioisotopes in the materials. Special detectors and analyzers then scan the materials for radiations that are interpreted for such things as moisture and ash content in coal or ore or the individual components of a slurry. This on-line neutron activation analysis enables continuous operation without delays.

There are many other industrial uses of radioisotopes. They can, for example, be used to gauge the thickness of a material: a thick section of lead does not, for example,

let as much radiation through as does a thin section. Radiations from some isotopes are also used in the processing of foods—to destroy harmful microorganisms, to disinfect grains, to inhibit sprout formation in some vegetables, and in numerous other ways.

USE IN AGRICULTURE

The use of tracers in agriculture represents one of the earliest radioisotope applications. As early as the late 1930s, phosphorus-32 was being used for the study of fertilizer uptake in plants.

Throughout history plant diseases caused by fungi have been a major agricultural problem. The best way to fight the problem (aside from developing resistant strains of plants) is through chemical control of such diseases. With radioisotope tracers, such as sulfur-35, it is possible to measure chemical uptake in single spores and to follow the chemicals through the plant. We can study the life cycle of the microorganisms and learn how temperature and humidity influence this cycle. We can also find out the chemical changes within plant cells that make them vulnerable to the attacks of fungi.

Radioisotope tracers are also used to determine which insect predators devour the insect pests that eat crops. The pests are made radioactive. Traces of this radioactivity are found by a detector in the insects that devour the pests.

Radiation has also been used to sterilize certain pests—the screwworm, for example. The screwworm fly inhabits extensive areas of the southern United States, Mexico, and the Caribbean and is a menace to livestock in those areas, causing millions of dollars of damage. Fortunately a very effective way has been found to combat this pest. The male flies are sterilized in the pupal stage by low doses of radiation, and the sterilized insects reared in large numbers. The sterile males compete equally with normal males for mates. Females mate only once. If a female mates with a sterile male, her eggs will not hatch and she will never breed a new generation.

Researchers have also used radiation to change the hereditary makeup of plants.

A small percentage of these radiation-induced mutations have resulted in improved farm plants.

USE FOR POWER AND HEAT

A device that generated electric power through the heat supplied by a radioisotope was introduced in 1959. It was called SNAP, from the italicized capitals in "*S*ystem for *N*uclear *A*uxiliary *P*ower." It and similar devices have been put to use for various purposes, such as supplying power for artificial satellites, weather stations, navigational buoys, and offshore oil rigs.

As for the direct use of heat from radioisotope decay, it has been limited thus far. It is believed that the basic principle may be employed to heat small huts on trails between major stations in extremely cold regions and also to warm diesel-fuel oil pans in exceptionally cold climates.

USE IN THE SPACE PROGRAM

Radioisotopes have played a part, though a very small one, in the space program of the United States. The artificial satellite *Transit 4A*, launched on June 29, 1961, was the first to have a nuclear electric power generator. It must be said, however, that solar batteries have proved far more successful in the space program as a source of electric power.

Manned satellites operate in an essentially "zero-gravity" situation. Hence the usual types of fuel and oil gauges, which depend on gravity for their operation, are useless. An isotope transmission gauge can be used to give the average mass of fuel in a tank. Radiation is transmitted from a radioisotope source at one end of the tank to a detector at the other end. The radiation recorded by the detector will vary according to the amount of fuel or oil in the tank.

The idea of using carbon-14 to detect the presence of life on other planets was proposed many years ago. In July 1976 the idea was put to a test. The U.S. Viking spacecraft landed on Mars and scooped up soil samples. These soil samples were then analyzed using carbon-14. Any organisms growing in the soil will absorb the carbon-14. Like all living things—as we know them—they will then exhale carbon dioxide. Only in their case, the carbon dioxide will be radioactive, being made up, in part at least, of carbon-14 radioisotopes. However, the results of the Viking tests were inconclusive.

Fig. 2. Half-life of a radioactive substance can be illustrated in this manner. The half-life of the substance is three hours. Hence in three hours, from noon to 3 P.M., it loses half of its radioactivity. In the three hours from 3 P.M., it loses half of its remaining radioactivity; and so on. "Mc" stands for "millicuries," or "thousandths of a curie."

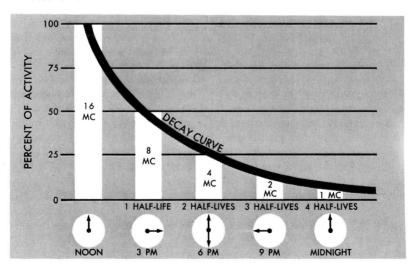

QUANTUM THEORY

by Lillian Hoddeson

When you climb up or down a ladder, only certain places or positions on the ladder—the rungs—can serve as stable footholds. You cannot rest your feet where there are no rungs. Thus you change your distance from the floor not in a smooth, continuous, gliding manner, as in sliding down a fire pole, but in a number of distinct, separate steps or acts: the acts of moving your feet from rung to rung.

There are many processes in nature where changes are made in distinct, separate steps. Each step consists of a definite increment or quantity. Because a definite quantity is involved in each step, we say the change is *quantized*. For example, in an atom, an electron moves in a distinct orbit. If it gains or loses energy, the electron jumps to another orbit. It cannot rest between orbits, just as you cannot rest your foot between two rungs.

The opposite of a quantized process is a process in which changes occur smoothly and continuously. In this process, any quantity or amount of change may be involved. For example, as a spacecraft travels from the earth to the moon, its distance from the earth increases in a smooth, continuous manner—not in a series of jumps or steps.

In a quantized process, energy is absorbed or emitted in a discontinuous manner rather than in a continuous manner. The energy is absorbed or emitted in bundles. One of these little bundles of energy is called a *quantum*. (The plural of quantum is *quanta*.)

The theory that underlies quantized processes is called the *quantum theory*. The theory is very important in science because it can, in principle, predict all the chemical and physical properties of a substance. In practice, however, scientists are not completely successful in making such predictions. The mathematical equations involved in the application of the quantum theory are so complex that, except for a few special cases, scientists do not yet know how to solve them. To obtain equations that are solvable, they make certain assumptions about the structure of an atom that are not really true. But these assumptions lead them to equations that are simple enough to be solved with known methods. The results calculated in this fashion are not really exact. They are only approximations to the truth. However, they are better than no information at all.

A NEARLY PERFECT BLACKBODY

During the end of the nineteenth century, most physicists were highly complacent about the state of their knowledge. They felt that everything important about physics was already known. They felt that no new major discoveries remained to be made. There was a certain amount of unrest, however, and some problems had not yet been solved. But it was generally believed that these problems would someday be solved within the theoretical framework then known, and no new tools would be needed. But as Hamlet said, "There are more things in Heaven and earth, Horatio, than are dreamt of in your philosophy."

One of the problems that baffled the physicists was the difficulty of understanding certain aspects of *blackbody radiation*. A blackbody, in the sense in which a physicist uses the term, does not necessarily mean an object that is black. Rather, it is an object that is a perfect absorber and a perfect emitter of light energy. Such a body completely absorbs all the wavelengths of light that fall on it. When it is heated it emits or radiates all wavelengths of light.

No known material is a perfect blackbody. But objects that are black in color approximate this type of behavior (hence the name "blackbody"). Black clothing, for example, absorbs most of the light energy that falls on it. This makes it warmer to wear than light-colored clothing, which reflects more light energy than it absorbs.

The work of these men led to development of the quantum theory. First row, left to right: Albert Einstein, Max Planck, Ernest Rutherford, Lester Germer. Second row: Clinton Davisson, Max Born, Niels Bohr, Wolfgang Pauli. Bottom row: Werner Heisenberg, Erwin Schrödinger, Louis de Broglie.

Art adapted from photos courtesy Niels Bohr Library/American Institute of Physics

The most nearly perfect blackbody is, of all things, a furnace with a single, very small opening. Light entering through the tiny opening is reflected back and forth between the walls of the interior. Since the opening to the outside is very small, very little light finds its way back out of the furnace. Hence, such a furnace traps virtually all of the light that enters it. In other words, it is a nearly perfect absorber of light and, therefore, a nearly perfect blackbody.

When such a furnace is heated to very high temperatures, it becomes a nearly perfect emitter of light. This means that it emits almost all the different wavelengths of light.

Figure 1 illustrates a wave. The *wavelength* is the distance between two successive peaks of the wave. It is the wavelength of a light wave that determines color. Of visible light, red light has the longest wavelength; violet light has the shortest. Other colors have wavelengths between these two extremes. There is an important principle we must remember: the shorter the wavelength, the greater the amount of energy

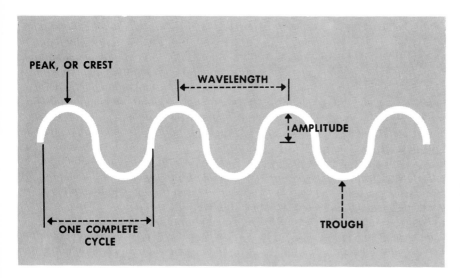

PEAK, OR CREST

WAVELENGTH

AMPLITUDE

ONE COMPLETE
CYCLE

TROUGH

Fig. 1. A typical wave. For a discussion of its characteristics, see the text. Waves and quanta are closely related phenomena.

associated with the light. Violet light (which has a short wavelength) contains more energy than red light (which has a long wavelength). We can express this another way. We can say that a quantum of violet light contains more energy than a quantum of red light.

Frequency is the number of cycles of the wave that pass a given point in one second. The greater the frequency, the shorter the wavelength; the smaller the frequency, the longer the wavelength. Which, then, has the greater frequency: a wave of violet light or a wave of red light? (You're right if you chose the violet light.)

THE WORK OF MAX PLANCK

Blackbody radiation is the radiation emitted from a blackbody when it is heated. It contains not just some but all wavelengths of light. If we look at Figure 1 again, we notice one other important characteristic of a wave: *amplitude*. Amplitude is one half of the vertical distance between a peak of the wave and a trough. The amplitude is related to the intensity of the radiation. When the intensity of radiation emitted for each wavelength is plotted on a graph, the result is called a *spectral distribution curve*. Such a graph is shown in Figure 2a. The curve shows us that a different amount of radiation is emitted at different wavelengths. The greatest amount of radia-

tion is emitted by those wavelengths in the "hump" of the curve. The amount of radiation falls off to zero at high and low wavelengths. As the temperature of a blackbody changes, the spectral distribution curve for its radiation also changes. The hump moves toward shorter or longer wavelengths. This is shown in Figure 2b.

In 1900, Max Planck, a then-unknown German physicist, decided to study blackbody radiation. He attempted to derive a mathematical equation that would account for the shape and position of the curves on a spectral distribution graph. Planck assumed that blackbodies contain tiny vibrating particles in their walls. He called these particles "little resonators." Furthermore, he assumed that these blackbody resonators emitted radiation continuously, in accordance with the laws of physics accepted at that time. These laws were all based on the fundamental laws of mechanics developed by Sir Isaac Newton. But Planck could not derive a formula on this basis.

When his efforts failed, Planck decided that the laws of mechanics that pertain to the workings of the inside of an atom are somehow different from Newton's laws. He decided to make up a new law. This was a daring thing to do, for he was challenging some of the most deeply respected and established laws of physics.

Planck started with a new assumption:

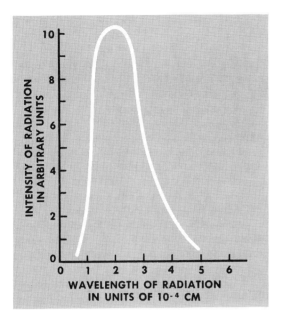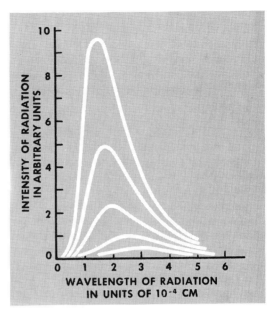

Fig. 2a-b. Blackbody radiation. Left: spectral distribution curve, where radiation intensity is plotted against radiation wavelengths (in 1/10,000 of a centimeter). Right: spectral distribution curves. Each higher curve represents a successively higher temperature.

the resonators in the walls of a blackbody do not absorb or emit energy in a continuous manner but, rather, in a discontinuous manner. The process, he said, is not smooth and gradual but stepwise. Each resonator, said Planck, absorbs energy in little bundles and, likewise, emits energy in little bundles. These bundles are called *quanta*, as we said before. When a bundle, or quantum, of energy is absorbed, the energy state of the resonator suddenly and abruptly jumps to one level higher on the energy scale. When a quantum of energy is emitted, the energy state of the resonator suddenly and abruptly falls to one level lower on the energy scale. A resonator cannot absorb $\frac{1}{2}$ or $2\frac{1}{4}$ quanta of energy at a time. It can absorb or emit only a whole number of quanta at a time. The energy content of a resonator cannot be increased by $\frac{1}{2}$ of a quantum any more than we can climb a ladder by moving up $\frac{1}{2}$ of a rung at a time. With this revolutionary hypothesis, Planck was able to derive a mathematical equation for blackbody radiation that did agree well with the experimental data. This equation, known as *Planck's blackbody radiation law*, shows

how the intensity of light emitted from a blackbody varies with the wavelength of light.

In deriving this law, Planck had to make an important assumption that can be written as follows:

$$E = h\nu$$

This states that the energy of a quantum (E) is equal to a certain constant value, known as *Planck's constant* (h), multiplied by the frequency (ν). In other words, the energy of a quantum is directly proportional to the frequency of the radiation. The greater the frequency, the greater the amount of energy in one quantum of the radiation. For example, since violet light has a greater frequency than red light, a quantum of violet light contains more energy than a quantum of red light. This is true for all types of electromagnetic radiation, not just light. Thus it is one of the most fundamental laws of physics.

Planck published his results in a highly respected scientific journal. But the significance of his work was not appreciated by the scientific community for some time.

The fact that recognition came slowly is not surprising, for his approach involved a completely new way of thinking. Recognition did come finally, and in 1918 Planck received the Nobel Prize in Physics for his work in quantum theory.

THE WORK OF ALBERT EINSTEIN

In 1905, Albert Einstein, soon to become famous, made an important application of Planck's quantum theory. He utilized the theory to explain the *photoelectric effect*. When light having at least a certain minimum frequency strikes the surface of some materials, electrons are emitted by the material. This phenomenon is called the photoelectric effect. All metals exhibit this phenomenon. So do some nonmetals, such as selenium, and certain compounds, such as cadmium sulfide. The photoelectric effect is widely used in electric eyes that operate electrical relays. Such relays open doors, set off burglar alarms, and cause photographic light meters to respond.

Surprisingly, the maximum velocity of the escaping electrons depends only on the frequency of the impinging light. One would expect the intensity of the impinging light to affect the velocity of the escaping electrons, but it does not. When the intensity increases, more electrons are emitted, but their maximum velocity remains the same. These facts had puzzled scientists for many years.

Einstein's explanation, using the quantum theory, was quite simple. He said that the least amount of energy that an atom at the surface of a material can absorb is one quantum. If an atom has not absorbed one quantum, an electron cannot escape from it. If it has absorbed one quantum, then an electron can escape. Since one quantum of energy has a certain frequency, electrons can escape if the frequency of the impinging light is more than this minimum value.

Increasing the intensity of the light increases the number of quanta hitting the surface of the photoelectric material. Thus it increases the number of electrons released by the material. But it does not change their maximum escape velocity. The maximum escape velocity is related to the energy of the impinging quanta, not to the number of quanta. However, if the frequency of the light is increased, then the energy of the quanta is increased and, hence, the maximum escape velocity increases.

Einstein's explanation of the photoelectric effect was very radical. It was not generally accepted for some time. But it did receive widespread notice among scientists when it was published in 1905, and it thus aroused considerable interest in Planck's previously ignored quantum theory.

In 1907 Einstein used the quantum theory to explain another phenomenon that puzzled scientists: the way that the heat

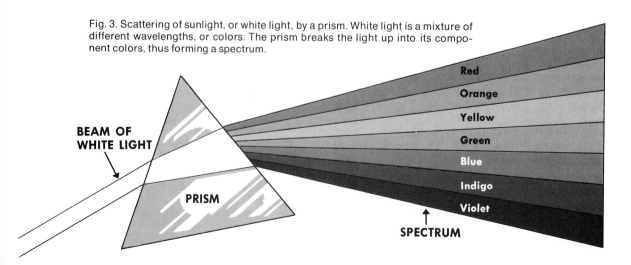

Fig. 3. Scattering of sunlight, or white light, by a prism. White light is a mixture of different wavelengths, or colors. The prism breaks the light up into its component colors, thus forming a spectrum.

Fig. 4. Characteristic spectral patterns produced by the dispersal of light emitted by chemical elements that have been heated to incandescence. Top: Spectrum of neon gas. Bottom: Spectrum of calcium vapor.

capacity of a material varies with temperature. The heat capacity of a material is the quantity of heat required to raise the temperature of one mole of that material by one degree Celsius. One mole of a substance is the total, in grams, of the atomic weights of the atoms that make up one molecule of the substance. One mole of carbon dioxide (CO_2), for example, is 44 grams, since the atomic weight of carbon is 12 and that of oxygen is 16 $(12 + (2 \times 16) = 44)$.

At room temperature the heat capacity of all simple bodies is the same. In 1872 measurements made on diamonds showed that the heat capacity varies with temperature—but only with temperatures below room temperature. More measurements followed and it was learned that the heat capacity of a substance varies from absolute zero ($-273.16°$ Celsius) to its limiting value at room temperature. Why, scientists wondered, doesn't it increase indefinitely?

Einstein explained the peculiar behavior of heat capacities by assuming that the absorption of heat by an atom or molecule is quantized. Starting with this assumption, Einstein was able to derive a mathematical formula that predicted the heat capacity versus the temperature curve for a substance fairly well. This was another achievement that drew attention to the importance of the quantum theory. In 1921, Einstein received the Nobel Prize in Physics for his work on the photoelectric effect.

THE WORK OF NIELS BOHR

In 1913, following Einstein, another phenomenon was explained by using Planck's quantum theory. Niels Bohr, a Danish physicist, applied the theory to the study of atomic spectra.

We are all familiar with the spectrum formed when white light is separated into its multi-colored wavelength components. The rainbow after a rain shower is an example of such a spectrum. This spectrum can also be produced by passing white light through a triangular glass prism (Fig. 3).

The white light from the sun is a composite of the spectra of all the chemical elements. The spectrum of an individual element is quite different in appearance. To produce the spectrum of an individual element, a specimen of this element is heated to incandescence. The light thus produced is passed through a triangular glass prism. In this case (the spectrum of a pure element all by itself), we do not get a full rainbow. There is no continuous band of color. Instead we get a pattern of sharp, discrete lines. The pattern of lines, or the atomic spectrum, is different for every element. It is unique and characteristic of that element. Thus the spectrum of an element can be used to identify that element (Fig. 4).

In a chemical analysis by spectroscopy, the specimen to be analyzed is heated to incandescence. The light produced in this manner is passed through a glass prism, and the pattern of lines that results is photographed. Then the photograph is matched up with the spectra of the known elements, just as the fingerprints of an unidentified person are matched up with those of known persons in a file. If more than one element is present in a specimen, the spectrum obtained is a composite of the separate spectra of all the elements present.

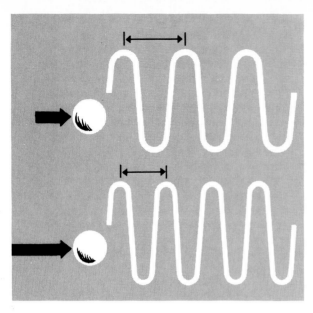

Fig. 5. The faster a subatomic particle, such as an electron, moves, the shorter the wavelength of its radiation. Top: slow electron. Bottom: fast electron.

Although spectroscopy had been used for chemical analysis since 1859, no one could explain the existence of the discrete lines in spectra. Then, in 1913, Niels Bohr proposed a new theory concerning the structure and behavior of the atom. This theory incorporated Planck's quantum theory and Ernest Rutherford's findings on the structure of an atom. In 1911, Rutherford, a British scientist, showed that the positive charge of an atom and nearly all of its mass were concentrated in a tiny, dense nucleus and that the negative charge—that is, the electrons—was outside the nucleus, surrounding it in some unknown manner. Bohr's theory attempted to account for the existence of discrete lines in the spectra of individual elements in a very clever and ingenious manner.

According to Bohr, the structure of an atom resembles a miniature solar system. There is a central nucleus containing protons and neutrons. Electrons revolve about the nucleus in orbits, as the planets revolve around the sun. When one of the electrons absorbs a quantum of energy, it jumps outward, to a higher or more distant orbit. When the electron emits a quantum of energy, it drops inward, to an orbit nearer the nucleus.

Bohr's theory makes it easy to account for the existence of the lines in the spectrum of an element. When a specimen of an element is heated, the electrons inside its atoms absorb energy. When the specimen is cooled, the electrons lose energy and fall back to a lower orbit, closer to the nucleus. When this happens, one or more quanta of energy are released in the form of light. The wavelength, and hence the color, of the light released depends on the energy content of the quanta released. Only the spectral lines corresponding to the energies of these particular quanta are present in the light emitted by the incandescent specimen. These lines, or wavelengths, are illuminated in the spectrum. Other wavelengths, absent from the specimen, do not appear.

Using the solar-system concept of the atom as his basis, Bohr derived a mathematical formula that made it possible to calculate the wavelengths of all the lines that appear in the spectrum of hydrogen. The calculated values were found to be in excellent agreement with the values obtained by direct experiment. This agreement between experiment and theory was taken by Bohr's contemporaries as evidence that his atomic-spectrum theory was essentially correct.

But, for elements more complex than hydrogen, Bohr's theory did not work well in predicting the wavelengths of spectral lines. Today the Bohr theory has been superseded by a more precise theory. But Bohr's theory was a major step forward, and for his contribution Bohr was awarded the Nobel Prize in Physics in 1922. Notice how Bohr, in his work, went one step further than Planck. Planck merely argued that blackbody resonators absorb or emit energy in a quantized manner. He did not specify what process or mechanism within the resonator was responsible for quantization. Bohr, however, showed precisely what process was responsible for this behavior: the quantum jumps of electrons.

PARTICLES AND WAVES

For many years there was a great deal of disagreement among physicists concerning another fundamental question: does light consist of particles or of waves? Some

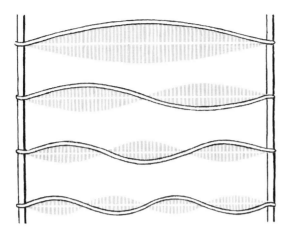

Fig. 6. Standing waves in a string. Similar waves occur in atoms. The wavelength of a matter wave is determined by the object's mass and velocity.

said light consists of tiny material particles. Others believed that light consists of energy waves. There was considerable evidence in support of each of these apparently contradictory points of view.

The first really important experiments on the nature of light, as far as we know, were those that Isaac Newton conducted in 1666. In one experiment, he allowed a beam of light, coming through a tiny hole in a window shutter, to pass through a prism and, when it emerged, to fall onto a viewing screen. Newton expected to find a single white spot on the screen. Instead, there were the colors of the rainbow: red, orange, yellow, green, blue, indigo and violet, in that order. Newton concluded, correctly, that white light is a mixture of all these colors. He also believed that light consists of rays or beams of extremely tiny particles moving at extremely high speeds.

The *corpuscular theory,* as Newton's concept of light was called, accounted for many of the observed properties of light. It accounted for the fact that light moves in straight lines and casts sharp shadows. It accounted for the fact that light is reflected from mirrors. It was argued that the light particles bounced off the surface of the mirror. It also accounted for the fact that light is refracted (bent) when it enters certain transparent materials such as water and glass. However, the theory failed to explain why two intersecting light beams do not interfere with one another. After all, shouldn't the particles collide? It also failed to explain why a beam of one color of light is refracted to a different extent than a beam of another color.

Twelve years after Newton's famous experiment, Christian Huygens, a Dutch physicist, proposed an alternative to Newton's corpuscular theory of light. Huygens proposed that light consists not of particles but of waves. This novel point of view explained why two beams of light can intersect without always interfering with one another. It also explained why light beams of different colors are refracted to different extents. However, Huygens' theory had its own weaknesses. It could not explain why light travels in straight lines or casts sharp shadows. It could not explain why light waves, unlike sound waves and water waves, do not pass around objects. And it could not explain how light can be transmitted where there is no medium for its propagation, as is the case in interstellar space.

For many years Newton's views prevailed, largely because of Newton's immense prestige. However, in 1801, Thomas Young, an English scientist, carried out an interesting experiment. His results persuaded most scientists of that time that Huygens was probably right after all.

Young showed that, under certain conditions, light can form bands of interference patterns, a phenomenon characteristic only of waves. Moreover, from his data, Young was able to estimate the wavelengths of light waves. He found that red light, at one end of the spectrum, has a wavelength of about 0.000075 of a centimeter. Violet light, at the other end of the spectrum, has a wavelength of about 0.000039 of a centimeter. These wavelengths are obviously very short. This is quite significant. It explains why light waves, even though they are waves, travel in a straight line and cast sharp shadows. Light waves will only be able to pass around particles smaller than the waves themselves (for example, virus particles, atoms, and molecules). The objects familiar

Fig. 7. Small physical particles, such as flour or rice, when dumped into a strainer, fall through and form a heap with indefinite borders. Compare with Fig. 8 below.

to us in everyday life are too big for light waves to pass around them. Thus Huygens' *wave theory* of light gained favor because Young's findings had given it considerable credibility.

THE LAWS OF QUANTUM MECHANICS

Could light be, simultaneously, both a particle and a wave? This seemed totally incomprehensible. In the 1920s, however, scientists found a link between these two seemingly contradictory characteristics.

The first breakthrough came in 1923:

French physicist Louis V. de Broglie postulated that all matter has waves associated with it. He suggested that what was true for light was also true for electrons and other subatomic particles and even for objects big enough to see: all have both particles and wave properties. Electrons, baseballs, planets, stars—any moving body—have wave as well as particle properties. But why can't we see the wave properties associated with, say, the motion of a baseball?

Objects large enough for us to see have extremely short wavelengths. They are so short that they cannot be detected by any known method. De Broglie explained the relation between the wavelengths of a particle and the particle itself in this equation:

$$\lambda = \frac{h}{mv}$$

This tells us that the wavelength of the wave properties (λ) is directly proportional to Planck's constant (h). It is inversely proportional to the mass of the object (m) and to the velocity of the object (v). In other words, the larger the object or the faster it is moving, the shorter the wavelength of its waves. If the object is very small—for example, an electron—the wavelength associated with it is comparatively long.

The waves associated with electrons and other particles of matter are called *De Broglie waves,* or *matter waves.* There is no direct way for us to see a matter wave. But a simple analogy will help describe the waves that accompany an electron as it orbits the atomic nucleus. Imagine a heavy

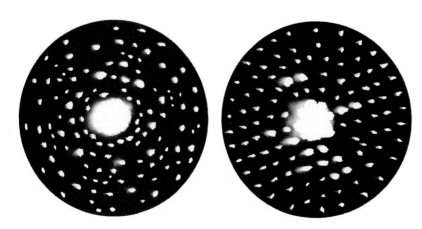

Fig. 8. Electromagnetic waves (far left) and electrons (near left), after they pass through crystals, form patterns of bright and dark spots. The waves are scattered by the molecules of the crystals, which act somewhat as the sieve in Fig. 7 above.

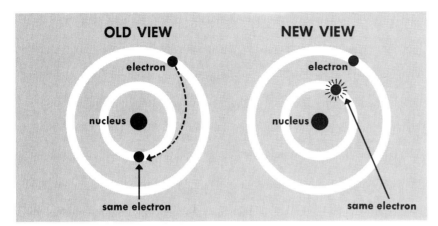

string attached horizontally between two poles, like a long clothesline. The string is attached so that it has just enough tension to be straight. What happens if you pluck the string, the way you would pluck a guitar string? The string will fill up with waves that cancel out in a short time. If the string has too much tension, the waves will travel so fast that you will not see them.

If you repeat this experiment, using a violin wire and a much shorter distance between the poles, you will notice that only certain size wavelengths are present. You will also notice that these wavelengths have a fixed pattern that looks very much like the diagrams in Figure 6. The waves in such a fixed pattern are called *standing waves*.

A similar thing takes place in an atom. Each electron that orbits the nucleus is accompanied by a standing wave that vibrates along the path, or orbit, of the electron. The wavelength of the standing wave obeys De Broglie's equation. In other words, the greater the speed of the electron, the smaller its corresponding wavelength.

SCHRÖDINGER'S CONTRIBUTION

At this point, a German physicist, Erwin Schrödinger, enters our story. He said that the orbit of an electron must contain a whole number of wavelengths. An electron cannot be accompanied by a standing wave of $2\frac{1}{2}$ or $6\frac{3}{4}$ or $7\frac{2}{3}$ wavelengths as it makes one orbit around the nucleus. Schrödinger expressed this thought in the equation:

$$\frac{\text{circumference of orbit}}{\text{wavelength of electron}} = \text{whole number}$$

If the circumference of an orbit divided by the wavelength of the electron does not equal a whole number, the electron cannot maintain itself in that orbit. For example, as an electron gains energy, its wavelength decreases. The orbit no longer contains a whole number of wavelengths. The electron jumps to another orbit.

In 1926 Schrödinger developed a mathematical equation, the *Schrödinger wave equation*, that can be used to calculate the wavelengths of wave properties. In the same year, Max Born, another German physicist, showed that the probability of finding an electron at a given point in space is proportional to the amplitude of the electron's matter wave. The electron, in other words, will probably be found where its wave has the greatest amplitude.

ELECTRON DIFFRACTION

In the following year, 1927, U.S. physicists Clinton J. Davisson and Lester H. Germer did an important experiment. They discovered De Broglie's matter waves for electrons. They did this by showing that electrons, like X rays, can be diffracted by crystals. When a beam of electrons is sent through a crystal, the beam is deflected and spreads out in the same way that light waves are deflected and spread out when they pass through thin slits. In crystals, the spaces between the atoms act like slits.

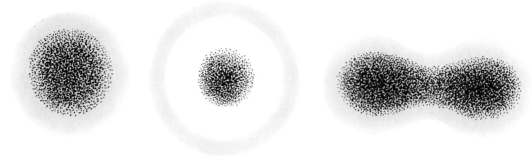

Fig. 10. Electron orbital regions of the hydrogen atom (left) and of hydrogen molecule (far right), consisting of 2 atoms. Electrons occur more often in the darker areas.

Diffraction is a property that only waves can possess. There was no longer any question that electrons had waves associated with them.

However, measuring these waves and measuring the particle aspects of an electron at the same time poses a problem.

ATOMIC AND MOLECULAR ORBITALS

In 1927 the German physicist Werner Heisenberg proved that it is impossible to determine, with exactness, both the position of an electron and its momentum. Any attempt to measure the position of an electron, Heisenberg showed, would disturb the electron. This would cause an uncertainty in the value of the momentum. Conversely, any operation intended to measure with exactness the momentum would disturb the electron. This would give rise to an uncertainty in the location of the electron. This is known as the *Heisenberg uncertainty principle*. It holds for the combination of time and energy, just as for momentum and position. Thus at a given time it is not possible to determine the exact energy of the electron. Thus, it is incorrect to ascribe definite, well-defined orbits to electrons, as was done in the Bohr theory.

The Bohr theory held that an electron occupies a definite, fixed orbit in its movement about the nucleus, like the orbit of a planet about the sun. Quantum mechanics, based on the theoretical and experimental discoveries we have just discussed, says that it is impossible to ascribe a fixed orbit to an electron within an atom. The best we

can do, quantum mechanics tells us, is to calculate the mathematical probability that a given electron will be found within a certain region of space.

Bohr depicted the hydrogen atom as consisting of a tiny nucleus surrounded by an electron that moved about it in a fixed orbit. Quantum theory depicts the hydrogen atom quite differently. A spherical, diffuse cloud of negative electric charge surrounds the nucleus. The sphere traces the region of space in which there is a 95 percent probability that the electron will be found. Or we can say that this sphere delineates the region of space in which the electron is found 95 percent of the time. Such a region of space is called an *orbital*. There are atomic and molecular orbitals.

The shapes of hydrogen and other atomic orbitals are shown in Figure 10. These shapes were determined by mathematical calculations based on the Schrödinger wave equation. Diagrams of this sort are used by chemists to help predict and interpret the chemical and physical properties of various substances.

UNDERSTANDING THE PERIODIC TABLE

One of the most useful tools of science is the periodic table (see The Periodic Table on page 33 in this volume). All the known elements are arranged into horizontal rows (called periods) in accordance with their atomic number, or number of protons. They are arranged in vertical columns according to related groups that have similar properties.

An interesting consequence of quantum mechanics helps us understand the arrangement of the periodic table. Each electron in an atom has associated with it a set of four *quantum numbers*. The first two of these numbers were discovered by Bohr. But he couldn't explain why they exist. Schrödinger showed that the numbers occur as constants in the solutions to his wave equation. Later, the other two quantum numbers were discovered, and they, too, were shown to be such constants.

The quantum numbers indicate the energy state of an electron. They can also be used to indicate the region in space, or orbital, where the electron is likely to be found. This region of space may be thought of as the "home" of the electron. Each electron in an atom is confined to its own home, but it may be found anywhere within that home. No two electrons occupy exactly the same home, or region of space, because two electrons in one atom cannot have identical sets of quantum numbers. This idea was presented by the Austrian physicist Wolfgang Pauli in 1925. It is called the *Pauli exclusion principle*. The four quantum numbers belonging to an electron are the:

(1) *Principal quantum number, n.* This may equal any positive whole number: 1, 2, 3, 4, and so on. The n's are called mainshells. Scientists call the mainshell in which $n = 1$ the K shell. The mainshell where $n = 2$ is the L shell, where $n = 3$, the M shell.

(2) *Subsidiary, or azimuthal, quantum number, l.* This may have any value from 0 up to $n - 1$. It cannot have a negative value. Thus, for the K shell, in which $n = 1$, the only value l can have is 0. In the L shell, in which $n = 2$, l may equal 0 or 1. The l's form subshells. In the K shell, where l has only one value, there is only one subshell. In the L shell, where l can be either 0 or 1, the shell contains two subshells.

(3) *Magnetic quantum number, m.* It may have the values of l and also the negatives of these values. Thus if $l = 2$, m could $= -2, -1, 0, +1$, and $+2$.

(4) *Spin quantum number, s.* This may have only two values, $-\frac{1}{2}$ and $+\frac{1}{2}$, for each value of m.

These numbers determine how many elements occur in each row of the periodic table. Each horizontal row of the periodic table represents a mainshell. The first row represents the first mainshell, the K shell, the second row the L shell, and so on.

We know that the K shell $n = 1$. Since l cannot have a value greater than $n - 1$, it must equal 0. If l equals 0 then $m = 0$; s will have two possible values, $-\frac{1}{2}$ and $+\frac{1}{2}$, for the one value of m. We can write the possible sets of quantum numbers for the K shell:

$$
\begin{array}{ll}
n = 1 & n = 1 \\
l = 0 & l = 0 \\
m = 0 & m = 0 \\
s = -\frac{1}{2} & s = +\frac{1}{2}
\end{array}
$$

Fig. 11. Tabulated values of the quantum numbers n, l, m and s for the second, or L, shell of an atom. Eight sets of values occur.

QUANTUM NUMBER	SET 1	SET 2	SET 3	SET 4	SET 5	SET 6	SET 7	SET 8
n	2	2	2	2	2	2	2	2
l	0	0	1	1	1	1	1	1
m	0	0	−1	−1	0	0	+1	+1
s	−½	+½	−½	+½	−½	+½	−½	+½

Thus there are two, and only two, distinct sets of quantum numbers possible for the first mainshell. There are only "homes" for two electrons in this shell. If the atom contains only one electron, only one home is occupied; this is true for the element hydrogen. If there are two electrons in the atom, both homes are occupied; this is true for helium. These are the only two elements that are found in the first row of the periodic table.

The situation for the second row of the periodic table, which corresponds to the L shell, is more complicated. Here $n = 2$. Therefore l can equal either 0 or 1. If l is 0, m must be 0. If $l = 1$, m can be either -1, 0 or $+1$. And for each value of m, s can be either $-\frac{1}{2}$ or $+\frac{1}{2}$. All the sets of quantum numbers that are possible for the L shell are shown in Figure 11. As you see, there are 8 sets of numbers. Consequently, the second row of the periodic table will contain 8 elements. In these elements, the first mainshell (K) contains 2 electrons; the second mainshell (L) contains from 1 to 8 electrons, depending on the element.

Can you determine how many homes, or sets of quantum numbers, are possible for the third mainshell? The answer is 18. But if you look at the periodic table, the third row contains only 8 elements. In these elements, the K and L shells are filled with electrons. The third mainshell (M) contains from 1 to 8 electrons (see Figure 12). These M-shell electrons are in the first 2 subshells. But there are 3 subshells here (since l equals 0, 1, or 2). Why aren't there any electrons in the third subshell, then, as this would seem to suggest?

Scientists suggest that this results from an overlapping of subshells as one gets farther away from the nucleus. The outermost subshell of M overlaps with the innermost subshell of N. Because of this overlap, scientists suggest that an electron will enter the inner subshell of N rather than the outer subshell of M.

Quantum mechanics has other applications in the physical sciences and has proven to be extremely important in analyzing and explaining the behavior of atoms. The laws of quantum mechanics are also of interest to philosophers. This is particularly true of Heisenberg's principle of uncertainty, which tells us that we cannot always measure things precisely. For example, we cannot tell exactly where an electron will be in an atom—and we never will be able to determine this. This implies that there is some randomness to events in nature. Einstein found this hard to accept: "I can't believe that God would choose to play dice with the world," he said. But neither he nor anyone else has disproven Heisenberg's principle that there is a certain amount of inherent uncertainty in our universe. This uncertainty is particularly evident at subatomic levels.

| | NUMBER OF ELECTRONS IN | | | | | ATOMIC NUMBER |
	$1l^1$	$2l^1$ $2l^2$	$3l^1$	$3l^2$	$3l^3$	
Sodium	2	8	1	—	—	11
Magnesium	2	8	2	—	—	12
Aluminum	2	8	2	1	—	13
Silicon	2	8	2	2	—	14
Phosphorus	2	8	2	3	—	15
Sulfur	2	8	2	4	—	16
Chlorine	2	8	2	5	—	17
Argon	2	8	2	6	—	18

Fig. 12. Numbers of electrons in the atomic shells and subshells of elements in the third row of the periodic table. The letter "l" designates a subshell. $1l^1$ means the first, or K, shell, having only 1 subshell; $2l^1$ and $2l^2$ indicate the second, or L, shell, with 2 subshells; $3l^1$, $3l^2$ and $3l^3$ mean the third, or M, shell, which has 3 subshells (the third subshell having no electrons). Atomic number is simply the total number of electrons = number of protons in an atom.

PARITY

The law of the conservation of parity has been considered very important in modern physics. This law states, in effect, that for the absolute universe of space and time, distinctions between left and right and up and down, between an object and its image, and so on have no real significance. These directions mean something only to an observer or in spatial relations among objects in various locations.

It is true that in our familiar everyday world, as opposed to the universe at large, distinctions between right and left assume great importance. We find a difference between right and left arms, right and left legs, right and left shoes and gloves. The structures of various organisms tend to be oriented to the right or to the left, more toward one than the other in different groups (the coiling in the shells of snails, for example). Thus, one direction may be preferred over another. Even the sun, planets, and moons mainly rotate and revolve in a prevailing direction. And yet, is it not possible to imagine a system like ours in all respects, but constructed on a pattern where directions are the reverse of ours? Except for this one basic difference, would it otherwise differ markedly? Probably it would not.

REAL AND MIRROR WORLDS

Consider a pair of shoes. Except for the difference between rightward and left-ward directed structures, no one doubts that both shoes are identical in style, size, and function. Suppose you have a pair of shoes, a and c, as shown in Figure 1. If you hold shoe a to a mirror, the reflection, b, will be identical with shoe c, the other shoe of the pair. If we use symmetrical kidney beans instead of shoes in the demonstration (Figure 2), we see that bean b, the mirror image of a, is exactly the same as bean c. We also observe that bean c is exactly the same as bean a turned upside down. In this case, then, there is no basic difference between the object (bean a) and its image (bean b).

PARITY IN THE SUBATOMIC WORLD

If we could directly see the particles that compose the atoms of matter and the waves of energy, we would notice, in certain situations, that particles spinning or moving in one direction tend to be balanced by identical particles spinning or moving the opposite way. To illustrate this aspect of parity, hold a plain ball, a, up to a mirror (Figure 3). Ball and image are absolutely identical and cannot be distinguished at all. Now spin the ball between thumb and forefinger along an axis parallel to the plane of the mirror. You will notice that the image of the ball in the mirror spins in the opposite direction. Reflection b is identical with a real ball, c, rotating in the same direction. Ball a corresponds to c turned upside down. The introduction of motion into the concept of parity here is of the greatest importance, as we shall see shortly.

Another important factor to consider

Fig. 1. Shoe b is the reflection of shoe a in the mirror. Shoe b is identical with shoe c, the other shoe of the pair.
Fig. 2. Bean b is the mirror image of bean a.
Bean b is also the same as bean c, which is simply a turned.

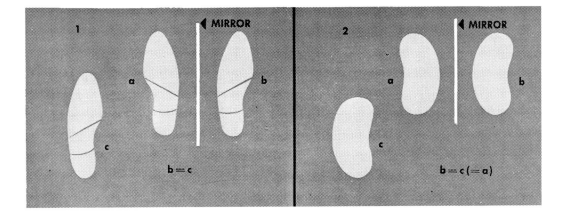

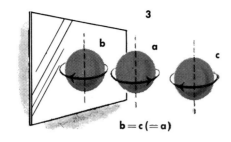

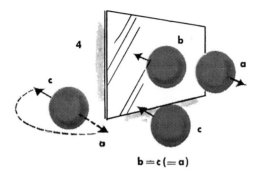

Fig. 3. Ball *b* is the mirror image of ball *a*; *b* is identical with ball *c*, which rotates in the same direction.

Fig. 4. Ball *b* is the mirror image of ball *a*; *b* is identical with ball *c*, which is *a* turned 180°.

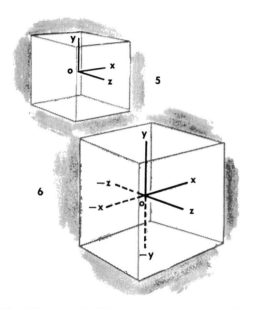

Fig. 5. Three straight lines, *x*, *y*, and *z*, meet at right angles at O and thus define a three-dimensional space.

Fig. 6. When the same axes, *x*, *y*, and *z*, are extended beyond the O point, the extended portions receive minus signs.

in balanced motion is symmetry. We illustrated mirror-image parity for the sphere by rotating it, in Figure 3. Suppose now that, as in Figure 4, we move the ball *a* in a straight line directly *away* from the mirror. Its image, *b*, of course, will move in the opposite direction. If we move another ball, *c*, directly away from ball *a* (which moves in the opposite direction), ball *c* will correspond exactly to the mirror image *b*. These motions may be considered as symmetrical or balanced. In Figure 4, balls *a* and *c* may be said to constitute a pair. If we turn *a* 180° to the left so that it is upside down in relation to its former position, it corresponds perfectly to *c*, even including the arrows. Therefore, *a*, *b*, and *c* are identical.

We can detect the nature of particles of subatomic size only indirectly, since they are too minute and move far too rapidly for us to see. We observe how they interact with other particles and waves, and we measure these interactions. But this system of measurement necessarily disturbs the particles, so that we cannot get an exact picture of what and where they are at a given moment. However, using statistics as a guide, physicists can guess about the probable positions of particles.

These bodies travel and may also spin, or rotate, on their axes as well. They are affected by electrical and magnetic fields. They react with each other, giving off or forming different particles, as in radioactivity. Some particles, such as electrons, have negative electrical charges. Others are positive, such as positrons and protons. Still others are electrically neutral, such as neutrons and the far lighter neutrinos and antineutrinos. Certain particles move freely through space, while others are located in atoms.

To locate such a particle in three-dimensional space, we construct three straight lines, called axes, at right angles to each other (Figure 5). The position of the particle is related to distances marked off on the three axes, or co-ordinates, which are given arbitrary symbols or letters — usually *x*, *y*, and *z*. Suppose this system of axes is reflected in mirrors (Figure 6). We could call the reflections negative, labeling them $-x$, $-y$, and $-z$, to distinguish them from the positive axes, x, y, and z, that are being reflected. Axes $-x$, $-y$, and $-z$ might also be called left-handed.

The actual picture in the world of subatomic particles is a bit more complicated than we have drawn it so far. Under certain conditions, groups of particles behave like waves in their motion. Physicists have made a careful study of these wave motions. They have determined that the probable positions of particles can be derived by a formula involving the square of what is called the *wave function*. If the axes which determine the position of a particle are reflected and the wave function becomes negative as the axes do, that particle is said to have *odd parity*. If the wave function still remains positive after reflection, however, the parity is called *even*. It is as if, despite the extension of the axes into the negative region, the particle's image is still positive. In a system with a given number of particles interacting, parity is not supposed to change from the beginning to the end of the process. It is either even or odd, but not both: parity is theoretically conserved.

IS PARITY REALLY CONSERVED?

Certain scientists began to express doubts about the law of parity during the 1950s. It was noted that certain particles, called mesons, during their transformation, or decay, into other kinds of mesons and various particles, had different parities, although they were absolutely alike in all other respects. One type of meson decays into two other mesons of different mass, each of the two having odd parity, so that the over-all parity is even. This is so because, when two negative quantities are multiplied together, the result is positive. In other words, two odds make an even. Another kind of meson decays into three mesons, each of the latter identical with the products of the first-described process and each one also having odd parity. In this case, the over-all parity is odd. The reason is that when three negative numbers are multiplied together, the result is negative; three odds make an odd. The two mesons at the beginning of decay are alike except for their parity — one odd, the other even. It is as if identical objects have different mirror

Fig. 7. Ball *b* is the mirror image of ball *a*; *b* is identical with ball *c*, which is ball *a* turned upside down. Figs. 8. and 9. Ball *b* is the mirror image of *a*, but is not equivalent to *c* in either case.

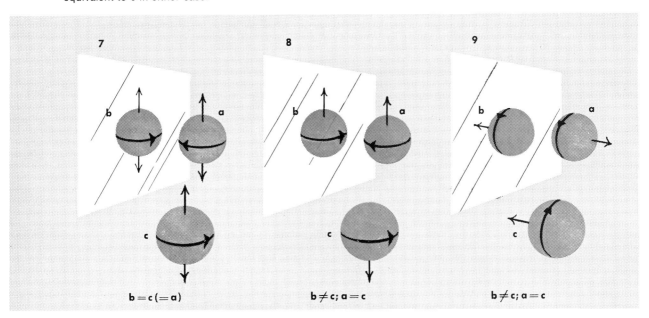

reflections. Even if we assume the same parity for both, then somehow parity has changed in one of the two decay processes above. It would appear that the principle of parity has been violated. Certain reactions, called *weak interactions,* in the subatomic field do not appear to be symmetrical and there appears to be a difference between right and left and between particle and image.

A STARTLING IDEA

In 1956, C. N. Yang and T. D. Lee, physicists working at the Institute for Advanced Study, in Princeton, New Jersey, proposed that parity is not conserved in weak interactions. They suggested that several quantities could be measured to confirm this theory. For their tradition-breaking theory, Yang and Lee received the 1957 Nobel Prize in Physics.

The predicted violations of parity were tested experimentally by C. S. Wu. She and her colleagues, E. Ambler, R. W. Hayward, D. D. Hoppes, and R. P. Hudson, set up an experiment. Wu believed that parity conservations could be tested by observing the direction of the electrons emitted in beta-decay of a spinning nucleus. If electrons were observed to travel in a direction parallel to the nuclear spin, rather than, say, antiparallel, then parity is not conserved. By supercooling the radioactive material cobalt-60 to diminish the effects of molecular motion and thereby enhance observation, Wu and her associates were able to determine that more electrons left in an antiparallel direction. In other words, parity is not conserved in this decay process.

MORE VIOLATIONS OBSERVED

Further experimentation supported the work of C. S. Wu and her group.

To test the parity law for mesons, physicists at Columbia University in New York City and at the University of Chicago conducted a number of experiments. The Columbia group of M. Weinrich, L. Lederer, and K. Garwin, for example, bombarded a target in a cyclotron with protons. As a result, the target emitted a stream of me-

sons, known as *pi mesons.* The latter then decayed into so-called *mu mesons* and into extremely light particles with no positive or negative electrical charge—*antineutrinos.* This process was carried out in a magnetic field, which lined up the mu mesons' axes of rotation in one direction. It also made these mesons spin in a single direction. This was important, because the movements and positions of the reacting particles could be observed and measured. The mu mesons were then collected in a carbon block, where they in turn decayed, giving off electrons, antineutrinos, and neutrinos—light particles with no positive or negative charges (unlike the electrons). It was noted that more electrons were emitted in one direction along a mu meson's spin axis than in the opposite one. According to the law of parity, the electrons should have come off equally in all directions or in equally opposed ones, as in Figure 7 (indicated by the arrows). That is, they were supposed to come off each end of each rotation axis. Parity in this case would have been preserved; note that image *b* is identical with *c;* *a* and *c* constitute a pair, and *a* is *c* turned upside down, so that *a = b = c.* (Compare this with Figures 1–4.)

The reader should note that each sphere in Figure 7 is symmetrical in that its upper half, with its arrow, is a mirror image of the lower half, with its arrow. That is, imagine a mirror inserted through the equatorial region of each ball, perpendicular to the axis of spin. Figure 7 shows what the situation would be if the law of parity held true.

The actual experimental results are shown in Figures 8 and 9. The meson *a* emits electrons off one end of its spin axis. Its paired opposite, meson *c,* spinning in the opposite direction, emits electrons off the opposite end of its axis. We see that *a* is *c* turned upside down. Is *a*'s reflection, *b,* identical with meson *c?* It is not, of course. In Figure 8, *b* spins in the same direction as *c,* but its electrons stream off its axis in the opposite direction. In Figure 9, electrons move in the same direction in *b* and *c,* but the spin directions are opposed. It is as if *a* has two different "mirror" images: *b* and *c.*

ALBERT EINSTEIN

by Aaron B. Lerner

Albert Einstein—for many of us, saying his name is the same as saying "modern science." He is one of the best-known scientists who ever lived, and for good reason. His work changed the way in which we understand the universe. We no longer think of time and space in the same way we did before he developed his theories. He changed fundamental concepts that had prevailed since the days of Isaac Newton. He can be called the main founder of theoretical physics in the 20th century.

EARLY YEARS

Einstein was born on March 14, 1879, in Ulm, Germany. One year later the family moved to Munich, where his sister Maja was born. Einstein remained in Munich for 15 years. His love of music, which became a necessity for him, took shape early when he began violin lessons at the age of six.

Although he was not especially good in the routine work of the classroom and was slow in talking, Einstein excelled in self-education. By the age of 12 he had taught himself calculus, and by 13 he had read Immanuel Kant's *Critique of Pure Reason.* He did not read light literature but was absorbed in books on mathematics, physics, and philosophy.

Einstein left Munich at 15 to join his parents in Milan, Italy, and a year later he sought to enter the Federal Institute of Technology in Zürich, Switzerland. Because he had no diploma from high school, he had to take an entrance examination. He failed in modern languages, zoology, and botany—subjects he had not completed in Munich. However, he performed so brilliantly in mathematics and physics that he was allowed to enter after a year of study. His performance in college was excellent.

For a time after his graduation in 1900, Einstein was not able to get any kind of job or assistanceship. He spent two years in a variety of minor teaching posts and finally got a job as a technical expert in the Swiss patent office in Berne.

INTERNATIONAL FAME

Einstein was a rapid worker and had ample free time to work on his own projects and complete requirements for a Ph.D. degree. During his seven years at the patent office he published several important papers. In 1905 alone he brought out four, any one of which would have made him famous.

One of the papers set forth his special theory of relativity, the importance of

Albert Einstein, the greatest scientist of his era and probably of the 20th century. His theory of relativity changed man's concept of the universe.

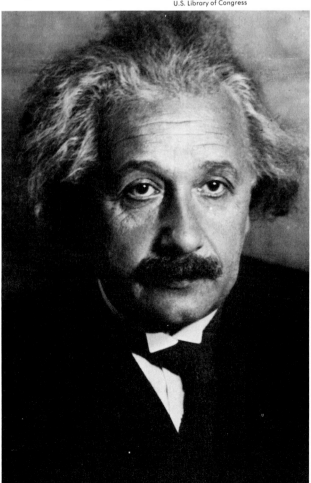

which is discussed below. Another paper came to the conclusion that radiation must be considered as traveling in small "bundles," or quanta. This paper also accounted for the photoelectric effect—the emission of electrons by metals when exposed to light of certain wavelengths. It was primarily for this that Einstein was awarded the Nobel Prize in 1921. The third paper showed that the Brownian movement—the zigzag movement of small particles suspended in an otherwise motionless liquid—was direct evidence for the existence of molecules. The fourth paper concluded that matter and energy are equivalent to each other.

Even though his work gained Einstein worldwide fame, he did not obtain an academic position until 1911. In the meantime, in 1907, he published a paper using the quantum hypothesis to account for heat phenomena at very low temperatures, and, in 1916, a paper that led to his general theory of relativity. In 1913, Einstein gained a research post in the Kaiser Wilhelm Society in Berlin, combined with a professorship at the University of Berlin. He published several other important papers during his years there.

LIFE ABROAD

Einstein had many invitations to lecture in other countries. He was in the United States when Hitler came to power in 1933. Since Einstein was Jewish and an outspoken opponent of the Nazis, they took over his property and denied him his German citizenship. He accepted a post at the Institute for Advanced Study in Princeton, New Jersey, and in 1941 he became a U.S. citizen.

Einstein did everything he could to support the war against the Nazis and for that effort was criticized by pacifist groups that he had long endorsed. He suggested to President Roosevelt that the United States begin development of the atomic bomb, fearing that the Nazis were also engaged in such research. After the war, however, he took a leading part in efforts to gain international agreement on the control of nuclear energy.

During his later years, Einstein worked mainly on trying to perfect a unified theory that would link together all the forces in the universe. These forces range from those at work within and between atoms to the force of gravitation that holds the planets in our solar system and the stars in their galaxies. How far he succeeded in this has yet to be determined, but his work raised many problems and gave rise to new theories of modern physics and cosmology. Einstein died in Princeton on April 18, 1955.

Einstein liked simplicity in everything he did. He was modest and direct, cared nothing about honors or formal affairs, and was suspicious of authority. He was known for his wit and joking, much of his humor being directed against himself. The most impressive of Einstein's characteristics was the lasting zeal with which he sought out the laws of nature.

A view of Einstein's study at the Institute for Advanced Study at Princeton University, the way he left it on the day of his death—April 18, 1955.

EINSTEIN'S IMPORTANCE

The scientific world was facing major problems in theory by the turn of the twentieth century. The concepts of space and time and motion developed by Isaac Newton in the seventeenth century, which dominated physics thereafter, were found no longer able to account for observations that had been made in recent years.

The Newtonian universe was a fixed framework of space and time. It was thought that a body's motion could be described in terms of this "absolute" framework, without reference to any other body. When it was found that light is an electromagnetic wave that travels at a certain velocity, scientists tried to explain how it travels through space. They hypothesized that a material, an "ether," filled all space and that it carried the waves.

Among the expected effects of this hypothesis was that the observed velocity of light should be different according to whether one made the observation "downwind" or "upwind" while traveling through the ether. Yet no such differences could be found in actual experiments. These and other problems had the scientific world in a quandary.

Einstein answered all this by making some very radical assumptions. He assumed that there is no such thing as "absolute" motion in a fixed universe. Instead, any motion of a body can be described only in relation to motions of other bodies. The velocity of light alone is an absolute. It does not vary, no matter what the relative motion of the observer. Time is also relative. Clocks run faster or slower depending on the relative motion of the observer. This was a fundamental change in our understanding of the nature of the universe.

Einstein also found that the mass of a body is relative to its velocity. The faster an object moves, the more massive it becomes. The limiting velocity is that of light. From this came his conclusion that matter and energy are equivalent. His famous equation, $E = Mc^2$, states that energy is equal to matter times the square of the velocity of light. Previously, electromag-

Einstein acquired a love of music from his mother and found enjoyment in pursuing the interest throughout his life.

netic radiation had been thought of as a disturbance in the "ether." Now it had to be thought of as having as definite a physical reality as matter.

Einstein also made basic contributions to quantum theory, which was first developed by Max Planck in 1900. Quantum theory, which underlies all modern physics, holds that radiation must be considered as being emitted or absorbed in "bundles," or quanta, of energy, as well as being wavelike. Einstein showed that this concept explains various puzzling physical phenomena, thus bringing the theory into general acceptance.

Among Einstein's many other contributions to science was his development of the concept of the process known as the stimulated emission of radiation. This process is now used in the devices known as lasers.

Besides all this, Einstein was a public figure and humanitarian who played an important role in history. He worked for the creation of modern Israel and once turned down an invitation to become its president. His philosophical writings, his love of music, his wit—even his physical appearance in later years, with his great shock of white hair and his bushy mustache—made him a unique figure of the twentieth century.

THE THEORY OF RELATIVITY

by Serge A. Korff

Certain theories, very few in number, have so greatly influenced scientific thinking that they stand out above all the rest. Among them are the hypothesis of Copernicus concerning the motions of the planets; the doctrine of universal gravitation, advanced by Sir Isaac Newton; the cell theory of Schleiden and Schwann; and Darwin's theory of organic evolution. We certainly must also include the theory of relativity, formulated by Albert Einstein and developed further by Einstein himself and several other scientists.

This theory has revolutionized our thinking. It has altered our ideas of space, time, mass, energy, motion, and gravitation. It has provided a new approach to the study of the universe. It is made up of a special theory and a general theory. Both rest on a solid mathematical foundation, and both have been confirmed by experimentation and observation.

FRAMES OF REFERENCE

The theory of relativity arose from the need for a *frame of reference*—a standard that scientists could use in analyzing the laws of motion. The need for such a standard is obvious, for the moment we attempt to analyze motion, we must ask ourselves: "Motion with respect to what?"

Suppose that a pilot is flying a plane far above the earth in a jet stream (a swift current of air) flowing at 300 kilometers per hour. The plane itself is flying, say, at 300 kilometers per hour in the same direction as the jet stream. The pilot will be moving at a 300-kilometer-per-hour speed with respect to the jet stream, but at 600 kilometers per hour from the viewpoint of an observer stationed on the earth. If the pilot flies against the jet stream, to the earth-bound observer he will seem to be hovering motionless in the air.

To an observer on the planet Mars, a speed of 300 kilometers or 600 kilometers per hour would seem insignificant compared to the nearly 30-kilometer-per-second speed of the earth as it revolves around the sun. It is moving at many kilometers per

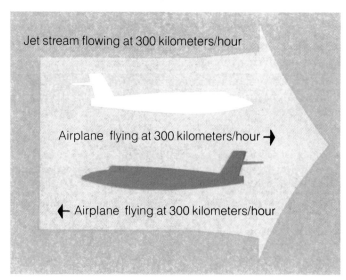

Jet stream flowing at 300 kilometers/hour

Airplane flying at 300 kilometers/hour →

← Airplane flying at 300 kilometers/hour

All motion is relative. The upper plane, flying at 300 kilometers per hour, is traveling in a jet stream which is flowing at 300 kilometers per hour. The plane is moving in the same direction as the jet stream. It is traveling at 300 kilometers per hour with respect to the jet stream, but at 600 kilometers per hour from the viewpoint of an observer on earth. The lower plane is flying at 300 kilometers per hour against the jet stream. To an earth-based observer, it will appear to be motionless.

All illustrations: Clifton Line

second around the center of the Milky Way, the galaxy of which our solar system forms a very small part. The Milky Way itself is moving with respect to other galaxies. Where, then, is our frame of reference?

At one time, men of science believed in a luminiferous, or light-bearing, ether that might serve that purpose. The ether was supposed to be a fluid, or a very elastic solid, occupying all the space between the atoms of which matter consisted. Light, it was thought, was transmitted in a series of waves through the ether—waves at right angles to the direction of motion. Some scientists believed that the ether itself remained motionless as the light waves passed through it. They compared the passage of light through the ether to the movement of ocean waves in the sea. It is the ocean-wave form, or pattern, they said, that moves forward. The water particles themselves do not move to any marked extent. If scientists could show that the ether had no motion of its own and that the heavenly bodies imparted none to it as they passed through space, we would have a reliable starting point for a general analysis of motion.

In trying to prove that the ether was stationary with respect to the earth, one would have to consider that from our viewpoint the earth is standing still. Therefore, the ether would appear to be rushing past us "like the wind through a grove of trees,"

to quote the English physicist Thomas Young. In the same way, the sun seems to us to be moving around the earth, while it is really the earth that is revolving around the sun.

ETHER CONCEPT DOUBTED

Two U.S. scientists, Albert A. Michelson and Edward W. Morley, tried to solve the ether problem by making delicate measurements of the speed of light. Let us suppose, they said, that the ether is motionless as light is transmitted through it. If the earth were also motionless in space, light would always seem to have the same velocity when measured by an observer on the earth. But the earth moves through space. Hence the velocity of light would appear to change according to the direction in which a light beam would be flashed—that is, in the direction of the earth's motion, or in the opposite direction, or at right angles to it.

The two scientists analyzed the differences in the arrival times of the light beams parallel to the earth's course and at right angles to it. No matter how many measurements they made, the light appeared to travel just as rapidly in the direction of the earth's motion as against it or at right angles to it. The experimenters concluded that "if there be any relative motion between the earth and the luminiferous ether, it must be quite small."

The man on the motorcycle sees the ball fall straight down. An observer on the side of the road, though, sees the ball fall in an arc. This illustrates the relativity of motion: Different people see the same event in different ways.

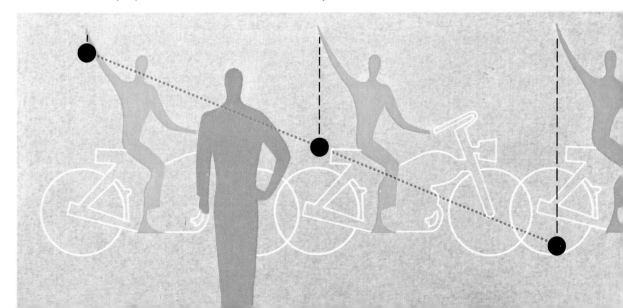

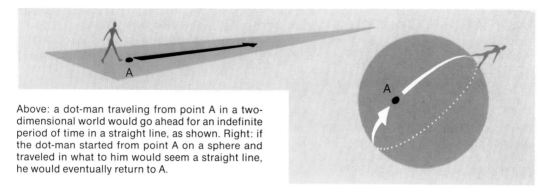

Above: a dot-man traveling from point A in a two-dimensional world would go ahead for an indefinite period of time in a straight line, as shown. Right: if the dot-man started from point A on a sphere and traveled in what to him would seem a straight line, he would eventually return to A.

The Michelson-Morley experiments caused scientists to doubt the existence of the luminiferous ether. What was even more important, they showed that the speed of light is independent of the motion of the observer. Remember that as we try to measure light, we are moving with the earth through space. In other words, the velocity of light is a constant, which remains the same under all circumstances. This concept was to form the basis of Einstein's *special theory of relativity.*

SPECIAL THEORY OF RELATIVITY

According to this theory, which was formulated by Einstein in 1905, it is impossible to measure or detect the absolute motion of a body through space. However, we can accurately determine its relative motion by using the speed of light as a basis. This speed comes to about 300,000 kilometers per second.

The special theory modified the ideas of classical physics with respect to space and time. According to these ideas, particles move in a three-dimensional framework in space and at specific times. The three dimensions are length, breadth, and thickness, or height. This concept is a familiar one. We know, for example, that a room has length, breadth, and height.

We can determine the position of any point in space. Suppose that we have located two points, P_1 and P_2. We want to trace the movement of a particle from P_1 to P_2. Ordinarily, we would take a time reading when the particle is at P_1 and another when it is at P_2—or rather when we would see the particle arrive at P_1 and P_2. Time

readings of this sort would not be absolutely accurate, because it would take time for light to travel from a light source to the particle and to be reflected from the particle to our eyes.

The discrepancy between the apparent and actual times of arrival of a particle at a given point would not be significant ordinarily because of the great velocity of light. Even if the particle's speed were one-tenth the speed of light, the discrepancy would be so small that it could be detected only with the most refined measurements. But at a speed of one-half that of light—150,000 kilometers per second—the discrepancy would begin to be apparent. It would be great at nine-tenths of the speed of light, and enormous at ninety-nine-hundredths of that speed.

If we were to chart accurately the progress of a particle at such great speeds, we would have to apply the special theory of relativity, which takes into account the time needed by light to traverse space. We could make the necessary calculations with accuracy. For the speed of light always remains the same, regardless of the particular position in which the observer may happen to be located.

FOUR-DIMENSIONAL UNIVERSE?

According to classical physics, there are two separate factors we must consider in charting the progress of a particle in space. These factors are *space* (the three dimensions of length, breadth, and height) and *time.* According to the relativity theory, we cannot consider time as something apart. It represents a fourth dimension,

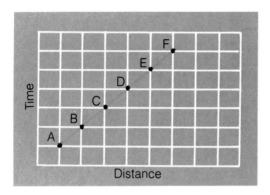

A dot-man, traveling in a two-dimensional world, goes ahead in a straight line from A to B; from B to C; from C to D; from D to E; from E to F. The line AF would represent part of the world line of the dot-man, the record of his movement.

which must be added to length, breadth, and height. We should refer not to space and time, but to space-time. As a famous German physicist, Hermann Minkowski, put it: "From now on, space by itself and time by itself are mere shadows and only a blend of the two exists in its own right."

We can represent four dimensions mathematically, but we cannot draw more than three dimensions in a diagram. Further, the fourth dimension—time—should not be thought of as a spatial dimension. In mathematics, a dimension is simply a quantity that must be specified in order to define a situation.

Most people find it hard to grasp even the idea of a fourth dimension. It would be just as hard for a dot-like man living in a world made up of two dimensions—length and breadth—to conceive of a third one.

If our dot-man traveled from point A (see diagram) in his two-dimensional world, he would go ahead for an infinite period of time in a straight line. Suppose he were transported to a three-dimensional world—a sphere of enormous size. The dot-man would not be aware of any change in the surface, since the curvature, from his point of view, would be slight. But if he started from point A on the sphere and advanced in what for him would be a straight line, he would be mystified to find he had returned to point A. Our dot-man could not visualize

a sphere, since he would be familiar only with a two-dimensional world. However, a dot-man mathematician could explain the structure of the sphere in terms of mathematics. He could show by means of mathematical symbols and equations just what had happened to the dot-man who had returned to the place from which he had started out. But unless the dot-man traveler were familiar with the language and methods of mathematics, the explanation would not clear up his doubts. Many humans would find it as hard to follow an explanation of the fourth dimension.

SPACE-TIME

The space-time of Einstein is called a *four-dimensional continuum*. It is also called a *space-time continuum*. The point of space and the instant of time at which any event occurs represents a single point in the continuum. The interval between any two events would be represented by a finite line (one with definite limits). A succession of finite lines of this kind would make up a record of the successive positions occupied by a particle in space-time. A complete record of this sort of the total course of the particle from the first to the last moment of its existence would be called the particle's *world line*.

To understand the principle involved, let us suppose that the two-dimensional world of the dot-man is marked out in equal squares, formed by intersecting lines (see above). The dot-man is taking a walk. He starts at point A, at the intersection of two of the lines and goes straight ahead. At the end of an hour he reaches point B, where two other lines intersect. At the end of two hours he reaches point C, and so on. If we draw a line connecting all these points, it will represent part of a world line based on the two dimensions of length and breadth and the third dimension of time. It will provide information concerning a series of events—the arrival of the dot-man at one point after another.

WARPING SPACE-TIME

Matter will have an effect on the network of the space-time continuum. Matter

According to relativity, an object shortens as its speed approaches that of light.

warps, or distorts, the network. Without matter, the continuum would be Euclidean, and go on forever in straight lines in any direction. We can show this warping in three dimensions by imagining a network of rubber bands, criss-crossing like a fly screen. Without matter nearby, the rubber bands would be flat in two dimensions.

Now if we drop a billiard ball into this net, the net will be distorted, and more distorted near the ball. In the same way, matter can be thought of as distorting a four-dimensional continuum. A small ball, free to roll, might fall toward a central, large billiard ball, or might circle it, depending on its speed and direction. This would be similar to a meteor coming in from outer space; or, in the case of the small ball circling the larger, to the earth revolving around the sun.

FITZGERALD-LORENTZ CONTRACTION

The special theory of relativity is indispensable in the study of objects traveling at very great speeds, approaching that of light. Interesting things happen to such objects. For one thing, they contract in the direction of their motion. This is called the *Fitzgerald-Lorentz contraction,* from the names of the two physicists who advanced the idea in the 1890s. It was proposed independently by the Irish physicist G. F. Fitzgerald and the Dutch physicist Hendrik Antoon Lorentz. At the same time that the object contracts, its mass increases and its time slows down.

If a rocket travels at a speed of 30,000 kilometers per hour, there will be no ap-preciable change in its shape. The reason is that its speed is negligible, compared to the velocity of light. Suppose now that the rocket whizzes past an observer on the earth at a speed that begins to approach that of light. If the observer would be capable of making out details, he would note that the rocket would have contracted greatly in the direction of its motion. If there were a watch inside the rocket, its time would run slow compared with that of a watch held by the observer on the earth.

As the speed of the rocket came nearer to that of light, its length in the direction of its motion would approach zero. It would turn into a flat disk. As its speed increased, its mass would approach infinity and its time would slow down more and more. Since it would require an infinite force to accelerate an infinite mass, we conclude that no physical object will ever move with quite the speed of light. It could approach that speed closely, however, if enough force were available.

Can we prove that when objects approach the speed of light, they contract in the direction of their motion? Can we prove also that their mass increases and that their time slows down? The answer to all three questions is "Yes."

The contraction of objects in the direction of their motion has been demonstrated by the study of cosmic radiation. When primary cosmic ray particles collide with nitrogen and oxygen nuclei at high levels in the atmosphere, they give rise, among other things, to particles called mu mesons. These particles are extremely penetrating, making their way through the layers of atmosphere to sea level. They can be readily counted with radiation counters and can be positively identified in cloud chambers. They have a half life of two microseconds—that is, two millionths of a second. In two millionths of a second, half of the mu mesons should have decayed into ordinary electrons. But in a millionth of a second a particle traveling at even the speed of light could move only 300 meters or so. How could any mu mesons reach us, since they are formed mostly at the 20,000 kilometer altitude level? They should all have decayed by the time they reached sea level.

One reason is that the "clock"—that is, the half life—of the mu mesons has slowed down by a very large amount. In 200 microseconds of our elapsed time, the "clock" of the mu mesons will have moved ahead only two microseconds.

The slow-down of clocks at speeds approaching that of light is the basis of the famous "clock paradox". A 20-year old astronaut takes off in a space ship, destination Arcturus, 36 light-years from earth. He travels at a speed close to the speed of light. His twin brother stays home. When the astronaut returns to earth, 72 years later, he will have aged one day. His twin brother, however, will have aged 72 years.

SOME PROOF

The German physicist A. H. Bucherer, around 1910, first demonstrated experimentally the increase of mass with velocity. He made mass measurements of beta particles—electrons emitted from naturally radioactive substances. He found that the mass increased with velocity, just as the special theory had predicted.

The time variation at velocities approaching that of light was demonstrated by the U.S. physicist H. E. Ives about 1925. He showed that hydrogen atoms, emitting an accurately known frequency of radiation, changed their frequency if they were accelerated to speeds approaching that of light, by exactly the amount Einstein predicted.

Other experiments have verified the variations of mass, length, and time with speed, as predicted by the special theory of relativity. Physicists now consider that these variations have been demonstrated beyond the shadow of a doubt.

The law of addition of velocities: Each car travels at 65 kilometers per hour; they approach each other at 130 kilometers per hour—65 + 65 = 130. However, the addition of velocities of two space ships travelling near the speed of light, cannot exceed the speed of light so that, 270,000 + 270,000 = 297,000.

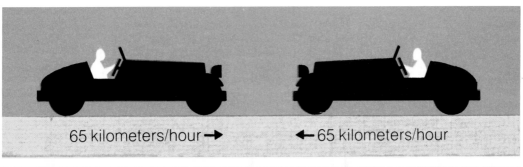

65 kilometers/hour → ← 65 kilometers/hour

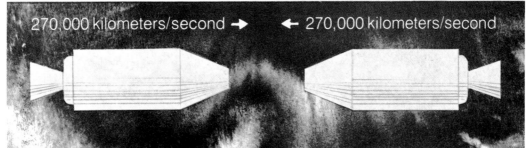

270,000 kilometers/second → ← 270,000 kilometers/second

VELOCITIES NEAR THE SPEED OF LIGHT

Velocities comparable to that of light might seem to be purely theoretical, but that is not the case. The speeds of particles in the beams of large accelerating machines, such as cyclotrons, synchrotrons, and betatrons, are already up in this range. In working with such particles, scientists use the calculations of the special theory of relativity. In fact they often refer to accelerated particles traveling at speeds that are close to the speed of light as *relativistic particles.*

A proton or other particle in a synchrotron reaches a speed substantially equal to that of light. As one continues to accelerate it, it does not move faster but it becomes heavier. Scientists have suggested that such big particle accelerators should be called *ponderators* (from the Latin *pondus,* meaning weight), or weight-bestowers, since they increase the mass of the particles they accelerate.

In the 1940s, engineers had to use relativistic formulas to calculate correctly the trajectories of electrons in the beams of fast cathode-ray oscilloscopes. In this age of space flight, scientists are devoting more and more attention to the curious phenomena that take place at speeds approaching that of light.

PECULIAR EFFECTS

The application of the principles of special relativity to the measurements of speeds produces some peculiar results. Suppose, for example, that two rockets are approaching each other at very high speeds, almost equal to that of light. Let us say that, with relation to the earth, each is traveling at nine-tenths the velocity of light. What is the relative speed of these two rockets with respect to each other?

One might imagine that we would use the same type of calculation in this instance as we would in considering the case of two automobiles speeding toward each other. If each of these cars is traveling at 65 kilometers per hour, they are approaching each other at the rate of 65 plus 65, or 130, kilometers per hour. On this basis, since each

of our rockets is traveling at nine-tenths the speed of light—that is, approximately 270,000 kilometers per second—their speed with relation to each other would be two times 270,000, or 540,000, kilometers per second. Actually, the combined speed of the two rockets would come to less than the speed of light—that is, less than 300,000 kilometers per second.

For one thing, the speed of light is the ultimate speed, which no combination of speeds can possibly surpass. Also, we must remember that at speeds approaching that of light, time is slowed up. Speed is a quantity measured in units of length per unit time. The time of our two speeding rockets would slow up sufficiently so that their combined speeds would amount to some ninety-nine hundredths the speed of light—approximately 297,000 kilometers per second.

MASS-ENERGY EQUIVALENCE, or $E = mc^2$

The special theory of relativity also

The clock paradox. One twin goes on space trip at relativistic speeds. The other twin stays home.

concerns itself with the transformations of energy. Nineteenth-century physicists knew that there are different forms of energy: kinetic energy (energy of motion), electrical energy, chemical energy, thermal (heat) energy, potential energy, and so on. They also realized that one form of energy can be transformed into another. In the twentieth century scientists came to realize that energy can be transformed into mass, and mass into energy. Mass and energy are regarded as equivalent.

In his special theory, Einstein worked out the equation for this transformation. This was the famous mass-energy-transformation equation: $E = mc^2$, in which E represents the energy in ergs, m the mass in grams, and c the velocity of light in centimeters per second. Modern nuclear physics has proved beyond the shadow of a doubt the correctness of this equation. It has played an all-important part in the development of nuclear energy.

The sum total of energy, or its equiva-

The traveling twin returns 40 years later, to find his brother has aged—although he has not.

lent in mass, always remains the same in the universe or in any closed system. A nucleus that is about to undergo radioactive decay has, locked up in itself, energy that is in the form of mass. This turns up later in the form of the kinetic energy that is released as the particle disintegrates. If we weigh the particle before and after decay, weigh the products of decay, and measure the kinetic energies of the alpha particles and gamma rays (if any have been emitted), we can show that nothing of the original particle has been lost. Generally only a part of a given particle is converted into kinetic or electromagnetic energy. However, there are several examples of a complete transformation of mass into energy. When, for example, a negative electron collides with a positive electron, or positron, the entire mass of both electrons is converted into gamma rays. This process is called "annihilation."

The special theory of relativity is valid because it correctly describes the important physical properties of any system. It has not caused the older theories of Galileo, Newton, and others to become obsolete. The differences between the calculations of classical physics and those of the special theory, when applied to bodies traveling at speeds far less than that of light, are insignificant and we can ignore them. But when we deal with speeds approaching that of light, we find that the calculations of classical physics will not serve. We must apply the special theory of relativity if we are to avoid serious error.

GENERAL THEORY OF RELATIVITY

This theory, advanced by Einstein in 1915, deals with fields of force. The theory may also be regarded as a geometrical representation of gravitation. To show how fields of force operate, let us imagine that an observer is shut up in a box. He cannot see out of the box and therefore can make no observations of the exterior world. However, he is amply provided with instruments with which he can measure various forces.

We shall suppose, first, that the box is resting upon the surface of the earth. By

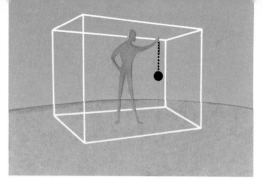

This man is standing on the bottom of a box that rests on the surface of the earth. When he lets an object fall, it is drawn toward the earth's center by the force of gravitation.

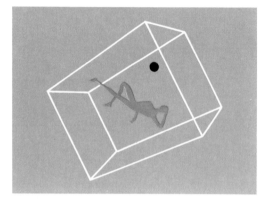

The box is now floating freely in space, far above the earth. The man himself and any objects in the box would float about aimlessly, not being subject to earth's gravitational force.

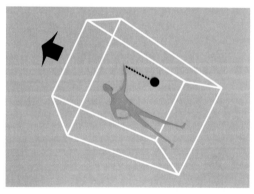

If any outside agency applied a force to the box, causing it to have a uniform acceleration in a given direction, the objects in the box would act as if subjected to a gravitational force.

using a pendulum, the observer inside the box would discover that he is in a gravitational force field. He would notice that all objects in the box are attracted toward a point, in this case the center of the earth.

Next, suppose that the box, with the observer still in it, is transported to some distant point, far from any mass exerting gravitational force. If the box were stationary or if it were moving at a constant speed, there would be no measurable forces in the interior. The observer himself and any objects in the box would float aimlessly, since they would have no weight.

We must now imagine that some outside agency applies a force to the box and causes it to have a uniform acceleration in a given direction. All the objects in the box would then be acted on by this force and they would acquire weight. They would act in exactly the same way as the objects that undergo the effects of gravity. The man in the box would not be able to tell whether it was stationary and affected by a gravitational field or whether it was being accelerated uniformly in free space.

Suppose we connected the box by a cable to some central point and caused it to revolve about this point. The centrifugal effect (the force acting away from the center) would then produce the same effect as that of gravitation or of uniform acceleration in one direction. It is believed that the space stations of the future, established 1,500 kilometers or so above the earth, will be based on this principle. These stations will be wheel-shaped. The wheel will spin slowly, and centrifugal force will be substituted for the force of gravity.

According to the general theory, it is impossible to distinguish between gravitational fields and the other force fields we have just described. This gravitational field or any field of force produces measurable effects because of the presence of matter in the universe.

TESTING RELATIVITY

When Einstein analyzed these effects in his general theory, he found that they differed from those predicted by Newton's law of universal gravitation and other laws of classical physics. For one thing, according to the concepts of classical physics, the path of a beam of light is a geodesic—that is, the shortest distance between two points. Einstein held that this would be a perfectly straight line only in a universe

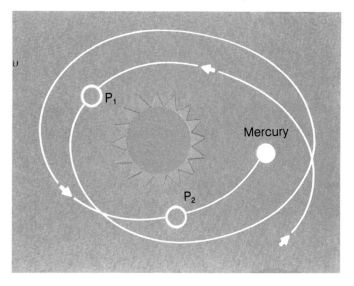

As the planet Mercury revolves in an elliptical orbit around the sun, the ellipse itself revolves in the direction of the planet's motion. P_1, in the diagram, stands for "perihelion, first position," P_2, "perihelion, second position." The perihelion of a planet is the point in its path nearest the sun.

containing absolutely no matter. Actually, he said, light would be bent from its path whenever it passed a massive object. The geodesic would be curved.

Scientists put this theory to the test. First, they photographed a part of the sky when the sun was not in that vicinity. They later took another photograph of the region when the sun was there. Of course they had to take this photograph during a total solar eclipse, since otherwise the stars near the sun could not be seen at all. If the sun actually bent the light rays, the position of the stars would appear to be slightly altered on the second photographic plate. Such was indeed the case. The amount of the alteration was what the general theory had predicted.

The general theory helped to solve a problem that had puzzled astronomers for many years. They had noted that as the planet Mercury revolves in an ellipse around the sun, the ellipse itself rotates in the direction of the planet's motion. As the perihelion (the point at which the planet is nearest to the sun) moves around the sun with the shifting orbit, it describes a sort of secondary orbit.

Astronomers found that the perihelion showed a greater advance in the direction of the planet's motion than could be accounted for by the gravitational attraction of the other planets. This excess came to about 40 seconds of arc per century. Einstein showed that the general theory accounted for the discrepancy. At the perihelion, the planet would be closer to a massive body—the sun—than at any other

point of its orbit. Its path, therefore, would be altered slightly, by just the amount indicated by the calculations of the general theory.

Another proof of the general theory came from its prediction of a shift in the frequency of spectral lines emitted by atoms in an extremely high gravitational field. There are fields of this sort around the so-called white dwarf stars, extremely dense bodies with very small diameters. Shifts in the spectral lines of atoms in these fields followed general-theory predictions.

Another interesting outcome of the general theory of relativity is the prediction of "black holes". Suppose a star should collapse into an extremely small volume, without losing much mass. Then it would become very dense and concentrated. The attraction of gravity in its immediate neighborhood would be very strong. Now the velocity needed to escape from any body is called the "escape velocity." If the concentration of matter is great enough, the escape velocity might even exceed the speed of light. In that case, light cannot escape from the collapsed star and as it will not shine, it will appear black.

Anything falling directly toward the black hole, such as a light beam or an object, would be accelerated to very high speeds and could not escape. Yet the black hole would still have a gravity field surrounding it, and a planet or a second star could circle it just as the earth circles the sun.

Astronomers now believe that black holes exist in the universe. The theory of relativity does not say that black holes exist, merely that they could.

THE GENERAL THEORY AND COSMOLOGY

"Cosmology" is a word that means a study of the structure and geometry of the universe. Today, many different cosmologies are known. Observational astronomers try by various measurements to find which one most closely fits the actual universe. It is not known conclusively today which comes nearest. In fact, it is not known which type of geometry gives the "best fit."

The geometry may be "closed," as we illustrated above with spherical (or elliptic) curvatures. It may be "open," like hyperbolic ("saddle-shape") geometry, and extend to infinity. Both solutions are consistent with general relativity. Whether the universe is open or closed depends on the average density of matter. This is not well enough known today to permit a definite statement to be made.

In his early papers, Einstein proposed a theory of the universe based upon his general theory of relativity. He pointed out that if space were entirely empty, it would extend to infinity in all directions. In such space, light would travel in a straight line. However, we know that space is not empty. It contains matter, very tenuous in most areas, highly concentrated in others. In regions where matter is concentrated, the world lines for the path of light would describe curves rather than straight lines. The average density of matter in space is so low that the net effects of the presence of matter would be small. However, in view of the immense extent of the universe, Einstein thought it likely that if a beam of light could be kept moving for thousands upon thousands of millions of years, the curved course it would follow would cause it to return at last to its starting point.

Leading theoretical physicists like Steven Hawking of Britain are attempting to explain how all the fundamental forces of nature acted during that first millisecond of creation. Such a "Grand Unification Theory" would go one step further than Einstein's theories. For although Einstein produced a model of how the universe grew out of the Big Bang, he was unable to explain how subatomic forces came into being. Indeed, many modern physicists approach cosmology from the standpoint of what happened before the Big Bang.

Top: The position of a star, from the viewpoint of an observer on earth, when the sun is not in this particular part of the sky. Bottom: When the light from the star is bent by the sun, the position of the star will appear to be changed, as far as the observer is concerned.

Real position of star

Observer on earth

Apparent position of star

Real position of star

Earth

© Bill Pierce/Time Magazine

A magnet floats above a superconductor. Levitation is among superconductivity's promising applications.

SUPERCONDUCTORS —HOT AND COLD

by Benedict A. Leerburger

One of the most important technological developments of the twentieth century has been the use of electrical energy. Look around you: there is hardly a man-made item in view that in some way doesn't rely on the flow of electricity. Your digital watch, telephone, computer, automobile, microwave oven, and hearing aid all need electric current. However, as our increased use of electrical devices becomes more and more sophisticated, we are constantly searching for ways to make better, cheaper, and safer use of electrical energy.

In the middle of the twentieth century physicists announced a pair of major developments that may affect the way electricity is transmitted: plasma and superconducting materials. Ironically, although both developments may lead to far more efficient electrical conductors or, in the case of plasma, a source of plentiful, low-cost energy, plasma involves extremely high temperatures while superconducting materials involve extremely cold temperatures.

PLASMA

When a physician mentions plasma he is referring to blood; when a physicist mentions plasma he is referring to a gaslike collection of charged particles. Thus they show many features not displayed by ordinary gases, whose particles are electrically neutral. Plasma can be made by one of two methods: by ionizing an ordinary gas by means of an electrical discharge or by heating a gas to an extremely high temperature.

Every atom of a gas consists of a positively charged nucleus surrounded by a negatively charged cloud of electrons. Its net electrical charge is zero. When an electrical field is applied to a gas, a few electrons are stripped off the atoms. Every atom affected in this way becomes a positively charged *ion*. The stripped electrons, which are negatively charged, are free to move about. As they pass through an electrical field the negatively charged electrons speed up and collide with other atoms, ion-

The color of a plasma (gas atoms stripped of their electrons) depends upon the kind of gas.

USIS

temperatures matter exists only in a plasma state—considered the fourth state of matter, a state distinct from solids, liquids, and gases. Our sun and the stars, glowing at temperatures rarely achieved on earth, are in a plasma state.

Fortunately, the fusion reaction in the sun goes very slowly, so that it will supply heat and light for a long time. The sun supports a self-sustaining fusion reaction. As yet, no self-sustaining controlled fusion reaction has ever been achieved on earth. The stumbling blocks are the twin problems of creating a plasma at ultrahigh temperatures and then holding it long enough for it to react. It seems clear that the sun's huge size and mass are the keys to its success. At the sun's birth, its intense gravitational field compressed and heated its matter to the thermonuclear ignition point. This same gravitational field keeps the hot gases from escaping into outer space. On earth, we have as yet achieved sun-like densities and temperatures for only a fleeting millionth of a second or so, in the hot core of the atomic bomb.

DESIGNING A PLASMA BOTTLE

There are two major problems in plasma research. The basic problem is designing a container to hold matter at temperatures that would melt any known material. However, from the physicist's point of view there is an even greater problem. Plasmas generated under laboratory conditions for fractions of a second are of a very low density. The total pressure of an ordinary gas or a plasma is proportional both to its density and to its temperature. The greater the density or the higher the temperature, the greater the pressure. To keep the pressure within workable bounds, the density of a high-temperature plasma must be kept extremely low—typically less than 1/10,000,000th the density of ordinary solid matter. The heat content of even a very hot plasma is quite small. The problem facing physicists is how to keep plasma particles from losing energy as they bury themselves in the walls of a container. Any loss of energy would instantly cool and dissipate the plasma. This would happen in a

izing them and freeing up still more electrons. This is called a *cascade process*. Through it, an electrical discharge develops, and an electrical current can now flow through the gas, which was formerly an insulator. In a particularly powerful discharge, the gas may become completely ionized by the cascade process, becoming a plasma. Plasmas are produced in nature by lightning bolts in just this way.

The second method of producing plasma involves heating a gas. As the gas temperature increases, the energy of all the particles of the gas also increases. The hot particles bound about, colliding with one another. If the temperature could be raised high enough, the collisions would be so violent that all the electrons would be knocked loose from their atomic nuclei, resulting in a state of complete ionization.

The temperatures required to achieve this kind of *thermal ionization* are measured in terms of thousands of degrees Celsius. However, the plasma physicist is particularly interested in the behavior of ionized plasmas at even higher temperatures—millions of degrees Celsius. At such fraction of a millionth of a second under ordinary conditions. Such a brief period would be far too short to create and heat a suitable plasma. It was necessary, therefore, to develop some kind of nonmaterial "insulation" with which to line the chamber walls so that plasma will not strike them.

The solution, initially suggested in the mid-1940s, was that a magnetic field might provide the nonmaterial insulation between the hot plasma and the surrounding material walls of the vacuum chamber in which the plasma is created. It is known that a strong magnetic field can have a powerful effect on the motion of a charged particle, causing it to move in a helical, or coilspringlike, orbit.

Shaping the bottle, or container, to hold both the plasma and the magnetic field

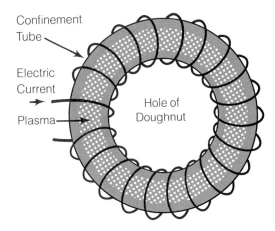

Torus shape for a magnetic bottle. If we bent the confinement tube shown in the previous diagram back on itself to form a torus, or doughnut-shaped structure, the device would have no ends. Theoretically, the particles making up the plasma would be trapped. Note, however, that the turns of wire are more crowded together at the "hole" of the doughnut than at the outer walls. There would be a greater concentration of electric current at the inner surfaces of the torus. Hence, the magnetic field inside the tube would be stronger near the inner wall than near the outer one. Particles would drift toward the outer wall, where the magnetic field would be weakened. The magnetic bottle would spring a leak.

Theoretical structure of a simple magnetic bottle, which is really a tube. Electric current is sent through the coils and a magnetic field is set up. Supposedly, the magnetic field within the bottle would serve to confine the plasma. Unfortunately, the particles of which the plasma consists would make their way to the ends of the tube, and the plasma would be dissipated.

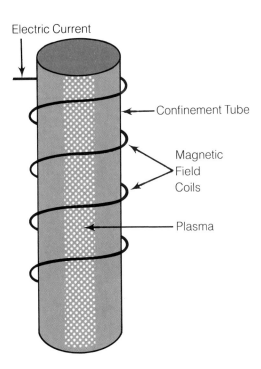

was another matter. The initial suggestion was to design a torus, a doughnut shaped container. Since the device would have no ends, the plasma particles, supposedly, could not escape to the chamber walls. Unfortunately, it was quickly discovered that when magnetic coils are wound around a doughnut they become more crowded together as they pass through the hole in the center than they are as they pass around the outside portion. Thus the greater concentration of electric current on the inner surfaces distorts the motion of the particles and causes them to cut across the magnetic field.

In 1951 physicist Lyman Spitzer of Princeton University suggested a new type of torus, one bent into the shape of a figure 8. Thus, like a Mobius strip, a particle traveling along the inner part at one end of the figure 8 will go along the outer edge of the other end. The drifts in the two curves will tend to cancel each other out. Spitzer's figure 8 torus is called a *stellarator*.

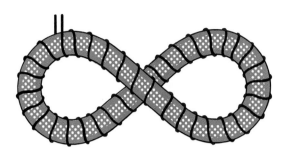

The basic principle of the stellarator. The torus shown in the preceding diagram is bent into a figure 8. What is the inside wall in one curve becomes the outside wall in the other one. The particle drifts occurring in the two curves of the device will tend to cancel each other out.

A further improvement on the design came from the Lawrence Livermore National Laboratories near Oakland, California. Scientists at Lawrence Livermore built a Mirror Fusion Test Facility (MFTF). Instead of using a doughnut-shaped chamber, they tested magnetic mirrors that caused escaping plasma to bounce back into a long tubular chamber as it reflected from a conventional mirror. The facility features a steel cylinder larger than a Boeing 747 fuselage; at each end is a "yin-yang" magnet as tall as a two story house and weighing almost 400 tons.

Physicists at Princeton University had a different idea. They designed a torus called a *tokamak,* whose plasma-containing tube is fatter and whose doughnut hole is smaller. A larger radius means that particles take longer to leak out because they have to travel a greater distance. The Princeton tokamak (Princeton Large Torus or PLT) uses three magnetic fields to hold a plasma.

INERTIAL-CONFINEMENT FUSION

While many plasma physicists have concentrated on building a better "magnetic bottle" to confine the plasma, others have been pursuing a different approach, called *inertial-confinement fusion* (ICF). Rather than confining a diffuse gas by using magnets, ICF calls for enclosing the fuel, composed of a mixture of deuterium and tritium, as a frozen ice within a glass sphere the size of a grain of sand. This microsphere is then zapped with an intense and powerful laser beam. The momentary burst focuses on the outer layers of the sphere, heating and compressing its inner layers. Thus, for some trillionths of a second, the center of the sphere reaches the conditions found at the center of a star. In 1984 scientists at Sandia National Laboratories in Albuquerque, New Mexico, an-

Technicians install the vacuum vessel and coil assembly of the Tokamak Fusion Test Reactor at Princeton.

High-speed "maglev" trains will become a reality once scientists develop materials that exhibit superconductivity at ordinary temperatures. Propelled to speeds exceeding 483 kilometers per hour (300 mph) by an alternating superconducting magnetic field, the train will streak along 6 inches above the tracks by means of superconducting magnets.

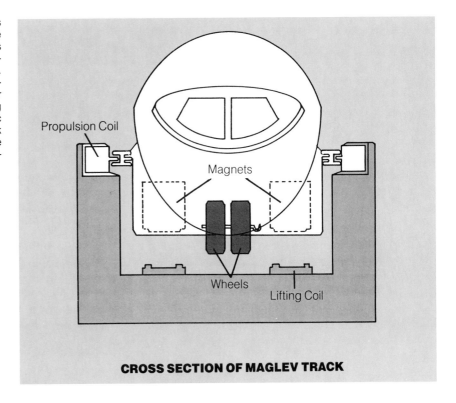

Propulsion Coil

Magnets

Wheels

Lifting Coil

CROSS SECTION OF MAGLEV TRACK

nounced that they had succeeded in focusing a beam of ions so sharply that it could be used to drive fusion targets. Meanwhile, scientists at Lawrence Livermore Laboratory designed and built Nova, the world's most powerful laser, for test shots on ICF targets.

In 1987 President Reagan endorsed the construction of a $4.4 billion particle accelerator with a circumference of 53 miles. It will use more than 10,000 powerful magnets to keep streams of protons on course.

Because the development of a successful and possibly commercial fusion capability has such great significance, scientists worldwide are striving for breakthroughs. Successful fusion means not only a major scientific breakthrough but also the possibility of a source of almost endless energy. Elements such as hydrogen are plentiful, and a tremendous amount of energy is released during fusion. Moreover, unlike the product of nuclear fission when an atom is divided, no radiation is generated as a byproduct when atoms are melded.

SUPERCONDUCTORS

Imagine yourself driving from New York City to Detroit. You cross the Hudson River, get on Interstate 80, and accelerate to the legal speed limit. Then you reach down, turn off your engine, flick a switch on the dash, and coast to Detroit without slowing down. Or, picture yourself aboard a streamlined train leaving New York. You feel and hear the train pull out of the station. Then, almost magically, you sense the train settle a few inches and "float" above the tracks all the way to Washington at a speed of more than 300 miles per hour. Such "fantasies" were in the realm of science fiction only a few years ago. However, thanks to one of the great scientific breakthroughs of the twentieth century, these dreams of yesteryear could well become reality in a few decades. A "floating car" has already been developed in the laboratory and a prototype levitating train exists in Japan. What makes these dreams possible is the creation of giant

CURRENT FLOW AND RESISTANCE

INSULATOR
In materials with extremely high resistance, such as rubber or glass, electrons are tightly bound to atoms and cannot be jostled loose to sustain a flow of current.

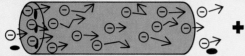

CONDUCTOR
In materials with lower resistance, some electrons are loosely bound and form a current when voltage is applied. Resistance is a measure of the energy lost in the form of heat from electron collisions.

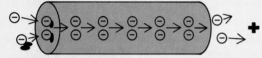

SUPERCONDUCTOR
When materials become superconductive, all resistance disappears because electrons are bound into pairs, which move in step with each other, avoiding collisions. Current flows with no energy loss.

magnetic fields created by superconducting materials never thought possible only a few years ago.

To understand superconductivity, it is necessary to understand conductivity. Around the turn of the current century scientists began to understand the nature of matter. A new designator, "nuclear physicist," described the men and women who attempted to define atoms, neutrons, electrons, protons, molecules, and the many ingredients that composed subatomic particles. Prior to their many discoveries, scientists believed that electricity was a weightless, or massless, "fluid" that flowed through matter. The nuclear physicists, however, showed that electricity is the movement of tiny charged particles having energy and a small amount of mass. Physicists called these negatively charged particles *electrons*. It was also discovered that the atom is not the smallest part of matter, as previously thought, but is in actuality composed of many parts, of which the electron is but one. The electrons form a series of loose shells about the dense, positively charged atomic nucleus. Electrons orbit about their nucleus very much the way planets orbit their sun.

CONDUCTORS: GOOD AND BAD

Materials, such as most metals, in which the outermost electrons are loosely held "in orbit" by an atomic nucleus are usually good conductors of electricity. Their electrons are easily freed from the positive attraction of the nucleus to produce an electric current. The free electrons "float" in the spaces among the atoms of a metal. A solid metal has its atoms arranged as a three-dimensional repeating pattern in a structure known as a *crystal lattice*. Visualize the free electrons as a cloud or gas of negative charges filling the interatomic spaces of the lattice. When a metallic conductor is connected to a source of electrical power, such as a generator or a battery, electrons are disturbed or "shoved" by the resulting voltage. This movement of electrons through a circuit is called an electric current.

Electricity flowing through a conductor, such as a copper wire, is hampered by the natural resistance of the conductor and by temperature. At ordinary temperatures, electrons cannot move too freely. Their paths are deflected by the atoms. More exactly, the motion of the atoms in their lattice interferes, or resists, the passage of the electrons. Atoms in a crystalline substance are free to vibrate around their fixed position in the lattice. The warmer the crystal, the greater the vibrations of its component atoms. Resistance is a measure of the energy lost in the form of heat from electrons colliding with each other. Materials with extremely high resistance, such as glass or rubber, are called *insulators*. Their electrons are tightly bound to their atoms and cannot be jostled loose to sustain a flow of current. This phenomenon explains why resistance to electricity builds up with an increase in temperature. Conversely, extremely low temperatures reduce the atomic vibrations in crystals to a minimum.

In dealing with the relationship between heat and work, physicists use the Kelvin scale, in which absolute zero is equal to $-273.16°$ Celsius. (The size of the degree on the Kelvin scale is the same as the size on the Celsius scale. Thus, to con-

A magnified view of a crystal of yttrium-barium-copper oxide, a high-temperature superconducting material.

vert degrees Celsius to degrees Kelvin, add 273.16°.) At absolute zero all atomic and molecular motion comes to a halt and thermal energy vanishes. Physicists discovered that at absolute zero a gas which obeyed the known laws of thermodynamics would have no volume and so, theoretically, could never be achieved. The lowest temperature reached in the laboratory, however, is only 0.0000005° Celsius above absolute zero.

In the early 1900s the American physicist Josiah Willard Gibbs described how all materials can exist in one of three "phases": liquid, solid, or gas. His "phase rule" described the thermodynamic effect on a material when it changes phase. To liquify a gas, for example, it must be cooled. The temperature at which a gas changes phase to a liquid varies with the material.

DISCOVERY OF SUPERCONDUCTIVITY

In 1908 the Dutch physicist Heike Kamerlingh-Onnes became the first person to liquefy helium gas at −269° Celsius or 4.2° Kelvin. One of Kamerlingh-Onnes' prime interests was the ability of metals to conduct electric currents at very low temperatures. Three years later he froze pure mercury and measured its resistance to the passage of electricity at increasingly lower temperatures.

The electrical resistance of the mercury dropped. At 10° above absolute zero its resistance was only 1/100th of what it was at room temperature. But when the temperature reached 4.2° above absolute zero Kamerlingh-Onnes received a surprise. Rather than reaching a low point and leveling off, the mercury's resistance suddenly disappeared. Electricity was continually conducted through the solid mercury without the least amount of interference. Kamerlingh-Onnes called this phenomenon *super-conductivity*. He tested other metals and alloys and found that many, but not all, displayed this property at different low temperatures. Above these temperatures resistance to electricity develops and increases as temperatures increase.

To explain superconductivity three American physicists—John Bardeen, of the University of Illinois, Leon N. Cooper of Brown University, and J. Robert Schrieffer of the University of Pennsylvania developed the *BCS theory* in 1957. Their theory, which won them the Nobel Prize, bases superconductivity on the ability of electrons to associate in freely moving pairs at extremely low temperatures.

One of the problems facing scientists was dealing with a physical law which states that no two electrons in a current can have the same velocity. In handling this law imagine a rule which says that all cars on a road must travel at different speeds. Under these conditions, it was hard to imagine how electrons could flow in any coordinated manner. Schrieffer and Cooper solved the problem.

First Cooper suggested that when a metal's lattice vibrates it does not unify the entire current. Rather, it creates pairs of electrons, called "Cooper pairs," that act as a single unit. Schrieffer then stated that since each member of a pair must have a different velocity, both from its partner as well as from every other electron in the stream of current, a pair can have the same collective velocity as every other pair. The pairs of electrons flow in unison as if attached together on a string. "Imagine soldiers crossing a field," says Brian Schwartz of the American Physical Society. "If they link arms in formation, one can step in a rut, the others will hold him up and everyone will keep going."

The result is a high concentration of positive charge behind the first electron. This charge attracts the next electron, which follows the first one, thus creating a Cooper pair. When the pairs move along, they are thought to cause a momentary "buckling" or "puckering" of the lattice as they pull on the nearby metal atoms. This enhances the propelling effect of the atoms' positive charges on the electrons. Thus, the atoms now "cooperate" in moving the electrons, instead of interfering with or resisting them, as in ordinary conduction.

The temperature at which a conductor becomes a superconductor, and vice versa, is called the *superconductive transition temperature*. It varies with different superconductors. For decades the highest known transition temperature for a pure metallic element, technetium, was 11° Kelvin. Until 1983, the highest transition temperature for an alloy, germanium and niobium, was 23° Kelvin. Then suddenly the world of superconductivity research was dramatically altered.

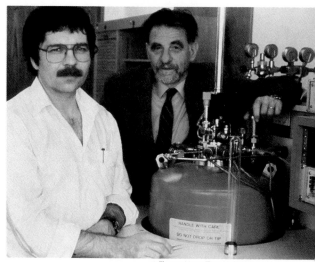

© Thomas Meyer/Time Magazine

Physicists Karl Alex Müller and Johannes Georg Bednorz won the Nobel Prize for their pioneer work in superconductivity.

THE GREAT BREAKTHROUGH

At IBM's Zurich Research Laboratory in Switzerland, physicist Karl Alex Müller decided to take a different approach. Rather than test the superconductivity of various metallic compounds and alloys, he turned his attention towards compounds of metals and oxygen, or metallic oxides, called *ceramics*. Müller and his colleague, Johannes Georg Bednorz, tested hundreds of different oxides. Then, almost by trial and error, they mixed a compound of barium, lanthanum, copper, and oxygen and were startled to find that the transition temperature jumped to 35° Kelvin, the highest yet achieved.

Müller and Bednorz published their findings and were surprised that they generated little interest among their colleagues throughout the world. Several physicists, however, recognized the fact that if a material could be found whose transition temperature was high enough it would revolutionize the world of physics. At AT&T's Bell Laboratories, for example, a team led by physicist Bertram Batlogg and chemist Robert J. Cava reproduced Müller and Bednorz's work, then developed another compound that bettered IBM by a single degree Kelvin. Similar work was being undertaken in China, in Japan, and in a small laboratory at the University of Houston headed by Paul C. W. Chu.

RAISING THE TRANSITION TEMPERATURE

Chu wondered what would happen if he pressurized the IBM ceramic. "Using known theories," he says, "you don't expect the transition temperature to go up rapidly under pressure, but it shot up like a rocket. It suggested to us that there might be some new mechanism involved." Chu discovered that when the IBM compound was subjected to from 10,000 to 20,000 times normal atmospheric pressure it remained a superconductor up to 52° Kelvin.

In the spring of 1987 Chu attended a meeting of physicists in Houston, Texas,

Paul C. W. Chu led the University of Houston team that discovered the yttrium-barium-copper oxide compound that becomes superconductive at 93°K.

© Dan Ford/Picture Group

and shocked his associates by announcing that by subjecting a new substance, composed of several rare earth elements combined with an oxide, to a series of heat and chemical treatments he achieved a transition temperature of 93° Kelvin. When he duplicated his experiment before the Houston group he bettered his thermal record by 5°. Other scientists, both in the United States and abroad, duplicated Chu's feat and also achieved the 98° Kelvin number.

At a meeting of scientists in the summer of 1987 Chu told the assembled group that his research team "is now able to readily reproduce materials that are superconducting at about 130° Kelvin." He also stated that he was testing a material "that showed a loss of electrical resistance at 225° Kelvin" but that other researchers had not, as yet, been able to reproduce the results.

The achievement of developing a superconducting material at a temperature approaching 0° on the Fahrenheit scale has caused a major stir in the usually complacent world of physics. "If we can reach 98° Kelvin why not a material that is superconductive at room temperature?" reason many scientists. In laboratories throughout the world physicists are trying to achieve just that goal. Some scientists compare the importance of these advances in superconductors to the invention of the transistor. Jack S. Kilby, co-inventor of the integrated circuit, thinks even that comparison is an understatement. "This is much broader," he says. "It could impact almost everything."

The BCS theory which worked so well for all previous superconducting materials immediately became questionable. According to Bardeen, his theory can explain superconductivity up to around 40° Kelvin. But at 90° Kelvin, he says, "I think it's highly unlikely. We no doubt are going to need a new mechanism." Says Schrieffer, "Superconductivity may turn out to have as many causes as the common cold." Also, scientists have yet to discover why the oxides are such superior superconductors. "It may be several years before we know what's going on," adds Schrieffer.

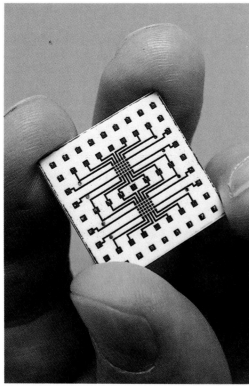

A ceramic chip spray-painted with wires formed of a superconductor shows how superconductors might be used in a computer's integrated circuits.

"There may be no theoretical limit to how high the temperature may go."

One of the first benefits of raising the transitional temperature of materials was the opportunity to switch to a far less expensive coolant in driving a metal's temperature down to the transition temperature. Certain superconductors such as those used in today's "atom smashing" cyclotrons and magnetic resonance imaging machines, used for medical imaging, must be cooled with costly liquid helium. New superconducting materials can be cooled with plentiful, cheap liquid nitrogen.

PRACTICAL APPLICATIONS

Although the first commercial applications are not expected until the mid-1990s, scientists and industry observers are convinced that Chu's quantum leap could lead to dramatic new technologies in fields as diverse as electronics, transportation, and power. To be practical, superconductors must be fashioned into wire, magnet cores, and the microscopically thin coatings that are the foundation of computer circuits. Says AT&T's Robert J. Cava, "It's a long road between discovery and the use of the devices." However, steps along that road have already been taken. Scientists at several laboratories, for example, are developing processes to lay strips of superconducting film just a few hundred atoms in width on computer chips.

Engineers have already found that new superconducting materials may make possible the most powerful electromagnets ever built. Tests at Westinghouse Electric Company and AT&T indicate that the new superconductors can withstand magnetic fields up to 10 times greater than those possible with the previous standard, niobium.

With more powerful, more efficient electric motors we may see the day when gas-powered vehicles are obsolete. Although electric-powered vehicles have been used as recreational vehicles and for industrial use, tomorrow's superconducting electric motors could be efficient enough to power an automobile at highway speeds. And unlike a golf cart, they would scarcely need recharging.

One of superconductivity's most promising applications may be in the field of electric power transmission. Copper wire is the conductor used to move electricity from a power plant to homes and industry because it carries current more economically than any other material. However, copper is inefficient. As much as 15 percent of the electricity generated must be used solely to overcome the material's resistance. Large utility companies waste hundreds of millions of dollars each year because of this loss. According to John K. Hulm, director of research for Westinghouse Electric Company, "If these wires were superconducting, there would be no resistance. And that opens up a lot of tremendous possibilities. The power you'd save would reduce the need to build more plants. And because you could transmit power great distances with no loss of en-

High-temperature superconductors have the potential for
profoundly advancing many
aspects of technology, including transportation, computers,
power transmission, and medical diagnostic equipment. Left:
a magnetically levitated train.
Advances in superconductivity
may soon lead to the development of practical, full-sized
maglev trains.

© Kaku Kurita/Gamma-Liaison

ergy, you wouldn't need to keep your generators close to your customers, so you
could build nuclear power plants away
from population centers."

With the development of more powerful electromagnets we may see high-speed
trains crossing the country suspended, or
levitated, inches above the ground. A magnetically levitated train, called a "maglev,"
may not be that far in the future. Such a
train, powered by low-temperature superconducting electromagnets, was tested in
Japan in 1979 and reached a top speed of
321 miles per hour. The principle behind a
"maglev": opposite magnetic poles attract;
like poles repel.

In Japan's model, eight superconducting electromagnets are built along the base
of each train car and thousands of metal
coils are set into the floor of a monorail-
shaped guideway. When the train moves,
the electromagnets on the sides of the cars
induce electric currents in the guideway's
coils, which then become electromagnets.
With an increase in power, the opposing
sets of magnets repel each other, lifting the
train four inches above the guideway.
Other magnets on each wall of the guideway repeatedly reverse their polarity from
positive to negative, pushing or pulling the
magnets built into the train so that the train
moves forward on a cushion of magnetic
force. Without wheels or rails, there is little
friction to slow the train and no tracks from
which the train can derail.

The Japanese train required liquid helium to supercool the electromagnets, as
well as large compressors in each car to re-
liquefy the helium. With the next level of
superconducting materials, inexpensive liquid nitrogen and much smaller compressors
may make the "maglev" a practical reality.

GENERAL BIOLOGY

Left: Ames Co., Division Miles Laboratories, Inc.
Above: Peg Estey/PR

Life—what is it? How did it start? How does it continue? How, for example, does a baby tortoise (above) develop from an egg laid by its mother months earlier? A complex molecule—DNA—may hold the answers. The determination of the structure of DNA, a plastic model of which is shown on the opposite page, was a giant step toward an understanding of "what is life?"

WHAT IS LIFE?

Of all the wonders of the universe, the most striking, perhaps, is life itself. Just what is life? There is no direct answer. We know, of course, that some things are living and others are nonliving. A man, a lion, a fish, an oak tree, and a rosebush are certainly alive. Rocks, icicles, and such man-made objects as tables and steel pillars are just as certainly not alive. But some things are not easily classified.

If we leave a loaf of bread and a bar of iron exposed to the open air, mold will form in time on the bread and rust on the iron. How do we tell whether the mold and the rust are living or nonliving? We cannot say simply that they are alive because they are "like men or trees," or that they are not alive because they are "like rocks or icicles." If we only knew in what respects living things differ from nonliving things, we could classify a mold or rust as living or nonliving. Actually, we know a good deal about such differences.

CHARACTERISTICS OF LIFE

One important distinction is that prac- tically all living things are made up largely of a complex substance called *protoplasm,* which is arranged in units known as *cells.* To study protoplasm is to study life. All the activities of living things take place in this substance. However, certain organ- isms called viruses do not have proto- plasm and cellular structure, yet they are classified as living things.

IRRITABILITY

A second point of difference between living and nonliving things is that living things display *irritability*. By this we do not mean that men, or lions, or trees are quick-tempered. We simply mean that they respond to changes in the environment.

Let us suppose that a grain of sand and a seed lie buried side by side in the ground. The grain of sand may remain there indefi- nitely, or it may be brought to the surface by some animal—an earthworm, perhaps. It may be exposed to heat, moisture, or cold, but it will still remain a grain of sand. What of the seed? It, too, may lie inert for a time. But suppose its environment begins to change. Suppose the earth is warmed by the sun and moistened by rain water or melting snow. The seed will respond: it will begin to sprout. Before long, its stem will push its way up out of the ground. In the course of time, the tiny stem will become a tree. The seed, unlike the grain of sand, has displayed irritability: it has reacted to environmental changes.

Irritability is expressed in movement. Of course, movement is by no means confined to living things. The water in a river moves, yielding to external forces. Gravity causes it to flow from a higher to a lower level. External forces also bring about movement in living things. When a parachute jumper leaps from an airplane, gravity will act upon him just as surely as it will upon the water in a river. This is not what we mean by movement as an expression of irritability. What we have in mind is rather such irritability movements as the swelling of the germ within the seed in response to sunlight or the closing of some flowers in response to darkness.

Living things respond to an internal stimulus as well as to external forces. When a child hastily withdraws his hand after touching a hot stove, it is because of an impulse from within—an impulse transmitted from the injured area along the nerves to a reflex center and from this to the muscles of the arm. There is nothing corresponding to this in nonliving things.

GROWTH

Living things, again, differ from non-living things because of the manner in which they grow. Things that are not alive are capable of growth only by accretion—that is, by adding to their existing bulk a mass made up of the same substance. In this way, a river increases its size as brooks flow into it. An icicle becomes bigger as more and more water is added to the core of ice and is frozen in its turn. But the water of the enlarged river and of the enlarged icicle has not changed in any way—it is still water. On the other hand, a living thing, like a child, develops by taking into its body food that is chemically different from itself, transforming this food chemically and making it a part of itself. A growing child owes his increased weight to his diet of milk, eggs, bread, and meat. But these foods have been so transformed that you would seek in vain for little particles of food in the child's biceps muscle or the calves of his legs.

REPRODUCTION

Finally, living things have the ability to reproduce their kind. Men, snakes, and insects all have young and these young develop into adults. Trees yield seeds and in the course of time these seeds become trees. Nonliving things do not have this property.

Long ago, it was believed that certain nonliving objects could give rise to living things. Learned men used to teach that aphids, or plant lice, came from dew in plants; that maggots were formed from decaying meat; that horsehair turned into

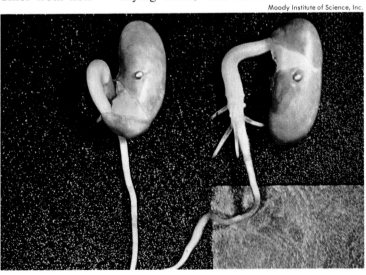

Living things show irritability. These beans have been pinned to a wet blotter (blue). As the roots grow, the ones belonging to the bean on the right touch the dry blotter (red). The root has turned toward the moisture—moving in response to its need for moisture.

Another example of irritability—the tendrils of this morning glory are very sensitive to touch. As the tendril touches the rope, it begins to spiral and wrap itself around the rope.

worms; that the mud of the Nile gave forth fishes. We know now, however, that aphids, maggots, worms, and fishes are all produced from eggs that are laid by the adult female. In other words, life begets life, as far as our present knowledge goes. Only living things are able to reproduce themselves.

Living things, therefore, differ from nonliving things because (1) they generally contain protoplasm; (2) they display irritability; (3) they are capable of growth; (4) they can reproduce their kind. By applying this fourfold test we can determine whether a mold or a speck of iron rust is living or nonliving. The mold (1) contains living protoplasm; (2) displays irritability; (3) is capable of assimilating food and growing; (4) can reproduce its kind. Therefore, it is alive. The speck of rust on the iron does not fulfill any of these requirements for life; therefore, it is not alive.

FORMS OF LIFE

Generally speaking, we can establish clearly the characteristics of living things. However, there are living forms which do not exhibit all these characteristics. They are the viruses.

VIRUSES

Viruses were discovered by the Russian scientist Dimitri Ivanovsky in 1892, while he was investigating the plant disease called tobacco mosaic disease. The general belief at that time was that the disease was caused by bacteria—tiny organisms which can be seen only under the microscope. It was known that bacteria can be removed from the liquids in which they live by means of a very fine filter made of porcelain or compressed earth. The liquid will pass through the filter, but the bacteria will not.

Ivanovsky showed that he could produce tobacco mosaic disease in healthy tobacco plants by inoculating them with the juice of diseased plants even after the juice had been filtered in order to remove bacteria. He came to the conclusion that certain diseases are caused by agents that can pass through the finest filters.

Other scientists confirmed Ivanovsky's findings. The newly discovered disease agents were called viruses. *Virus* is a Latin word meaning "poison." They are sometimes called filterable viruses because they can filter through any porous mass. They are also known as ultramicroscopic viruses because they cannot be seen through the ordinary microscope. They naturally grow and reproduce in the living cells of more complex organisms, where they may cause disease. Bacteria, higher plants, animals, and humans are affected. In humans, viruses produce such diseases as smallpox, measles, colds, influenza, rabies, encephalitis, and certain cancers.

It is possible nowadays to filter viruses through a special type of membrane. We still cannot see them through the ordinary optical microscope. But the electron microscope enables us to make shadowgraphs of viruses and these shadowgraphs show tiny particles that have a definite size and shape. The viruses of a particular disease have the same size and shape.

Detailed studies by means of extremely powerful electron microscopes and by other methods have shown that most viruses have complex structures. They consist of various nucleic acids, proteins, and other compounds, such as sugars. Some viruses are able to hook onto a living cell of some organism, dissolve their way into the cell, and use the protoplasm to reproduce themselves. The cell is usually destroyed as a result.

Certain viruses can take over the heredity mechanism—the genes—in the nucleus of an invaded cell. They are then

Only life can give rise to life. These photos show cells before dividing (1) and through the stages of division (2–5) to the formation of complete new cells (6).

J. P. Suzzoni

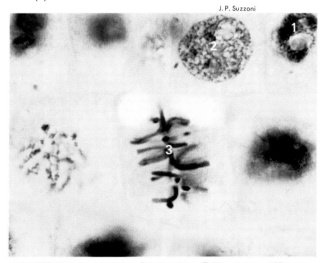
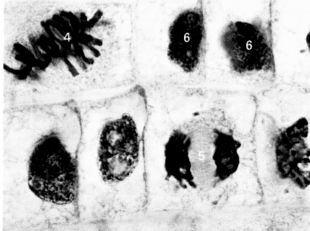

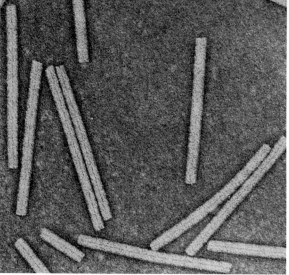

Grant Heilman

R. Mernaugh

Viruses—are they only half living? They do not exhibit all the characteristics of life. Left: tobacco plant infected with mosaic virus. Right: the tobacco mosaic virus.

able to regulate the activities of the cell for viral growth and reproduction. In fact, the virus is much like a gene in composition and structure. Some authorities consider a virus as a "gene without a cell."

Viruses can be transformed into ordinary-looking crystals that resemble inorganic matter. The tobacco mosaic virus, responsible for a disease of tobacco plants, can be crystallized in this way. Yet these crystals can be dissolved and injected into a tobacco plant to cause the mosaic disease, many times over.

Apparently, there are living things smaller than viruses. Called *subviruses,* they cause certain serious nerve diseases in man and lower animals. At present, very little is known about subviruses.

LINKS BETWEEN VIRUSES AND BACTERIA

Certain microorganisms seem to stand somewhere between the true viruses and the bacteria in the life scale. For example, the microbes responsible for psittacosis (parrot fever) and various related diseases are often classified as large and rather complex viruses. They resemble the latter in appearance. Like viruses, they can grow and reproduce in living cells only. Psittacosis organisms undergo a cycle of development by dividing as small spherical bodies into still smaller granular particles. In the process, they disintegrate the in-

fected cell. They are somewhat like bacteria in the stages of their life. They can also be stained in the same way as bacteria.

The rickettsiae, named after their discoverer, H. T. Ricketts of the University of Chicago, are more like bacteria in appearance and habits. They are minute, spherical, or elongate bodies, which divide in reproduction. Like bacteria, the rickettsiae can be stained and then examined with an ordinary, or optical, microscope. However, ricketts organisms can reproduce only within living cells, where they may cause serious diseases, such as typhus and spotted fever, among others. These microbes are usually carried by arthropods (insects, spiders, crustaceans, and related animals), in which they grow and develop. The arthropod that harbors rickettsiae injects them into humans or other mammals by means of its bite.

PLANTS AND ANIMALS

Among subviruses and viruses, the distinction between plant and animal is meaningless. These organisms are parasites, low in the life scale. But higher living things are themselves divided more or less definitely into two separate groups—animals and plants. This division holds true particularly in the case of the more highly developed living things.

Perhaps the most obvious difference between animals and plants is that most animals move about freely and spontaneously because they possess nervous and muscular systems, while the vast majority of plants, including practically all of the higher forms, are fixed in one place. There are exceptions, however. Many of the microscopic one-celled plants known as diatoms make their way freely through the fresh or sea water in which they live. On the other hand, the tiny sea animals known as true corals, or madrepores, often form colonies that are attached to rocks or reefs. The individual animal stays here during its lifetime. After it dies, its skeleton remains. Coral reefs are formed by successive generations of madrepores attaching themselves to the skeletons of their predecessors.

Animals differ from plants in the outer structure of their cells. Most plants have a fairly rigid and inelastic cell framework made up chiefly of cellulose. Cellulose is a carbohydrate—a chemical compound consisting of carbon, hydrogen, and oxygen. It

is not found in animals, with the exception of a lowly group called the sea squirts, or tunicates. In the case of animals, the external cell "walls," which are called pellicles, are much softer and more elastic and are made up chiefly of protein substances.

Finally, animals do not obtain their food in the same way as plants. Most plants manufacture their own food by a process called *photosynthesis*—a process that has no counterpart in animals. It is possible because the plants possess a green pigment, chlorophyll. In the presence of this pigment and of sunlight, the plant produces glucose, a kind of sugar, from plain water and from the carbon-dioxide gas found in the air. This glucose then combines with other substances within the plant to build up a long line of products—sugars, starches, fats, and proteins. Part of these products are used up by the plant. Part of them are stored up within the plant and become available as food for animals.

Animals move freely; plants do not. Here are two exceptions to the rule: (left) animals—corals—that are permanently attached to a rocky substrate and (right) a microscopic plant—a diatom—that moves about freely in a watery environment.

Cell types. Top: we see the one-celled animal known as the ameba. The ameba is capable of locomotion, enveloping and assimilating food, and other essential activities of animal life. Middle: a series of gland cells grouped around ducts which open into the human intestine. These cells are highly specialized. They produce a secretion which flows into the intestine and aids in digestion. Bottom: a cross section of cork cells. Note the well-defined cell walls characteristic of plants.

Upper photo: Hugh Spencer; middle photo: National Teaching Aids, Inc.; lower photo: Armstrong Cork Co.

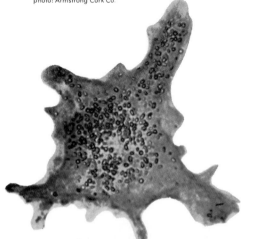

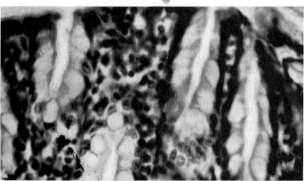

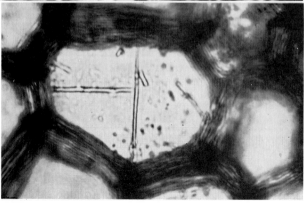

Animals do not manufacture their own food. They obtain it from the living things — plants or animals or both — that they eat. They digest this food; that is, they break down the complex substances of which it consists into simpler substances that they can assimilate more easily. A few plants, it is true, also devour living things. The Venus's-flytrap, a plant that grows in bogs, traps and digests insects; so does the pitcher plant. But such exceptions are rare.

Most living things may be definitely classified either as animals or as plants. Science recognizes the fact by dividing biology — the study of living things — into the two branches of *zoology,* the study of animals, and *botany,* the study of plants. However, a few groups defy such classification. For example, the tiny green organism called *Euglena* is like a plant in that it contains chlorophyll and can manufacture its own food by utilizing the radiant energy of the sun. Yet it has a mouth and a gullet and can swallow and digest food like an animal. It strikingly suggests an animal, too, as it propels itself through the water. Many spirochetes — microscopic disease-producing forms — are likewise on the borderline between plant life and animal life. Some scientists have proposed that such single-celled organisms should be grouped separately — in the kingdom Protista.

THE CELL — BASIC UNIT OF LIFE

All living things more advanced than viruses, whether animals or plants, have one thing in common — that is, their basic unit is the cell, made up of the life-giving substance protoplasm.

This protoplasm may have a jellylike consistency or it may be in the form of a sticky liquid. It contains somewhat less than thirty chemical elements, including oxygen, carbon, hydrogen, nitrogen, phosphorus, potassium, sulfur, chlorine, sodium, calcium, magnesium, copper, and iron. These elements are combined in various ways to form complex chemical compounds — proteins, fats, and carbohydrates — and certain other substances.

Water makes up from 60 to 90 per cent of protoplasm. Water is such a common substance that we take it more or less for granted. However, without it life as we know it would be impossible. It has many valuable properties. Among other things, it can dissolve substances, and it retains heat very well.

Next to water, proteins are the most abundant substances in protoplasm. Proteins are exceedingly complex compounds. Each of their molecules contains hundreds, and in some cases thousands, of atoms. Proteins have been called the building blocks of living matter. They are necessary for growth and for the maintenance of life.

The carbohydrates and fats in protoplasm are combinations of carbon, hydrogen, and oxygen. They are sources of energy for the activities of the cell. Most cells contain fat in the form of droplets. The higher animals may store quantities in the cells of certain tissues. Carbohydrates are generally stored as solid particles in the liver and muscle cells of animals. The starch grains of plant cells and the cellulose of which the cell walls are composed are two forms of carbohydrates; another is sugar.

Mineral salts usually make up a very small percentage of the cellular protoplasm, but they are very important nevertheless. They are essential elements of the complex substances that the plant manufactures; they also serve as chemical regulators in the manufacture of these substances. They are just as essential for the formation of bone.

Finally, protoplasm contains the substances known as enzymes, hormones, and vitamins. They serve to speed up the chemical reactions that take place within the protoplasm, without being affected themselves by these changes.

Most cells can be seen plainly with the ordinary microscope, but few of them can be made out with the naked eye. Among the largest ones are the egg cells of fishes and birds—the portion of the egg that is known as the yolk. Cells have a wide variety of shapes: they may be round, oval, flat, elongated, spindlelike, or cylindrical. The cell is a complicated affair with many intricate structures.

As we have pointed out, a living thing may consist of a single cell, in which all the functions of life take place. However, the one-celled arrangement is not ideal. Animals and plants that possess many cells have certain advantages. For one thing, they are not so easily affected by the environment. Bacteria are constantly buffeted by the movement of the molecules of the water in which they are to be found. But a fish swimming in the same water is not disturbed by this bombardment. A one-celled animal or plant cannot, of course, survive the destruction of its one cell. On the other hand, the destruction of even a considerable number of cells may not affect a many-celled animal or plant at all.

In many-celled organisms, cells are grouped together to form *tissues,* which serve certain special purposes. The cork that makes up the outer bark of the stems and roots of woody plants is a tissue. It guards the inner bark against injury and it prevents excessive evaporation. The epithelial, or surface, tissue of animals also serves as a protective covering.

Tissues in turn are combined to make up *organs*. An organ is an important part of a living thing. It performs a single task or a series of closely related tasks. An animal's heart is an organ. It keeps up the circulation of the blood by alternately contracting and expanding. The plant organ known as the root absorbs water and mineral salts from the soil, conducts these materials into the stem, anchors the plant to the soil, and sometimes stores food.

Organs are grouped together to form *systems*. The digestive system in man, for example, includes the mouth, the esophagus, the stomach, the small and large intestine, and the rectum. All of these organs contribute to the same general function: the utilization of food. A somewhat similar system in plants, the vegetative system, made up of the roots, stems, and leaves, absorbs raw materials, manufactures food, and uses the food.

Finally, the systems of an animal or plant are combined to form an individual.

CELLS

Practically all living things, from the very smallest to the greatest—from the microscopic ameba and single-celled algae to huge whales and redwood trees—are made up of *cells*. Many animals and plants consist of one cell, which is the whole individual. Others have passed far beyond this *unicellular,* or one-celled, stage. They are made up of a great number of cells that are joined together. Such animals or plants are called *multicellular,* or many-celled. The exact significance of "many" in this case can be seen from the fact that more than six million cells may be found in a drop of human blood.

What do we mean by a living cell? Ordinarily, when we speak of a cell, we have in mind a compartment—say a prison cell or the cell of a beehive—which may or may not be empty. The walls are the essential factors in compartments of this kind. The prison cell would be just as much of a cell even if there were no prisoner in it. The cell of the honeycomb would still be a cell, even if it were not filled with honey.

EARLY IDEAS

When Robert Hooke, an English experimenter of the seventeenth century, examined cork under magnifying lenses, it seemed to him to be made up of just such compartments as those of a bee's hive or the cells of a prison. It is natural, therefore, that he called these plant compartments cells. Later investigators were impressed by the strong and thick walls of the plant cells. They were impressed, too, by the fact that while some of the cells seemed to be filled with something, others appeared to be empty. They came to the conclusion, therefore, that the cell wall was the important thing. They did not realize that if certain cells appeared to be empty when they were viewed through the microscope, the reason was simply that the contents had oozed out while the cells were being prepared for microscopic investigation.

Still later, it was discovered that cells could be found in animals, too. But the biologists of that period had no idea of the wide distribution and importance of cells.

It was not until the nineteenth century that scientists came to realize the true significance of cells. In 1838–39, two German physiologists, Theodor Schwann and Matthias Jakob Schleiden, working independently, advanced a new and revolutionary cell theory. They held that living things, from the simplest to the most complex, are made up almost entirely of cells and that these cells play an important part in all the activities of life. Later, it was found that not only do the bodies of all the higher animals and plants consist of cells but that each one of these living things has sprung from a single cell.

CELLS ARE LIVING PROTOPLASM

Modern science has discarded the old idea that the wall is the cell and that the contents are mere stuffing, so to speak. We realize now that the important part, the truly living part, of each cell is the contents, to which we give the name *protoplasm.* We know that the protoplasm is the scene of the chemical changes that bring about digestion, absorption, muscular activity, and all the other activities of life. To be sure, we still use the word cells in referring to the units of living organisms. But when specialists in *cytology,* or the study of cells, speak of cells, they have in mind not only the enclosing walls or membranes, but also the living matter inside.

According to the modern conception, the cell is a mass of protoplasm enclosed in a membrane, known as the *plasma membrane,* and often by a more or less durable wall. The protoplasm is not uniform throughout but shows certain well-defined areas. A *nucleus,* or core, lies in a more or less central position. Some cells have more than one nucleus, while some mature cells have no nucleus. The protoplasm that surrounds the nucleus is called the *cytoplasm.* The nucleus is also enclosed in a membrane, called the *nuclear membrane.* Both the cytoplasm and the nucleus contain various important structures, which we shall discuss later.

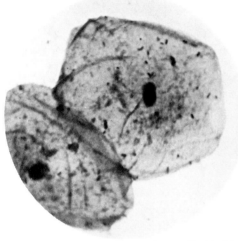

Plants and animals are made of cells. These cells come in a variety of sizes and shapes. Top: magnified onion-skin cells. Bottom: epithelial cells from the human cheek.

The cells of multicellular organisms form *tissues*—nervous, muscular, connective, and the like. Tissues combine to form *organs* such as the heart, intestines, and pancreas. The tissues and organs simply coordinate the various activities of the individual cells so that the organism as a whole can function properly. We can think of the individual cells as having a life of their own. They absorb various substances.

They burn fuels derived from food, and in so doing release energy for their many activities. They synthesize, or build up, complex substances. They manufacture and secrete hormones and enzymes, which control vital body processes. They eliminate waste products. They reproduce by undergoing division. Thus the activities that engage an organism as a whole are accomplished by the protoplasm contained in an individual cell.

COMPOSITION OF PROTOPLASM

The composition of protoplasm varies considerably, depending on the particular plant or animal and the kind of tissue. The water content averages roughly 75 per cent. There may be only a small proportion of protein in a cell, or else there may be as much as 30 per cent, as in certain muscle cells. The other constituents of protoplasm, including fatty substances, carbohydrates, and minerals, generally make up only a small proportion. In bone, however, the mineral content may be nearly 50 per cent.

Water. Some of the water in protoplasm is bound chemically with the proteins of the cell. The rest of it exists in a free state as the water molecule, H_2O, in which two atoms of hydrogen (H) are combined with one atom of oxygen (O). The free water in protoplasm is a most effective solvent. In it, the inorganic substances called salts ionize — that is, they break up into electrically charged atoms or groups of atoms called ions. Organic substances are either dissolved or dispersed in the water of protoplasm. The result is that chemical reactions are speeded up. The chemical exchanges between dissolved molecules and ions proceed faster than if these substances remained in solid form.

Generally, the greater the water content of cells, the more intense the vital activities that take place. For example, the cells making up fatty tissue have relatively little water, consume little oxygen, and release little energy. The cells composing the nervous tissue of the spine and brain contain much more water and consume much more oxygen. They provide the energy for the conduction of nerve impulses.

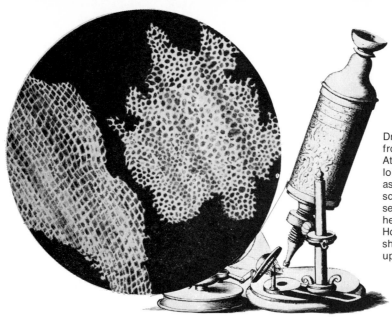

Drawings by Robert Hooke from his book *Micrographia*. At the far left is a very thin longitudinal section of cork as seen under the microscope. Next to it is a cross-section of cork. We also see here the microscope used by Hooke. He was the first to show that plants are made up of cells.

(cork) The Bettmann Archive
(microscope) N.Y. Academy of Medicine

Proteins. The proteins contained in protoplasm are essential building blocks of this living substance. Among other things, they are important constituents of *enzymes* —substances that speed up chemical reactions within the body. Associated with proteins in this vital activity are the *vitamins*. These substances, which are present in minute quantities, serve as *coenzymes*— that is, as partners in enzyme systems.

Proteins are formed from varying combinations of chemical substances called *amino acids*. There are more than twenty of these acids. Each contains an amino group, $-NH_2$ (a compound of nitrogen and hydrogen), and a carboxyl group, $-COOH$ (a compound of carbon, oxygen, and hydrogen). Each of the amino acids forms a link in the chain of which the protein molecule consists. There may be more than a thousand links. Some proteins are made up entirely of amino acids. They are called simple proteins. Other proteins are more complex. They consist of simple proteins to which are attached various other substances, such as carbohydrates and phosphorus.

Since the amino acids can form all sorts of combinations, the number of entirely different kinds of proteins is exceedingly great. Different cells have different kinds of proteins. The proteins in a man's kidney cells differ from those in his muscle cells and from those in the kidney cells of a dog.

However, it has been found that the cells of corresponding tissues in different animals have a certain affinity for each other. Cells from the kidney tissue of mice and chickens have been grown together in special solutions and have yielded strange hybrid "pseudotissues." Such research may well lead to medical repair of human tissues and organs with other animal tissues.

Lipids. The fatty substances in the cell are called *lipids*, from the Greek word *lipos*, meaning "fat". They form part of the protoplasmic structure. They also serve as a reserve energy source and compose many of the cell pigments. Some of the lipids are completely insoluble in the water of the protoplasm. They disperse themselves throughout in the form of tiny droplets. In other cases, certain chemical groups attached to the lipids apparently bind these fatty substances to the water molecules.

Lipids may be either simple or complex. The alcohol known as glycerol (glycerine) and the so-called fatty acids are combined to make up the simpler lipids. Weight for weight, the latter release more energy when burned in the cell than do any of the other substances in the protoplasm. However, since they are not so easily burned as are the sugars, the simple lipids are stored in cells as reserve sources of energy. They are found in abundance in the cells of *adipose*, or *fatty, tissues* of animals.

The simple lipids of plant and animal cells include cocoa butter, olive oil, lard, and fish oils.

The complex lipids include the *phospholipids, steroids, carotenoids,* and *lipoproteins.* Phospholipids are fatty substances containing phosphorus and nitrogen. They occur principally in the cellular structure of nervous tissue. Some steroids are produced by the liver and are a part of the bile. Certain steroids (called sterols) serve as *hormones,* regulating various activities of the body. The carotenoids are represented by several important cell pigments, including the carotenes, found in carrots and grass. Lipoproteins are lipids linked to protein molecules. They form part of the nucleus and the membranes of the cell.

Carbohydrates. The carbohydrates are made up of the chemical elements carbon, hydrogen, and oxygen. The proportion of hydrogen atoms to oxygen atoms in carbohydrates is always 2 to 1, just as it is in the water molecule, H_2O. A carbohydrate, therefore, may be considered as "watered carbon." This accounts for the name "carbohydrates," for "hydrate" is derived from the Greek word *hydor,* meaning "water."

One of the most important of the carbohydrates is the sugar known as *glucose.* It has the formula $C_6H_{12}O_6$, indicating that the glucose molecule contains 6 carbon (C) atoms, 12 hydrogen (H) atoms, and 6 oxygen (O) atoms. It is formed in the cells of plants by the process of *photosynthesis,* which we describe later in this article, and it is used by these cells as an immediate source of energy. Animal cells, too, burn glucose to provide energy. Glucose is a *monosaccharide*—the simplest form in which a sugar can exist.

When two monosaccharides join together, they form a *disaccharide. Lactose, sucrose,* and *maltose* are all disaccharides. Lactose is found in animal cells. Sucrose and maltose are found in plant cells. The disaccharides can be broken down to give energy. They readily dissolve in the cell's protoplasm.

When a number of monosaccharides combine chemically, the resulting substance is called a *polysaccharide. Starch* and *cellulose* are plant polysaccharides. Starch serves as a food reserve. It exists in the form of granules in the protoplasm. Cellulose goes into the plant cell walls and other supporting structures forming the plant's skeleton. *Glycogen,* or animal starch, is a polysaccharide found in animal cells. It is an energy source. It occurs, dissolved in the protoplasm, in a variety of cells, but most abundantly in those of the liver and the muscles.

Carbohydrates often unite with protein or mineral elements, such as sulfur and phosphorus. Various important complex substances are formed in this way. Pentoses—sugars whose molecules contain five carbon atoms—link with phosphoric acid and nitrogen-containing groups to form *nucleic acids.* One of the nucleic acids, called *ribonucleic acid,* or RNA, is concerned with the building up of proteins in the cell. Another one—*deoxyribonucleic acid,* or DNA—occurring in the nucleus, is associated with the chromosomes, the hereditary matter of the cell. We discuss nucleic acids elsewhere.

Inorganic materials. The inorganic substances in protoplasm include various salts and mineral elements. In the watery protoplasm, the salts are broken up into ions. Thus common salt, or sodium chloride (NaCl), separates into positive sodium and negative chloride ions. The numerous ions must be present in the right amounts in the cell so that a balance, or equilibrium, may exist among them. For instance, the protoplasm must have calcium ions, in minute quantities, in order to balance sodium, potassium, and magnesium ions.

We find about the same relative concentrations of salt ions in the body fluids that circulate around the cells as we do in sea water. It is especially true of the ions mentioned in the preceding paragraph. This is a significant fact, since most biologists think that life probably originated in the ocean.

It is important to note that the various ions in protoplasm remain as free ions as long as they are in solution. Thus, instead of referring to the common salt in protoplasm, it is more accurate to refer to the sodium and chloride ions into which the salt is broken up in solution. In the case of the other salts, too, we are particularly interested in the ions they form in protoplasm. These salts exist only in the form of ions as long as the protoplasm retains its water content—that is, as long as the cell is alive.

Sodium ions are more abundant than the other ions of cells. They play an important part in controlling the diffusion of substances into and out of the cell through the plasma membrane. Together with potassium, sodium increases the viscosity, or sticky consistency, of the cell's internal protoplasm.

Potassium ions seem to play an important part in the conduction of nervous impulses by nerve cells and the contraction of muscle cells. Magnesium, which exists in protoplasm in small quantities, forms an essential part of the green plant pigment, *chlorophyll*. Magnesium and calcium ions tend to make the protoplasm more fluid.

The presence of calcium ions in the cell lowers the permeability of the plasma membrane, so that only small amounts of water and a few dissolved substances can pass through. Calcium ions are vitally important in speeding up the action of various enzymes. They are essential, too, for nerve-cell responses and the contraction of muscle cells.

Iron, copper, and zinc occur as part of various enzymes. Sulfur is found in proteins and is an essential constituent of many enzymes. Phosphate forms a bond with lipids, proteins, and sugars. This bond is often extremely important as a source of energy.

CONSISTENCY OF PROTOPLASM

The components of protoplasm, ranging from relatively simple water molecules and salt ions to the complex molecules of carbohydrates, lipids, and proteins, occur in various forms. Salts and some carbohydrates are dissolved in the water of the protoplasm. Other carbohydrates and the proteins and lipids are suspended in water as aggregates (clumps) or as macromolecules (large or long molecules). Though they are too small to be seen through the ordinary microscope, they are too large to pass through membranes that are permeable to ions and small crystalline substances. They belong to the group called the *colloids*. The vital chemical activities of the cell take place at the surfaces of the colloidal particles.

The consistency of protoplasm can vary widely in different cells. It even changes in the same cell from a *sol*, or rather freely flowing liquid, to a fairly firm *gel*, or jellylike structure. The gel condition is apparently due to the fact that the dispersed colloidal particles are attracted to one another and are linked together to form a three-dimensional network. In the gel condition, water fills the spaces in the network. When the network breaks down, the colloidal particles are dispersed and the protoplasm becomes a sol.

Many colloidal particles, such as various proteins and enzymes, may clump together in crystalline form. There is some evidence that the structure of protoplasm, especially in the gel state, has an inherent crystalline pattern.

The living cell may possess a skeleton in the form of complex internal membranes stretching throughout the cytoplasm. These membranes make up what is called the *endoplasmic reticulum*, a network occurring within the protoplasm. It may be a continuation of the outer cell, or plasma, membrane. The endoplasmic reticulum is probably involved in the vital process of metabolism.

CELL SURROUNDINGS

The protoplasm of the cell is usually enclosed by walls or membranes of some sort, to protect it from the environment and to regulate what passes out of and into the cell proper. In plants, each cell is often surrounded by a more or less rigid *cell wall* of cellulose, a woody carbohydrate material. The animal cell is sometimes contained in a more flexible envelope, called a *pellicle*, or *cuticle*, which is made up of protein. Both the cellulose walls of plant cells and the pellicles of animal cells permit water and substances dissolved in water to pass through quite freely.

Just inside the cell wall or pellicle at the surface of the living cytoplasm is the plasma membrane, much thinner than the surrounding cell wall or pellicle. It is known as a *semipermeable membrane*. It permits certain substances—liquids, gases, or dissolved solids—to pass through and restricts or prevents the passage of others. The diffusion of a liquid through such a membrane is known as *osmosis*. It involves movement from a region of higher liquid concentration on one side of the membrane to a region of lower liquid concentration on the other side. Suppose that the concentration of water molecules inside a cell is less than that outside. Water will then flow into the cell through the plasma membrane. The cell will increase in size and will become more or less turgid, or swollen. If the water concentration in the inside of the cell is greater than that on the outside, water will flow out of the cell, and the cell will shrink.

Water is not the only substance that flows through the plasma membrane. Glucose, amino acids, salt ions, oxygen, carbon dioxide, and other substances essential for the proper functioning of the cell also pass through. The passage of materials such as these is not a constant process. It varies with the concentration of these materials inside and outside the cell and with the condition of the plasma membrane.

Biologists are learning much about the structure of the plasma membrane. We know that it is exceedingly thin and slightly elastic, and that it is apparently constructed of lipids and proteins. Lipid molecules are arranged with their long axes perpendicular to the plane formed by the surface of the protoplasm. The proteins form a network at both sides (outer and inner) of the lipid layer, parallel to the surface.

It is thought that pores of various sizes perforate these protein-lipid layers and that water and water-soluble substances (such as glucose and amino acids) pass through by way of these pores. Certain pores seem to be electrically charged. Salt ions pass into the protoplasm from the outside by being attracted to these pores. Substances that are soluble in lipids get into the cell by being dissolved in the lipid layer.

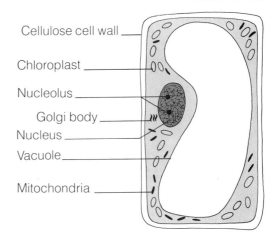

PLANT CELL

Cellulose cell wall
Chloroplast
Nucleolus
Golgi body
Nucleus
Vacuole
Mitochondria

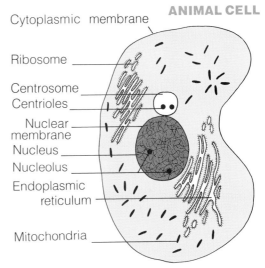

ANIMAL CELL

Cytoplasmic membrane
Ribosome
Centrosome
Centrioles
Nuclear membrane
Nucleus
Nucleolus
Endoplasmic reticulum
Mitochondria

Top: a typical plant cell. Bottom: a typical animal cell.

The cell may engulf food and other materials outside itself. The plasma membrane and the cytoplasm wrap themselves about these materials, forming around the latter a clear space called a *vacuole*. The vacuole ultimately detaches itself from the cell boundary and moves into the cytoplasm, where its contents are absorbed. This phenomenon often occurs in one-celled animals.

THE CYTOPLASM

We saw that cytoplasm makes up the

bulk of the cell—all of it outside of the nucleus. The cytoplasm contains accumulations of protein and carbohydrate, drops of fat, pigments, small crystals, and secretion granules. It also has the specialized structures called *organelles* which are distinctly visible through the ordinary microscope. The organelles include the mitochondria, lysosomes, ribosomes, Golgi apparatus, centrosomes, and plastids.

The *mitochondria* are small structures made up of proteins and lipids forming a stable and rigid gel. They are often cigar-shaped but may have other shapes. Internally a mitochondrion consists of a series of membranes. Apparently mitochondria can divide. Also, the threadlike mitochondria break up into granules, and the granules unite to form filaments. Mitochondria are vital in respiration; in the breaking down of carbohydrates, proteins, and fats; and in the release of energy. Mitochondria contain DNA, which is vital to the synthesis of proteins. The numbers of mitochondria in different cells vary enormously.

Lysosomes are vacuolelike bodies that secrete enzymes to digest food substances. Also, at the death of a cell, lysosomes release substances that disintegrate the cell.

The *ribosomes* are granular masses associated with the endoplasmic reticulum. They contain RNA and are the sites of protein synthesis in the cell.

The *Golgi apparatus* consists of an irregular network of rodlike, globular, or granular bodies in animal cells, often concentrated around the nucleus. Much Golgi matter occurs in gland and nerve cells, but little in muscle cells. For a given type of cell, Golgi bodies are uniform in shape. The chief function of Golgi material in gland cells is collecting protein secretions from the endoplasmic reticulum and passing them outside the cell. Golgi bodies add carbohydrates to these secretions to stiffen them into material called *zymogen granules*.

The *centrosome,* a rather dense area of protoplasm, lies close to the nucleus. In the middle of the centrosome are two small dot-like, rod-shaped, or V-shaped bodies called *centrioles*. They play an important part in cell division, as we shall see.

Left: plant cells with a low concentration of water. Right: plant cells after water has diffused through the membrane. The cell vacuoles are now fluid-filled, and the cells have become swollen, or turgid.

Nuridsany

Nuridsany

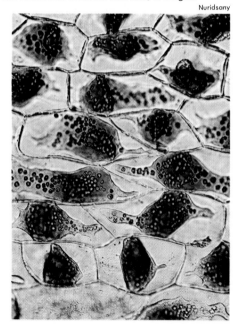

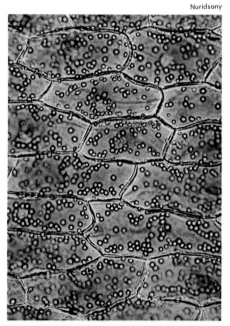

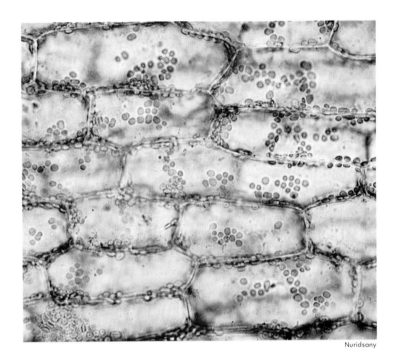

Cells in an elodea leaf. The round green bodies are chloroplasts. These contain chlorophyll, a pigment that is essential for photosynthesis.

Plastids, bodies of varying shape found in plant cells, are made up of lipids and proteins. Plastids synthesize fats, proteins, and starch. They vary in number in different cells. There are at least three different kinds of plastids — *leucoplasts, chromoplasts,* and *chloroplasts.* Leucoplasts, which are colorless, occur in sex cells and in storage cells of roots and underground stems. They form starch grains. Chromoplasts carry pigments that color parts of plants. Chloroplasts contain chlorophyll, the green substance that converts carbon dioxide and water into sugar in sunlight.

Microtubules are long, thin tubes that have been detected in many kinds of cells, both plant and animal. They consist of protein. Their chief function is to maintain the shape of a cell and to direct the movements of cell components. Microtubules also assist in mitotic cell division.

Vacuoles, more common in plant than animal cells, are each separated from the cytoplasm by a membrane, which may be like the plasma membrane. Vacuoles are filled with water fluid — *cell sap* — which contains food, cell secretions, and wastes.

THE NUCLEUS

The cell nucleus is a comparatively large, spherical, egg-shaped, or irregularly shaped structure, surrounded by the cell's cytoplasm. There is generally only one nucleus, though some kinds of cells possess two or more. The nuclear membrane encloses a watery fluid, the *nuclear sap.* Scattered throughout this fluid are filaments, granules, or flakes linked by a fine lacy network. These particles form the substance called *chromatin,* which carries the *genes,* or determiners of heredity. One or several spherical bodies called *nucleoli* (singular, *nucleolus*) are enmeshed in the chromatin network.

The nucleus is composed mainly of *nucleoproteins* — that is, combinations of proteins and nucleic acids. Many enzymes are present, making the nucleus an active center. From the nucleus, across the nuclear membrane and into the cytoplasm, flow substances that are connected with gene activity, and protein synthesis. When the cell divides, the nucleus undergoes great changes, as we shall see.

COMMON CELL FUNCTIONS

One-celled animals and plants can do, in the single cell that forms their bodies, all the things that are necessary to keep alive. Multicelled organisms, such as mammals and trees, also face the problem of remaining alive. They meet it by a division of labor among the cells that compose their bodies. Thus there are plant cells that serve as conductors of food materials, as strengthening units, as food-storage compartments, as centers for absorbing food from the soil. In animals, certain cells are responsible for making bone and cartilage, for shortening in muscle contraction, for carrying nervous impulses, for lining cavities, for engulfing bacteria, for absorbing food from the gut. Yet all cells have certain functions in common. They all must absorb food and obtain energy from it.

Bone cells of a young cat. Each cell's nucleus, cytoplasm, and plasma membrane are clearly visible.

Nuridsany

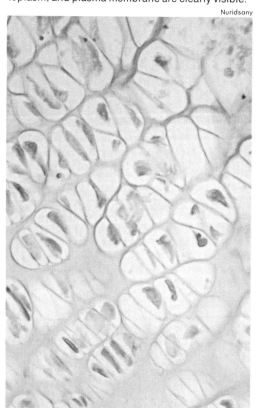

CELLS NEED FOOD

In the first place, all cells have to obtain essential food elements. Plants absorb water and mineral ions through their root systems and transport these materials by way of conducting vessels to the cells. Carbon dioxide gets into plant cells after it has made its way through the *stomata,* or pores, of the leaves. In animals, solid foods are broken down, or digested, in a cavity (alimentary canal or intestine) within the body. Digestion is speeded up by the action of enzymes. Protein food is broken down into amino acids, carbohydrate into simple sugars, fat into glycerol and fatty acids. These products then pass through the walls of the cavity where digestion has taken place and enter nearby cells or are carried to distant cells by the blood.

The simplest way food material gets into cells is by *diffusion.* Like water molecules, the molecules or ions of a dissolved substance flow, or diffuse, from a region of higher concentration to one of lower concentration. Amino acids, for example, pass into the cell when there are more amino acids on the outside than on the inside. The concentration of food molecules and ions is usually lower within the cell, because, as soon as these enter, they are built up into more complex compounds or are passed on to other cells. Food molecules and ions, therefore, continue to diffuse into the cell from the greater concentration outside. There is much evidence that protoplasm takes an active part in absorbing food materials. Salt ions from the outside are absorbed into the cell by combining with protein material of the protoplasm. Colloids in the protoplasm attract certain ions.

Once food substances get into the cell, they are converted, or assimilated, into a variety of other substances, some of which form the essential parts of protoplasm itself. In the cells of animals, assimilation is the reverse of digestion. The same type of enzymes are involved. Proteins are built up from amino acids, fats from fatty acids and glycerol, carbohydrates from simple sugars. When new protoplasm is made from these substances, it is not stable, but undergoes

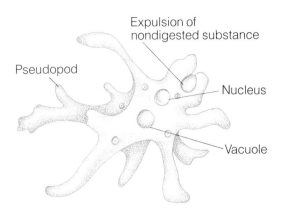

Pseudopod

Expulsion of
nondigested substance

Nucleus

Vacuole

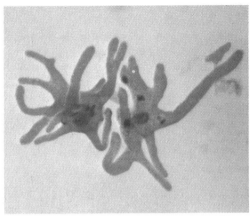

Fotogram

Amebas display one type of protoplasmic movement. They actually change
shape, pushing forth projections, called pseudopodia, into which their protoplasm flows. Above left: diagram of an ameba; right: photo of two amebas.

continuous changes. The nitrogen atoms of proteins, for example, are constantly re-placed by new nitrogen coming into the cell. The atoms composing fats also change continually. The simple sugars are some-times immediately burned to produce ener-gy before they can be built up into more complex carbohydrates.

The cells of green plants can assimilate much simpler materials than animal cells can. Plants, for instance, use ammonia and other simple compounds to synthesize the amino acids that form proteins. Carbon dioxide and water are transformed into a sugar by photosynthesis. In this process, the water molecule (H_2O) is split—by sun energy trapped in the pigment chloro-phyll—into hydrogen and oxygen. The hy-drogen then combines with carbon dioxide, in a series of complicated steps, to produce a simple sugar, glucose. The light energy that was trapped by the chlorophyll now exists as chemical energy in the sugar. Many of the sugar molecules combine with one another to form starch or other poly-saccharides, which are stored as reserve material or used in the structure of the plant. Some of the sugar is burned in the process of respiration.

CELLULAR OXIDATION

Respiration, or cellular oxidation, is another essential activity of the cell. It causes chemical energy to be released from fuel foods. Carbohydrates are important cellular fuels. Fats and proteins are also used. The oxygen that serves in respiration is absorbed by diffusion in most plant and animal cells. At one time, it was believed that respiration invariably involved the combination of cellular fuel with oxygen, in the process called *oxidation,* and the result-ing release of energy. We now realize that the loss of hydrogen from food material is also involved in respiration. The sub-stances that give up hydrogen are called hydrogen donors. Those that take up hy-drogen are known as hydrogen acceptors.

Even in *aerobic respiration,* or respira-

Unicellular paramecia display a different kind of motion. These cells are covered with hairlike pro-cesses, called cilia, that undulate or vibrate, caus-ing the cell to move.

Fotogram

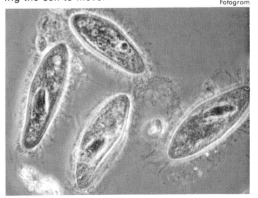

tion in the presence of oxygen, the giving up and taking up of hydrogen play an important part. A chain of reactions is involved. First, a hydrogen donor gives up its hydrogen to a hydrogen acceptor. This hydrogen acceptor, acting now as a donor, hands the hydrogen on to another acceptor and so on through a number of steps. Finally, oxygen, as a hydrogen acceptor standing at the end of this chain of reactions, takes up the hydrogen, and the combination of oxygen atoms and hydrogen atoms yields water (H_2O). A great deal of energy is liberated when hydrogen and oxygen combine. Should this happen at once, the energy would be freed in such great quantity that much would be wasted. In the hydrogen donor-hydrogen acceptor series, only small amounts of energy are conveniently released in steps so that the energy can be fully utilized in the different activities of the cell.

When food material is oxidized in the presence of oxygen, water and carbon dioxide are end products. We have seen that water is formed from the combination of oxygen and hydrogen, but from where does the carbon dioxide come? This is the generally accepted explanation. Starch, glycogen, and other carbohydrates are decomposed to the simple sugar glucose, which has six carbon atoms. In a series of complicated reactions, glucose is converted to a three-carbon substance, pyruvic acid, with an accompanying release of energy. The pyruvic acid then enters a cycle, called the *citric-acid cycle,* in which a chain of reactions keeps repeating itself, releasing energy and throwing off carbon dioxide. Various other reaction systems are connected with the citric-acid cycle. In these systems, proteins, fats, and other substances are first built up and then broken down, producing further energy.

In the case of *anaerobic respiration,* or respiration without oxygen, pyruvic acid does not go into a citric-acid cycle but is converted to lactic acid. Energy liberated in anaerobic respiration is rapidly available to the protoplasm, for there is no need of a continuous oxygen supply. Muscle cells take advantage of this quick energy to per-

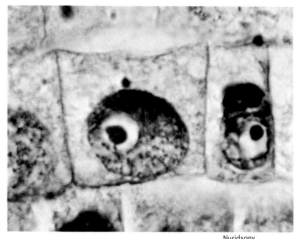

Nuridsany

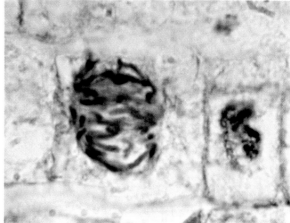

Nuridsany

Mitosis occurs in several stages, some of which are shown in the photos above and on the opposite page. During interphase (upper left) the nucleolus is distinct, but individual chromosomes are not. The chromosomes become shorter and thicker and split. By early metaphase (upper right) they are lined up along the equatorial plane of the spindle. They then begin to move apart (lower left) until at telophase (lower right) they are at each pole. Then, a cell membrane develops, dividing the cell into two new cells.

form their work. Once they are at rest, the muscle cells use oxygen to oxidize some of the lactic acid. The energy produced by this is used to synthesize the rest of the lactic acid into carbohydrate.

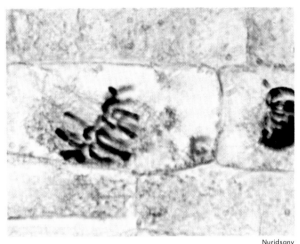

Nuridsany

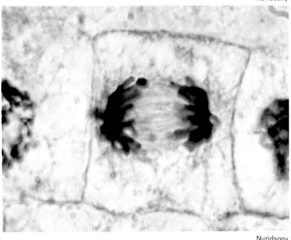

Nuridsany

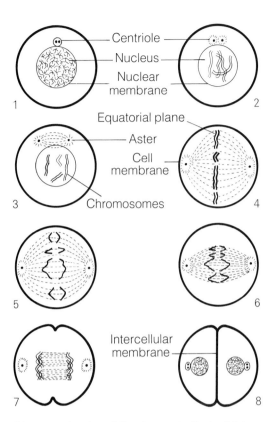

Diagrams showing all the stages of mitosis. 1) Interphase. 2) Prophase. 3) Early metaphase. 4) Later metaphase. 5) Early anaphase. 6) Later anaphase. 7) Telophase. 8) Interphase.

Labels in diagram: Centriole — Nucleus — Nuclear membrane — Equatorial plane — Aster — Cell membrane — Chromosomes — Intercellular membrane

CELLS USE ENERGY

The energy released by respiration is used for the building up of complex materials and, in animals, for muscular work. Protoplasm does work, or exerts energy, when it absorbs certain substances, as we have seen. Energy is also used in several kinds of protoplasmic movements. One of these is *cyclosis*—the streaming of protoplasm round and round within many plant cells and a few animal cells. Another is the so-called *ameboid movement*. In this the cell actually changes shape by pushing forth projections, called *pseudopodia* (false feet), into which protoplasm flows. The cell can move from place to place as a result of this process. The single-celled amebas display ameboid movement, as do white blood cells and various connective-tissue cells. In certain cells, we find another kind of movement—the vibratory lashing or undulating of hairlike processes known as *cilia*. Cells of this type are called *ciliated cells*. They line respiratory passages and alimentary canals, are spread over the gills of fishes and various other water-dwellers, and may cover the entire outer surface of certain animals.

Some of the chemical energy of cells is converted to heat, some to electricity—especially in the organs of specialized fish, such as the electric eel—and some to light, in the luminous cells of fish, worms, insects, and certain plants.

Another energy-expending activity is that of *secretion,* by which substances absorbed into the cell are chemically transformed into new materials and forced out of the cell. Cellular secretions serve to protect the organism, to stimulate or inhibit (check the activity of) other cells, or to act chemically on various substances. Mucus, cellulose, nacre (mother-of-pearl), hormones, auxins (growth hormones of plants), and digestive enzymes are good examples of these secretions. Substances to be secreted are often visible in the cell in the form of granules, vacuoles, or droplets. The cell eliminates its secretions in various ways. It extrudes, or pushes out, solid materials, such as plugs of mucus. It diffuses or actively ejects fluids and dissolved particles.

Various waste products are formed in the cell. Carbon dioxide and water result from the breakdown of fats and carbohydrates. When proteins are broken down, the waste materials may be ammonia, urea, uric acid, and other nitrogen-containing compounds. Usually wastes leave the cell through the plasma membrane by diffusion. Since the concentration of these products is greater inside the cell, they make their way to the outside. No energy is exerted by the protoplasm, as a general rule, in the process of excretion. In certain special cases, however, energy is expended to excrete water and salts.

CELL DIVISION

As the cell synthesizes some of its absorbed food materials into new protoplasm, it grows in volume. But this increase is definitely restricted. After a cell attains a certain size, it generally divides into two cells, which later divide in their turn and so on. As the different cells multiply, the organism as a whole grows and develops. Apparently growth and division bring about chemical and physical changes that prevent the protoplasm from growing old and dying. It should be pointed out, however, that certain cells, such as muscle and nerve cells, do not divide after reaching the limit of their growth, and yet they continue to carry on their activities.

The process of cell division is called *mitosis*. In this process, the nucleus of the cell divides into two daughter nuclei. The cytoplasm also divides, so that in place of one cell there arise two daughter cells, each with its own nucleus. The stage between two successive cell divisions is termed *interphase*. It is during interphase that the cell performs most of the functions that we have described.

Mitosis is a continuous process, but certain definite stages can be distinguished

Nuridsany

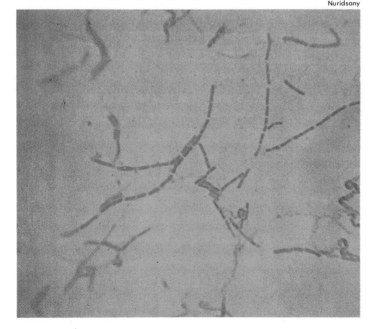

Bacteria are also unicellular organisms. They vary in shape. One common shape is rodlike as in the anthrax bacteria shown at left. These cells frequently form long chains.

in it. These are prophase, prometaphase, metaphase, anaphase, and telophase.

At *prophase,* the chromatin material of the nucleus begins to shorten and increase in thickness to form *chromosomes.* It would be more accurate, perhaps, to say that the chromosomes exist as independent bodies at all times, but that during interphase they are elongated and intertwined, like thread in a tangle of yarn. There is always the same number of chromosomes in the cells of individuals belonging to the same plant or animal species. In the first part of prophase, they become shorter and thicker, acquiring a rodlike shape. Then each chromosome splits lengthwise, so as to form two parts known as *chromatids.*

We mentioned previously the two small bodies, called centrioles, that lie in the cytoplasm close to the nucleus. At prophase, certain fibers, called *asters,* or *astral rays,* radiate from each centriole. Their function is unknown. As the chromosomes become distinct, the centrioles migrate in the cytoplasm to opposite ends of the nucleus, so that they are antipodal, or opposite one another. Between the two centrioles arises a bundle of delicate filaments, forming a structure called the *spindle.* The cells of most plants lack centrioles. A spindle forms between two points called *polar centers.* During prophase, the cell becomes more spherical and turgid.

At *prometaphase,* the nuclear membrane seems to disintegrate. The nucleolus disappears. Some of the nucleolus material attaches itself to the split chromosomes. The rest appears to diffuse into the cytoplasm. The spindle can now be made out more clearly. Some of the filaments stretch from one centriole or polar center to the other. Others make contact with the chromosomes.

During *metaphase,* the divided chromosomes are arrayed for a moment in the equatorial, or central, plane of the spindle. Then the two chromatids of which each chromosome consisted begin to move away from each other. They have become perfectly identical daughter chromosomes.

The cell now enters upon the *anaphase* stage. Each of the daughter chromosomes

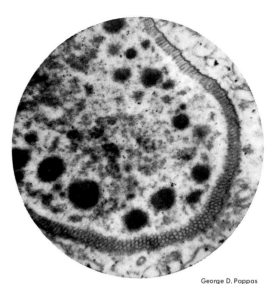
George D. Pappas

All cell activities depend on the passing of material through semipermeable membranes—the cell membrane itself or the nuclear membrane surrounding the nucleus. Above a photo of the nuclear membrane of an ameba.

formed from a given chromosome migrates to a different centriole or polar center. Finally, they all arrive at their destinations.

At *telophase,* the daughter chromosomes are bunched at the poles of the spindle, and the spindle begins to disintegrate. The daughter chromosomes begin to elongate and to intertwine. The tangled chromatin network appears. A new nuclear membrane is developed around each clump of chromatin. The nucleoli are re-formed. Two daughter nuclei have now developed. Meanwhile, the cytoplasm has been divided into two roughly equal parts by a partition formed at the equatorial region of the spindle. The original cell has given rise to two daughter cells, each with a nucleus. In plant cells a new rigid cell wall is formed between the two daughter cells. The organelles and particles contained in the cytoplasm of the old cell have been more or less equally divided, so that each new cell gets its share.

In effect, then, cell division brings about an equal separation of chromosome material and a nearly equal division of cytoplasm and its constituents. Each daughter cell then enters the interphase stage in which it grows until ready to divide again.

VIRUSES

by C. Arthur Knight

Standing before the Russian Academy of Sciences on February 12, 1892, a Russian botanist named Dmitri Ivanovski presented a short paper. It was entitled *On the Mosaic Disease of the Tobacco Plant.* Toward the end of the article, he commented on an experiment he had carried out: "I have found that the sap of leaves attacked by the tobacco mosaic disease retains its infectious qualities even after filtration through Chamberland filtration candles."

The device he referred to was a porcelain filter. Its fine pores permitted the passage of liquids, but held back bacteria. Since the disease agent passed through the filter, it must have been smaller than ordinary bacteria. Oddly enough, Ivanovski did not accept this conclusion until years later.

Unaware of Ivanovski's research, a Dutch botanist by the name of Martinus W. Beijerinck later conducted a similar experiment. He then advanced the idea that tobacco mosaic disease was caused by a strange, previously unknown infecting agent—a "contagious living fluid" or "virus."

In the early nineteenth century, the word "virus" meant any poisonous substance, including snake venom. Later it was also applied to bacteria that cause disease. When it was discovered that certain disease agents could pass through bacterial filters, they were called "filterable viruses."

Unlike ordinary chemical poisons and venoms, viruses are able to reproduce. They are much smaller than bacteria, microscopic one-celled organisms. With few exceptions, viruses are too small to be seen with an ordinary, or optical, microscope. Viruses first became visible when the electron microscope was developed in 1940. The electron microscope magnifies thousands of times.

VIRAL DISEASES

The roll call of viral diseases in man, animals, and plants is long. In human beings, viruses cause: smallpox, yellow fever, poliomyelitis (infantile paralysis), influenza, rabies, dengue, mumps, measles, rubella (German measles), chicken pox, a form of encephalitis, hepatitis B (or serum hepatitis), herpes, warts, perhaps some cancers, the common cold, and AIDS (Acquired Immune Deficiency Syndrome). Virus-like elements have been implicated in such brain disorders as Alzheimer's disease, kuru, and Creutzfeldt-Jacob disease.

Animals are also prey to numerous viral infections. These include cowpox, rabies, pseudorabies (mad itch), Newcastle chicken disease, dog distemper, and horse encephalomyelitis. The names of plant virus infections are picturesque and generally describe the symptoms. Some examples are sugar-beet curly top, tomato bushy stunt, tobacco mosaic, tomato spotted wilt, and tulip break.

A virus may not harm one type of host, but may injure or kill another. The so-called "healthy potato virus" has almost no effect on potato plants. Transferred to a tomato plant, the same virus causes noticeable symptoms.

A mild viral disease may grant its host immunity, or resistance to related but more serious viral infections. For example, cowpox makes a person only slightly ill, but it also makes him immune to smallpox. However, a new race, or strain, of a virus may arise suddenly, through a biological process called *mutation*—a change in the genetic material of the virus. People who are immune to the older strain may not have resistance to the new one.

Viruses may also cause the wild growth and reproduction of cells, rather than their direct destruction. The result is a malignant tumor, or cancer. Viruses definitely cause certain cancers in animals. These include kidney tumor of the leopard frog, several fowl sarcomas (deep-seated cancers), rabbit tumors, leukemia (cancer of white blood cells) in mice, and breast cancer

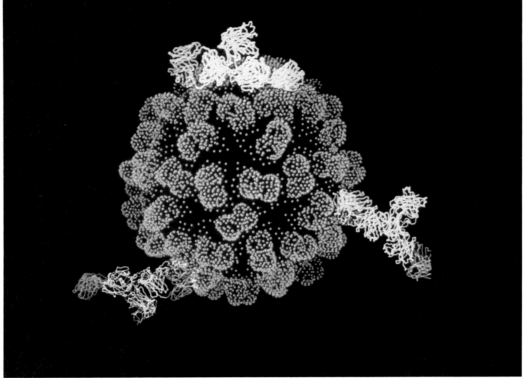

In this computer-graphics model, antibodies (yellow and orange) recognize a plant virus (blue) and bind to it.

in mice. In humans a cancer of the T-lymphocytes (white blood cells) is caused by the RNA virus HTLV-1. The Epstein-Barr virus has been implicated in Burkitt's lymphoma, the herpes simplex virus in cervical cancer, and the hepatitis B virus in liver cancer.

HOW VIRUSES SPREAD

Viruses pass from one host to another by direct and indirect contact. Among humans and animals, coughing, sneezing, and touching transmit viral diseases such as the common cold, influenza, and mumps. Viruses can penetrate cells simply by coming in contact with them. It is suspected that warts and fever blisters are transmitted in this way. Some viruses, such as the poliovirus, are excreted in feces. Others are carried by flies to food or drinking water. The rabies virus is transmitted through the bite of infected mammals. Contaminated inanimate objects, such as dishes and silverware, may also carry infectious viruses.

Some plant viruses are easily spread through contact. Simply by brushing against a diseased plant, a human being (or

a tool he is using) may become a carrier of viruses from a bit of sap from an infected plant. If the roots of different plants of the same species grow together, a bridge from a healthy to an infected plant may be formed.

Serious viral illnesses are often spread by insect *vectors,* or carriers, in the act of feeding. The vector itself is seldom affected. Among men and animals, the most common vectors are flies, ticks, and mosquitoes. Particularly deadly are mosquitoes which transmit yellow fever, dengue, various types of encephalitis, and Rift Valley fever. Plant viruses may also be carried by insects.

SIZE AND STRUCTURE

All viruses are extremely small. They are only a few millionths of a millimeter in diameter. The vaccinia virus is one of the largest. It can just be made out under a powerful optical microscope. The influenza virus is of about average size: 500 of its particles can rest on the point of a pin.

The shapes of viruses are as varied as their sizes. Viruses come as rods, threads, spheroids, polyhedrons (many-sided figures), and tadpolelike structures. Individual viruses may combine a number of shapes. For example, one type of virus has a polyhedral "head" attached to a rodlike "tail."

Protein, the chief structural material of a virus, determines its shape and structure. Hundreds or thousands of similar protein molecules together produce the structure typical of a given virus. The protein, or shell, protects the important nucleic acid inside from the environment. In some viruses, the enveloped viruses, the outer coat contains fat and carbohydrates as well as protein, as a kind of membrane.

CHEMISTRY OF VIRUSES

All reproducing systems in biology have one or both of two chemical types of a natural compound called *nucleic acid:* DNA (deoxyribonucleic acid) and RNA (ribonucleic acid). Cells generally have both DNA and RNA but a virus usually has either one or the other, not both, and is thus classified as a DNA virus or a RNA virus. The information for the structure and function of cells is carried from generation to generation by the exact duplication of the DNA. This DNA is located in a central, dense body, the nucleus. Thus a newly formed cell will look the same as, and work just like, the old one. In most cells, RNA acts as a messenger between DNA and protein-manufacturing sites (ribosomes) located outside the nucleus. A virus particle, however, is little more than a package of genetic material. It lacks the energy-containing chemicals, raw materials, and machinery for making things. It has only the blueprint for making more virus and must use the invaded cell for protein-synthesizing machinery. When a virus or its genetic material gets inside a cell, it causes the cell to produce messenger RNA, which, when carried to the cellular ribosomes, results in the production of viral protein.

The protein coat that most viruses have helps the virus to get into cells, which it must do in order to multiply. Once viral DNA or RNA has gotten inside a cell, more RNA or DNA must be made in order for more virus to be made. This requires the help of substances called *enzymes,* proteins that speed up reactions. Often, special enzymes coded for by the viral nucleic acid are required for production of more viral nucleic-acid strands. The larger viruses may code for several hundred proteins, while smaller viruses must rely on the host cell enzymes to manufacture protein and DNA.

The size, shape, and number of strands of DNA or RNA vary among different viruses. A simple virus may have only a single strand of DNA or RNA. In more complex viruses, the nucleic acid may be double-stranded or looped into a closed circle. In other cases, the nucleic acid may occur in several pieces, all of which are packed in one particle or, in some cases, in two to four particles.

MECHANISMS OF VIRAL INFECTION

Outside the host's living cell, a virus is inactive. It is a parasite that requires the chemical machinery of the living cell for its multiplication. How the virus takes over cellular machinery, uses chemicals and energy present, and affects the health of the cell are of great interest in *virology.*

There are certain viruses that attack bacteria only. They are called *bacteriophages*—a name that means "bacteria eaters." Because bacteriophages and their bacteria-hosts are easy to grow in test tubes, the bacteriophage, also called *phage,* has been useful in helping virologists study the method by which a virus invades its host and takes over the cell's machinery.

Some phages, such as one called *T4,* infect a species of colon bacteria, *Escherichia coli,* common in the human colon, or large intestine. The method by which this virus enters the bacterial cell and assumes control of its biochemical activities is swift and efficient. In shape, this phage resembles a little tadpole. First the phage attaches its tail to the surface of the bacterial cell. Then the phage, somewhat like a hypodermic syringe, injects its nucleic acid into the cell. Once the viral nucleic acid has entered the host cell, the virus takes over the cell's genetic machinery, forcing it to serve the virus's needs.

Under the direction of the virus, the cell begins to produce certain important enzymes, necessary for the synthesis of more virus material. One of them may destroy the host's DNA. Virus particles themselves have few or no enzymes. Instead of producing cell parts, the host cell, using the enzymes, begins to produce parts for new viruses: phage DNA, protein sheath, and tail parts. After about twenty to sixty minutes, new virus particles are completely assembled. Suddenly, the infected cell bursts and falls apart—a process termed *lysis*. The new phages are then freed and are ready to infect other healthy bacteria.

Sometimes the introduction of phage nucleic acid results in a guest-host relationship, rather than that of master-slave. Instead of destroying the bacterial DNA, viral DNA may be inserted into it. This process is termed *lysogenization*. A strand of DNA in the host bacterial cell is broken, and the DNA of the phage is then made part of the strand. This connected, or integrated, form of viral and bacterial DNA is not an active virus particle, or phage. In fact, the viral DNA reproduces only when the bacterial DNA reproduces. The resulting bacterial cells thus contain both DNA forms.

However, if this combined form of bacterial and viral nucleic acid is exposed to radiation, such as ultraviolet light, or is treated with certain chemicals, the virus may be reactivated. First the viral DNA is released from its bond with the bacterial DNA. Utilizing the cell machinery, the viral DNA begins to produce complete new phage particles. Eventually, the helpless bacterial cell bursts, just as in ordinary lysis, and the virus particles are set free. Thus a virus, which has remained inactive within an infected cell, suddenly becomes active and goes about its destructive work producing new phage particles.

During the release process another interesting event may occur. The virus may act as a genetic "delivery service." As the phage DNA is released from the bacterial DNA, it may take along some of the bacterial nucleic acid. When the new virus particles, carrying bacterial nucleic acid, infect another bacterial cell, they may introduce

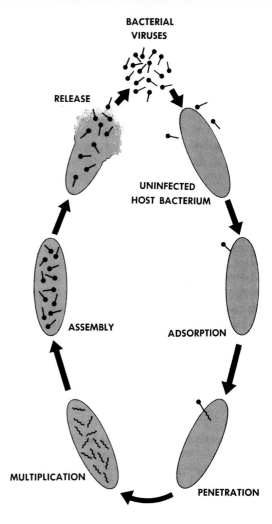

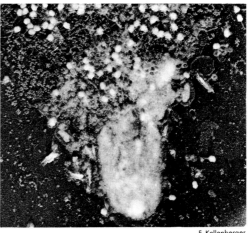

E. Kellenberger

Top: life cycle of a bacteriophage. A virus attaches itself to the uninfected bacterium. The nucleic acid in the viral head penetrates the bacterium. Parts of the virus are assembled within the bacterium. They multiply and assemble to form new bacteriophages. Finally the bacterium bursts, releasing the new phages to infect new hosts. The micrograph shows an infected bacterium bursting.

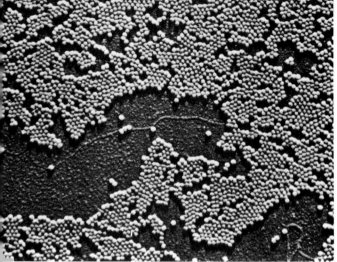
March of Dimes

The ball-shaped virus that causes polio is very tiny. It would take more than a million of these placed side by side to equal one inch.

some of the foreign genetic material of the first bacterial cell into the DNA of the new bacterial host, thereby changing its hereditary characteristics. Through the medium of the virus, the second bacterial cell has acquired traits of the first cell. This transfer by viruses of bacterial traits from one bacterium to another is called *transduction*. Biologists make use of the phenomenon of transduction to investigate heredity.

Other types of viral infection resemble phage attacks on bacteria, but there are important differences. Animal and plant viruses, for example, lack tails and do not inject their nucleic acid into cells. Instead, most animal and plant viruses invade cells as entire particles, usually after attaching specifically by their coat proteins to certain sites on the cell membrane. Some viruses, such as poliovirus, can penetrate cells directly after attachment. Commonly, however, the cell itself "swallows," or engulfs, virus particles, much as a white blood cell engulfs bacteria, or as an amoeba closes around bits of food. Once inside a cell, the virus loses its protein coat and, under the direction of its nucleic acid, gets down to the business of reproducing itself.

The viral genes insert themselves into the genetic material of the host cell. They then use the machinery of the cell to manufacture proteins that may affect the growth of the cell, sometimes resulting in neoplastic transformation of the cell, or cancer. The host cell also serves as the site

of production of more viruses, which bud off from the host cell to infect new cells.

RNA viruses operate in a different manner. They must first synthesize DNA before they can integrate into the host cell's genetic material. The RNA viruses do this by means of an enzyme called reverse transcriptase. Then the newly made DNA becomes part of the host cell chromosomes and is called the provirus. At this point the viral genes begin protein synthesis. A group of RNA viruses called *retroviruses* includes the agents responsible for certain tumors, leukemia, and AIDS.

CREATING NEW VIRUSES

Additional experiments with viruses are shedding more light on the chemical activity of genes. For example, when bacteria and other organisms are infected with a mixture of two different viruses, new types of viruses often appear, in addition to the original types. Each new virus recombines certain separate traits, and hence portions of nucleic acid, of the original viruses in the mixture. As a result, new viral types are formed, often with greater infective power. An animal or a plant may be immune to attack by one strain of a virus, but be vulnerable to a mutant form of the same virus.

Because of these experiments, scientists have been able to map the location of genes in certain viruses. They have also been able to study the effect of enzymes on the behavior of nucleic acids.

Genetics experiments have been conducted with viruses in test tubes as well as in living cells. This great breakthrough has simplified some of the problems of viral research by eliminating the complex environment of the cell. In such experiments, virologists have made an artificial RNA of a virus that normally infects a bacteria, as well as a synthetic phage DNA.

Studies of lysogenizing bacterial viruses—those that join their DNA to that of the host cell—have explained another important genetic mechanism observed in all forms of life. Once the viral and bacterial DNAs combine, many viral genes are inactive: the virus does not synthesize new phage particles.

Biologists have found out why. A viral gene may be inhibited or held back by an *inhibitor,* or *repressor,* substance. The inhibitor is a protein that combines with viral DNA at a certain site and keeps it from expressing itself as long as the inhibitor remains attached to the DNA or its host. The inhibitor itself is controlled by a gene in the combined viral and bacterial DNA.

Viruses are now being used by researchers as packages for introducing new genetic information into cells—one of the techniques of recombinant DNA research. These viruses are called vectors. In the case of one circular DNA virus commonly used as a vector—simian virus 40—a region of the virus DNA is replaced with new DNA—a cloned gene—that is then transfected into the target cell. If the transfer is successful, the new gene begins to function in the host cell. Genes that have been inserted in this way include those for growth hormone, β-globin, and insulin. RNA viruses are particularly good vectors because of the wide variety of cells that they infect.

CONTROL OF VIRAL DISEASES

As the result of recent important advances in the study of viruses, scientists have begun to develop new drugs and vaccines for curing and preventing viral diseases. These substances can stop the growth and reproduction of a virus at almost any stage of its life cycle.

The drugs that can cure or arrest some viral infections are either synthetic chemicals or natural substances derived from microorganisms other than viruses. Some of the most promising compose a group of natural proteins called *interferons.* Secreted by most cells in vertebrates, including man, interferons are so named because they interfere with viral development. One interferon can inactivate many types of viruses.

VACCINES AGAINST VIRUSES

Vaccines work just the opposite of interferons. Vaccines, which may be swallowed or injected, prevent viral diseases before they occur. A given vaccine is usually effective against one type of virus. But the immunity granted to a human being or an animal by one or a few injections of vaccine may last a lifetime.

There are three types of antiviral vaccines: (1) *Killed virus vaccine,* such as the Salk polio vaccine, is inactivated chemically. Once in the body, the dead virus stimulates the production of antibodies, proteins in the blood that will then guard against invasions by the same type of virus. (2) *Live virus vaccine,* such as the Sabin polio vaccine, consists of virus that is changed genetically so that it is harmless. The surface protein of the virus remains unchanged. Thus, when the vaccine is injected or scratched into the skin, it causes a mild infection and the production of protective antibodies. (3) *Subunit vaccines* are created by recombinant DNA techniques. Selected pieces of the virus, especially the protein coat, are used to construct the vaccine. Vaccines for hepatitis B and childhood meningitis are made by this method. These vaccines often produce only a weak immunological response, and large amounts are needed for vaccination.

A vaccine against influenza has been developed by recombinant DNA tech-

Herpes simplex virus, magnified 100,000 times. Viral particles have been crystallized inside the cell. Each particle has a spherical core and a surrounding membrane.

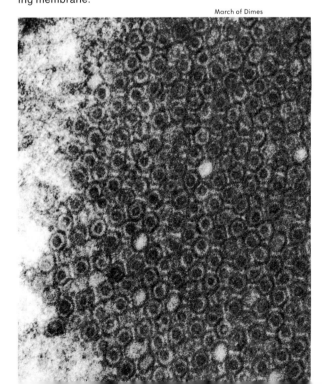

niques. It is constructed of two surface protein genes from the flu virus and six harmless genes from the laboratory version of the virus. Administered as nose drops, this promising vaccine is now being tested.

Despite their faults, vaccines remain powerful weapons in medicine's war against viruses. They combat such human viral diseases as yellow fever, rabies, influenza, polio, smallpox, measles, mumps, and rubella. Vaccination also controls a number of viral infections in domestic animals.

Vaccination, however, cannot be used to curb viral diseases in plants, which do not produce antibodies. One important way to prevent viruses from injuring crops is to breed plants that are resistant to viruses.

VIRUSES AND AIDS

One particularly devastating disease, AIDS, or Acquired Immune Deficiency Syndrome, is caused by infection with the human immunodeficiency virus (HIV). The disease was first identified in 1981, and the virus was identified in 1983. AIDS destroys the body's immune defenses, and victims usually contract rare forms of pneumonia and cancers such as Kaposi's sarcoma.

The disease is transmitted in body fluids and is particularly prevalent in homosexuals and intravenous drug users, although the rate of infection among heterosexuals, especially women, has been steadily on the increase. In the United States, it was estimated in 1991 that more than 1.5 million people were infected with the virus. By the end of 1992, experts estimate that 360,000 more Americans will have developed AIDS and of those, 260,000 will have died. Programs have been instituted to test blood intended for transfusion for the AIDS antibody, and researchers are working to produce a vaccine. However, because of mutations in the coat protein of the virus, progress has been slow. Experimental drugs are being produced in the hope of killing the virus once it has infected a victim.

VIRUSES AND CANCER

The first tumor-causing virus was discovered in 1910 by Peyton Rous in a Plymouth Rock hen from an upstate New York farm. He found that the tumors caused by this virus, Rous sarcoma virus, could be transplanted from one chicken to another and also that extracts of the transplanted tumor could induce new tumors when injected into uninfected chickens.

The direct causation of human cancers by viruses has been difficult to establish. The first established cancer-causing virus in humans was HTLV-1, which causes cancer of the T-lymphocytes (white blood cells).

There are four types of tumor viruses: (1) papovaviruses, which are DNA viruses such as simian virus 40; (2) retroviruses, a group of RNA viruses; (3) adenoviruses (DNA); and (4) herpesviruses (DNA). These viruses can affect the growth of cells in two ways. In the first method, called insertional mutagenesis, the viral DNA is integrated into the host genetic material. The activity of the viral DNA then results in transformation of the cells. Transformation is the process by which the structure of the cell and the factors that control its growth are changed. The cells begin to divide uncontrollably, resulting in tumor growth.

The second method involves oncogenes, or cancer genes. These consist of genetic material carried by viruses that when inserted into the host cell are directly responsible for neoplastic transformation. More than 20 of these cancer genes have been identified in viruses so far. These oncogenes seem to be derived from normal, cellular genes, or proto-oncogenes. The proto-oncogenes are also believed to have tumor-producing abilities and at least 20 of them have been identified.

Oncogenes, once activated, seem to produce proteins that interact with growth factors, hormones that prompt a cell to divide. Once the normal regulation of cell growth is disturbed, the result is uncontrolled growth, or a tumor. Active oncogenes have already been isolated from lung and bladder tumor cells and from leukemic cells. Oncogenes have been activated experimentally by carcinogenic chemicals and by chromosome fusion.

HOW TO DO
AN EXPERIMENT

by Philip Goldstein

Doing an experiment is an exciting experience. True, it takes up lots of time that might be spent in other activities. It requires careful planning and hard work to carry it out. But it is worth the time and effort. The purpose of this article is to give you a little guidance in conducting an experiment. If you read this article carefully and follow the advice it gives, you may save yourself a little sweat and a good many tears when you set out to do an experiment.

Curiosity often leads to the discovery and study of problems. Why does the stick appear bent when pushed under the water?

An experiment is *a planned effort to answer a question by making something happen under controlled conditions.* Usually a single experiment answers only a small, simple question that in itself may seem insignificant. But an accumulation of many experiments, each testing a little question, may ultimately give rise to something important. "Each adds a little to our knowledge of nature," said Aristotle, "and from all the facts arises grandeur."

How does an experiment begin? It begins in your mind. It begins when you ask yourself a question or become curious about something. Curiosity and a desire to know lie at the bottom of all experiments.

HOW ARE PROBLEMS RECOGNIZED?

Every experiment begins with a problem. Everyone encounters puzzling situations in the course of everyday life. Few of us, however, recognize that a problem exists. Even fewer of us are willing to do anything about solving the problem even if we do recognize its existence.

If you want to do an experiment, you must give your curiosity free sway. Ask yourself questions about what you read, what you see, what you learn in school. Ask yourself why things happen, how they happen, and under what circumstances they will happen again.

Sir Ernest Rutherford, the Nobel Prize-winner in chemistry, often told how, when he was a youngster, he pushed a stick into a pond. The stick appeared to be bent. When he pulled it out it was straight again. He pushed it back, it was bent. He pulled it out, it was straight. "This sort of thing makes you think," he commented. This story illustrates how curiosity leads to the discovery of scientific problems.

Don't worry if the problem seems to be a little one. Even an unimportant question can give you a start. If you set up an experiment to solve a little question, you are almost sure to be led into bigger and more important ones. It is a well-known truth in science that every experiment raises more questions than it answers. Your little experiment most likely will not be an exception to the rule. Let us consider a case to see how this works.

In biology you learned that a geranium leaf has tiny pores called stomata. Your teacher explained that these pores allow the leaf to exchange gases with the outside atmosphere. Immediately questions begin to arise. What gases are exchanged through these pores? How do the pores work? Are they all the same size? What happens to a geranium leaf if its pores are clogged?

The last question interests you, and you decide to set up an experiment to find the answer. You realize that you must first find out where the stomata are located. Are they equally distributed over the whole leaf surface? Are there equal numbers of stomata on the upper and lower surfaces?

As you begin to work on this question, you think of other interesting questions. Both sides of the geranium leaf are exposed to the air. In a water lily, however, only one side of the leaf is exposed. Does this in any way affect the distribution of stomata? What arrangement of stomata is found in a plant such as elodea, which grows under water? And if a plant grows partly below and partly above the surface of the water, is the arrangement of stomata the same in both parts?

This line of questioning can lead you toward some exciting problems. Can environmental conditions affect the number and arrangement of stomata in a leaf? In other words, do plants adapt to changing environmental conditions by producing new stomata? Or by losing existing stomata? Or by rearranging them? Is there a seasonal change in the number and arrangement of stomata in a given plant?

This is only the beginning. The questions spread to cover wider and wider areas, like the circles in a pond when a pebble is dropped into the water. How do stomata open and close? Is there a daily cycle? A seasonal cycle? Does light affect their behavior? Sunlight? Moonlight?

SELECTING A PROBLEM

If you ask questions about the things you see and hear, your mind will surely be bombarded with potential problems. But which problems should you select for your very own? Perhaps the following guidelines will help you make a decision.

1. Does the subject interest you? Pick a question that really excites you—one to which you are eager to learn the answer. Don't let anyone talk you into choosing a topic that bores you.

2. Is the proposed experiment suitable to your present level of maturity and knowledge? It is far better to select a relatively simple project that you can bring to a successful conclusion than to begin a complex experiment too difficult to complete.

3. Do you have available the equipment, apparatus, and laboratory space that your investigation requires? Obviously a ninth-grade student who is interested in astronomy could hardly hope to investigate the radio signals reaching earth from quasars lying far out in the cosmos. Such a student does not have access to a radio telescope with which to detect them.

4. Do you have the technical skills that the project requires? For example, you may be interested in synthesizing DNA from its component parts. But do you have the chemical skills required to carry out such a complex chemical operation?

5. Can the experiment be completed in the time available to you? Remember, this experiment is not going to be your life's work. You probably have a rather limited amount of time during which you must conduct your investigation. Therefore choose something that can be completed in a reasonable length of time.

National Teaching Aids, Inc.

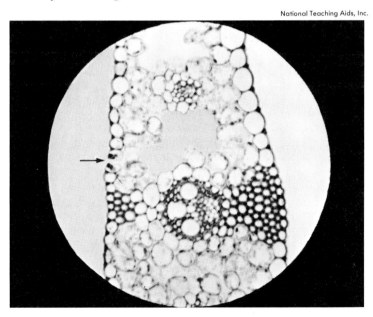

This is a thin slice across a green leaf. The arrow points to a small opening in the underside of the leaf. This opening is called a stoma. What is its purpose? Do all leaves have stomata? How can you learn the answers to these questions?

6. Does the proposed experiment entail any danger to you? If it does, forget it. You should not consider an experiment dealing with pathogenic bacteria, explosives, dangerous drugs, or anything else that may do you injury.

7. Will your experiment cause harm or pain to the animal subjects you use? You are not an accomplished surgeon who can operate on an anesthetized animal under humane conditions. Thus you should avoid experiments that are apt to injure an animal or cause it pain. There are many possible experiments with small invertebrate animals or with plants where these factors do not come into play.

STATING THE PROBLEM

When you first become interested in an area of research, you have a generalized idea that must be narrowed down and brought into sharp focus. Precisely what do you want to find out? You should begin by formulating a clear, concise statement of the specific problem that you wish to investigate. Let us consider an imaginary situation to see just how this works.

During a class discussion on adaptation, you learn that organisms are constantly adjusting to their environment. Various questions flash through your mind. Are organisms really aware of small changes in their surroundings? Can they detect a gradual change? Do they do anything when such a change occurs?

You decide to plan an experiment to test whether organisms respond to a changing environment. This is a fine idea for research, but obviously it is too broad. You would be wise to narrow your efforts to the effect of one environmental factor on a single species. You decide to use guppies because you are familiar with them, and you have many of them in your aquarium. But which environmental factor should you study?

At this point you may recall a recent television program. The commentator said that the increase in water temperature around a nuclear electric plant was killing fish in that area. He called the increase in water temperature "thermal pollution." You decide to study the effect of water temperature on guppies. More specifically, you want to see if guppies will move to that part of the water where the temperature is best for them. How do you state the problem?

How does "A study of adjustment to the environment" strike you? Or "How fish respond to changes in their environment"? Would either of these make it clear to any-

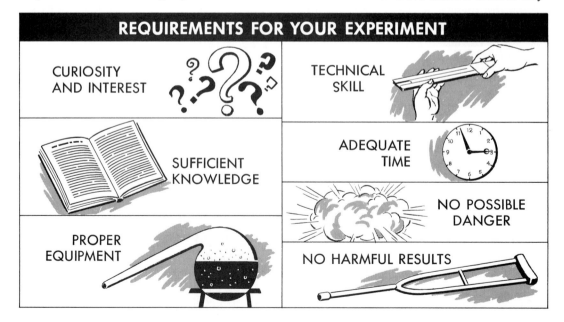

REQUIREMENTS FOR YOUR EXPERIMENT

CURIOSITY AND INTEREST

TECHNICAL SKILL

SUFFICIENT KNOWLEDGE

ADEQUATE TIME

NO POSSIBLE DANGER

PROPER EQUIPMENT

NO HARMFUL RESULTS

one what you are trying to find out? Wouldn't it be better to include the name of the organism and the condition being tested? For example, "Thermal pollution and guppies." This is better, but not yet good enough. At last you come up with the following: "How do guppies react to a gradual change in water temperature?" This is a fine statement of the problem. It tells clearly that you intend to subject guppies to a gradual change in water temperature in order to see how they react. You are now ready to proceed to the next step.

LAYING THE GROUNDWORK

Before you begin the actual experiment, it is wise to take certain preliminary steps to lay the groundwork for your research.

1. Collect all the equipment and materials you will need. Answer questions like the following: If the school cannot supply a certain chemical, can I get it from another source? Can I afford to buy it? Does the school own such and such a piece of equipment? If not, can I build it?

2. Practice any new techniques that your experiment will require. The time to develop skills is before the actual experiment begins, not after it is under way.

3. Find out what other researchers have done in this and related fields. Long ago, Francis Bacon, one of the earliest experimenters, commented, "When a person prepares himself for discovery, he first obtains a full account of all that has been said on the subject by others." Bacon's advice is as good today as it was in 1620.

Reading about the work of others provides an insight into the best line of procedure. It gives you the opportunity to take advantage of earlier successes and failures. It tells what pitfalls may lie ahead, and what errors are to be avoided. It gives information on special apparatus and techniques that may help you solve your problem.

Where do you find such information? The American inventor Thomas Edison once said, "When I want to discover something, I begin by reading everything that has been done along that line in the past. That's what all those books in the library are for."

Before beginning an experiment, read about the work of other researchers who have studied similar problems.

This is sound advice. The library has many sources of background information: college textbooks, magazine articles, original articles by research scientists, review articles, publications of government agencies, and so on. In your search for such materials, enlist the help of the librarian, who can direct you to many aids that help locate materials bearing on your project.

Your school library is almost certain to have *The Readers' Guide to Periodical Literature.* This is an index, arranged like a dictionary, that lists the articles appearing in over one hundred magazines. Other, more technical, publications index special science articles. There are also periodicals like *Biological Abstracts* and *Chemical Abstracts,* which give brief summaries of important articles from the current science literature. These specialized journals are not likely to be in your school library, but they can be found in the libraries of almost all colleges and universities. (Most such institutions allow high-school students to use their library facilities.).

GIBBERELLINS — discovery

Gibberellins occur in several forms, of which gibberellic acid is one of the most common.

Discovered in 1926 by Japanese botanists who noticed that rice seedlings attacked by the fungus *Gibberella fujikuroi* grew 2 to 3 times taller than non-infected seedlings.

Cronquist, Arnold, INTRODUCTORY BOTANY, Harper + Brothers, New York, 1961, pages 657-658

R. E. Hudson

One method of recording what you learn from your reading is by placing the information on index cards. Summarize the material clearly and precisely. Label the cards and arrange them in an orderly fashion so that you can easily refer to the information.

4. Record what you learn from your reading. Prepare yourself with a large notebook or a batch of file cards, and make your records as complete as possible.

If you find something in a book, summarize it in your notebook or on an index card. Be sure also to record the full name of the author, the complete title of the book, the date of the publication, the name and address of the publisher, and the pages on which the information is found.

If your information comes from a magazine, note the full name of the author, the complete title of the article, the name of the magazine, the date of the issue, the volume number, and the pages on which the article appears. One again, you should make a good summary.

5. Finally, prepare a preliminary plan of your experiment. Before you begin the actual experiment, you must have a clear picture of what you are going to do and how you are going to do it. Let us now see what goes into the planning of an experiment.

PLANNING THE EXPERIMENT

Whenever you are faced with a problem, it is logical to plan a line of attack. You should be certain that you have a clear picture of precisely what it is you are attacking—exactly what question you hope to answer.

If you are wise, you will think about the question for a while. You will turn it over in your mind and ask yourself what the answer might be. You should propose an answer that seems to make sense. Maybe you will even think of several possible answers. You should weigh them against each other, and form an opinion as to which is likely to be the most probable answer.

Being asked to guess at the answer may strike you as an unscientific procedure. But every research scientist makes such guesses. A guess of this type is called a *hypothesis*. It is an educated guess based on knowledge, background, and experience. The task of the experimenter is to plan a test that will tell him whether his guess is reasonable and correct, or incorrect and unacceptable. Such a test is called a *planned experiment*.

A planned experiment is an attempt *to make something happen under controlled conditions where all variables except one have been eliminated or neutralized.* You hope that whatever happens in the experiment can be attributed to the single, remaining variable. Let's consider an imaginary case to see how this works.

You have read that growth substances called "gibberellins" cause certain plants to grow to two or three times their normal height. You decide that it would be interesting to try these substances on bean plants. You frame a clear question: "How do gibberellins affect growth in bean plants?" You realize that there are several possibilities. The gibberellins may make the plants grow faster and taller. They may have little or no effect. They may inhibit growth, so that the bean plants are dwarfed. They may kill the plants. You conclude that the third and fourth hypotheses can be eliminated because they do not seem logical in view of what you have read about gibberellins. This leaves two possibilities for testing.

How do you set up the test? How do you make bean plants grow under controlled conditions that will provide the answer to your question? You decide on the following plan:

Place a bean seed in a pot of good soil in a location favorable for growth. Moisten the soil daily, but instead of using plain water, use a solution containing five parts per million (5 ppm) of gibberellins. As the seed sprouts, make daily observations and accurate measurements. Count the number of leaves and measure their length and width. Measure the height of the flower stalk and the size of the flower when the plant blooms. In addition to recording these observations accurately in a notebook, make a series of dated photographs showing the plant in various stages of its development.

This sounds like a fine plan, doesn't it? You have made the bean grow under care-fully controlled conditions, and you have made accurate observations. Therefore you should get an answer to the question. But no, this experiment cannot provide a definite answer. In certain respects, it is poorly designed. What is missing is a basis for comparison.

Without another bean plant for comparison, how can you be sure what caused the experimental plant to grow the way it did? Was it something in the soil? Some peculiarity of location? The gibberellins you added? Or was it simply the normal growth pattern of your particular bean seed?

NEED FOR A CONTROL

Your plan would be improved if you added another part for comparison. This second part is called the *control* or *check*. The object is to make the experimental part and the control part of your experiment *identical in every respect except one*. The one difference is the variable you are testing.

This, then, is your new and improved experimental design: use two identical bean seeds planted in identical pots containing equal portions of the same soil. Put

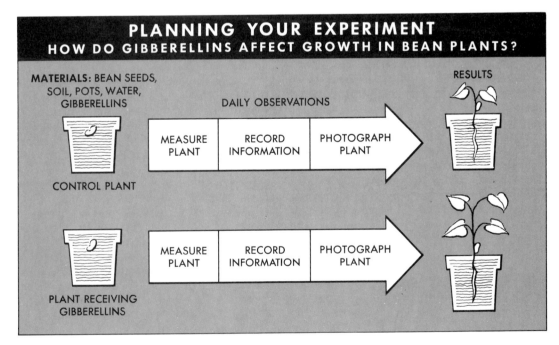

PLANNING YOUR EXPERIMENT
HOW DO GIBBERELLINS AFFECT GROWTH IN BEAN PLANTS?

MATERIALS: BEAN SEEDS, SOIL, POTS, WATER, GIBBERELLINS

RESULTS

DAILY OBSERVATIONS

MEASURE PLANT — RECORD INFORMATION — PHOTOGRAPH PLANT

CONTROL PLANT

MEASURE PLANT — RECORD INFORMATION — PHOTOGRAPH PLANT

PLANT RECEIVING GIBBERELLINS

Before beginning an experiment, you should learn the scientific skills needed for your project. The microscope is an important tool in many experiments.

both pots in the same desirable location. Label one "CONTROL" and the other "GIBBERELLINS." Moisten the control pot daily with plain water. Moisten the gibberellin pot daily with a solution made from the same water, but containing 5 ppm of gibberellins. Be sure to give the same amount of water to each plant.

Again make daily measurements of growth and record them accurately in your notebook. Make side-by-side photographs of the two bean plants, so that any differences in growth pattern are clearly visible.

Now you surely have a good plan. The beans, the soil, the pots, the location, the water are all identical. There is only one variable: one pot was treated with gibberellins, the other was not. Surely any difference in growth must result from this one variable.

A LARGE SAMPLE

But there is still another factor to consider: the question of sampling. Your experiment is not testing all beans, but only two. How do you know that the two you happened to choose are representative? How do you know that they will react like normal, average bean plants should?

The only way to be reasonably certain that your results are representative is to test as many beans as possible. You are in a better position to determine the behavior of the average bean plant if each half of your experiment has 10 plants or 50 or 100 or 1,000. The more times the experiment is repeated and the more subjects you have in your sample, the more reliable your conclusion will be.

Thus a well-designed experiment has two parts — experimental and control — with a single difference between them. And there must be many subjects in each part. This describes a planned experiment in its simplest form. But there are many other, complicating factors in the design of an experiment. Let us go one step further.

Suppose you want to know what concentration of gibberellins causes the greatest increase in the growth of beans. You decide to test three concentrations: 5 ppm, 25 ppm, and 50 ppm. Must you set up three parallel experiments, each having its own control? In each case the control would be a group of plants moistened with plain water. One control can therefore serve all three experimental parts. Thus the experimental design has four parts: three are experimental (5, 25, and 50 ppm gibberellins), and the fourth is the control (plain water).

It is impossible to cover all the complications of experimental design. The basic points that have been examined should be sufficient to give you a start. However, there is one more bit of advice. Before you begin work on the actual experiment, prepare a written plan. Such a plan should describe the experiment as you have worked it out in your mind. It should include one or more educated guesses as to the answer to your question, and proposals for testing them. It should detail the information you will need to determine whether your hypothesis is correct or incorrect, and how this information will be collected and analyzed. Once this plan of action is complete, you are ready to begin the actual experiment and collect the data.

COLLECTING THE DATA

When you undertake an experiment, it is necessary to maintain complete and accurate records of all your observations. One way to do this is to keep a dated record of everything you do and everything that happens. Since you can never tell what may turn out to be important, it is wise to record every observation, no matter how trivial it may seem.

Observation is simply the use of the senses to discover what is happening. You see things, hear them, taste them, smell them, feel them, sense them in a dozen different ways. Making observations is part of life itself. But observation becomes doubly important during an experiment: it provides the evidence on which you will decide whether or not your hypothesis is correct.

Although man has a variety of amazing sense organs, their power is comparatively limited. The environment is filled with things that our unaided sense organs cannot detect. For example, we cannot see the millions of tiny bacteria and viruses that exist all about us. Our universe is crowded with heavenly bodies too distant to see with the unaided eye. The atmosphere is filled with radio waves, but the human body has no way of detecting them.

But microscopes reveal the tiny bacteria and viruses. Telescopes bring the distant galaxies into view. Radio receivers change the radio waves into audible sound waves. These are but a few of the remarkable devices that man has invented to increase and extend the powers of his sense organs.

In addition, man has developed measuring devices with which he can substitute objective numbers for subjective observations. The human senses can be fooled into giving inaccurate and distorted impressions. Measurement can make observations accurate and objective. Since accurate observation is an absolute must in science, measuring devices are basic tools of the researcher.

The simplest measuring devices include rulers to measure distance, scales to measure weight, thermometers to measure temperature, manometers to measure pressure, voltmeters and ammeters to measure electricity, and so on. These tools have been refined in a hundred ways to make the measurements as accurate as possible.

AVOIDING PITFALLS

Observation can be a very tricky business during the conduct of an experiment. There are many traps and pitfalls to be avoided. One of these is illustrated by the gibberellin experiment. The experimenter planned to see if treated plants grew larger than untreated plants. But what is meant by larger? Is a larger plant one that has more or larger leaves? One that is taller or thicker around the base? Is it one that weighs more? One with a taller flower stalk and a

One of the most important phases of your experiment is keeping accurate records. These must be made at the time you make your observations. Drawings, charts, and photographs are also important. Keep your records in a notebook or binder.

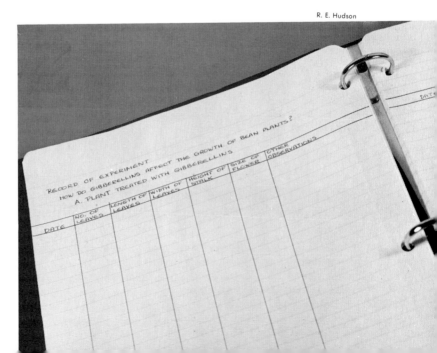

M. J. Bukovac, Michigan State University

These photographs show a time sequence of control and gibberellin-treated bean plants. (A) Control plant after 48 hours. (B) Gibberellin-treated plant, 48 hours after it was treated. (C) Control plant after 96 hours. (D) Gibberellin-treated plant after 96 hours. How has gibberellin affected the plant?

larger flower? There are many different aspects to being "larger."

It is not enough therefore to note that one plant is larger than another. You must tell in what way it is larger, how much it is larger, and how you determined the degree of largeness. Between which two points did you measure? From the base of the stem to the top of the leaf? From the beginning of the flower stalk to the top of the flower? Obviously the comparison of two measurements is meaningless if a different basis was used for each.

Selection of the proper measuring instruments also requires good judgment. Would you use a meter stick to measure a salt crystal? An analytical balance to weigh out 100 grams of a chemical? Obviously these would be foolish procedures. A meter stick is much too coarse a measuring device for a tiny salt crystal. And the analytical balance is intended for weighing hundredths of a gram rather than a hundred grams.

Even when you use an appropriate measuring device, how accurate can you be? To how many decimal places can you measure? As a general rule, it is unwise to estimate more than one place beyond what the device measures accurately. Let us consider an example:

When you measure the length of a line with a ruler marked off in centimeters, you find it to be 9 centimeters long with a little left over. You estimate the extra piece to be just over half a centimeter, so that you write the observation as 9.6 centimeters. You can hardly say it is 9.58 centimeters or 9.62 centimeters. Your powers of estimation are not that good. But if you used a ruler marked off in millimeters, you might find that the line measures 96 millimeters plus a little extra that you estimate as two tenths of a millimeter. Now you can say the line is 96.2 millimeters (9.62 centimeters) long.

BLIND AND DOUBLE-BLIND

There are also certain psychological factors that can distort your powers of accurate observation. It is easy to see only those things that favor your hypothesis and to overlook things that are opposed. Any person who consciously does this is dishonest. But often it can be a completely unconscious act on the part of the experimenter. Without being aware that he is doing so, he can favor observations that he hopes the experiment will produce. This is well illustrated by a research scientist studying the problem of lung blackening in coal miners. He said:

"One of our problems is the need to take elaborate precautions to prevent those examining the lungs from knowing where they came from. I don't care how dedicated a pathologist is, if he knows the history of the tissue he is examining, his findings are going to be influenced by that history. In our study those looking at the tissues know them only by a number. About 50 per cent of our subjects are miners. The other 50 per cent serve as controls."

This approach is known as doing an experiment "blind." Usually the term *blind experiment* describes a situation where the subjects are not aware whether they are in the experimental or control group.

Suppose a doctor is testing whether drug X causes blood pressure to rise. He divides his subjects into two equivalent groups. The subjects in the experimental group get a pill containing drug X. The controls get an identical pill without drug X. (In medical terminology this is called a "placebo.") To his great pleasure, the doctor finds that, on the average, blood pressure rose in the subjects getting the drug, but not in the controls. He concludes that the drug is effective.

However, this conclusion is not really valid. Blood pressure is affected by many psychological factors. Is it possible that the subjects reacted psychologically when they knew that they were getting drug X? Did their pressures rise merely because they knew that this drug was supposed to raise pressure, and not because of any action of the drug?

Experimenters try to avoid this problem by doing the test blind. The subjects are not divided into two groups. Each subject gets an identical-looking pill with a code number. Half the pills contain drug X, the other half do not. The subjects do not know what they are getting. Only the experimenter knows.

Even this is not the final word. In this experiment the subjects were "blind," but the experimenter was not. He knew which subjects were receiving drug X and which were not. Perhaps he can still influence the outcome unconsciously by the way he looks, what he says, or what he does when he gives the subject the pill. This is avoided if neither the subject nor the experimenter knows which sample is which. The coded samples are prepared for the experimenter by someone else. The outcome of such a "double-blind" experiment cannot be influenced psychologically by either the subject or the experimenter.

ARRANGING THE DATA

Let us return to the collection of data. In the planning stage of your experiment, you decided what observations and measurements you would require. You should also decide what method you will use to record the data. Choose a practical arrangement that is both convenient and understandable.

One practical device is a table or a chart. A table summarizes on a single page all the data that you collected during your experiment. It is much easier to see the figures all at once if they are organized in a table than if they are scattered over the

Careful observation is important in all scientific work. These are the chromosomes of a human. How many can you count in the photograph?

Dr. J. H. Tjio

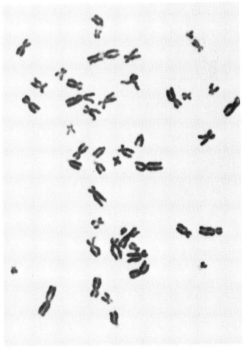

many pages of your record book. By looking at the table, you can get a good idea of where your experiment is heading. But you can only do this if you decide in advance what arrangement of columns will make the table most useful. When your experiment is complete, and you have collected all the data, you are ready to begin analyzing the results.

ANALYZING THE DATA

The data that you originally collected and recorded in your experiment are often referred to as the *raw data*. This means that they are the original measurements without treatment or processing of any kind. Your next task is to prepare the raw data in such a way that they tell a meaningful story.

Sometimes a graph can be used to summarize the data. The most common types of graphs are bar graphs, pictographs, line graphs, and circle graphs. Each type has its own advantages and disadvantages.

At other times it is necessary to analyze the data by applying a form of mathematics called statistics. Statistics has been called the art of drawing inferences from the raw data of observation. An inference is something derived by logic and reasoning from a collection of evidence. In your experiment, your observations and measurements are the evidence. What you decide the evidence is telling you is your inference.

Statistics is a very useful tool for examining evidence. It can make a diffuse mass of data become a good deal more meaningful. While this brief article cannot possibly cover the science of statistics, there are certain fundamental ideas that you really should understand. Four of these statistical concepts are explained here.

1. If you measured or weighed a large number of specimens, you might need a figure that is representative of the whole group. For this you use a *measure of central tendency*. There are three to choose from: the mean, the median, and the mode.

The *mean* is simply the arithmetic average. To determine the mean, you add the separate measurements and divide by the number of measurements. The *median* is the middle measurement of your series. To

find it, you list all the measurements in order from the highest to the lowest and locate the one in the middle. There are always as many cases above the median as there are below. The *mode* is the measurement that occurs most frequently. It indicates that there are more cases at that level than at any other level.

2. If you need to know how variable or consistent your measurements are, you use a *measure of variability*. Such a measure tells you how far away from a central point your data may spread. There are three such measures: the range, the average deviation, and the standard deviation.

The *range* is the spread from the highest measurement to the lowest. This tells you the upper and lower limits of your data. The *average deviation* is the average of the individual deviation from the mean. To determine the average deviation, you must find the difference, or deviation, between each individual measurement and the mean. Then you add the sum of the deviations and divide by the number of measurements. The mean, plus and minus the average deviation, always includes 50 per cent of the measurements in the series. Thus if the mean is 9 and the average deviation is 2.5, then 50 per cent of all the cases lie between 6.5 and 11.5. The *standard deviation* is a similar figure obtained by a slightly different calculation. The mean, plus and minus the standard deviation, includes about $2/3$ of the measurements instead of $1/2$.

3. To find whether there is a relationship between two series of events you use some measure of *correlation*. Two things are said to be correlated if they occur together. This does not mean, however, that one of these things is the cause of the other.

Correlation can be positive, negative, or zero. A positive correlation exists when a subject that is high on one series of measurements also tends to be high on a second series. A negative correlation exists when a subject that scores high on one series of measurements tends to score low on the second. Zero correlation is represented by the situation where there is no recognizable relationship between two events or two series of measurements.

© Bill Hubbell

Young scientists try to find out if absorbent cotton makes a good heat insulator.

4. An experimenter is always interested in knowing whether his or her results are accurate, and whether they would come out the same way again. For this he or she uses a *statistical measure of reliability*, such as the probable error. This is really a way of applying the "laws of chance" to the results. In simple words, it says, "the chances that a similar result will come out again are 100 to 1, or 10 to 1, or 1 to 1, or some such ratio."

Statistical reliability depends on two factors: the number of cases in your experiment and the variability of the sample. An experiment that includes a large number of cases is more apt to be reliable because it is more likely to be a true sampling of the total population. If the sample you use is variable, it may give unreliable results even though you use a large number of subjects.

FORMING A CONCLUSION

The conclusion is a final answer to your question. Early in the planning process, you proposed a hypothesis as a possible answer to the question. Now you must see whether the evidence produced by your experiment does or does not support this hypothesis.

Don't be upset if your hypothesis is unsupported. You will not be the first to face this situation. The nineteenth-century English scientist Thomas Huxley made the point when he said, "The tragedy of all inquiry is that a beautiful hypothesis may be slain by ugly facts."

Your conclusion should be stated simply and clearly. It should be a natural outgrowth of your observations and the analysis of the data. Don't let your conclusion be tinted by your opinions or by what you had hoped to find. However, in discussing the conclusion, there is nothing wrong with offering an opinion as long as you label it such. It is often necessary to state that you are offering a tentative conclusion subject to further checking.

Your experiment was set up under

prescise and controlled conditions, and your conclusion is valid only within the limits of these conditions. Be careful not to state a conclusion not supported by the data. Mention if your conclusion is limited by factors you did not or could not control.

Accept the facts you discovered, even if they disagree with what you read. State your conclusion as the experiment indicates — not as authorities say it should be. Here is an actual case where an important discovery grew out of observations that ran counter to authority.

UNEXPECTED FINDINGS

Until 1956 every biologist accepted 48 as the number of chromosomes in a human cell. This figure appeared in every biology textbook and in every medical book. In that year Joe Hin Tjio and Albert Levan claimed that the number should be 46. Their claim was based on chromosome preparations made with new and improved techniques. When the claim was checked and varified by other biologists, the figure of 46 was accepted, and Tjio and Levan were credited with the discovery.

Several years earlier, however, a team of biologists working with human liver cells had kept getting chromosome counts of 46. Instead of following up their curious observation, they dropped the whole project on the assumption that their technique was bad. They accepted the voice of authority over their own observations, thus losing the opportunity of being credited with making an important discovery.

NEGATIVE RESULTS AND FAILURES

What kind of conclusion can you reach if your results are negative? In science it is just as important to know what does not happen as to know what does happen. So state your negative result — it may be very worthwhile.

Perhaps your experiment fails completely, and you do not get any results. The same thing has happened before — even to great scientists. Like most human beings, scientists are more apt to advertise their successes than their failures. Thus, you are more likely to hear about experiments that worked than about those that failed. However, you can still form an honest conclusion under these circumstances. You can say that the experiment failed to answer the question for which it was proposed.

Once the conclusion is reached, whether positive or negative, you are ready to report the results of your experiment. This may be done in writing or orally. In either case, remember to make things clear and understandable for the benefit of people who don't know as much about the subject as you do. One of the values of reporting your results is that it gives other people the chance to repeat your experiment. A good experiment must be reproducible. If it cannot be reproduced, there is something wrong.

SUMMARY

An experiment begins when you try to solve a problem. You would be wise to begin with a relatively simple problem that interests you. Be sure that you have the necessary time, space, materials, equipment, and technical skills. Under no circumstances should you choose an experiment that is dangerous.

Your first step is to frame a concise statement of the problem, which gives a clear picture of the purpose of the experiment. Next, you must take care of the preliminaries. These include background readings, developing skills, gathering equipment, and preparing a plan of action.

In its simplest form, the well-planned experiment consists of two parts that are identical except for the single variable that is being tested. There should be many subjects in the experimental and control parts of the experiment. All observations and data coming out of the experiment should be collected and recorded.

The raw data collected from the experiment must be organized and analyzed with the help of graphs and statistics. A simple, clear-cut conclusion is drawn from the inferences growing out of this analysis.

When it is completed, the experiment can be reported in the form of a written report or by a speech before an audience. If the experiment is valid, it is reproducible.

EMBRYOLOGY

by Lynn Lamoreux

Life goes on and on. By definition, living things are those that are able to reproduce themselves. Rocks and oceans do not have offspring. But people, butterflies, bacteria, plants, and other forms of life produce new generations.

Scientists, of course, have often wondered where babies come from. It may surprise you to learn that we do not yet have the full answer to that question. But we do know that the babies of many living things, including the vertebrates—animals with backbones, such as amphibians, reptiles. birds, and mammals—begin life when the ovum, or egg cell, of the mother is fertilized (joined) by a sperm cell from the father. The sperm activates the ovum to develop from a one-celled fertilized egg into a baby. The baby later is either born alive from the mother or hatched from the egg after it has been developed within and released from the mother's body.

Obviously a great deal of growth and change must take place for a microscopic egg cell to become a much larger baby, or infant, or, in plants, a seed. After the egg is fertilized, and before it is born or hatched, it is called an *embryo.* The embryo is simply an unborn infant creature or plant. In higher animals, an unborn infant, particularly in its later stages of development, is also known as a *fetus.* Scientists who study the growth and development of embryos are known as *embryologists,* or *developmental biologists.*

We shall discuss here some of the most important principles of animal embryology, particularly among vertebrates. We shall concentrate on the major changes that take place in the earlier stages of embryonic growth, when the ground plan, so to speak, of the future adult is laid out. We cannot go into all the details of the changes that take place. However, an understanding of the general principles is enough for us to grasp why these things happen and what they mean.

THE GENETIC CODE

Each of the many forms of life on the earth has some means for reproducing its own kind. It may be by *asexual reproduction,* in which only a single parent is necessary. But many living things, especially the more complex species, need a pair of parents—a male and a female—in order to reproduce. This is *sexual reproduction.*

Common to both forms of reproduction is the passing on, from one generation to the next, of the so-called *genetic code.* This is a coded message that guides the development of an embryo, making it into a member of a given species. The genetic code is built into the structure of the genes. *Genes* are chemical molecules attached to each other end-to-end. They make up larger bodies called *chromosomes.* The chromosomes are located in the nuclei of advanced cells. Genes and chromosomes, in the suitable environment supplied by the cell, are able to reproduce themselves by dividing.

The mother and the father provide the genes and chromosomes, which are contained in living cells known as the *egg,* or *ovum,* and the *sperm.* Each ovum and sperm contains a half set of chromosomes and genes. When egg and sperm unite, in the process that is known as *fertilization,* their respective sets of genes pair up with each other. The fertilized egg then has a full set of genes, complete with the genetic code. The code will then be the blueprint for the growth and development of the resulting embryo.

After the embryo has developed, it is born or hatched. It grows and becomes an adult, ready to reproduce should it survive. It will, like its parents, pass the genetic code on to its offspring.

The code passes on from generation to generation. The chain of life presumably extends from the beginning of life on earth. If the chain is broken at any point, the life of a given species ends there.

COMPARISON: SPERM AND EGG FORMATION

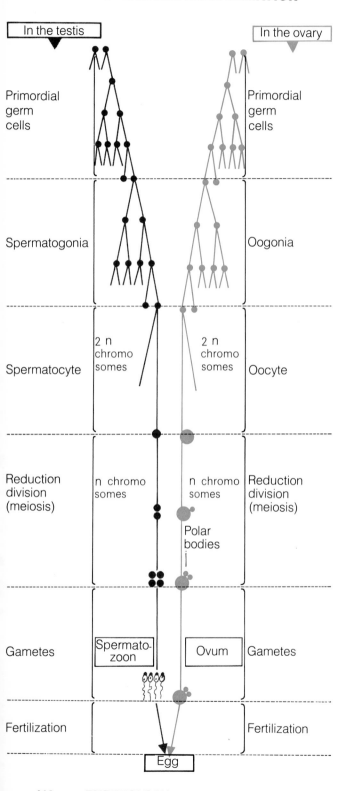

In the testis

In the ovary

Primordial germ cells

Primordial germ cells

Spermatogonia

Oogonia

Spermatocyte

2 n chromo somes

2 n chromo somes

Oocyte

Reduction division (meiosis)

n chromo somes

n chromo somes

Reduction division (meiosis)

Polar bodies

Gametes

Spermato-zoon

Ovum

Gametes

Fertilization

Fertilization

Egg

If the chain of living cells with their codes goes back unbroken to the origin of life, why is life now so marvelously varied and so different from its microscopic beginnings? Surely a one-celled bacterium, a butterfly, and a mouse cannot all belong to the same ongoing chain of life? Perhaps not, but they may all have sprung from the same origin, hidden in the distant past.

Over the ages, scientists say, the genetic code has been passed from generation to generation of living things. During that time, chemical changes have occurred now and then in this code, just as they do today. In this way, different blueprints have been provided for the development of different branches of the plant and animal kingdoms. This process may explain the differences and the similarities among bacteria, plants, butterflies, mice, and all other organisms.

Scientists accept the theory of *biological evolution*—the change of one form of life into another. Many of them are interested in embryology. Some scientists think that certain steps in embryonic growth are similar to those that may have taken place during evolution. Thus, a mammal seems to go through successive embryonic stages in which it resembles a creature without a backbone, and then an animal with gills and a backbone, before it finally become recognizable as a mammal.

ENVIRONMENT OF THE EMBRYO

The genetic code is so powerful a factor in embryonic development that, at certain stages, it is even more important than the environment of the embryo. One might think that the mammalian egg would develop according to what kind of mother it is growing in. It would not surprise you to find that a mother mouse produces baby mice

The genetic code—the blueprint for development—is passed from generation to generation in the germ cells—sperm in the male, eggs in the female. The sperm and eggs are formed in the parent's bodies through a special process of cell division, which results in each germ cell having only one half of the normal "genetic package." When the germ cells unite in fertilization, the resulting fertilized egg cell has a complete "genetic package." It then begins to divide.

and a mother elephant a baby elephant. We know, however, that a fertilized egg has its own destiny, regardless of the particular mother it is in.

The genes, however, must still get the necessary food and other material from their environment—the cell containing them.

The genes are located, as we stated, in the nucleus of the fertilized egg and, in fact, in the nuclei of the cells. The cell matter surrounding the nucleus—the *cytoplasm*—may have considerable influence on how the genes carry out their coded message.

The egg and the embryo must live and function within their external environment —the mother's body, in the case of the mammal. Embryonic development thus depends, to a great extent, on this external environment. Bad factors in the environment may cause birth defects, which are not always due to mistakes in the genetic code. For this reason pregnant women should take proper care of their health and avoid smoking or taking any drugs or medications without a doctor's advice.

STATUS OF EMBRYOLOGY

Because we ourselves are mammals, we have a natural curiosity about mammalian embryology. In the past it has been very difficult to study the mammalian embryo, because it is so protected and nurtured by the mother's body. Although embryology has had a long and productive history, most of the work had been done on embryos of amphibians (frogs, salamanders, toads), birds (chickens), and other species less closely related to mammals.

But many investigators prefer the embryo of the mammal. It provides a glimpse into our own development, a glimpse that leaves us searching for more understanding. Other scientists react in a more practical way. Mammalian embryology can answer questions about birth defects and many other questions of medical and biological value to man.

EARLY DEVELOPMENT OF THE EMBRYO

Soon after fertilization, the visible development of the egg into the embryo be-

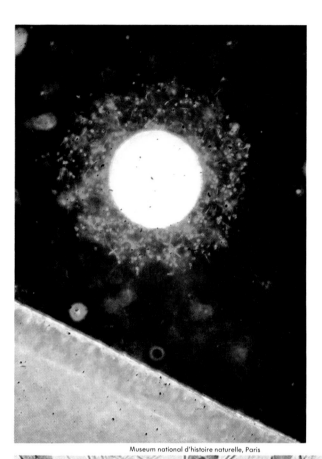

Museum national d'histoire naturelle, Paris

C. Nuridsany

Top: an ovum, or egg cell, surrounded by sperm. Normally, only one spermatozoon will penetrate the egg cell to fertilize it. Bottom: photomicrograph showing numerous sperm. Each spermatozoon has a head and tail region.

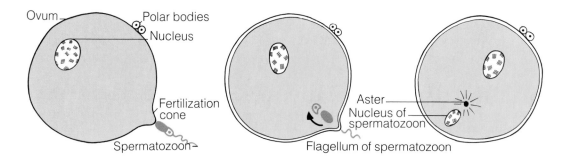

Ovum — Polar bodies
Nucleus
Fertilization cone
Spermatozoon —

Aster
Nucleus of spermatozoon
Flagellum of spermatozoon

A spermatozoon approaches an egg cell and a fertilization cone develops on the egg membrane.

The spermatazoon has penetrated the protective layers of the egg an activated it.

Spermatozoon nucleus starts to approach egg nucleus and preparation for cell division begins.

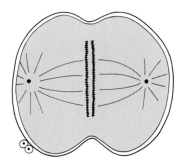
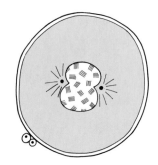
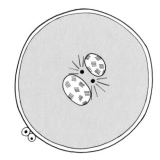

The two nuclei line up near each other in preparation for fusion and cell division.

The two nuclei fuse. The new cell has a complete "genetic package."

The single fertilized cell now starts its first division in the process of cleavage.

gins. The fertilized egg cell starts dividing into more cells. As in all typical cell division, the chromosomes in the nucleus are first reproduced, or replicated. Then the rest of the cell divides. In this way, each daughter cell receives a full set of chromosomes, which provide the information for normal development.

After the first cell division, each daughter cell divides once more. The embryo then consists of four cells sticking together. Next, cell division follows on cell division, until the former single egg cell has become a mass of many tiny cells.

The embryo has not yet grown larger, but has made hundreds of copies of its own unique set of genes, each set inside one of the cells. This cell division is known as *cleavage,* and the embryo produced in this way is called a *blastula.* The blastula is a ball-shaped embryo. Its center is a hollow —the *blastocoel.* The thick walls of the blastula itself are made up of the many small cells.

Up to this point you would not notice much difference between an elephant embryo, say, and a mouse embryo. But differences among mammal, bird, and amphibian embryos have already appeared. At present it is difficult to explain how or why the blastula develops from the fertilized egg.

The change from the blastula to the next stage of embryonic development, the *gastrula,* is known as *gastrulation.* Just before gastrulation in the amphibian, a slight difference becomes evident between the ends of the blastula. One end, the so-called *animal pole,* will develop into the baby animal. The other end, or *vegetal pole,* provides food for the embryo in the form of

The early stages of development of an amphibian illustrate embryonic processes common to many higher animals. The fertilized egg begins to divide. One cell becomes two, two cells become four, and so on in the process of *cleavage*. The embryonic ball of cells, or *morula*, develops a central hollow. The hollow is known as the *blastocoel* and the embryo at this point as a *blastula*. Immediately after the blastula stage, there are shiftings and rearrangments of cells. This stage, known as *gastrulation*, leads to the formation of three main layers of cells: an ectoderm, a mesoderm, and an endoderm. The central cavity at this point is called the *archenteron*, or primitive gut. As the differentiation of the three primary layers continues, the beginnings of organs and organ systems, including the intestinal cavity and notochord, appear. At this stage, the embryo is often known as a *neurula*. The notochord forms a supporting rod that will later become part of the vertebral column, or backbone, of the animal.

somewhat larger cells, which are filled with yolk. We may compare the shape of the blastula to the shape of the earth, with the animal and vegetal poles corresponding to the north and south poles.

In the amphibian embryo, gastrulation proper takes place when cells migrate from the surface of the blastula toward the inside. The result is a hollow ball of cells with a hole at one side, through which the cells move. This hold, or *blastopore,* is located between the animal and vegetal poles. Thus the animal itself will develop around, and above, its food supply at the vegetal pole. This pole will form the ventral, or belly, surface.

After cell migration to the interior of the gastrula has been completed, the embryo is still a hollow ball of cells. But it now has two layers of cells making up the walls: an inner layer and an outer layer. The inner consists of cells that were once on the outer surface of the embryo and later flowed to the inside through the blastopore. The blastocoel has been pushed out of existence, and the new inner space is known as the *gastrocoel*, or *archenteron*.

Gastrulation may not seem like a great step forward. For all the migrations and rearrangement of its cells, the embryo still remains a tiny, rounded object with a central cavity. But at this stage certain cells become destined to develop along more specialized lines. Gastrulation, therefore, is

STAGES OF DEVELOPMENT OF AN AMPHIBIAN

Cleavage

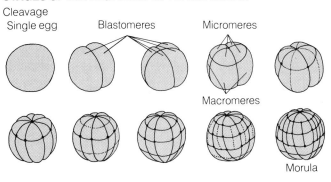

Blastula

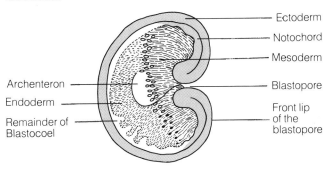

Gastrulation

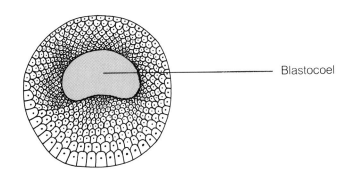

Neurula

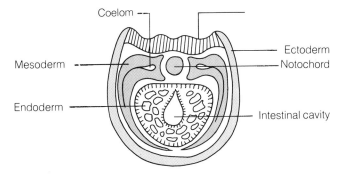

a critical period in embryonic development. During gastrulation, development may easily be disrupted, leading to abnormality or death of the embryo.

LATER DEVELOPMENT OF THE EMBRYO

After gastrulation, further normal development conforms to a pattern that is much easier to follow. But exactly how it works, and why, remains largely mysterious.

The cell lining the gastrocoel will form *endodermal,* or inner-layer, *tissue*. The *endoderm* will eventually give rise to certain glands and to all the tissues that will later come to line the inner surfaces of the body.

As times goes by the gastrula becomes longer, and the endoderm grows to line the entire inside of its cavity. A new opening forms at the end opposite the one that is already present. When the time comes, the embryo will have a gut and two openings— a mouth (the new opening) at one end and an anus, or waste opening, at the other.

Meanwhile, two other layers of embryonic tissue are forming: the *mesoderm,* or middle layer, and the *ectoderm,* or outer layer. The ectoderm is on the outside, and the mesoderm lies between the ectoderm and the endoderm. The mesoderm will eventually develop into muscle, bone, heart, and the *germ cells*. The germ cells produce new sperm or egg cells for the next generation. The ectoderm will finally become the outer layer of skin, the pigment, or color cells, and the nervous tissues, including nerves, spinal cord, and brain.

PROGRESSIVE DETERMINATION

The manner in which all these tissues and organs take shape, and the way they interact, are now well understood in the amphibian and the bird. The embryos of birds and amphibians can be watched at every step of their development. Experimental changes can be made in these embryos. In these ways, embryologists have made truly important discoveries about the basic principles underlying all biological development. They have begun to build concepts of how genes control embryonic growth.

The first important concept was discovered by embryologists working with amphibians. They found that groups of cells gradually lose more and more of their ability to develop in various ways. At very early embryonic stages, if you were to take away one of the two or four cells of the cleaving embryo, that cell would equally well divide again to form a complete, normal embryo, under the proper environment.

However, if you were to take a cell from, say, the ectoderm of an older gastrula, you would find that its capacity to develop had become restricted. It might go on to form nerve, skin, or pigment tissues only. It would no longer be able to express the full possibility of its genetic code, in the

In birds, gastrulation, or the formation of the three main tissue layers, takes place as cells migrate through and around a thickened band of cells known as a *primitive streak.* Below: a chick embryo of 36 hours with primitive streak forming an early notochord. Right: series of photographs showing the development of a chick. From top to bottom: embryo at 6 days, at 10 days, at 13 days, and at 21 days.

J-P Suzzoni

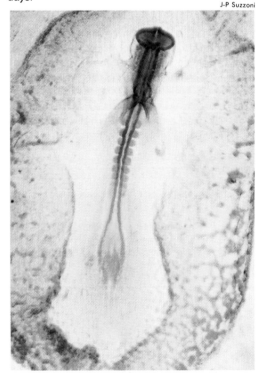

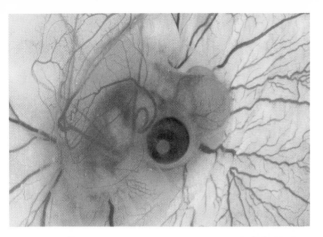

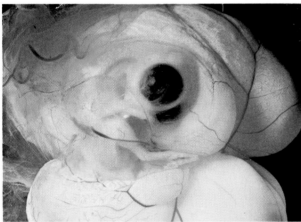

all photos, G. Whiteley, Photo Researchers

way a fertilized egg cell can. Evidently, some of the genes in the ectoderm cell are no longer working.

At a still later stage of development most of the body cells would be completely specialized to do the jobs required of them in the newborn and adult creature. They would be brain cells, skin cells, pigment cells, and so on. If these cells were to divide, they could produce only cells of their own type.

This increasing specialization of cells is know as *progressive determination*. After a cell's future is determined, it normally cannot go back to the immature stage in which it was able to develop in any one of a number of specialized directions. As time passes and the embryo grows, each cell type becomes progressively more fixed in an ever narrower specialization.

How does a cell become determined? We are not entirely certain. We do know, however, that for progressive determination to take place, there must be interaction between different tissues at particular times and places during development.

For example, in the amphibian embryo, during gastrulation the mesoderm comes in contact with the ectoderm at the animal pole. This new contact between mesoderm and ectoderm induces, or stimulates, the ectoderm to form brain tissue at that spot. If the contact does not occur at the correct time and place, devlopment of the brain does not proceed correctly. This contact is known as *induction* of the ectoderm by the mesoderm. It is also known as *inductive interaction* between the mesoderm and ectoderm. Embryologists discovered it by experimentally manipulating the amphibian embryo.

DEVELOPMENT IN BIRDS

In a chicken egg, unlike an amphibian egg, the yolk is enormous compared with the embryo itself. It you were to look at the egg during the original cleavage, you might just be able to see the tiny future chick as a whitish speck on the surface of the yolk.

In the bird egg, as in the amphibian egg, the yolk and the embryo are surrounded by a jellylike layer, or egg white,

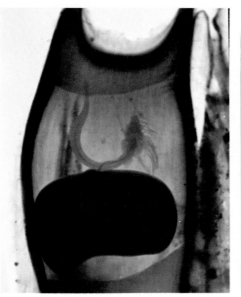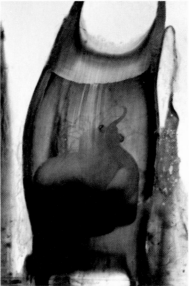

Skates—harmless cousins of sharks and rays—undergo their embryonic development in special egg cases, known as "mermaid's purses". Three stages of development are shown above: (left) wormlike form above yolk; (middle) embryo further matured with future eye region apparent in head area; (right) embryo beginning to have characteristic rounded form of a skate.

provided by the mother's body. This white provides protection and other advantages necessary to life; it is not a true part of the egg itself. The chicken egg has a protective shell necessary for survival on land.

The same basic development occurs in the chick as in the amphibian. But the chicken egg is so unusual, with its huge yolk, that its embryonic growth at first seems different. The tiny animal pole rests on top of the yolk in such a way that it cannot form a hollow ball of cells with the yolk enclosed, as is the case in the amphibian. Instead, the chick blastula is a flattened disk, which is actually the animal pole itself. Under this pole is a space—the blastocoel. The yolk, hundreds of times larger than the embryonic chick is, forms the vegetal pole, underneath the animal pole.

Next, a long *primitive streak*, rather than a round blastopore, forms, and numerous cells migrate through it. Thus the chick gastrula takes shape. As in the amphibian, three tissue layers—ectoderm, mesoderm, and endoderm—develop. But in the chick, these layers are flattened upon the surface of the yolk, with the endoderm next to the yolk. Through progressive determination and inductive interaction, the tissue layers develop into the baby.

DEVELOPMENT IN MAMMALS

The egg of the mammal is much smaller than that of the amphibian or the chicken. Since the mammalian embryo implants itself into the *uterus,* or *womb,* of its mother, it has no need of egg yolk for food and white for protection. At about the blastula stage, the small embryo sinks into the soft tissues of the uterus, which then close around it.

The embryonic development of the mammal has more similarities to than differences from the development of amphibians and birds. New evidence suggests that the mouse embryo, however, may not go through the massive cell migrations that are necessary for gastrulation in the chicken and the amphibian.

According to this new theory, the mouse gastrulates primarily by means of cell division. Also, the internal cavities of the blastula open by a simple pulling apart of the tissues. There may be little rearrangement of tissues during gastrulation. Major cell movements, if any, occur at later stages, during the formation of the organs.

This, if true, would be a surprising difference in the embryonic growth of the mammal compared with that of the chicken or the amphibian. The difference very likely may be due to the fact that mammals do not need large amounts of yolk in the egg. The mammalian embryo has attached, instead of yolk, a tissue called the *placenta*. The placenta holds the embryo to the inside of the uterus and also serves as a path for food and oxygen from the mother's bloodstream to reach the embryo. The placenta also removes the embryo's wastes. Mammalian development follows the pattern of all the vertebrates: the formation of ectoderm, mesoderm, and endoderm, and the organ systems derived from them.

We may therefore assume that the embryo mammal develops through progressive determination and inductive interaction, as directed by the genetic code. The code is present in the set of genes inherited from each parent. But the evidence for this is still only indirect. Embryologists and geneticists have now developed the techniques with which to study the mammalian embryo intensively.

Mammalian embryo research has led to several practical developments for women who experience difficulty in conception. In the early 1980s, for example, scientists successfully developed an *in vitro* (outside the body) fertilization technique in which a ripe egg from a woman's ovary is fertilized with her partner's sperm in a laboratory dish filled with nutrients. Then, after several days growth, the developing embryo is inserted into the mother's womb. In 1983, a technique was pioneered at UCLA in which a fertile ovum is transferred from one woman to another. By the late 1980s, a number of different techniques were successfully used to transfer fertilized eggs or developing embryos into "foster" wombs in humans as well as other mammals.

Embryo manipulation can also lead to curious results. For example, scientists at the Institute of Animal Physiology in Cambridge, England, mingled embryo cells from both sheep and goats. The crossbred embryos were then placed in the wombs of sheep and goat mothers and allowed to grow to full term. The hybrid offspring, called a *chimera*, has characteristics of both parents. Scientists believe such research may make it easier to rear embryos of endangered species in the wombs of other species.

In many invertebrates, embryos hatch prematurely as feeding larvae that then undergo a series of changes, known as metamorphosis, which produce an adult form. Art below shows the development of a bee from the time the queen bee lays the egg (1); through the feeding of the larval form by worker bees (2); and the gradual change of the larva into a form known as a nymph (3,4); to the emergence of a young bee (5,6); 22 days after the egg was laid.

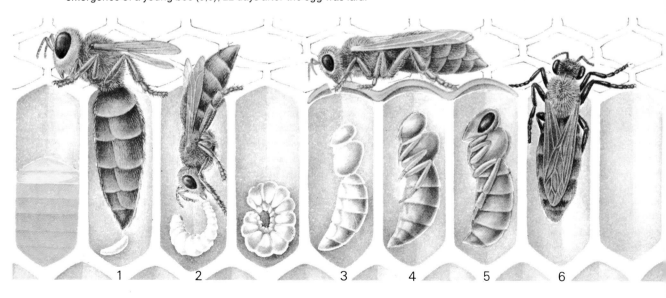

BIOCHEMISTRY

By Stephen N. Kreitzman

© Jane Burton/Bruce Coleman

Biochemists seek to unravel many of the puzzles of life. One recurring question is why a normally pigmented animal like the guinea pig above gives birth to a litter containing both normal and albino young.

J. P. Suzzoni

It was inevitable that biologists, who were carefully studying plants and animals, should begin to ask questions that could only be answered by chemists. How do our bodies get energy from foods? What is a gene made of? What is the nature of protoplasm? What is the chemical composition of fat and bone and brain tissues?

To try to answer these questions, scientists combined the concepts and skills of biology and chemistry. Thus they created the science of *biochemistry:* the study of the chemistry of life.

In very recent years, biochemists have begun to obtain exciting answers to their questions. More and more scientists, sensing this excitement, have become interested in biochemistry.

Perhaps we can share in this excitement by discussing some of the things that biochemists have discovered.

Each of us began life as a single cell. This cell divided and became two cells. Each of these cells divided, and the process continued until there were many millions of cells. Then—at some signal—some of the cells began to form liver cells; others began to resemble muscle cells, and so on. This process continued until the mass of cells began to look like a baby. At birth the baby may have weighed three kilograms or so, but he or she quickly grew larger, from infancy through childhood and into adulthood. All this change, from a single cell to an adult, came from food. Food provided the sugars, proteins, vitamins, minerals, and fats needed for growth.

WHAT ARE HUMANS MADE OF?

A human being is made up of sugars, fats, minerals, and proteins, the same as

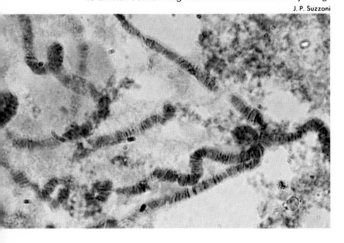

Chromosomes carry the hereditary information from generation to generation. Biochemists have succeeded in tracing many of the steps involved as this information is translated into the complex processes of development. Left: fruitfly chromosomes.

food. Every body cell needs the chemicals in food to grow and divide. The story of biochemistry is the story of how the body and its cells change these chemicals into many different forms so that they can be used for living.

Early in the history of their science, biochemists determined the chemicals in foods and in body cells. They measured the amount of carbon, hydrogen, oxygen, nitrogen, and phosphorus in the living cells. They quickly discovered that special groups of atoms—called *amino acids*—were very important to the cells.

The scientists found that the proteins in food are made of about twenty different amino acids. They learned that these amino acids are the chemicals that human cells use to make their own kinds of proteins. The fats in food are composed of *fatty acids*. These fatty acids are used by living cells for energy. Starches and sugars in food are made of *simple sugars*. There are many uses for these simple sugars in a cell.

STRUCTURE OF PROTEINS

Biochemists discovered that it does not make any difference in which order amino acids occur in a specific protein of food. However, the total amount of food eaten by a person must contain enough of each amino acid to satisfy the cells of the body. These cells recombine the amino acids into human proteins, each of which has a specific job in the cell. This job is determined by the number and order of the protein's different amino acids.

Thus proteins are not all alike. They are all strings of amino acids, but the order in which the amino acids are strung together is different for each type of protein.

Biochemists call the unique number and order of amino acids in a protein the *primary structure* of the protein. The special functions of proteins and the ability to carry out these functions are determined by the primary structure of the protein and the way in which the string of amino acids folds and twists. The twisting, which in many cases is a spiral like a spring, is called the *secondary structure* of the protein.

The folding, however, may be even more complicated. The spiral itself may twist around to form very complex shapes, with some parts of the spiral exposed and some parts hidden. This extra complicated folding is called the *tertiary structure*. It is this special folding of the long chain that determines what function a protein performs. When the protein unfolds so that it has a different shape, it is *denatured*. The primary structure is unchanged, but the function of the protein is lost.

One of the most important functions of

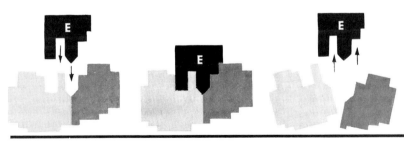

Catabolism: Here an enzyme breaks a large molecule into 2 smaller molecules. Note that the enzyme's shape is specifically tailored to enable it to attach to the large molecule.

Anabolism: In this process, an enzyme builds 1 large molecule from 2 smaller molecules. The enzyme is not changed during this activity but can be used over and over again.

proteins in a cell is their ability to act as *enzymes*. Enzymes are special protein molecules that allow a chemical reaction to take place at body temperatures, reactions that would otherwise occur only at very hot temperatures. Enzymes may also regulate the speed of chemical reactions, enabling them to occur slowly or very rapidly. Every kind of chemical reaction in the cell has a specific enzyme protein helping it. In other words, each enzyme protein has its own kind of chemical reaction to control. It cannot affect another kind of reaction. The specific action is determined by the order of the amino acids, and the special way that the protein chan folds upon itself.

METABOLISM: BUILDING UP AND BREAKING DOWN

The term *metabolism* refers to all the chemical reactions that occur in a cell. Some of these reactions result in the breakdown of large molecules. Other reactions take small units, and combine them to make new large molecules. When large molecules are broken down into smaller ones, the process is called *catabolism*. Rebuilding the complex molecules from the smaller units is called *anabolism*. Both anabolism and catabolism involve changes in molecules that occur one at a time in assembly-line fashion. It is almost as if the molecules are traveling along a conveyor belt and at each stage an enzyme was present to add something or take something away. Many kinds of molecules are being changed in the cell. For each kind, there is a specific conveyor belt, or *metabolic pathway*.

During catabolism large molecules that come into a cell are taken apart. A great deal of energy is released when molecules are catabolized. This energy is released in a controlled manner, to be used later for the processes of anabolism. Much of the energy that was used in plant and animal cells to put together the food molecules we eat is released by the processes of catabolism. This energy originally came from the sun and was used by plants to make sugars during the process of photosynthesis.

When food molecules, especially sugars and fats, are catabolized, large amounts

of energy can be released. This energy can be captured by a cell to be used at a later time. The cell forms an energy-storage molecule called *adenosine triphosphate* (ATP).

ATP has often been described as a type of energy money. Before the use of money, people exchanged goods or services, and the exchange had to be made directly between two persons. If one person wanted some eggs, and the other person wanted some cabbage, they could exchange some of one for some of the other. This was not always convenient, however: at times only one person had something the other wanted. With the invention of money, it was possible to convert goods and services into money which could be spent later. ATP is formed when some reactions of catabolism give off a lot of energy. Then it is like money. The energy can be spent later in another reaction that requires energy.

Most of the energy for the cell comes from the catabolism of starches, sugars, and fats. Each has its own metabolic pathway, but eventually many of the starches and sugars and fats are changed into the same compounds. This allows the cell to take the energy from a large number of different chemicals found in foods and convert them into molecules that the cell needs.

There is one metabolic pathway where most of the energy is produced. It is very famous and important. Almost all the metabolic reactions that occur in the cell, whether anabolic or catabolic, begin or end with one of the stages of this pathway. It is called the *Krebs citric acid cycle*.

The Krebs cycle is named after the German biochemist Hans A. Krebs, who discovered it. It consists of a series of chemical reactions that occurs in all living cells. When foods are partially broken down in a cell, acetic acid is produced. The chemical reactions in the Krebs cycle change this acetic acid to carbon dioxide and water. During the process, ATP is formed.

SIZE AND SHAPE OF MOLECULES

In our earlier discussion of proteins, we indicated that the way a molecule folded

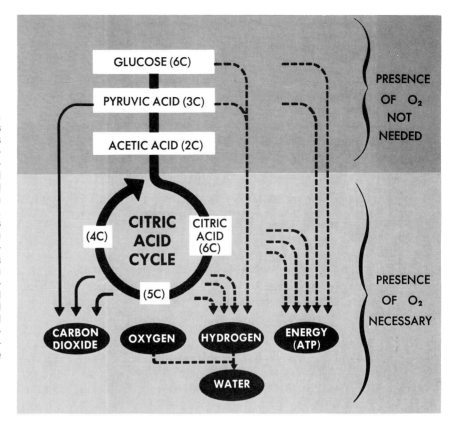

GLUCOSE (6C)

PYRUVIC ACID (3C)

ACETIC ACID (2C)

CITRIC ACID CYCLE

CITRIC ACID (6C)

(4C)

(5C)

PRESENCE OF O₂ NOT NEEDED

PRESENCE OF O₂ NECESSARY

CARBON DIOXIDE

OXYGEN

HYDROGEN

ENERGY (ATP)

WATER

The Krebs citric acid cycle. In the cell, sugar is in its simplest form: glucose. It is broken down into two molecules of pyruvic acid. Carbon dioxide is formed and removed, leaving acetic acid which enters structures in the cell called mitochondria. Here the acetic acid joins with oxalacetic acid to form a six-carbon compound, citric acid. This undergoes a series of reactions in which hydrogen and carbon dioxide are released. At the end of the cycle, oxalacetic acid is reformed, ready to be used again. During the entire process, large amounts of energy, in the form of ATP, are released.

upon itself in three dimensions was an important factor in determining the function of the protein. Nucleic-acid molecules, which we will soon discuss, are also quite large and have a great deal of information in their size and shape. Biochemists have used various methods to determine the structure of these *macromolecules*.

One of the most important ways that biochemists can learn about the size and shape of macromolecules is based upon a very simple principle. If you put some particles of sand in a glass of water, the particles will sink to the bottom. They are pulled to the bottom of the glass by the force of gravity. If the glass were taken to the moon, the particles would sink much more slowly. If we could increase the force of gravity on the sand particles, they would sink more quickly.

Biochemists use this principle to learn about macromolecules. They add some macromolecules to water that has been made syrupy by the addition of sugar. At normal gravity the macromolecules sink very slowly—if at all—in this syrupy water. But if the force of gravity is increased, the macromolecules will sink more quickly.

A force equivalent to gravity can be produced by simply spinning the glass containing the macromolecules very fast. And spinning the glass even faster will increase the equivalent force. A machine that can spin such a solution very fast is called a *centrifuge*. As the force is increased in the centrifuge, the macromolecules are pulled toward the bottom of the tube. Not all of the macromolecules sink at the same rate, however.

Even more exciting, perhaps, is to know exactly where each and every atom of the macromolecule is located. This is very difficult to do, but biochemists are now learning how to calculate the exact position of the atoms of macromolecules. Calculations are based on the manner in which a crystal of the macromolecule reflects a beam of X rays. Such calculations have now been worked out for several proteins, most recently for two different proteins that have similar enzyme activities. The results indicate that these two proteins—the enzymes elastase and chymotrypsin—have similar shapes (which biochemists had expected because of the substances' similar activities).

MEMBRANES AND COMPARTMENTS

A *membrane* is a divider. The wall of your room is a membrane. It separates one room from another and therefore everything within that room from everything within the adjacent room. There are many membranes in a cell. One membrane completely surrounds the cell to keep all the contents contained. Another membrane encloses the nucleus, keeping it separate from the rest of the cell.

While the wall of your room may separate you from the people in the next room, a cellular membrane allows certain molecules to pass freely through, back and forth. Water is allowed to pass through a cellular membrane. Other chemicals are able to pass through some membranes but not others. It was once thought that a membrane acted like a strainer and let only small chemicals through. This is not completely true. Some very large molecules can pass through a cellular membrane, but they have to be carried across by a special mechanism that acts as a ferry service. This ferry service is very selective and will transport only certain substances. ·

The chemical components of a cell are located in very specific places within the cell so that they may carry out their particular job. There are many membranes within the cell. These set off compartments which contain different kinds of molecules. For example, *mitochondria* are small cigar-shaped bodies in the cytoplasm of the cell. In the mitochondria are Krebs-cycle enzymes and other molecules necessary for harnessing the energy of catabolism.

HOW ARE PROTEINS MADE?

Proteins and other macromolecules that have specific functions in the cell need to be made exactly right in order to carry out their functions. An enzyme that has a specific reaction to carry out must have the exact amino-acid sequence necessary to do the job. One of the most important advances in our understanding of the chemistry of living things came with understanding how proteins are made in the cell.

The master code for the construction of almost all proteins in the cell is found in long, spiral molecules called deoxyribonucleic acid, or DNA. Understanding how information is coded in DNA molecules is one of the outstanding achievements of modern science.

A molecule of DNA consists of a long chain of repeating, similar units. The repeating units are called *nucleotides*. Each of these units in turn consists of three parts: (1) a phosphate, (2) the sugar deoxyribose, and (3) compounds called bases.

In the DNA molecule, two chains attach at the bases, forming an H-like shape. The vertical parts of the H are made of deoxyribose. The horizontal part is made of two bases. One H-shaped unit is connected

The structure of DNA. A molecule of DNA consists of two long chains of repeating, similar units. Each unit has three parts: a phosphate, a sugar, and a base. The chains are attached to each other at the bases.

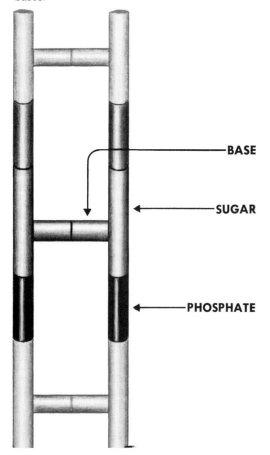

BASE

SUGAR

PHOSPHATE

to another H-shaped unit by connecting bars made of phosphate (see diagram).

The DNA molecules differ among themselves only in the bases that make up the horizontal parts of the H's. The exact order in which the H's are strung together determines what information will be contained in the DNA. That is, the order of the repeating units determines the order of the amino acids in a protein.

The information contained in the DNA is called the *genetic code* and eventually allows specific proteins to be constructed. The fascinating mechanism for reading the code and turning it into a sequence of amino acids has been worked out.

The DNA contains almost all the instructions for the cell and therefore must be protected against damage. It remains safely in the nucleus of the cell, away from all the machinery that is busily making new proteins. However, the information coded in the DNA must be brought to the parts of the cell making the new proteins. This is done by means of a messenger molecule composed of nucleotides, also, but the sugar in the nucleotides is ribose and not deoxyribose. The molecule is therefore called ribonucleic acid, or RNA.

The RNA is formed in the nucleus, apparently under the influence of DNA. The RNA messenger molecule contains the same information as the DNA chain except that it is expressed in a series of ribose nucleotides rather than deoxyribose nucleotides. The messenger RNA travels from the nucleus to the cytoplasm. There it attaches to workbench-type molecules called *ribosomes*. It is on the ribosomes that proteins are made. With the coded information, in the form of RNA, now attached to the ribosomes, amino acids can be assembled in the sequence to form proteins.

The RNA code has now been completely deciphered. Every three nucleotides on the messenger RNA molecule indicate a specific amino acid that is to be placed in sequence and then zippered up to form part of the completed protein. For example, if the nucleotide base sequence on the messenger RNA was adenine-cytosine-adenine, then the amino acid threonine would be put into place and zippered into the growing protein chain. The next three nucleotides would determine the next amino acid of the protein, and so on.

The protein therefore is determined by the sequence of nucleotides in the DNA of the nucleus. A complete "genetic-code dictionary" has now been compiled, and, as a remarkable example of the uniformity of nature, the code seems to be the same in almost all living creatures.

It has long been suspected that there is some way to transfer information from one generation to another so that the species would be continued. Scientists believed this information was held in the *genes*.

We have learned a great deal about genes. We know, for example, that the genes are, in fact, the long polymers of DNA—the master code within the cell's nucleus. When a cell is ready to divide, a copy of its DNA is made. This identical copy of the master code is given to the new cell.

BIOCHEMISTRY APPLIED

The understanding of the processes by which cells naturally add, delete, and transfer genetic information necessary to protein production opened the way for scientists to carry out genetic manipulations. These methods are collectively called recombinant gene technology, or genetic engineering. Important advances in the field were made in the 1980s and 1990s, and genetic engineering has already begun to revolutionize medicine and agriculture.

Biotechnicians have succeeded in producing genetically altered bacterial cultures that can synthesize human insulin and human growth hormone cheaply and in large quantities. By using bacteria to synthesize only the protein coats of viruses, a number of safe vaccines—totally free of viral DNA or RNA—are now being produced. *Rennin*, an enzyme used in cheesemaking, is being made from genetically engineered bacterial cultures, as is *cellulase*, an enzyme that can convert indigestible cellulose into glucose. Agricultural plants are being genetically altered for more efficient photosynthesis and nitrogen fixation, to increase their productivity.

Left: *Mycena* toadstools photographed in daylight; right: same toadstools at night.

BIOLUMINESCENCE

by E. Newton Harvey

If you have ever taken a walk through deep, damp woods, you may have been startled when you came upon a shining stump or tree. This natural glow, sometimes called fox fire, is caused by the mycelia, or thread-like processes, of a luminous fungus that grows on wood. The fungus threads eventually form a fruiting body—a mushroom, which also glows. Quite often one finds clumps of such mushrooms on stumps of trees, emitting their eery light.

The emission of light from a living thing is called *bioluminescence* ("living glow"). There are many instances of it in nature. For thousands of years seafarers have known that the sea sometimes glitters with light when a boat passes through it or when an oar strikes it.

Benjamin Franklin was inclined at one time to accept the theory that sea light represented an electrical display. He quickly changed his mind, however, when he visited the seashore and was able to experiment. He found that a bottle of luminous sea water glowed brightly when it was shaken, but that after a while it did not glow at all. He knew that certain well-known marine animals could produce light. He came to the conclusion that the glow of sea water was due to much smaller living things. Their presence, he thought, could be detected only by their trail as they passed through the water. Franklin was quite right.

The glow one sees in the eyes of a cat in the dark certainly looks like biolumines-cence. However, the "glow" is simply the reflection of weak external light. If the animal's eyes shine in the dark, you may be sure that there is at least some illumination, however feeble it may be. Put the cat in a completely dark room, and the glow will vanish.

LIGHTS IN THE OCEAN

By far the greatest variety of light-producing species are found in the oceans. Among the microscopic protozoans, there are luminous radiolarians and flagellates. The flagellates, especially the species called *Noctiluca miliaris,* give rise to the spectacular displays of sea light in which a ship seems to move through liquid fire.

Certain sponges, a number of jellyfish and siphonophores, and many comb jellies (ctenophores) emit light. There are luminous hydrozoans and sea pens. They are attached to rocks or grow in the sand. Various marine worms, nemertines, brittle stars, nudibranchs, and marine snails produce light as they crawl over solid surfaces.

The bivalve mollusk, *Pholas dactylus,* bores into the sea floor and secretes a luminous slime that it squirts out into the sea water through a tubular organ, called a siphon, whenever it is disturbed. The Romans considered this mollusk a delicacy and served it at their feasts. "The pholades," wrote Pliny the Elder in his *Natural History,* "shine in one's mouth as one chews them; they shine in one's hands. If any drops of their fatty liquid fall on one's garments or on the floor, these drops also shine."

Of the free-swimming marine animals, many squid, deep-sea shrimps, and prawns produce light. A considerable proportion of deep-sea fishes are also luminous. In some cases a luminous organ is located at the end of a long process extending in front of the fish. In other fish, there are rows of lantern-like lights along the sides of the body. Sometimes only a few of these "lanterns" are lit; sometimes all of them are aglow, providing an amazing display.

Usually, luminous marine organisms are found at considerable depths. However, certain marine animals living on the surface also produce light. Among them are the plankton, tiny forms of life that drift about or swim very slowly. They include protozoans, jellyfish, and comb jellies, which we have already mentioned. A few additional surface forms, such as the transparent snail *Phyllirrhoë* and the tunicate *Pyrosoma,* are also luminous. They account for the prominent individual flashes of light that one can observe on the sea at night.

BACTERIA PROVIDE LIGHT FOR FISHES

Certain fish produce light not because they themselves manufacture a luminous material, but because they harbor luminous bacteria in certain structures of the body. The bacteria are always present in these

R. Rowan, Audubon/PR

Some species of coral are luminescent. The specimens above are in an aquarium.

structures. They receive food substances from the fish, and in return supply the fish with light. They do not harm the host at all. This is a remarkable example of symbiosis — a condition in which two dissimilar organisms live together to the mutual benefit of both.

Occasionally a living animal will emit light when it is infected by luminous bacteria. These spread through the body and cause a disease that is usually fatal. An infection of this sort is common among sand fleas, those lively little creatures that hop about on piles of seaweed washed up on a sandy beach. Infected sand fleas are very active and very luminous at first. In a few days, however, as the infection spreads, the insects become sluggish. Finally they die and their bodies no longer emit light. Luminous bacteria also infect freshwater shrimp, mole crickets, May flies, caterpillars, midges, and sow bugs.

LIGHT ORGANS

Luminous organs vary in structure. The light may come from within a single cell, as in a bacterium or flagellate, or a group of cells, as in the lighting region of the firefly. In some cases the light is directed by reflectors or lenses. This type of structure, called a *photophore*, is like a highly complicated lantern. It is found in squid, shrimp, and fishes.

Light may also be produced outside of a cell or cells. In this case it is a glandular secretion that makes its way to the outside of the body. Secretions of this kind are found in the mollusk *Pholas* and in certain crustaceans, earthworms, and centipedes.

Generally speaking, we can be reasonably sure that a living animal that produces a continuous light has been infected by luminous bacteria. Luminous bacteria and fungi emit light at all times. Luminous animals emit light only when they are stimulated. The firefly flashes when an impulse passes along its nerves and acts upon its light organ. The mechanical disturbance caused by the tossing of the waves or the passage of a boat causes various marine forms to luminesce.

COLD LIGHT

Light has generally been associated with heat in human experience. For example, the incandescent lamp is based on the general law that all substances give off light when they are heated to a certain degree. Incandescence could not possibly account for the production of light by living things. The heat that would accompany the light would soon prove fatal to any organism. Obviously bioluminescence must be due to so-called cold light—a light that produces little heat.

There are various kinds of cold light. Researchers have found that in a certain type of chemical reaction practically no heat is produced. The energy released by the reaction is largely converted into visible light. This is called *chemiluminescence*. It accounts for the glow of phosphorus, an element discovered nearly three centuries ago. And it accounts for bioluminescence.

An infinite number of low-temperature chemical reactions are continually taking place in living cells. They supply the energy for movement, for secretion, for cell division, and in fact for all vital processes. It is not surprising, therefore, that for the production of light in living things a chemical reaction of the chemiluminescence type should have been selected during the course of evolution.

Raphael Dubois, a French physiologist, was the first to point out the chemical nature of bioluminescence. In 1887, he suggested that the light of the boring mollusk *Pholas dactylus* was due to a reaction involving a substance he called *luciferin*. He maintained that luciferin combined with

Many fish that live at great depths have special light-producing organs. Note the row of light organs on the underside of this almost-transparent fish.

The bioluminescent sea anemone, *Anthopleura artemisia,* is found in Monterey Bay, California.

oxygen (was oxidized, a chemist would say) in the presence of an enzyme that he called *luciferase,* and that the reaction produced light.

Dubois's hypothesis was not confirmed until years later. It was found that luciferin is not a single compound. Actually many kinds of luciferins occur in luminous animals or plants. These luciferins are not necessarily closely related in the chemical sense.

SOME STUDIED IN DETAIL

In recent years, researchers have made very careful studies of the light-producing processes of three different organisms—luminous bacteria, fireflies, and the crustaceans belonging to the genus *Cypridina.* It was found that the luciferins in these three groups have little in common.

The luciferin found in luminous bacteria is a combination of organic compounds. It is made up of reduced flavin mononucleotide plus an aldehyde. When these compounds and bacterial luciferase are brought together in solution in the presence of dissolved oxygen, they produce a bright light. This lasts until the luciferin is used up. The end product of this oxidation process is acted on by another compound—reduced diphosphopyridin nucleotide—and luciferin is reformed. It is oxidized, reformed, and so on. All this results in a continuous light emission, which, as we have seen, is characteristic of bacteria.

The firefly has a particularly complex system for producing light. In addition to firefly luciferin, oxygen, and luciferase, a biological compound called ATP (adenosinetriphosphate) and magnesium ions are necessary. When all these substances are mixed together in a test tube, light will be produced. No light will appear if any one of them is missing.

Cypridina is responsible for much of the sea light observed in the Far East. The luciferin and luciferase are manufactured in separate gland cells and poured into the sea water from openings near the mouth of the animal. The luminescence appears in the water as a puff of light.

Usually a luminous animal or plant emits light of only one color. The colors most frequently produced are blue, green, and yellow. One remarkable insect—the "railroad worm" of South America, emits light of two colors—yellowish-green and red. This worm is a grublike luminous female beetle, belonging to the genus *Phryxothrix.* The male is winged and nonluminous. The railroad worm is about five centimeters long. It shows a bright red luminescent spot in the head region and a yellowish-green light on each side of eight segments of the abdomen. At night it offers a truly startling spectacle. The lights on the abdomen look like lighted windows. The red on the luminous spot of the head suggests the rear light of a train. As the worm crawls along on the ground, it looks like a tiny, fully illuminated train backing up slowly.

EVOLUTION OF BIOLUMINESCENCE

Bioluminescence is scattered in haphazard fashion throughout the animal kingdom. There is no indication that light production arose at a particular time or in a particular group and that it then followed definite evolutionary pathways. The first appearance of luminescence in a living organism was probably a chemical mutation, or sudden variation, resulting in a luminous spot. Once such a spot appeared on an animal or plant, the light may have been of value and may have persisted by natural selection.

The reverse process also occurs. A luminous organism such as a bacterium may suddenly produce nonluminous mutants. This type of mutation is due to the loss of one or another of the substances necessary for light production. The cell suddenly loses its ability to manufacture the substances in question. In the luminous bacteria, three kinds of dark mutants are known. One type lacks luciferase; another, the flavin mononucleotide; the third, the aldehyde. If the missing substance is added,

light will appear. Suppose, for example, that a colony of nonluminous mutants lacks aldehyde. If one blows the vapor from an open bottle containing aldehyde over the colony, it will suddenly begin to glow.

USES OF BIOLUMINESCENCE

It is doubtful whether the ability to emit light benefits the simple species of animals or plants. Of what use would light be to a luminous bacterium or fungus? How would it help a marine flagellate, with no nervous system and little power of locomotion? Such light probably represents a chance mutation, which persists for no apparent reason.

In the case of animals with a nervous system and definite behavior reactions, we can suggest various possible uses for bioluminescence. It is definitely known that the light flashes produced by the firefly serve as a signal to attract the sexes to one another.

In deep-sea forms, living in perpetual darkness, the function of light organs is not known. Many luminous marine forms live in regions of the sea where light penetrates. In such cases the animals may luminesce at night in order to lure prey or to elude natural enemies.

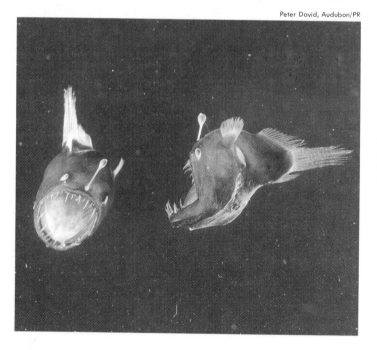

The angler fish uses a luminous bulb as bait. Smaller fish are attracted to the light and are swallowed by the angler.

Kenneth Lucas at Steinhart Aquarium

The light of the flashlight fish is produced by luminescent bacteria that live in beanshaped organs beneath the fish's eyes.

The deep-sea squid *Heteroteuthis* has often been cited an an example of a marine animal that uses luminescence in order to escape from its foes. This squid is found in the Strait of Messina between Italy and Sicily and is sometimes brought to the surface by the strong currents in that area. It is readily caught and can be observed in an aquarium. *Heteroteuthis* squirts a large amount of luminous fluid into the water whenever it is disturbed. The ordinary squid ejects black ink to divert the attention of a pursuing foe. It is believed that *Heteroteuthis* uses its luminous secretion for the same purpose. A pursuer would be confused by the glittering "liquid fire" and the squid would make good its escape.

The characteristic rows of lights on the sides and belly of deep-sea fish may serve to keep a school of fish together. Perhaps the fish has a complex signaling system, like that of the firefly. However, we can only guess about such uses of luminescence at the present time.

Within comparatively recent years investigators have begun to penetrate the depths of the sea in bathyspheres, benthoscopes, and bathyscaphes and have used the television camera to pry into the secrets of the deep. With such devices they will be able to make detailed studies of deep-sea animals in their natural environment. Some day they may be able to solve the fascinating mysteries of bioluminescence.

These members of the shrimp crustaceans excrete a luminescent cloud when chased by predators. They live in the ocean between 50 and 750 meters.

Peter David, Audubon/PR

BIOLOGICAL RHYTHMS AND CLOCKS

by Frank A. Brown, Jr.

The physical environment of living things on this planet follows certain rhythmic patterns. Night succeeds day as the earth rotates on its axis, turning first one side and then the other toward the sun. The day-night cycle represents the solar day of 24 hours. The solar day corresponds to the time that elapses from one appearance of the sun at the meridian to the next appearance. The actual length of the solar day ranges from about 23 hours, 45 minutes to 24 hours, 15 minutes. The 24-hour value represents the average length of time between one appearance of the sun at the meridian and the next. There is also a lunar day, marking the period from one time when the moon reaches the highest point in the sky to the next time. The lunar day is about 50 minutes longer, on the average, than the solar day. Since the rhythmic ebb and flow of the tides are effected chiefly by the gravitational pull of the moon, ebb and flow occur on the average about 50 minutes later each day. The sun also exerts a gravitational pull upon the earth. When sun and moon add their influences to one another, we have the highest, or spring, tides. When their joint action is at the minimum, we have the lowest, or neap, tides. Here, too, we have a series of rhythmic variations.

While the earth is steadily rotating on its axis, it is continuing in its annual course around the sun. Since the axis of rotation is tilted slightly, the earth's Northern Hemisphere is directed either more or less toward the sun at different points in its annual orbit. The same is true of the Southern Hemisphere. This gives rise to the rhythmic variations that we call the seasons, with their gradually changing lengths of daylight and of heating by the sun.

RHYTHMICAL NATURE OF LIFE

One of the most notable aspects of life on earth is its remarkable adaptation to the rhythmic characteristics of its environment. Like the environment, animals and plants display rhythmic patterns of activity. On the basis of such patterns, they tend to fall into two major groups—*nocturnal* (nighttime) and *diurnal* (daytime). Nocturnal animals, such as rats, mice, cats, owls, cockroaches, and moths, forage actively at night and rest secluded by day. Diurnal animals, including most songbirds, bees, butterflies, lizards, and generally, people,

Plants and animals generally fall into two groups— diurnal, or day-active, and nocturnal, or night-active. Below: some day-active species—a bee pollinating a flower and an iguana.

Schwitter

Devez—Jacana

use daylight for their foraging. Plants, too, show nocturnal and diurnal activities. Certain species bloom by day; others, by night. Plants carry on photosynthesis only during daylight. They usually grow most rapidly at night.

At the seashore, some animals feed when submerged in water during high tide, and protect themselves from exposure to the air at low tide. They include oysters, clams, barnacles, and many snails. Other animals, such as shore birds and fiddler crabs, feed actively upon the beaches exposed at low tide.

The foregoing patterns of activity are geared to the day-night cycles or to the ocean tidal cycles. Yet they are not simply direct responses to changes in light or tides. When living things are deprived of every obvious cue to the time of day, they commonly go on with their rhythmic daily patterns of behavior much as before. When an animal with a tidal rhythm is taken from a beach and placed in a tideless environment, its activity patterns may continue to match the tidal changes of its home beach.

BIOLOGICAL CLOCKS

The term *biological clock* is applied to the means by which living things adjust their activity patterns, without any obvious cue, to the time of day, or the month, or the year. The biological clocks seem to be beautifully adapted to the needs of living things. They are affected but little, if at all, by drugs, chemicals, or wide temperature differences—factors which may alter substantially the rates of all ordinary processes of the body.

The nature of the biological clocks' mechanism is still a mystery. Two quite different theories have been advanced to account for them. According to the first of these theories, each individual contains its own independent timing system. This is believed to have evolved, aided by natural selection, as an adaptation to the rhythmic environment. It has now become independent of the environment. According to this view, the clocks are not perfect timers. They require regular corrections by the natural light and tide cycles and the chang-

R. Tereafs—Jacana

Atlas Photo

Owls and bats are normally nocturnal animals. Their internal biological clocks schedule most of their activity for darkness.

Rives/C.E.D.R.I.

The activity of clams is regulated not so much by light and dark but rather by the cycle of the tides. They normally feed at high tide.

ing lengths of the day throughout the year.

The other theory holds that living things react continuously to their rhythmic physical environment. The setting of their biological clocks, therefore, involves a constant adjustment to subtle environmental forces. If this view is correct, the basic living clocks are potentially perfect timers.

Biological clocks appear to be everywhere in living things—even in individual cells or parts of cells. But the search for the specific timing system has been futile thus far. Despite the careful study of many rhythmic phenomena and even of detailed chemical variations between cells, there is no evidence that any one of them is the clock-timer itself. Not only has no independent timing system ever been discovered, but there has not yet been even a plausible guess as to its nature.

SOLAR-DAY CLOCKS AND RHYTHMS

In both plants and animals, there are persistent daily rhythms, based on the 24-hour day-night environment. In plants, these rhythms involve such phenomena as the movements of leaves and petals, the ability to carry on photosynthesis when light is offered, the growth rate, and the influence of lowered temperature. The daily rhythms of animals apply to such things as rest and movement, changes in skin color, sensitivity to light, and the composition of the blood. In fact, there is reason to believe that persistent daily variations occur in any organism or in any part of an organism. This is true of such fundamental phe-

nomena as the synthesis of proteins in accordance with the biological code supplied by DNA.

Perhaps the most characteristic property of clock-timed daily rhythms is the fact that they can be reset in such a way as to have the rhythmic events occur at another time of day. For example, a mouse or a cockroach, which is naturally active during nighttime, can have its rhythm reset so that it will show activity during the hours of daylight instead. This is done simply by subjecting the animal to a few inverted 24-hour light cycles, in which there is darkness by day and illumination by night. Thereafter, if the animal is deprived of the daily natural light changes, it will show a persisting rhythm set to this different time of day. Indeed, this is the principal means by which any plant or animal that is moved to new time belts of the earth can reset its daily rhythms to conform to the new times of day and night. Investigators have reported that living things can also use daily temperature variations to reset rhythms. In this case, lower temperature is treated like darkness and higher temperature like light.

Clock-timed daily rhythmic variations may also be employed as a clock for time-training. For example, one may teach honeybees to come to a feeding station at the same hour each day, or to arrive each day at two different hours. But training is possible only on a 24-hour schedule. Bees cannot be taught to arrive, say, at 19-hour intervals. Bean seedlings have been subjected to quite the same sort of time-training. Beans growing in continuous dim light will respond to briefly increased light by an extra brief elevation of their leaves. If there is no further stimulation, this extra elevation will recur for the next few days at the time of day when the increased light had been previously supplied.

These two examples from widely different kinds of organisms show how the daily cycle of change may be modified. Perhaps such modifications may contribute to the closely-knit animal and plant communities. This will depend on an adaptive daily-clock programming of the activities of each member of the community.

Evidence indicates that the setting of the daily rhythm to an altered experimental 24-hour light-cycle or to a new time belt of the earth depends upon a daily rhythm of light response in the organism. Both dawn and dusk contribute to the final accurate adjustment. In one part of the daily cycle of response, the rhythmic pattern is adjusted forward in time. At another part, the pattern is adjusted backward in time. The final adjustment is attained when the cycle is fitted to the external daily light variation so that the two responses just balance one another. Probably much the same sort of thing happens in rhythm-resetting by temperature cycles.

Here we might mention a phenomenon that is quite common. When any one of a wide variety of animals or plants is placed in unvarying light and temperature, its normal continuing daily patterns of change slowly drift toward progressively earlier or later hours of the day. The rate and direction of the drift varies from one individual to the other. This results in periods that are either a little shorter or a little longer than the 24-hour day. The daily rhythms that change their pattern a little in this way are called *circadian*, from the Latin *circa diem:* "about a day." The observed period of a circadian rhythm usually lies somewhere within the range from 19 to 29 hours. It varies not only from one individual to another but also with the previous treatment of the organism. It also varies with the levels of light and temperature and even, from time to time, for no obvious reason at all.

TIDAL AND MONTHLY CLOCKS AND RHYTHMS

Animals that live in the intertidal regions of the beaches, as do fiddler crabs, are provided with a tidal clock. This adjusts their periods of activity and rest to the appropriate times in the cycles of ebb and flow of the tides. Fiddler crabs on the beaches feed most actively just before low tide, and normally rest quietly in their bur-

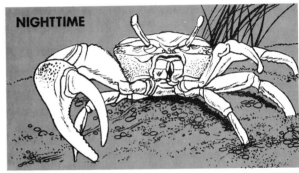

Fiddler crabs (right) offer a fine example of the functioning of the biological clock. These animals change their color in accordance with the 24-hour cycle of light and dark. They have a pale coloring at night and a darker coloring during the daytime. In the diagram below, the changes in color have been plotted for several days.

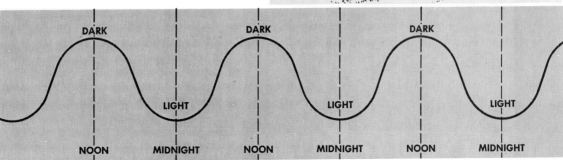

rows at high tide. When the crabs are removed from the beaches and the tides, their activity may still remain timed to the tidal variations of their former beach.

Careful studies have revealed that the tidal rhythms in crabs do not have a uniform period, any more than the tides themselves do. It is true that the tides, on the average, rise about 50 minutes later each day. But at some times during the month, the tidal cycles occur up to 1¼ hours later each day. At other times they occur only a half hour or so later. When crabs are taken away from the tidal changes of their beaches, they may follow in an uncanny manner the systematic variations in the tidal cycles. To do so, they probably, as do the tides themselves, depend upon the cooperation of solar-day and lunar-day rhythms. These systematically supplement one another to the greatest extent at half-monthly or monthly intervals.

The presence at one and the same time of tidal clocks and 24-hour clocks provides living things with an effective means for timing longer lunar cycles. In many kinds of animals and plants in the ocean, there are semimonthly and monthly rhythms in swarming for breeding purposes or in the production and liberation of eggs and sperm. One of the most famous examples is

that of the fishes called grunions, which swarm onto the sandy beaches of California to breed just as the highest tides of the month reach their greatest height.

CLOCKS AND NAVIGATION

Rhythms timed by biological clocks are used in direction finding by a number of different creatures, ranging from birds, turtles, and fishes to insects and crustaceans. Many of these animals seem to employ the azimuthal angle of the sun as a compass reference. The azimuth of a body in space is measured westward along the horizon from the south point to the vertical circle of the body—an imaginary circle passing through both the body and the zenith. The azimuth angle of the sun is the sun's azimuth, measured in degrees, minutes, and seconds. But as the earth rotates, the sun's position steadily moves from east to west. To maintain a fixed geographic direction, the angle of the animal relative to the sun must change continuously throughout the day. This ability is in some manner dependent upon clock-coupled rhythms. This has been demonstrated by subjecting a bird to light-dark cycles set to earlier or later times of day than the natural ones. When the bodily rhythms are thus set experimentally to a 6-hour earlier time, the path of animals using the sun as a compass is 90° to the right of what it would otherwise have been. When set 6 hours later, it is 90° to the left.

It has been reported that some birds use the stars or constellations as compass references for their nighttime migrations. To maintain an unvarying geographic direction under these circumstances, still another kind of daily "clock" is required—a

This photograph, taken at night by flash on a sandy beach of southern California, shows the remarkable breeding activities of the fishes called grunions. They emerge onto the sand between waves at the high-tide peak of the lunar cycle. The female burrows into the sand, tail first, and lays her eggs. The male wraps himself around her and releases sperm, which fertilize the eggs. The fishes return to sea with the next wave that reaches them. All this takes place only during the high tides that occur at night. The young that hatch from the eggs go to sea about two weeks later with the next high-tide peak.

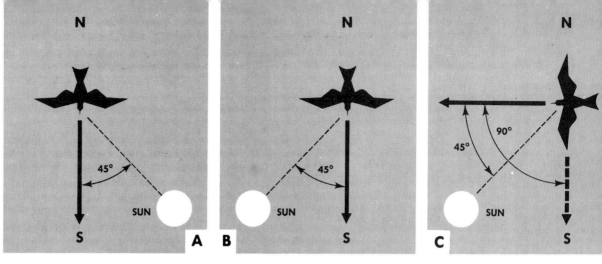

Experiments have shown that a bird's biological clock affects its sense of direction. A starling in an outdoor cage during the fall migrating season will keep fluttering its wings while pointing in a southerly direction. It takes its direction from the sun. If the sun is in the position in *A* at 9:00 A.M., the bird heads south keeping 45 degrees to the right of the sun (right, as the bird sees it). In *B*, it is 3:00 P.M.; the sun has shifted its position as shown. To keep on course due south, the bird maintains a 45-degree angle to the left of the sun. Suppose now that the bird is subjected to a series of artificial days in a laboratory, in which a man-made source of light is substituted for the sun. Each of the artificial days is made to begin and end six hours later than in the natural cycle. After the bird's biological clock has been reset in this way, it is brought outdoors at 3:00 P.M. *(C)*. The bird will now act as if the sun were in the 9:00 A.M. position. It will flutter in a direction at an angle of 45 degrees to the right of the sun, and consequently it will be headed due west instead of south.

sidereal-day clock, measuring recurring periods of 23 hours and 56 minutes. The sidereal day is the interval between successive passages of a star across the meridian. As the earth rotates, the stars rise and set about four minutes earlier each day.

Celestial navigation clearly depends upon a truly complex clock system. We know that any given star rises at the same time of day once a year and that the moon rises at the same time of day once a month. Hence the ability of living things to navigate using the heavens shows that their clock systems can provide rather precise information as to the time of month and the time of year.

24-HOUR CLOCKS FOR SETTING ANNUAL CYCLES

We have seen that the coming and disappearance of light in the 24-hour day may set very effectively the daily patterns of behavior. Just as effectively, the naturally changing relative lengths of night and day set annual patterns. Such responses to the daily photoperiod (period of light) determine the times of year of bird migration, nesting, and breeding; of the emergence of moths from their winter pupal cases; of the growth and flowering of plants; of seed germination; of the start of dormancy; and numerous other phenomena. This is especially characteristic of the temperate zones. Here the great weather-related variations in temperature and intensity of illumination cause the time between about sunrise and sunset to be the most reliable indicator of the time of year.

Some "long-day" plants and animals are brought into the flowering or breeding state by response to the long days and short nights of spring. In such spring-bloomers and spring-breeders, a daily rhythm occurs in response to light. The pattern of responsiveness is set to the time of light onset in the morning. Maximum stimulation is reached a short time after the onset of light. Shortly after 12 hours have elapsed, there is a rise toward a second maximum in light-stimulating response. This commonly reaches its peak about 16 to 18 hours after the onset of light.

In contrast are the fall-blooming flowers and fall-breeders—the "short-day" organisms. Their sensitivity to the light that occurs 16 to 18 hours after the normal daily light onset causes their activity to be depressed or else reverses the influence of the morning stimulation.

It must be recalled that the same daylight period usually occurs twice each year. Therefore some additional means must be present to regulate the annual cycles of living things. In the case of birds, one of these other factors is the need for a period of "short days" in order to prepare the birds for ability to respond to "long days." There may also be an extended period of resistance to photoperiods after one annual period of activity. In many seeds and insect pupae, a period of exposure to low temperatures must occur between periods of light responsiveness.

ANNUAL RHYTHMS AND CLOCKS

The possession by living things of deep-seated periods paralleling those of the organism's geophysical environment extends even to the year. In the absence of any obvious clues as to the time of day or year and in the arbitrarily set temperature of a refrigerator, slugs display their annual breeding activity on schedule. Samples of seeds stored in constant conditions, even at $-22°$ Celsius or $+45°$ Celsius, exhibit an annual rhythm in their ability to germinate. In one experiment, bean seeds were stored for three years in darkness and in essentially uniform temperatures. They were then allowed to sprout and they were tested for metabolic rate. In the course of the next three years they showed significant annual variation. Weaver finches have been observed to go through two practically normal annual breeding cycles on schedule, despite the absence of any annual cues such as changes in temperature or photoperiod.

It is becoming increasingly evident, as investigations continue, that almost all, if not all, living systems possess a biological clock for yearly periods.

BIOLOGICAL RHYTHMS AND CLOCKS IN PEOPLE

Humans possess a biological-clock system like those of other animals and of plants. Most conspicuous is the daily rhythm of sleep and wakefulness. There are also a host of daily variations which are correlated with the sleep-waking cycle. The daily fluctuations in body temperature, heart rate, and blood pressure have long been known. There are daily variations, too, in the activity of the endocrine glands, the composition of the blood, and the excretion of different substances by the kidneys.

The human daily rhythms, like those of other living things, have been observed to be circadian under isolation in conditions of unvarying light and temperature. Under these circumstances, as in other living things, the rate of daily drift of various bodily processes in the same individual may differ one from another. This may affect the former harmonious relationships between these processes.

Such a situation occurs following long trips in eastbound or westbound airplanes. Man arrives at his destination after a trip of this sort with his rhythms still adjusted to his former location. Various daily rhythmic events cause his rhythms to be reset for the new day-night cycle. But different bodily processes will be readjusted at different times. Some may become completely reset in only a day or two, while others may require up to a week or so. During this period of readjustment, the diverse bodily processes are out of gear with one another. Hence the individual is not operating at maximum efficiency. This has become increasingly evident in the case of such persons as jet-plane pilots and stewardesses, who fly constantly back and forth through many degrees of longitude and latitude.

Experiments have been performed with humans in the Arctic in summer at latitudes where no significant differences in illumination occur between day and night. Different groups were given watches that had been altered to indicate false day lengths — 22 hours and 27 hours. Those taking part in the experiment were asked to use these watches to time their daily routines. Some bodily rhythms tended to follow the imposed artificial activity cycles. Others stubbornly retained their 24-hour period. The result, again, was a lack of harmony between the different systems of the body.

BIOELECTRICITY

Tell a friend that his body is seething with electrical activity and he will probably smile in disbelief. The word electricity generally calls to mind the house current that operates radios and washing machines or the terrific power of lightning. But the flow of electric current and the flash of lightning are only two phases of a broad term, applying to numberless displays of force and energy. Some of these are mighty, while others are insignificant. Some have to do with nonliving materials, and others with life processes. All are based on the properties of certain atomic particles.

The electrical impulses in living things can be detected in various ways. Our own bodies can be made to generate a feeble electric current under certain circumstances. If two electrodes are placed in contact with the body at appropriate places and connected with a copper wire or other conductor outside the body, current will flow through the conductor—current whose pattern and strength can be accurately analyzed. Electrical activity in living things is called *bioelectricity*.

ATOMIC BASIS OF BIOELECTRICITY

Modern electronics has contributed a great deal to the study of bioelectricity. Miniaturized transistors and other precision instruments can accurately measure the electrical currents produced by living tissues. These instruments are also used to create or change bioelectric currents in living tissues, and they are often implanted in a living animal.

All of these instruments measure the effects of unlike electric charges in the human body. What produces these charges in the first place? Our most useful starting point is the modern theory of atomic structure. According to this theory the atom has a nucleus, or core, made up of positively charged protons and, in most cases, uncharged neutrons, packed tightly together. Whirling around the nucleus are one or more negatively charged electrons.

Only the electrons in the outer shell of atoms are involved in chemical reactions, in which atoms combine with each other. They combine in several different ways. In some cases the atoms are joined together in a compound through electrical, or, as scientists would say, *electrostatic,* attraction. Ordinary salt, or sodium chloride, $NaCl$, furnishes a good example of this kind of bond.

IONS IN SOLUTION

Atoms and molecules that acquire an electrical charge, either positive or negative, are called *ions*. When a salt crystal is dissolved in water, the ions in the crystal

Muscle contraction is one part of the body's electrical activity. The rate of muscle contraction in this volunteer's forehead is being measured in a study of relaxation and its effect on muscle activity.

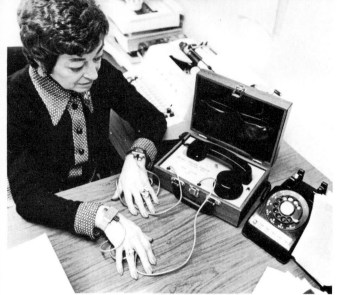

Cardiac pacemakers regulate the electrical activity of the heart. This woman with a pacemaker is obtaining instant evaluation. The electrical impulses are transmitted by a phone system to a physician.

are separated from one another. Instead of being bound together through the force of electrostatic attraction, they become free ions moving every which way in the liquid. The creation of free ions of this kind is called *ionization*. Various substances in solid, liquid, and gaseous form undergo ionization. Water itself, though it is the most effective of all solvents, hardly ionizes at all.

Neither water nor common salt in its crystalline, or solid, form can conduct electricity to any appreciable extent. They are good insulators. But when the salt is dissolved in water and is ionized, the solution conducts electricity very well indeed. Any substance which, like common salt, becomes a conductor of electricity when it ionizes in water or some other liquid is called an *electrolyte*.

There are three kinds of electrolytes: acids, bases, and salts. When an acid is dissolved in water, it yields positive hydrogen ions (H^+). The greater the concentration of H^+ in a solution, the stronger the acid.

When bases are dissolved in water they yield negative ions, made up of oxygen and hydrogen bound together. These ions, called *hydroxyls*, have the symbol OH^-. The greater the concentration of OH^- in a given solution, the stronger the base is.

The electric charges produced when electrolytes ionize are responsible for a vast number of chemical and physical phenomena. In this article we are particularly interested in the effects of ionization in the body. They account for a vast number of essential chemical reactions in the body. Fortunately for us, our bodies abound in electrolytes—acids, bases, and salts—and these substances can be readily ionized because bodily tissues are bathed in water, the most effective of all solvents. Much of the fluid in the cells—a large part of the cells' content—is essentially water.

ACID-BASE AND LIVING TISSUE

All living tissue contains hydrogen. And the hydrogen ion (H^+) is necessary for such activities as muscle contraction, nerve impulses, digestion, and respiration. Measuring the H^+ content of electrolytic solutions is standard procedure in much of the work of the chemist and the physiologist.

In the human body, life seems to be possible only within a very narrow acid-base, or pH, range. Yet acids and bases are constantly being produced in various processes. Some, of course, pass out as wastes. Nevertheless, excess amounts get into such fluids as the blood. To offset their effects, systems of *buffers* are present.

Given the fact that living cells are bathed in electrolytic solutions, in which ions are constantly being formed, it is hardly surprising to find that the surfaces of practically all cells have electric charges.

FUNCTION OF THE PLASMA MEMBRANE

Cell surfaces are usually called *plasma membranes*. They may be actual membranes or very thin films, though in many cases no cell wall is visible even under the most powerful microscope. Only in plant tissues do most of the cells have real walls. In many kinds of animal cells the plasma membrane does not seem to be much more than a boundary, or interface, such as is formed when a drop of oil is suspended in water. Nonetheless, the plasma membrane acts as if it were semisolid.

Within the cell there seems to be a similar arrangement, with the nucleus in a kind of envelope. Besides this, fine particles of protein are suspended in the fluid contents of the cell. The proteins are not dissolved but are colloidal particles, evenly distributed in the fluid. Apparently they are held in place because they all have like electric charges and so repel one another. Protein is, as you know, an important ingredient of protoplasm, the living matter of the cell.

If the cell is to be nourished and is to get rid of its wastes, food substances must pass in and waste products must pass out through the plasma membrane. At one time it was thought that the membrane has pores, but this idea has been largely discarded. Passage through the membrane occurs only when the fluids on either side are unequal in concentration — that is, one side must have more dissolved or suspended matter in it than the other. By what is called the process of *osmosis,* the two fluids tend to become equal in concentration, with that of less concentration moving toward the other, though there is also a slower and less noticeable movement in the opposite direction.

The plasma membrane is semipermeable. It will let some substances pass in or out and not others. In general, it tends to permit water, the solvent, to pass readily but is selective regarding dissolved substances, or solutes. As we have seen, the solutes are composed of ions. Some ions appear to pass in and out much more rapidly than others and some seem to be blocked altogether.

In any event, in living tissue a balance of ions on either side of the membrane is never reached. This would account for the difference in electric charges on the inner and outer side of the membrane — a *potential difference,* which would produce a current if an appropriate pathway were provided.

NERVE IMPULSES

The idea of a membrane with different charges on either side is the basis for the most widely accepted explanation of nerve impulses. At rest, the membrane of a nerve fiber is positively charged on the outside and negatively charged on the inside, which indicates that the membrane is keeping the

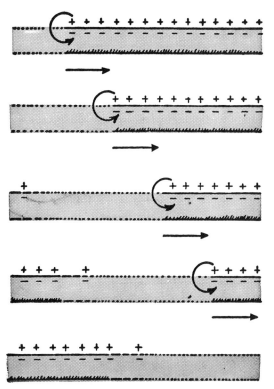

From "Machinery of the Body," by Carlson and Johnson, Univ. of Chicago Press

Transmission of an impulse along a nerve fiber, in five stages. As the membrane is penetrated (dotted area), positive and negative ions come together and neutralize one another. Ions from the adjoining section now make their way through the neutral gap. Another area becomes permeable. Thus the impulse makes its way.

ions on either side separated. Since no electrical activity is going on, we say that the area is inactive.

When a nerve impulse reaches any given place in the fiber it seems that the membrane is penetrated. The area then becomes active electrically. Its negative and positive ions come together and neutralize one another. Through the neutral gap thus made, ions from the next not-yet-active place along the nerve fiber penetrate. This makes the outer surface of this second area permeable. It, in turn, affects a third place, still farther along the nerve fiber, and so on until the impulse completes its journey. Meanwhile the charges and membrane impermeability of the "rest" state are restored, after a brief interval, to each area in succession. The passage of the nerve impulse is often likened to a spark traveling along a fuse.

This idea of the nerve impulse would also explain the fact that an injured nerve is negative because an injury would expose the inner side of the membrane, with its negatively charged ions.

Sodium, calcium, and potassium ions are important in the function of nerves. A complicated compound called *acetylcholine* that is produced by nerve tissue also enters the picture. The method by which nerve impulses are transmitted has been widely studied, and a complex picture of impulse transmission has been developed.

HUMAN BRAIN RHYTHMS

By recording the electric impulses generated by the brain and the heart it is possible to detect or pinpoint some disorders of these two important organs.

Of particular interest to the brain specialist are the waves generated by the cerebral cortex. This is the soft mantle of folded gray matter that covers the two hemispheres of the cerebrum, much the largest part of the brain. We think and feel and have knowledge of the world about us because of the cerebral cortex. Into its texture are woven thousands of millions of nerve cells. By means of the fibers that extend from each one, the cells communicate with one another and with other areas of the brain and these, in turn, with the rest of the body. The vast number of connections of nerve endings thus made possible staggers the imagination.

Because it lies just beneath the top of the skull, the cortex is, of course, the most accessible part of the brain. So it is possible to get quite close to the cortex by attaching electrodes to the scalp.

From the human cortex, three principal kinds of electrical waves may be detected. The *alpha wave* has a frequency or vibration, of 8 to 12 hertz (cycles per second) and a voltage of from 10 to 100 microvolts—millionths of a volt. It is picked up most easily from the occipital region (the lower back lobe) of the cerebrum when a person is at rest. With mental effort the rhythm tends to disappear.

The *beta wave* is faster—18 to 35 hertz—and is considered normal if the voltage does not rise above 50 microvolts. This rhythm seems to come most steadily from the frontal lobes and to be related to the motor-sensory system. Motor nerves carry messages from the brain to the muscles. Sensory nerves carry messages from the sense organs to the brain.

The *delta wave* is extremely slow, 6 hertz or less, and has a voltage of from 20 to 200 microvolts. It appears most often in normal sleep.

The *theta wave* is abnormal and has a frequency of 6 hertz. It sometimes indicates the presence of a deep tumor; and a "spike," an abnormal sharp wave that lasts no longer than one-tenth of a second and is an indication of epilepsy.

The pattern of brain waves differs with each person. It requires an expert to interpret an *electroencephalogram*—as the record of the pattern is called. Even then it is used only in the light of other information to diagnose a brain disorder.

An amplifier is used to magnify the weak brain waves. It may be harnessed in an electroencephalograph or in a cathode-ray oscilloscope. The *electroencephalograph* is the usual instrument for examining the human brain. The scalp is moistened with a salt solution and electrodes are attached to it, painlessly. They pick up the faint electric signals, which are then relayed through amplifiers and magnified perhaps millions of times. The voltages pass to an electromagnet which, in turn, impels a pen to write the pattern on a moving tape.

In the *cathode-ray oscilloscope*, the magnified voltage is directed against a beam of electrons moving in a cathode-ray tube. The beam is thrown on a fluorescent screen, where it appears as a quivering, luminous line. If the area of the brain being tested flares with increased electrical activity, the line pulses in rhythm with the variation in voltage. The length and height of the line's waves are measures of the electrical activity of the region.

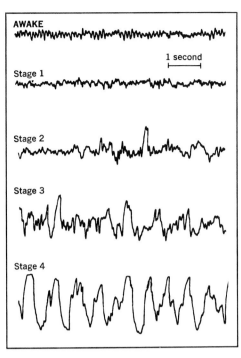

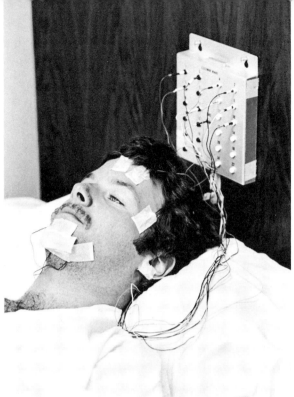

Electrodes are attached to a subject to record brain-wave activity, eye motion, and facial-muscle activity during different stages of sleep. Some brain-wave patterns typical of wakefulness and of progressively deeper sleep are shown.

CONTRACTION OF HEART MUSCLES

The electric impulses that accompany the beat of the heart—the contraction of heart muscles—are recorded by the *electrocardiograph.*

Normally the heart beats about 70 times per minute so that one beat lasts only about 0.86 second. Yet in that short space of time there is not one contraction but a series of contractions, called systoles, and periods of relaxation, called diastoles, which together are called the *cardiac cycle.* It begins in the right auricle, a little ahead of the left auricle. The auricular systole, by way of a bundle of tissues connecting auricles and ventricles, in turn stimulates the ventricles, which contract together. Each chamber relaxes immediately after its contraction is completed. Thus the auricles are relaxed at about the same instant that the ventricles begin to contract.

Just before each contraction in the cycle there is an electrical change. Active muscle fibers (like nerves) become negative to inactive fibers. Thus the electrocardiograph records the potential differences set up by each contraction in the cycle.

The charges are comparatively strong, and since the body's fluids are electrolytic solutions, the charges may be detected on the surface of the body by such a sensitive instrument as the electrocardiograph. Its electrodes are used in pairs and must be carefully placed so as to connect unlike charges. One may be placed on the hand and the other on the foot, for in the fraction of a second during which the upper part of the heart contracts, either hand will be more negative than either foot because the hands are closer to the active heart region.

The record, or *electrocardiogram,* of a healthy heart beat shows a low wave, P, at the beginning of the auricular systole. At the onset of the ventricular systole there is a high wave, QRS. A low wave, T, marks the end of the ventricular systole. Normally, the peak, R, of the highest wave has a value of from 0.70 to 1.80 millivolts—thousandths of a volt. Diseased heart conditions may be revealed by a departure from the normal pattern.

BIONICS

by Morley R. Kare

A sparrow swoops through the air. Adjusting swiftly for flight speed and wind, it lets down its landing gear just in time to stop precisely on a twig. How does this bird make all the calculations and accomplish the feat so deftly? Most airplanes, even with pilots' compartments filled with instruments, still need long landing fields.

If we were to place a piece of rotting meat in a stream, it would soon be covered with invertebrates such as *Planaria* that are directed to the meat by their chemical senses. Many fish, with taste buds on the outside of their bodies, can demonstrate even more impressively a sensitivity for chemical stimuli. How do these animals perceive the chemical, and how do they track it down?

These questions are of special interest to researchers in an area of scientific technology called *bionics* (*bio,* from the Greek word for "life," and electro*nics*). Bionicists attempt to solve technical problems by the application of mechanical and electrical mechanisms comparable to those found in nature. They design and develop machines and instruments that to some extent mimic, or simulate, the reactions and behavior of living organisms.

Although the name bionics was coined as recently as 1960, the subject to which it refers has interested man for hundreds of years. Early inventors of flying machines, for example, tried to imitate biological examples by building their unsuccessful flying machines with flapping wings. Some models even included a covering of chicken feathers. Levers such as crowbars and bottle openers are applications of mechanisms found in the skeletal systems of animals. More recently, an optical ground-speed indicator for airplanes was designed by mimicking a beetle's response to moving light patterns. And an understanding of the function of the visual system of the horseshoe crab has suggested a means of creating a television image with sharper contrast than

any previously produced television system.

THE SCOPE OF BIONICS

The field of bionics overlaps or approaches other comparatively new disciplines. It is closely related to *cybernetics* — the science of control and communication systems in animals and machines. It merges with *cyborg* (from *cyb*ernetic *org*anism) — the mechanical, electronic, and biomedical techniques used to improve the performance of the human body. Bionics is related to *robotics* — the design of robots, or machines that resemble men or animals or that perform some of their functions.

Finally, bionics is related to the abstract (not actual or mechanical) simulation of life processes by means of computers or mathematical models. For example, biomathematicians have made computer models of systems in the human body, including the respiratory, nervous, and circulatory systems. Information about a particular system is fed into a computer, which determines the relationships between the various components of the system, and analyzes the ways in which their actions are controlled. Mathematical models representing each of the theories are developed, and the computer determines which of these models best simulate the function of the system. Such abstract models may lead to bionic applications. For example, a simple mechanical model of a nerve has been developed from such a study. Based on a steel tube and water bath, it is able to simulate various nervelike phenomena.

In this article, we will restrict ourselves to mechanical, chemical, and electronic simulation of biological reactions.

THE BIONICIST'S APPROACH

A simple example will serve to illustrate bionic techniques. In building a submarine, designers wish to determine the most efficient shape for the hull. The bionic approach to this problem consists of study-

ing organisms that exhibit the desired characteristic of moving through water with the least amount of resistance. We know that fish meet this requirement and that sharks and dolphins are fast, efficient swimmers and could serve as models for the shape of our hull. The answer given to engineering by the biological model may not be simple and straightforward. For example, after a group of engineers determined the muscle power, size, and shape and calculated the theoretical speed of the fast-swimming dolphin, it was determined that this aquatic animal's observed speed was ten times its capacity to produce energy. The explanation is thought to rest upon the rubbery skin of the dolphin, which tends to reduce the drag of the water flowing along its surface. Thus it would seem that an actual imitation of the elastic covering of the dolphin might increase a submarine's velocity. This information has also been considered for airplane coverings and the inside coatings of pipelines.

The lesson to be gained is that engineers must do more than superficially copy nature. They must determine the crucial principles that account for the specific characteristic: the why and how in the construction of an organism. Of primary importance in the flight of a bird is the shape of the rounded leading edge and the thin, tapered trailing edge of the wing. The feathers are not so important.

Many of the basic characteristics of living organisms suggest that there are new and more fundamental approaches to old engineering problems. At present, for example, we use heat to generate steam. This steam drives a turbine that drives an electric generator. It would be more efficient to convert the heat directly into electrical energy. It is known that certain microorganisms will generate electrical potentials in the process of breaking down organic waste products such as garbage. Might a properly designed garbage dump yield significant amounts of electrical power?

STUDY OF ANIMAL SENSES

It is the sensory mechanisms of animals that have attracted the most interest among bionicists. In the past it was assumed that simpler forms of animal life had less complex sense organs than man. It is now abundantly evident that this is not always the case. Some animals with nervous systems less developed than man's have certain sensory capabilities that exceed our own. A dog can hear sounds in ranges beyond those discernible to the human ear. A bird has an acute sense of visual detail. It can better discriminate objects than can a human being. A butterfly, using sense organs on its feet, can detect minute concentrations of sugar. Man is unable to taste such small amounts of this substance.

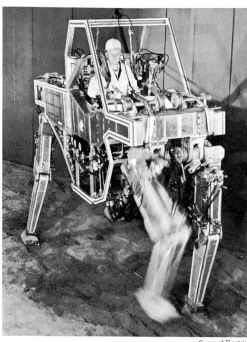

By means of an advanced control system, this four-legged "walking machine" mimics and amplifies the movements of its operator. The operator's arms control the unit's front legs. His legs control the rear legs of the machine.

General Electric

Not only do many organisms surpass our abilities in sensory modes similar to our own, but they possess types of perception unknown to man. An African fish, *Gymnarchus niloticus,* possesses special organs that generate a weak electric field around the fish. Any object that enters this field causes a disturbance. A change of $1/5{,}000{,}000$ of a volt is detected by the fish.

Perhaps the most studied of the senses not found in man is the echo-location system of the bat. This system is akin to the principle of radar and sonar. The bat emits short shrieks of sound, which sweep or scan over a range of frequencies, too high for a human to hear. These are reflected from obstacles and return to be heard by the bat. In natural surroundings, the bat locates moths and other insects by means of this echo ranging. Moths, in their turn, can detect the high-frequency shrieks with a pair of tiny hearing mechanisms, each consisting of only two cells. When the moth detects a batlike sound, it begins to fly erratically, rapidly changing directions and seeking ground or other cover. This behavior can be observed in the summer by attracting moths to a light and shaking a set of

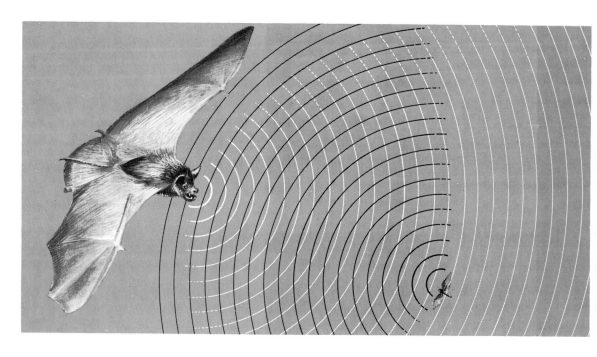

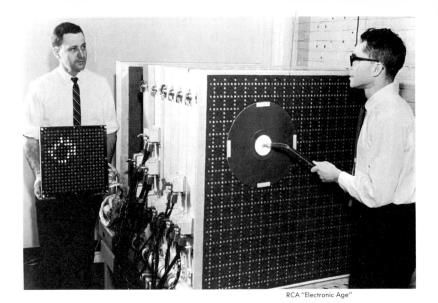

This electronic device "sees" like a real frog's eye. The display panel held by the engineer on the left shows what the simulated eye is seeing. Compare it with the disk held before the device by the man on the right. The location of the disk on the "frog's eye" is the same as on the display panel.

RCA "Electronic Age"

keys. Some of the metallic sounds of the keys mimic those of the bat, thus provoking the moth's defensive flight pattern.

While man's senses are often downgraded when compared with those of some animals, they are, nevertheless, of interest to bionicists. Under ideal conditions man can perceive a wire one third of a centimeter in diameter at a distance of almost one kilometer. He can detect some odors when they are present in only a fraction of a millionth of a gram. He can hear sounds so low that his eardrums vibrate through a distance of less than the diameter of a hydrogen atom.

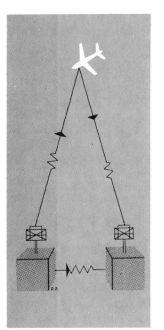

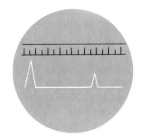

Tracking aircraft by radar, a bionic application of the bat's echo-location system. As a transmitter (left) sends out a searching pulse, it also sends out a locking pulse to the receiving set. The range indicator (below) shows a deflection from the locking pulse and a second, smaller deflection from an echo pulse.

BODY MECHANICS

Although bionicists have been most interested in sensory mechanisms, there are equally important engineering lessons to be learned from a study of other biological systems. There is the internal-gland cooling system of a fowl; the double-barreled pump in the proboscis of a mosquito; the bioluminescence of a firefly; the nutrient-distribution system in plants.

Studies of insect flight reveal that these animals are equipped with tiny hairs to sense speed. The leading edges of their wings are slotted and able to detect turbulent airflow. It was also found that insects obtain the greatest lift on the flat wing surfaces when they are close to stalling speed. Airline pilots flying between San Francisco and Hawaii, in efforts to conserve fuel during the 4,300 kilometer journey, discovered the same principle several decades ago.

APPLICATIONS

A number of mechanical and electrical principles found in living organisms have been used to improve man-made devices or to invent new ones. Pumps, levers, and bellows are inventions in which man has unknowingly simulated animal mechanisms. Applications in the design of airplanes, submarines, and other machines have already been mentioned.

The eye of a frog reacts primarily to small moving objects such as insects. But it is also very sensitive to the bluish parts of the light spectrum. It is believed that this adaptation enables the frog to jump toward water when danger threatens (water reflects blue light waves). Military and aerospace engineers have developed a radar system based on the optics of a frog's eye. It can detect planes, missiles, and spacecraft better than ordinary radar, which is too often confused by accidental intrusions such as clouds, rain, and birds.

Medical applications include a small, portable sonar apparatus built into a special cane for the blind. Using echolocation, the sonar detects obstacles in the person's path. The resulting changes in the sonar sound are heard by the person, thus enabling him to walk more freely.

Studies of the nature of muscular contraction and of the relationships between muscles and nerves have led to the construction of an electronic arm. This artificial arm operates through signals from the brain, in much the same manner as a normal arm. The arm is powered by a battery which, in turn, is activated by electrodes attached to the skin over muscles in the

This artificial arm simulates the actions of a real arm. When the man tenses certain body muscles, tiny electrical currents are produced. These currents activate the arm.

F. Ray Finley

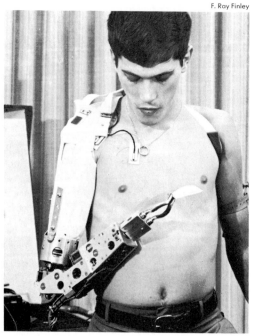

chest, back, and shoulder that control arm motion. If, for example, the person wishes to open a door, electrical impulses for the necessary movement travel from the brain to nerves in the muscles. Here the impulses are picked up by the electrodes, which amplify and transmit them to an electronic box in the arm.

In a system called Visilog, the distance-judging ability of an eye is simulated in an instrument within a moving vehicle. This device detects the vehicle's approach to a solid object by the apparent change in the texture of the solid's surface as it becomes larger in the field of view. The device can compute the rate of approach from these textural impressions. It then slows the vehicle accordingly.

In some cases, man's machines far surpass living things in power, speed, and adaptability. In other cases, they are woefully inferior. The biological cell, for example, is a miracle of complex functions in a microscopic package. Man's attempts at miniaturizing his devices are almost laughable in comparison. There are also limits on the usefulness of biological models for engineering-system designs. For example, no animal uses the wheel as a mechanism to propel itself.

The purpose of bionics is to look for and simulate systems, techniques, and devices in the world of plants and animals. The ultimate bionic development would encompass all the abilities of the nervous system of a living organism. It would be capable of modifying its operation to meet the demands of the environment. It would be capable of learning to cope with changes and have the capacity of self-repair. It would be able to receive, store, and intelligently use information.

As human society becomes more and more complex, new ideas for devices are needed in an ever-increasing number to keep in step with the changing mode of living. There is ample incentive to encourage engineers to study the self-adaptive and self-organizing systems found in nature. We can expect that the systematic combination of physical science and biology will be a fruitful union of benefit to all mankind.

CHARLES DARWIN

by Eldon J. Gardner

When Charles Darwin published *On the Origin of Species* in 1859, he set off a revolution in the biological sciences. The effects of that revolution are still being felt today. Although his theory of evolution by natural selection has been accepted by the scientific world, it continues to be a matter of controversy for many people. The idea that all living things and the physical world change with time was not new with Darwin, but his work gave it the authority of a logical, scientific structure. The idea has broad philosophical meaning as well as being central to biology. Thus Darwin's work has had an important influence on human thought.

EARLY LIFE

Charles Robert Darwin was born on February 12, 1809, in Shrewsbury, England. His father's father was the famous physician and philosopher Erasmus Darwin, and his mother's father was the celebrated potter Josiah Wedgwood.

Darwin's early education was obtained at a nearby boarding school. At the age of 16, complying with his father's wish, he went to Edinburgh to study medicine. He did not succeed there, however, and at 20 he transferred to Christ's College, Cambridge, to study for the ministry. His school-work was again neglected as he collected insects, hunted, rode horses, and socialized with friends.

Darwin met scientists there, however, who encouraged his interest in natural history. He was especially influenced by a professor of botany, Reverend John S. Henslow, who introduced him to current geological thinking. In fact, Henslow shaped the course of Darwin's life when he informed Darwin that a certain Captain Robert Fitz-Roy was seeking a naturalist to serve without pay on H.M.S. *Beagle*.

The *Beagle* was an Admiralty vessel scheduled for a five-year mapping survey of the South American coastline and a voyage around the world. Darwin had mixed feelings about such an adventure. When he made the decision to sail, he purchased shirts, slippers, Spanish language books, two pistols, a rifle, and a microscope. With this meager equipment, plus a few more books, he boarded the *Beagle* in 1831 for the long journey.

THE VOYAGE OF THE *BEAGLE*

Darwin had heard of seasickness, but he had no forewarning that the condition would plague him throughout the voyage and apparently contribute to his poor health in later life. Captain Fitz-Roy even considered returning him to England for the sake of his health.

Charles Darwin

However, the enforced idleness during his period of illness gave Darwin time to reflect deeply on the things he observed during the voyage. He had planned to use his spare time to catch up on languages, mathematics, and classical literature. Instead he spent as much time as possible on land, working mostly on geological studies.

His devotion to geology at first caused Darwin to neglect botanical and zoological observations. When awakened to his responsibility for studies of living things, he collected specimens industriously. His insufficient knowledge of anatomy and taxonomy — the classification of living things — caused his collections to be inadequate and incomplete, however. Some were so poor that no institution would accept them upon his return to England, and he was forced at times to borrow specimens from Captain Fitz-Roy's collections to make his own appear respectable.

In his first encounter with a tropical island — Sao Tiago of the Cape Verde Islands — Darwin observed and collected enthusiastically but carelessly. *Geological Observations,* published later, included descriptions of rock formations and a hypothesis for the origin of this volcanic island.

Moving inland at Bahia, Brazil, Darwin observed a vampire bat feeding upon a goat and examined many fossils and living animals unique to South America. The trip around the southern end of that continent left him very ill. At Concepción, Chile, his experience of an earthquake confirmed him in his theory of the role of earthquakes in altering land levels.

THE GALÁPAGOS AND HOMEWARD

The *Beagle* arrived at the Galápagos Islands on September 15, 1835. It was here that Darwin made his most significant observations and collections, starting him on a train of thought that ended in *On the Origin of Species,* published 24 years later. He saw such things as finches with beaks suited to a particular way of life on a particular island, and giant tortoises on different islands showing different shapes and colors. This led him to speculate on how species

can vary and what causes such differences.

The Galápagos Archipelago, which includes 10 major volcanic islands, had become the home of some animals from the South American mainland in the geological past. Isolation had allowed the animal populations on the different islands to develop their own mixture of genetic characteristics and to diverge from those of other islands. Comparisons of current populations provided Darwin with evidence of evolutionary change. He thought deeply about the mechanisms by which such change could occur and new species originate. Natural selection — that is, the "selection", by nature, of better-adapted species — became the basis for his theory.

The final years of the voyage were devoted to a cruise of southern oceans and then around the world. Darwin collected animals and plants of Australia and numerous islands. A study of coral reefs in the vicinity of Australia gave him material for a later work on the way in which coral reefs and atolls develop.

On his return to England in October 1836, Darwin was surprised to find that he was already well known in scientific circles. Letters he had sent to friends in England, describing his adventures and accompanied by samples from his collections, had been circulated among scientists, and some had been read before univeristy groups. He now busied himself with lectures, distribution of his collections to museums, and supervision of the publication of his official report.

PUBLICATION OF HIS THEORY

Darwin devoted the rest of his life to research and writing. Some 20 years were spent working over his theory of evolution by natural selection. In 1842 he prepared a short paper outlining his theory and sent it to biologist friends for criticism. A large book was planned in which all supporting data could be documented. In the meantime he undertook the writing of an abstract of his theory.

When the abstract was about half written, Darwin received for review a short manuscript by explorer-naturalist Alfred

Darwin and Bergh, founder of the Society for the Prevention of Cruelty to Animals. The Defrauded Gorilla: "That Man wants to claim my Pedigree. He says he is one of my Descendants. Mr. Bergh: "No Mr. Darwin, how could you insult him so?"

The publication of *On The Origin of Species* and of later works on evolution caused much controversy, some of which was depicted in cartoons of the day.

Russel Wallace, *On the Tendency of Varieties to Depart from the Original Type*. He found that Wallace and he had arrived at similar conclusions by different paths. With the help of friends, they made an agreement for a joint paper, which was published in the *Journal of the Proceedings of the Linnaean Society* for August 1858. Darwin never wrote the massively documented book he had planned, but the abstract— actually, some 350 printed pages—was published on November 24, 1859, as *On the Origin of Species*. The entire first edition was sold out on the very first day of publication.

A heated controversy immediately developed over Darwin's theory, largely because it ran counter to accepted theological opinion, which held that all species had been specially created and that new species did not develop from older ones. Also, many people simply found the idea of evolution repugnant and its implications about human beings and their relation to the animal kingdom disturbing.

However, Darwin was ably defended by supporters, particularly by the biologist and writer Thomas Henry Huxley, and Darwin's theory soon came to be accepted by many scientists around the world. He spent the rest of his life doing botanical and zoological research and publishing numerous books and articles, such as *Descent of Man, and Selection in Relation to Sex* in 1871. In his last book he demonstrated the importance of earthworms in keeping soils fertile. Darwin died on April 19, 1882, and was buried in Westminster Abbey.

HIS IMPORTANCE

Darwin's concept of evolution by natural selection provided a framework for the biological sciences. Basically, Darwin stated that individuals of a species are not identical but that they show variations. Some are better adapted to survive than others, and these variations have a better chance of producing offspring. Offspring resemble their parents by heredity, so the better-adapted ones come to outnumber those less well adapted. When populations of a species become separated from one another, the differences between them eventually become so great that the populations thereafter must be regarded and classified as distinct species.

What Darwin's theory lacked was the detailed mechanism for the evolutionary process. This was found later in Mendelian genetics. However, Darwin's work provided the basic theory for such later findings, and it was a source of organization for biological sciences in general. In addition, Darwin can be considered one of the principal founders of the modern sciences of ecology and ethology.

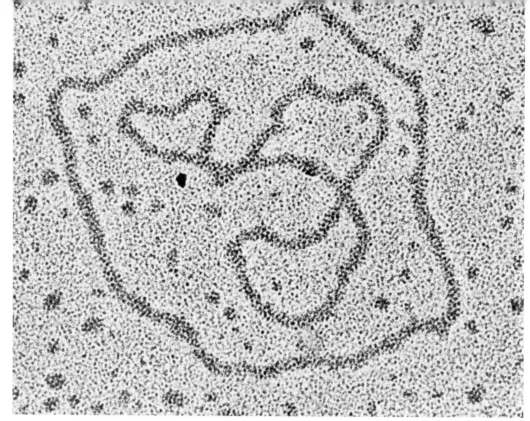

California Institute of Technology

Evolution depends on the action of natural selection on a population of reproducing organisms. The key to reproduction and heredity—and thus to evolution itself—lies in the genes of each organism. Above, an electron micrograph of the genetic material of a virus. Detailed study of such material increases our understanding of the diversity and evolution of life.

EVOLUTION

For well over 3 billion years, life has existed on earth. It probably began with fairly simple molecules consisting of carbon, hydrogen, oxygen, and nitrogen. Through eons it developed until today it is represented by hundreds of thousands of plant and animal species. Many living things have remained comparatively simple, though not so simple as they appear at first glance. They include viruses, bacteria, and protozoans, algae, mosses and fungi, jellyfish, and flatworms. Other forms have become highly complex—insects, spiders, vertebrates, ferns, conifers, and flowering plants. Each kind of animal and plant has a different body structure, different bodily processes, and a different way of behaving.

This extraordinary diversity has been brought about, according to scientific theory, by the working of organic evolution. The theory of evolution holds that existing animal and plant species have descended from preexisting organisms. The world of living things, as it exists today, is the current product of long lines of ancestry, extending far into the past.

The course of evolution depends upon the action of natural selection on a population of reproducing animals or plants. To understand this relationship we need to know certain facts concerning reproduction and heredity. The offspring of animals and plants more or less closely resemble the parents because hereditary material is passed on by reproduction from one generation to the next. This material is contained in structures called *chromosomes,* which are found in the nucleus of each cell.

REPRODUCTION

In the process of *sexual reproduction* a new individual is produced when a sperm, or sex cell from the male, fuses its chromosomes with those in a ripe egg cell of the female. This act is called *fertilization.*

Sex cells are produced by cell division in the sex organs of the parents. In each primary sex cell of the sex organ are two sets of chromosomes, the number depending upon the species of plant or animal. At the beginning of cell division, the matching chromosomes come together in the nucleus and become intimately entwined. Then they unravel but lie side by side in pairs. Each chromosome of each pair then splits lengthwise, so that there are now four.

The cell then divides, forming two cells. Each of these divides into two cells in its turn. The original primary sex cell has now produced four mature sex cells, or gametes, each having a single set of chromosomes. This type of cell division is called *meiosis.* When a gamete from the male (sperm) fertilizes a gamete of the female (egg cell), the male sex cell adds its single set of chromosomes to the single set of the egg cell. The fertilized egg, or *zygote,* will therefore have two sets of chromosomes—the original number of chromosomes present in the primary sex-organ cells of each parent.

The zygote, representing the new individual, undergoes another kind of cell division, known as *mitosis.* In this case, when the cell divides, the chromosomes are doubled. Therefore, each of the two new cells will have the normal number of chromosomes. The cells of the new individual, or offspring, divide again and again. Eventually they differentiate to make up various tissues.

Not all organisms reproduce by sexual means. Certain plants (bacteria, various algae, and others) and certain animals (including amebas and sea anemones) reproduce asexually. They divide in half or bud off a piece or pieces of the body. When an ameba, say, or another single-celled organism undergoes division, the chromosomes in the nucleus are doubled. The nucleus divides, and the body splits in two. The two offspring, or *filial cells,* have the same number of chromosomes as did the parent cell. Much the same thing happens when a parent organism buds off a filial organism. The cell or cells of the bud possess the same chromosome number as the parent.

Sometimes a species may reproduce either sexually or asexually. Sometimes these two modes of reproduction alternate in the life cycle.

By means of sexual reproduction, hereditary material contributed by two parents forms the new and distinctive hereditary material of the offspring. By asexual reproduction, the filial cell or bud has exactly the same chromosomal material as that of its single parent.

GENES

Thus far we have considered chromosomes to be the primary hereditary material. Actually, the fundamental units of heredity are the *genes,* which have been likened to beads strung on the threadlike chromosomes. There is evidence that certain genes exist outside of the chromosomes; they are called nonchromosomal genes and are located in certain cytoplasmic structures. They also play an important role in heredity.

Genes indirectly bring about the structure of an organism and the way that it functions. For example, eye color, the size of the brain, and the rhythm of the heartbeat are all the result of processes of development in which several to many genes are involved. Each gene has a distinct molecular pattern, capable of duplicating itself from other material in the cell. When the cell divides, either in mitosis or meiosis, the chromosomes of the filial cells duplicate exactly the distinct genes in the parent cell.

The genes consist of nucleoproteins—compounds of nucleic acids and proteins. The most important nucleic acids are DNA and RNA. The development of an organism depends not only upon specific genes

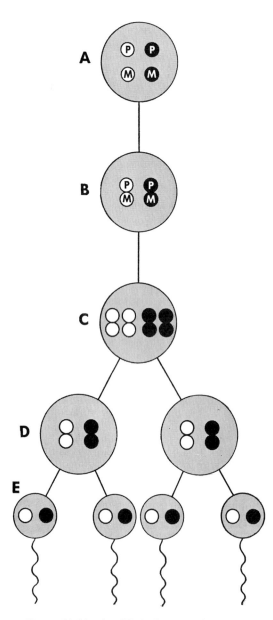

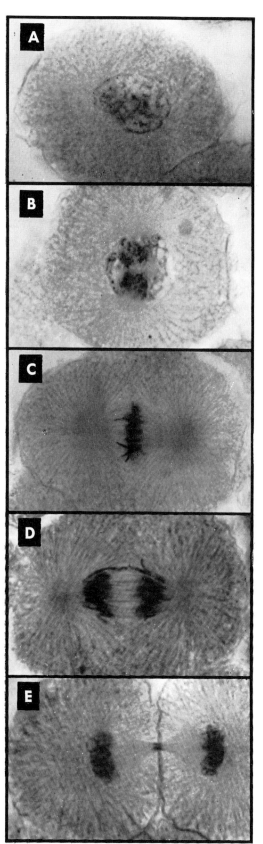

Above: highly simplified diagram of how a male germ cell gives rise to sperm by the process of meiosis. Assume that the germ cell (A) has two pairs of chromosomes—the white P matching the white M and the black P matching the black M. The chromosomes split, giving four pairs (C). Next the cell divides, forming two new cells (D). Then these two cells become four (E), each of which contains two chromosomes, one from each original pair. At right: a series of photos showing mitotic cell division, in which the chromosomes in a single cell (A) split, separate, and pull apart (B-D). At that time, a new cell membrane develops (E) and we have two cells, each with the same chromosome complement as the original cell.

acting alone, but also upon numerous genes acting together. Certain genes must lie next to one another in the chromosome to produce their specific effects.

Genes do not produce their effects in a vacuum. As a living thing develops, its genes interact with the environment. The amount of heat, light, humidity, air or water pressure, food (including minerals and water), oxygen, and carbon dioxide may all directly affect the development of the organism. The genes set limits—sometimes wide, sometimes narrow—on the kind of organism to be produced. Within these limits environmental factors are free to influence the molding of the finished product. A living thing, then, has a specific hereditary endowment with which it faces its environment; but its genes and environmental conditions must be in harmony if the organism is to live and develop.

EACH INDIVIDUAL VARIES

No two plants or animals, even of the same species, are exactly alike. This variation may be due to differences in genes, or to the varying effects of environmental factors, or to the combination of these two factors. Evolution takes place because such variations exist.

As we have seen, every plant and animal has a specific gene pattern; this is called its *genotype*. Under ordinary circumstances it remains the same throughout life and is passed on to the offspring. The genotype interacts with the environment to produce a definite kind of organism, functioning in a particular way. The physical appearance of a living thing and the way in which it functions are called its *phenotype* —"the type that shows" (from the Greek *phainein:* "to show" + *type*). Each individual's phenotype changes continuously from birth to death because of the effect of the environment. For this reason organisms with the same genotype will not have identical phenotypes.

For example, puppies from purebred parents may all have practically the same genotype. Yet each varies in phenotype from its litter mates. The variation may be considerable. There may be a runt, which is much smaller than the rest. The runt's phenotype is due probably to its not receiving enough food materials while in the mother's womb. It may be, too, that after birth the litter mates competed so strongly with the runt for the mother's milk, that the little creature could not get enough. The runt is dominated by its brothers and sisters and assumes a defensive attitude or runs away from them. It may not learn as quickly or as well. Yet if it breeds with one of its litter mates or with another individual of the same "purebred" stock, the runt's phenotype need not necessarily reoccur among the resulting offspring.

The case is different with a litter of mongrel pups. They will vary considerably from each other because they have different genotypes (inherited from parents whose genotypes differ). There will be, consequently, a noticeable variation in their physical appearance and manner of behaving. When these offspring mate, their young will also vary greatly.

VARIATION DUE TO RECOMBINATION

Variation is the indispensable material of evolution. Yet variation must be of a particular kind if evolution is to occur. As we have seen, animals with the same genotype may develop different phenotypes; however, these phenotypes do not serve as material for evolution. Certain organisms with different genotypes may show similar phenotypes—that is, will grow up to look and act more or less alike—because they develop in a similar environment. Evolutionary changes will not be brought about in this case by variations in genotype. It is only when both the genotype and the phenotype vary together—that is, when an altered genotype is reflected by a different phenotype—that evolution is served.

Alterations in the genotype can come about by the formation of new gene combinations. This may happen whenever sexual reproduction takes place, for the genes of the two parents are recombined to form the new genotype of the offspring. Suppose, for example, that a purebred curly-haired, erect-eared dog mates with a purebred straight-haired, floppy-eared individual.

There are genes for curly hair and erect ears in one parent and genes for straight hair and floppy ears in the other. Let us assume that the genes responsible for hair character and ear character are on different chromosomes; also that the traits curly hair-erect ear are dominant over the traits straight hair-floppy ear. If the matching genes on a pair of matching chromosomes are in some way different, one gene may be dominant over the other (the recessive). A dominant gene is one whose outward, or phenotypic, effect largely or entirely obscures that of the recessive gene.

The phenotypes of all the offspring of such a cross will show the dominant traits of curly hair and erect ears. The genotypes, however, will harbor genes for both curly and straight hair and for both erect and floppy ears. Therefore, the hereditary material of the offspring is a combination of the different genes found in each parent.

If these offspring, which are hybrids, mate with each other, their young will show the following combinations: curly hair-erect ear, straight hair-floppy ear, *curly hair-floppy ear* and *straight hair-erect ear*. Two new phenotypes—those indicated by italics—have been produced.

If the genes that regulate hair color and tail character are linked on the same chromosome, they will be inherited together under ordinary circumstances. Suppose that brown hair and straight tail are dominant over red hair and curly tail. When a purebred brown-haired, straight-tailed dog mates with one purebred for red hair and curly tail, all the hybrid offspring will have brown hair and straight tails. One of the matching chromosomes of each hybrid has the genes for brown hair and straight tail; the other has genes for red hair and curly tail. When two hybrids mate, normally most of their offspring will have brown hair and

Left: genealogical chart showing the first and second generations from "purebred" ancestors. The cross between a curly-haired, erect-eared parent and a straight-haired, floppy-eared one results in first-generation hybrid offspring having curly hair and erect ears. When these hybrids mate, four possible phenotypes may result.

Below: the phenomenon of crossing over takes place during meiosis. As a result, some of the genes (represented by small and capital letters) are exchanged between the sets of chromosomes, altering the offspring's genotypes.

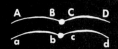

MATCHING CHROMOSOMES LINED UP PRIOR TO SPLITTING LENGTHWISE DURING MEIOSIS

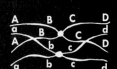

CROSSING OVER TAKES PLACE WHEN THE MATCHING CHROMOSOMES SPLIT

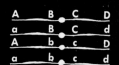

CHROMOSOMES POSSESSING CROSSED OVER SEGMENTS

RESULTING CHROMOSOMES

straight tails, and one or two will be red-haired and curly-tailed.

When gametes are formed by one of the hybrids, crossing over may take place during meiosis. This means that when the matching chromosomes are splitting, they exchange segments with each other. One chromosome now links together the genes for brown hair and curly tail. The other has genes for red hair and straight tail. If this hybrid mates with a "normal" hybrid, the offspring may be as follows: brown-haired and straight-tailed, *brown-haired and curly-tailed and red-haired and straight-tailed*. Crossing over at meiosis has created new genotypes (given in italics) because of the recombination of genes.

These examples have been purposely simplified. Actually, thousands of genes make up an organism's genotype. For example, many genes may determine the traits of hair character and color, ear character, and tail character. A purebred individual is really a hybrid because of the recombination of genes. When hybrids breed, an infinite number of recombined genotypes are possible. Usually hybrids do not look and function exactly like either parent, because a number of genes, some from one parent, some from the other, act together in regulating each specific trait that appears.

Organisms that reproduce asexually do not recombine genes because only one parent is involved in reproduction. The descendants from the parent are normally genetically alike and can be called purebred. When asexual alternates with sexual reproduction, however, as in the case of algae, ferns, and some other plants and many animals, including some insects, variations come about by way of the regrouping of genes during the sexual phase.

VARIATION DUE TO MUTATION

The recombination of genes alters the genotype to a more or less limited extent. Different genes from the two parents combine in the offspring's genotype, changing the genotype somewhat, but no new genes are produced. Only the process of mutation creates a radically different genotype.

A mutation is a fundamental change in a gene or genes, caused by extremes in temperature, cosmic radiation, ultraviolet radiation, radiation from radioactive materials, and possibly other environmental factors, such as certain chemicals. Genes of any cell may mutate, but only mutations in reproductive cells are significant for evolution. Mutations that effect the primary sex cells, the gametes, the fertilized egg, and the cells that develop into gamete-forming tissue can be passed on to the offspring. Of course, when one-celled organisms mutate, the mutation is inherited by the offspring during either sexual or asexual reproduction.

A mutation may be a change in the chemical pattern of a gene. After mutating, the gene duplicates itself in its altered form and is passed on by inheritance in this condition. A mutation may also involve a group of genes. Such mutations, which are also inherited, occur during meiosis, when gametes are formed. For example, during meiosis, the division of the chromosomes may get out of hand, producing a double or triple set of chromosomes in the gamete. The offspring derived from these gametes will have more than the normal number of all genes. Sometimes a chromosome section, containing one gene or a block of genes, is lost. The offspring will then be deficient in certain genes. On the other hand, a section of a chromosome may be repeated. The offspring receives a duplication of certain genes. Finally, genes may be rearranged on the chromosomes. These rearrangements may alter the offspring's phenotype, because development is often determined not only by the structure of the genes but also by their position on the chromosome.

Since genes control the development of living things, mutations change the structure of an organism and the way in which it functions. Animals and plants are normally pretty well adapted to their environment. Therefore, mutations that change the existing genotypes are generally harmful. The mutated organism may not develop at all or may develop poorly. However, some mutations are useful, and the mutated animal or plant is able to survive and reproduce. The accumulation of useful mutations through the ages has led to the creation of all the species of living things that we know today. It is an all-important factor in organic evolution.

VARIATIONS IN POPULATIONS

Mutations in individuals do not bring about evolutionary change. To be effective as an agent in evolution, mutations must act upon a *population*. A population is a group of organisms of one kind, existing together at a given time and spread out in an environment suitable for them. Each individual differs from the rest in one or several genes. Identical twins are an exception because their genotypes are exactly alike. The members of this collection of somewhat dissimilar genotypes have one characteristic in common—all are tolerably well adapted to the prevailing environment.

In a way, a population is a sort of superindividual, with a vast supply of genes—a supply called the *gene pool*. Each individual has a share in it. When it reproduces, it adds, through its offspring, its specific genes to the gene pool. When it dies or stops reproducing its genes are eliminated from the pool. Though the members of a population vary somewhat, there are more of the same kind of genes in its gene reservoir than in that of any other population. Consequently, the individuals of any given population look and function more or less alike.

Certain mutations are recessive to the normal genes. Hence the organism retains the normal phenotype. Harmful mutations that are recessive remain in the population's gene pool. However, as long as they do not appear in the phenotype, they produce no effect. If the harmful mutation appears in the phenotype of one of the members of the population, that member will be eliminated. But a change in the environment may cause the former harmful mutation to be a useful one. Organisms carrying the mutation in the phenotype will survive and will reproduce.

ROLE OF NATURAL SELECTION

If the conditions of the environment did not change, evolution would gradually lead to a condition in which all individuals would be perfectly adapted to their environment. But the environment does change. If the population is to survive, its genotype must adapt to the new conditions. Because of its genetic variability, a population can be molded to fit the changing environment. Natural selection is the agency that brings this about.

We can get some idea of how natural selection works by examining the artificial selection applied by man to the preservation and breeding of dogs. The domestic dog is derived from the wolf species *Canis lupus* or from another very similar species. Once "wolf-dogs" had been tamed, they were selected by man for particular traits of size, build, and behavior. Man systematically picked the best individuals ("best" from his viewpoint) to become the parents of the next generation. Any animal that did not measure up—that, in other words, was not adapted to man's desires—was destroyed or at least not allowed to breed with the select individuals. A great number of dog breeds were eventually developed, differing greatly from their wolfish ancestors. The Great Dane and the Saint Bernard are larger than wolves; the Pekingese and Chihuahua, much smaller. The greyhound runs much faster than the wolf; the dachshund, slower.

Under domestication, descendants of the wolf tended to develop sexual maturity at an earlier age and produced more offspring. Their behavior patterns were changed so that they became greatly attached to people. Thus by selection from an original wolf population, there has come about, in a comparatively short time, the evolution of many different types of wolf-like animals.

SURVIVAL OF THE FITTEST

Natural selection usually does not work so fast, but its effects in the long run are just as profound. It depends upon adaptation to the existing environment. Certain members of the population, carrying particular genes in common, will produce more surviving offspring than others in their given environment. Under equal conditions,

Brookhaven National Laboratory

Mutations cause basic changes in the hereditary make-up of animals and plants. Above: a mutation in sheep produces short-leggedness. The ewe in the center is a mutant— that is, it has undergone mutation. It is flanked by a mutant ram on the right and a normal ewe on the left. The photo at the right shows mutations caused by radiations from radioactive materials. They have brought about tumorlike growths in the tobacco plant on the left. A normal tobacco-plant stalk, which has been stripped of all its leaves, is shown for comparison.

more of these offspring will survive to reproduce and so carry on their genotypes. Consequently, a particular genetic pattern is preserved for this particular environment. Gradually, generation by generation, the gene pool of the given population fills with more and more of the survival-type genes and gene combinations. In other words, natural selection preserves the individuals that, because of their genotypes, are best adapted to the environment. The individuals whose genotypes never meet the mark or are no longer useful in a changing environment are eliminated. Since they do not survive, they do not reproduce their kind.

To adapt successfully, organisms must be able to meet every challenge of the environment. Plants and animals are affected by the climate and must respond in various ways to changes in temperature, humidity, wind velocity, and precipitation. They need a source of food. There must be space in which to live, a place where there is sun or shade, dampness or dryness; a place to build a nest, dig a burrow, extend roots, lay eggs, or bear young. An organism must adjust to other of its species and to organisms of different species. Sometimes it has to compete with them for space, food, sunlight, or water. If it belongs to a social species, it must cooperate with its species mates. It is affected by parasites and by natural enemies that would kill it for food.

Evolution is furthered, through the process of natural selection, when a population as a whole becomes better adjusted to the environment, when there is better care of the offspring, when conflict between the members of the population is eliminated, when there is more effective use of available food and space. Evolution is also furthered when some of the individuals of a formerly uniform population survive and reproduce successfully by using food and occupying space that the majority of the population cannot utilize.

Natural selection tends to be most effective in large breeding populations that are widely distributed. In such populations favorable genotypes will be gradually spread and unfavorable ones eliminated. In small breeding populations, on the other hand, the influence of selection is not so great. Evolutionary changes are usually at random, because there is no decided trend in the direction of one most favorable genotype. There is little variability in the population at any one time; its members, regardless of their genotypes, are eliminated at random. Almost always small populations become extinct. In rare instances, however, a small population may evolve rapidly into a new type if conditions are such as to favor a specific genotype. Since the population is small, the favorable genes would be spread rapidly throughout the gene pool.

LOCAL GROUPS

Usually a large population is not evenly distributed, but is broken up into a number of moderately sized *local groups*. Inbreeding within each local group is common. Frequently crossbreeding between adjacent local groups takes place, resulting in an exchange of genetic factors between them. Natural selection can operate rapidly in such a population if environmental factors demand it. If a new genotype is needed in an altered environment, the chances are that crossbreeding between the different groups will supply it. The number of the mutations in the population as a whole is large. By being transferred (by crossbreeding) between different local groups, a mutation may become fixed in various new combinations. If one of these combinations proves effective for a given group, that group will thrive. If the combination is disadvantageous, the group may become extinct without endangering other groups.

A population that fluctuates in size is particularly apt to undergo evolutionary change. For a certain length of time, the population will increase, then it will decline greatly. The cycle will repeat itself again and again. During the period of decline, organisms of the population are eliminated wholesale: only the best-adapted organisms survive. During the upsurge phase, selection is not so severe and the population increases its variability.

RATES OF DEVELOPMENT

Since genes establish the rate of development of an organism, mutations may change this rate. Any mutation that alters the general plan of early development is a radical one. If, for example, a change upsets the development of two-sided symmetry in the early stages of an animal, the organism probably would not survive. If it survived, it could not move about properly. As another example, any change that affected the yolk sac of a mammal embryo would also be a drastic one. Though the yolk sac has no nutritional significance, the first phase of development of the embryo's blood cells takes place in this organ. Any mutation that knocked out or altered the genes responsible for the yolk sac would destroy the embryo.

If, however, a mutation does not affect the very early development, the genetic change that it brings about may not be a harmful one. If may even result in causing the organism to be better adapted to a changing environment.

Sometimes mutations accelerate the development of both the reproductive organs and the other tissues of the body. This may result in a considerable simplification of the organism. We often find such simplification in parasites. The tapeworm and the crustacean *Sacculina* are good examples of this trend.

PAEDOMORPHOSIS

Sometimes mutations slow down the rate of development of body tissues relative to that of the reproductive organs. This

Some of the complex interrelationships of organisms to their environment are shown in the illustration on the opposite page. The two plant species compete with each another for sunlight, water, and soil nutrients. In order to grow, the smaller plant must not be shaded by the larger one. The larger plant produces oils that attract butterflies, which lay their eggs on its leaves. The caterpillars (larvae of butterflies) compete with aphids for the plant's substances as food. The smaller plant, through natural selection, has developed adaptations that protect it from these insect pests.

process, which is called *paedomorphosis* (from Greek *paid:* "child," and *morphe:* "form"), produces sexually adult larvae. Instances of this type of development are found among the salamanders, such as the mud puppy *(Necturus)* and the eel-shaped *Proteus.*

Paedomorphosis may be extremely important in the evolution of new forms. Young organisms are less sharply adapted to existing environments than older ones because their requirements are not so strict. Therefore, if mutations slow down the rate of body development, larval forms that are sexually mature may be plastic enough to explore new environments. They will adapt readily to new conditions and in this way establish a new kind of population.

There is reason to believe that the insects, which have three pairs of legs, originated by paedomorphosis from a millepede-like larva. We know that certain millepede larvae have only three pairs of well-developed legs and have the same number of head segments as an insect. Another apparent case of paedomorphosis is that of the flightless ostrich. The adult feathers of this bird resemble the nestling down of young flying birds. The ostrich seems to have evolved from a flying bird and lost its flight feathers in the process.

Paedomorphosis has been very important in the evolution of man's body. In a sense, adult men can be considered as a sexually mature embryo. For one thing, his erect posture is due to the retention of the cranial flexure that occurs in the development of the embryo. In the embryo of mammals, the long axis of the head is at a right angle to that of the trunk. This angle is the cranial flexure. In other mammals the cranial flexure straightens out during development, so that the trunk and the long axis of the head form a more or less straight line. The cranial flexure becomes a permanent feature in man.

Embryonic and newborn apes have rounded skulls, non-jutting jaws, and no brow ridges. These embryonic characteristics are retained in the adult human being. Man has hair only on his head at birth and grows little body hair throughout his life. Other primates, however, have scant hair covering only as embryos or when young. As adults they are more or less completely covered with hair. Man's teeth erupt later than those of the apes and the other primates.

The sutures (lines of union, or seams) between the skull bones of apes and other mammals close soon after birth, limiting increase in the size of the skull. In the case

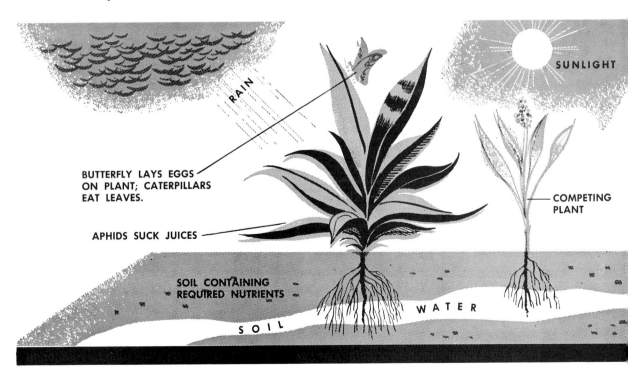

SUNLIGHT

RAIN

BUTTERFLY LAYS EGGS ON PLANT; CATERPILLARS EAT LEAVES.

APHIDS SUCK JUICES

COMPETING PLANT

SOIL CONTAINING REQUIRED NUTRIENTS

WATER

SOIL

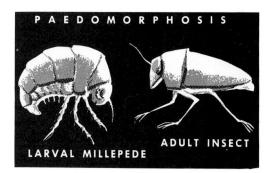

PAEDOMORPHOSIS

LARVAL MILLEPEDE   ADULT INSECT

The evolution of insects may have come about, through paedomorphosis, by early insectlike forms retaining as adults some of the larval characteristics of millepedes.

of man, the sutures do not close until much later. The skull increases in size and hence provides accommodation for a developing brain. In this way man's period of brain development has been extended into his early formative years, when he begins to use his hands, eyes, and other senses to explore his environment. The young human continues for some time to grow nervous tissue. New connections, or pathways, are made in the nervous system, and former pathways are changed.

DEVELOPMENT OF NEW SPECIES

A local group, or colony, of a population may diverge more or less radically from the population as a whole or from other scattered colonies. For this to happen, the diverging group must be isolated in some way from the rest of the population. If isolation persists for a long enough time and if the environment calls forth the building of a new genotype, the group gradually transforms to become a new species.

We can illustrate the effect of isolation by taking another example from the domestication of dogs. After the dog was tamed, human tribes quickly made use of the animal in many ways. Each tribe, which tended to be semi-isolated in its own geographical region, had its own population of dogs. This condition favored genetic changes in the tribe's dog population. The canine population was a small, isolated breeding one. By selecting those traits that were most

desirable, each tribe came to possess in a fairly short time its own type of dog.

Studies of dog skeletons associated with different North American Indian tribes show, apparently, that there were at least eight unique dog populations. The dogs tended to vary from tribe to tribe. Those associated with adjacent tribes were more alike than those belonging to widely separated tribes.

Different breeds, or genetic races, of dogs continue to exist because man prevents interbreeding among them. If interbreeding were not curtailed, there would be a rapid exchange of genetic material. The genetic constitution of one race would combine with that of another race or several races. In a short time, we would have a highly variable population of dogs. The mongrel is the product of such a process.

ISOLATING MECHANISMS

What man has done by maintaining various races of dogs in isolation is accomplished in nature by natural selection and by various isolating mechanisms. Isolating mechanisms hinder or prevent the exchange of genes between two or more segments of a once unified population.

The exchange of genes may be slowed up or entirely stopped because of *geographic isolation*. Seeds and spores of plants, larvae, and young forms of animals often develop to maturity at a distance from their parents. The next generation of offspring spreads out to a greater distance and so on. The individuals of the population as a whole, of course, are fitted for living only in a certain kind of environment. But this distinct environment may extend for tens, hundreds, or even thousands of kilometers. It need not extend uniformly. It may form pockets in the midst of other environments.

Once a colony, or segment, of a population establishes itself in one of these pockets, it may not easily interbreed with other colonies. A range of mountains, a body of water, a desert, or a forest may lie between colonies. It may stop the exchange of genes between them. In time the environment of a given pocket may change. The isolated colony will have to adapt to these changes if it is to survive. Natural selection will create new genotypes that will differ significantly from colony to colony.

If isolation is completed too soon, a race may become overly specialized to temporary environmental conditions. When these change, the race might become extinct because it could not alter its genotype quickly enough to fit the new situation. Until the environment becomes stable, there must be at least some genetic exchange between races in order to provide the necessary variability. Once the environment has become definitely stabilized, isolation can be complete.

Geographical isolation is not the only factor that limits or stops interbreeding and therefore brings about the formation of new species. Certain representatives of a variable population may harbor genotypes that will allow them to live under somewhat different conditions within the same general region. For example, they may be able to use different sources of food, build nests in somewhat different locations, tolerate greater or lesser amounts of sunlight, water, or heat. They may escape detection by enemies by somewhat different means, live in sandier or more clayey soils, withstand higher concentrations of poisonous substances. Mating or flowering periods may take place at different seasons. Differences in mating behavior may cause a lack of mutual attraction between certain males and females. Slight differences in the reproductive organs may prevent mating. If mating does occur, fertilization may not take place. Even if fertilization takes place, the hybrids may not survive, or, if they do, they may fail to reproduce. If they reproduce, their offspring may not live, or may not be able to adapt well to the environment.

POLYPLOIDY

New species may also arise as the result of a condition known as *polyploidy*. It happens, though rarely, at meiosis that the chromosome number is multiplied, so that each gamete will have a multiple of the normal number of chromosomes. It will have twice as many, or three times as many or even more. If reproduction occurs by self-fertilization, as in certain plants, or by the development of an unfertilized egg, as in some animals, the offspring may differ considerably from the parent. The offspring in such cases are called polyploids. If they can adapt to the existing conditions of the environment, they will be able to reproduce their altered genotype. In effect they will represent a new species.

Plant species arising in this way often differ from their ancestors in having longer stems and larger leaves and flowers. They may have differently shaped leaves and other organs, and may show slower growth and greater hardiness. *Datura,* a plant of the potato family, apparently came into existence through polyploidy. This condition is rare in animals. However there is good evidence that certain weevils, the shrimp *Artemia,* the sow bug *Trichoniscus,* and the moth *Solenobia* arose by polyploidy.

INTERBREEDING

Sometimes two species of plants may interbreed, producing a hybrid. The hybrid is almost always sterile. Occasionally, however, chromosomal doubling in the hybrid takes place, resulting in polyploid offspring that are fertile. For example, a species of radish, *Raphanus sativus,* can interbreed with a species of cabbage, *Brassica oleracea.* If the resulting hybrids undergo chromosomal doubling, they may yield a polyploid, *Raphanobrassica.* This is a new species. It breeds true with itself, but is infertile when crossed with either the radish or cabbage. The mint, *Galeopsis tetrahit,* arose in a similar manner. So, apparently, did the U.S. cultivated cottons.

SOME CLOSELY RELATED SPECIES

Depending upon environmental conditions and the rate of mutation, a few species or a considerable number may arise in a relatively short time. Some of these will remain conservative: that is, they will change little in a stable environment. Some may become extinct if there is a change in the environment and if they cannot adapt to it. Other species may change considerably by reconstructing their gene pool in response to a shifting environment.

They may exploit a new environmental niche not inhabited by another related species. They may adapt better to an already occupied habitat and so crowd out the former inhabitants.

Each surviving species may eventually be a parent to several offspring species. In this process, the parent species may survive as such, and may exist together with the new species derived from it. In some cases the parent species ceases to exist as such because it breaks up completely into new species.

As time goes on, certain species derived from a common parent species will tend to differ less from one another than they will from other species derived from the parent. They will form a distinct group, called a *genus* (plural: genera). There will generally be a number of such groups, all derived from one original species. The different genera will later be combined in different families; the families in orders, and so on.

DOES EVOLUTION FOLLOW A DEFINITE DIRECTION?

Some authorities have held that evolution proceeds in a fixed direction once a certain type of change has started. The development of the horse has been given as an example of this oriented evolution. It is maintained that *Eohippus,* a small horse living in the Eocene epoch, gradually but steadily evolved into our modern species, *Equus.* Among other things the horse became much larger in time, the number of its toes was reduced to one, and its teeth became more complicated.

Actually, this idea of steady advance in a single direction is exaggerated. Evolution did not proceed in a straight line in the case of the horse. There were a number of lines of development, called *adaptive radiations,* leading from *Eohippus.* A number of different kinds of horses evolved. Some of them led in the general direction of *Equus,* but many did not. The fossil record is filled with lines of horses that went just so far in their evolution and then became extinct.

Yet if the development of the horse did not follow a straight line, there is evidence of a distinct trend in a definite direction. In other words, there was a certain degree of orientation.

The tendency of evolving organisms to follow a trend is widespread. We see it in the development of the elephant's trunk, the wings of birds, the complex floral structures of flowering plants. In such cases as these, there may be many adaptive radiations. But there may be many blind alleys. However, viewed from the standpoint of the centuries or the millennia, there is evidence of a definite trend.

Some trends are favorable for the long-range development of a given organism. For example, the trend from *Eohippus,* a multi-toed forest browser, to *Equus,* a one-toed grasslands grazer, appears to have been an advantageous one. Certain trends, however, may ultimately prove to be harmful. The development of the huge antlers of the Irish elk furnishes a striking illustration. These structures increased in size and weight until they finally became most unwieldy. For a time, it is believed, the Irish elk's antlers did not usually handicap the animal, since it continued to thrive. However as the environment changed, the antlers became more and more of a drawback until the animal became extinct.

The different new types that are formed as the result of adaptive radiation may remain more or less firmly established, without showing any perceptible signs of a definite trend. Darwin's finches (subfamily Geospizinae), found in the Galápagos Islands, provide a remarkable example. Their ancestor was an unknown true finch, which came to the Galápagos from the mainland of South or Central America. In the course of time, the finches have split up into fourteen distinct species, forming a series of birds with quite varied habits. The main difference in structure is in the shape of the beak, which varies with the food habits. Some of the finches feed on seeds; others on insects and buds; still others on leaves. A warblerlike type with an elongated slender bill devours small, soft insects. Another Galápagos finch, the woodpecker finch, pries insects loose from bark by using a cactus spine. This striking series of radiations is due, in part at least, to the isolation afforded by the different islands of the group and to the absence of serious predators and competitors on these islands.

Adaptive radiation on a large scale took place in the evolution of the reptiles. Beginning about 280,000,000 years ago, the reptiles commenced to spread out into a variety of habitats and in time became quite different from one another. One group evolved into the modern turtles, another into the ichthyosaurs and other now-extinct aquatic reptiles. A third group led to the modern snakes, lizards, and tuatara, as well as to the crocodiles, flying reptiles (pterodactyls), birds, and all of the large and small carnivorous and herbivorous dinosaurs. A fourth group evolved to become the ancestors of the mammals. The mammals in turn radiated into a number of different types. One interesting radiation is represented by the marsupials on the isolated continent of Australia.

In the case of certain animals and plants we can find little or no evidence of adaptive radiation or an evolutionary trend. These organisms have survived virtually unchanged for millions of years, as if evolution had passed them by. The horse-tails *(Equisetum)* and club mosses *(Lycopodium),* for instance, are representative of land plants living 350,000,000 years ago. The ginkgo *(Ginkgo)* goes back at least 180,000,000 years. Among the oldest living animals are the lamp shell *Lingula,* the moss animal *Stomatopora,* and the bivalve *Nucula,* all of which originated in the Ordovician or Silurian periods over 400,-000,000 years ago. The pearly nautilus *(Nautilus),* horseshoe crab *(Xiphosura),* and an echinoderm, *Isocrinus,* survive from the Triassic or Jurassic periods, from up to 200,000,000 years back. The tuatara reptile *(Sphenodon* or *Homeosaurus)* of New Zealand has changed hardly at all since the Jurassic. The Port Jackson shark *(Heterodontus)* and cow shark *(Hexanchus)* have survived since the Jurassic some 150,000,000 years ago. Opossums have existed for well over 60,000,000 years.

These forms have managed to survive because they have continued to live under stable environmental conditions, to which they are well adapted. It is only when the environment changes and animals or plants fail to adapt accordingly that they become extinct.

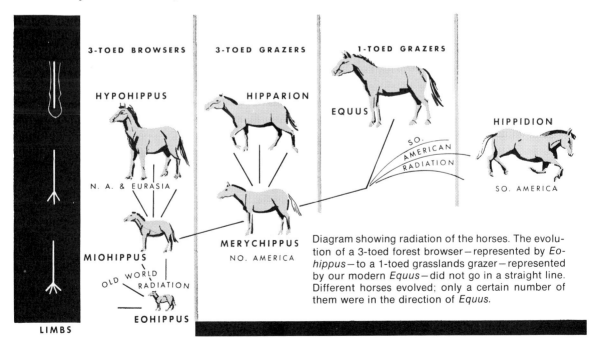

LIMBS

3-TOED BROWSERS   3-TOED GRAZERS   1-TOED GRAZERS

HYPOHIPPUS   HIPPARION   EQUUS   HIPPIDION

N. A. & EURASIA

SO. AMERICAN RADIATION

SO. AMERICA

MERYCHIPPUS
NO. AMERICA

MIOHIPPUS

OLD WORLD RADIATION

EOHIPPUS

Diagram showing radiation of the horses. The evolution of a 3-toed forest browser—represented by *Eohippus*—to a 1-toed grasslands grazer—represented by our modern *Equus*—did not go in a straight line. Different horses evolved; only a certain number of them were in the direction of *Equus*.

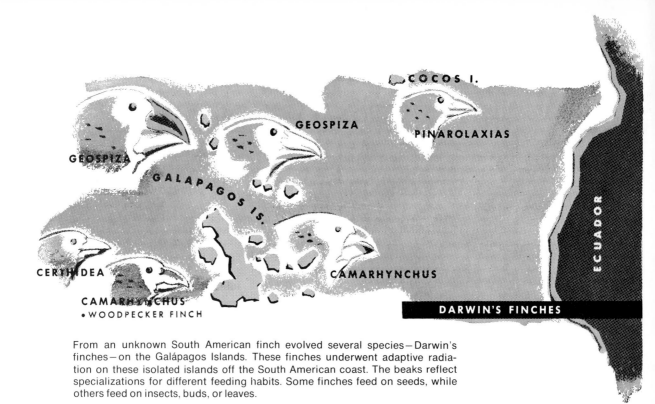

From an unknown South American finch evolved several species—Darwin's finches—on the Galápagos Islands. These finches underwent adaptive radiation on these isolated islands off the South American coast. The beaks reflect specializations for different feeding habits. Some finches feed on seeds, while others feed on insects, buds, or leaves.

DOES EVOLUTION REPRESENT PROGRESS?

The older evolutionary theories held that there was a definite progression from less perfect to more perfect forms of life. This idea is exaggerated. As living things evolve, there is apparent degeneration, as well as the development of more highly organized forms.

Degeneration is found among the plant and animal parasites. The dodder, for example, is a seed plant that absorbs food directly from other plants by means of modified roots. It has a slender twining stem, but lacks leaves and chlorophyll. Tapeworms and various parasitic crustaceans have degenerated to the point that, as adults, they lack almost all organs except those for absorbing food from their hosts and for carrying on reproduction. The barnacles, though not parasites, are degenerated crustaceans. They have given up a free-swimming existence for one in which they are firmly attached to some convenient rock or piling.

Can we say that adaptation is a sign of progress? Perhaps it is. As we have seen, however, every organism must be adapted to its habitat. Otherwise it would not survive. Is the development of a complex structure a sign of progress? Not necessarily, because a simplified structure may also evolve. For instance, the bony structure of the head of fish and amphibians is much more complex than that of the mammals, which evolved later in time. Are the more specialized plants and animals the most progressive? Not in all cases, for overspecialization may lead to a lack of variability. Birds are highly specialized, but their forelimbs impose a limitation on them for further evolutionary change. The forelimbs of mammals, on the other hand, are less specialized. It is believed that this feature has led to the exploration of more and varied environments.

In spite of the above examples, evolution, in many cases, does seem to represent progress. The most general kind of progress, in which all living things share, is the tendency to fill every available living place. There has also been improvement, in numerous instances, in the processes required for survival. As various kinds of plants and animals evolved, they developed greater mastery in their respective environments. They could move about better, reproduce more efficiently, give better protection for their offspring, and live in a greater range of habitats. Among the animals, there was a gradual improvement in the sensory organs, such as those for receiving light, air or water vibrations, touch stimuli, and stimuli from chemical substances. The nervous system became more highly organized. Animals with a well-developed nervous system could gather more and different kinds of information about their environment and react accordingly. They were able to cope with or adjust to a greater variety of factors in their particular environment.

As certain animals evolved, they developed a constant internal environment that was not affected by external conditions. To survive in freshwater and land environments, animals such as insects, spiders, and the vertebrates had to develop internal processes that would maintain the body in a more or less steady state. For example, the fluids bathing and entering the cells are normally much the same at all times in the higher animals, despite fluctuations in the environment outside the body.

HUMAN EVOLUTION

The species man, *Homo sapiens,* as a product of evolution, is a progressive type. Man's highly variable population spreads

A long-lived animal species. The horseshoe, or king, crab *(Xiphosura)* has lived as a species for some 180,000,000 years. The closest living relatives of the harmless, hardy horseshoe crab are probably the scorpions.

over all the inhabitable land masses of the earth. Humans are capable of adapting to a great variety of ways of life. Man has carried the process of developing the individual to its greatest extent. Humans will continue to vary because of the recombination of genes and because of mutations. We have no way of knowing, at the present, whether man will eventually evolve into a physically new type of human being.

Upon man's biological inheritance is imposed his cultural inheritance, by which he passes on his growing accumulation of knowledge to his descendants. Man's cultural development may be considered as a new kind of evolution.

Man's evolution has now reached a stage in which he has become a most significant part of the environment for other animal species and for plants. In effect, he is directly or indirectly a selective agent. Not only has he created a host of plant and animal breeds by artificial selection, but he has introduced his domestic animals and plants all over the world. In this way he has often drastically altered the environment of the native species.

Man competes against many animals for food and living space, exterminating them or driving them into isolated pockets. Human beings have brought about great changes in the face of the earth. The species of plants and animals that have not

Degeneration in evolution. The parasitic crustacean *Sacculina* has degenerated as an adult to a mere saclike body with a branching mass of tissue. This tissue grows throughout the body of *Sacculina's* host, which is a crab. *Sacculina* is highly specialized for a parasitic life and possesses only reproductive organs and tissue for absorbing food from its host.

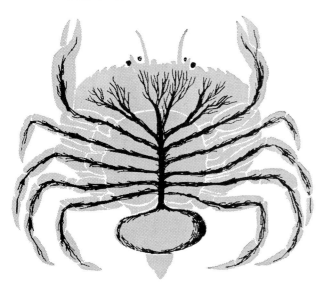

adapted to these changes have died out or have become rare. The use of insecticides and antibiotics has created new environments that have allowed new adaptive types of viruses, bacteria, and insects to appear. Humans now have the power to kill off unwisely almost any species they desire and the capacity to appreciate and protect the varied life about them.

Evolution is a continuous process of gradual change. It is going on today as it did in the past. Each living thing is a part of the constantly changing and developing pattern—a pattern that results from the relationships between living things and their environment. The concept of evolution shows how these relationships have developed in the course of eons in every part of the world.

HOW THE IDEA OF EVOLUTION DEVELOPED

The concept of evolution is associated most closely with the name of Charles Darwin, the great nineteenth-century English scientist. He and Alfred Russel Wallace, a contemporary, developed the scientific framework of the theory upon which all later work in the field has been built. However, the idea of evolution—that living things and the earth itself change and develop with time—has been around much longer than that. Some of the ancient Greek thinkers, such as Heraclitis and Anaximan-der, made the conjecture that the world is in a state of flux and that one form of life can develop, over a period of time, from another form of life.

While other thinkers over the centuries also speculated along such lines, the concept of evolution gained its solid basis with the work of zoologists and botanists in the late eighteenth and early nineteenth centuries. The evidence that they accumulated made the ultimate development of the evolutionary concept almost inevitable. The more different forms of life that scientists identified, the more certain they became that the wide variety of species were linked in some biological way—that an as yet unidentified "great chain of being" existed.

The first scientist of note who came out strongly for an evolutionary theory was Jean Baptiste Lamarck, a French naturalist. Early in the nineteenth century he published a work in which he presented his ideas, some of which came close to modern evolutionary theory. Lamarck also made some wrong guesses, although they were reasonable enough considering the state of biological knowledge in his time. A number of other biologists followed up on his work, but evolution was not yet accepted by the majority of scientists. It remained for the above-mentioned figures, Darwin and Wallace, to lay the groundwork for the theory as it is known today.

See also the article "Charles Darwin."

Humans have also evolved. Some human ancestors are shown below. Left to right: Homo erectus (Peking Man), Neandertal man, and Cro-Magnon man.

ADAPTATION

by F. L. Fitzpatrick

There is a sort of war that is being fought every single day in the vast arena of nature. It is a grim, unending war for survival among the various species of living things. Thousands of millions of individuals succumb each day in the fateful struggle. Sometimes entire species may be wiped out in the course of a few years.

The species to which we belong — *Homo sapiens* — has played an important part in this struggle. The advance of man's civilization has caused some species to be exterminated and has endangered the existence of others. On the other hand, man himself has fought a losing battle in certain regions of the world. The terrible tsetse fly, the carrier of a parasite that causes sleeping sickness in man, has made some areas of Africa almost uninhabitable. The plants of the jungle have overwhelmed human settlements again and again in such places as Yucatán and southeastern Asia.

In the continuous struggle for existence, each species is faced with various critical problems. Among other things, an animal must (1) obtain food, (2) defend itself against its natural enemies, and (3) provide for its offspring. To solve such problems, it must possess certain special structures, such as teeth, wings, or horns. It must also display certain forms of behavior, such as building nests or making long migrations in search of food.

We give the name of special adaptations to the structures and modes of behavior that enable a species to survive. There are literally thousands of such adaptations in nature, and it would take many pages simply to give a list of them. In this article, we shall tell you about a few of the more significant or striking ones.

ADAPTATIONS FOR FOOD-GETTING

We find in both animals and plants special adaptations for obtaining food.

Special teeth. Among the most striking adaptations are the teeth of animals. The

Among the most striking adaptations for finding food are the teeth of animals. Below: a tiger fish with teeth so sharp they can cut through a sheet of steel; a hippopotamus threatening and exposing its curved lower teeth; and a viper with fangs that it uses to strike prey and emit venom.

Vasselet/Jacana

Rigaux — Images et Textes

Chaumeton/Jacana

The anteater of Central and South America has a long tubular snout from which it thrusts out a sticky tongue and picks up ants and other insects.

teeth of carnivores, or flesh eaters—dogs, wolves, cats, hyenas, bears, seals, and the like—are especially adapted to seizing and rending prey. Not all carnivores, it is true, live by the hunt. Most bears, for example, are vegetarians, and it is not too often that a domestic dog has occasion to seize and devour prey. But most of the carnivores are hunters, and the possession of suitable teeth is for them a matter of life and death.

The four large and pointed canine teeth of carnivores are the fangs that seize a victim. Together with the sharp claws that are found in most species, the canines are the chief offensive weapons of the carnivores. Behind the canines are the premolars and molars. These teeth are provided with sharp cutting edges, suitable for shredding meat. As for the carnivores' incisors, or front teeth, they are so insignificant as to be almost useless.

We find adaptations of a completely different kind when we examine the teeth of the rodents, or gnawing animals, which include beavers, muskrats, rats, and squirrels. The large, chisellike incisors in the upper and lower jaws are fine for biting or cutting off food. The rodents grind their food with their formidable broad-surfaced premolars and molars in the back of the mouth cavity. They have no canine teeth at all—no great loss, since they are vegetarians for the most part.

A rodent's incisor teeth continue to grow throughout life. Thus a beaver that depends upon incisors to chisel off the bark that it eats need not fear that these teeth will wear down to useless stumps. As a matter of fact, it is probable that some rodents gnaw a good many things that they do not eat or use, and that this keeps the incisors worn down to a comfortable length. The unchecked growth of such teeth would be disastrous. If a rodent's lower incisor is broken off by accident, the upper incisor has no surface to wear against, and continues to grow unchecked. Such an upper incisor does not grow straight downward, but curves back into the mouth cavity and finally punctures the braincase at the base of the skull.

Snouts and bills. The great anteater, a mammal that lives in the jungle country of Central and South America, shows some remarkable adaptations for obtaining food. In its natural environment this animal feeds almost entirely upon ants, termites, and other crawling insects. An anteater's front feet are provided with powerful claws, which can be used to tear apart anthills. The animal has a long, tubular snout, at the end of which is a small, toothless mouth with a tiny slit as an opening. The long, sticky tongue can be thrust out far beyond the mouth. It can be used to pick up ants on the surface of the ground or in the passageways of ant colonies.

The bills or beaks of birds are very efficient devices for obtaining food. A beak consists of an upper and a lower mandible, or jaw. In a heron, both mandibles are long and pointed, and the beak is especially suitable for spearing small fish and frogs in shallow water. A pelican is also a fisheater, but its beak is constructed quite differently. The upper mandible is curved abruptly downward at its tip, and it is with this portion of the bill that the pelican seizes a fish. A fleshy, pouchlike sac extends between the two sides of the lower mandible. When a pelican has captured a fish, it tilts its head backward and works the fish down into the pouch. The fish cannot escape and the pelican can swallow it at leisure.

The powerful beak of a hawk, with its curving and sharply pointed upper mandible, is very effective in tearing up the flesh of prey. Hawks also have powerful talons, or claws, with which they secure a firm hold on their victims. The bills and claws of owls are similar to those of hawks. A woodpecker has a strong chisellike bill with which it cuts into the decaying wood of trees. It can thus get at boring insects and their wormlike larvae. Woodpeckers also have horny, lancelike tongues. These can be used to spear larvae when their tunnels beneath the bark have been opened.

Muscles, venom, and shock. The powerfully muscled boa kills by constricting, or squeezing, the birds and mammals upon which it feeds. The snake wraps several coils around the prey and begins to squeeze. As the coils tighten, the victim is unable to breathe. The action of the heart is stopped and death comes quickly. Because of the snake's mouth structure, prey several times larger in girth can be swallowed whole. The python, closely related to the boa, kills its prey in the same way.

Other snakes, such as the rattlesnakes, the copperhead, the water moccasin, and the cobras, subdue their prey by means of venom. The rattlesnakes of North and South America, for example, develop venom glands on the two sides of the head above the roof of the mouth cavity. Tubelike structures, known as ducts, extend from each gland to a pair of hollow fangs in the upper jaw. The fangs have openings a short distance from their tips and the venom spurts through these openings.

The fangs are attached to small bones hinged to long, rodlike bones of the upper jaw. Normally, the fangs lie flat against the roof of the mouth. When the mouth is opened in striking prey, the upper jaw bones slide forward, moving the fangs into

Birds' beaks are very efficient food-gatherers. Upper right: hummingbird with long narrow beak for obtaining nectar from a flower. Lower right: kingfisher with long pointed beak with which it seizes small fishes.

A boa kills its prey by coiling around it tightly, constricting the victim so tightly that it finally cannot breathe and dies.

477

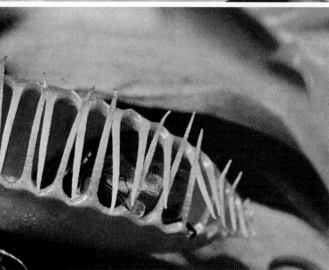

The Venus's-flytrap has trapped and firmly holds an insect in its specialized leaves.

sharks, store up electricity in organs on each side of the head. The discharge of the stored electric energy paralyzes or kills fishes in the vicinity and the ray pounces upon them. Even humans are affected. Many a swimmer has had his arms and legs numbed by the electrical discharges of the torpedo ray. The electric eel, found in the streams of tropical South America, has a similar shocking apparatus.

The mouth parts of the female mosquito are contained in a long sheath and represent a complicated adaptation. Sharp piercing structures cut into the victim's skin. A tube injects saliva into the wound, preventing the blood from clotting, while another tube sucks blood from the wound.

Plant adaptations. Striking adaptations for obtaining food are by no means limited to the animal world. There are plants that trap insects and obtain necessary proteins from their victims.

One of the best-known plants of this type is the Venus's-flytrap, which grows in moist places around freshwater ponds. The leaves of the Venus's-flytrap bear double-bladed structures with toothlike edges. These structures come together to entrap an insect that alights in the space between the blades. The victim is then partially dissolved by digestive juices, and the dissolved portions are absorbed by the plant tissue.

The pitcher plants form another famous group of insect-catchers. In this case, the leaves are like pitchers. They hold rainwater and form tiny ponds in which insects are drowned and digested. The pitchers of one South American species, thriving in an area of very high annual rainfall, have small holes near their tops. These holes permit excess water to be drained off. Hence the pitchers cannot overflow and the drowned insects contained in them cannot be floated off.

The adaptations of certain desert plants enable them to obtain food and to conserve water. Normally a higher plant bears many green leaves. Not only are they food-making centers but also through them water vapor is given off to the air. This sort of structure, however, is not at all suited to

the striking position. The rattlesnake, indeed, literally strikes, rather than bites, its victim. Muscles in contact with the venom glands contract so that venom, forced from the glands, passes through the fangs into the wound.

Certain fishes secure their prey by shocking it. The members of the torpedo-ray, or electric-ray, family, related to the

A porcupine with quills erect. Any enemy approaching will quickly find the quills embedded in its skin.

Hansle/Jacana

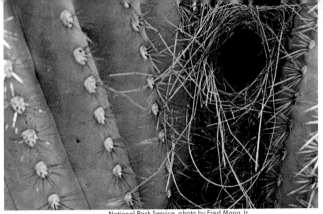

Closeup of part of a giant saguaro cactus. It offers the protection of its strong spines to some birds that nest among its stems.

National Park Service, photo by Fred Mang Jr.

life in the desert. Here water comes only from the rare showers or the melting of scanty winter snows, and it must be carefully hoarded throughout the dry months if the plants are to survive.

Members of the cactus group, such as the giant cactus of the North American Southwest, are well adapted to desert life. They have no broad leaves through which water may be lost to the atmosphere. The plants are therefore able to retain much of the water that is absorbed during or after a rain, and to conserve it throughout long periods of drought. At the same time, they are able to carry on normal food-making processes. Their food-making activities are carried on entirely in the green stems.

ADAPTATIONS FOR DEFENSE

Some adaptations serve for defense as well as for providing food. The rat often uses its teeth to defend itself when it is cornered. The electrical discharge of the torpedo ray is sometimes directed against larger fish which are pursuing it. The venom of snakes helps to ward off the attacks of stronger enemies, including man. Other adaptations are chiefly defensive.

Scent glands. The scent glands of skunks are a remarkably effective defensive weapon. Both the striped skunk and the little spotted skunk of North America possess such glands, just beneath the skin on either side of the anal opening. Skunks are able to discharge their ill-smelling secretion in the form of a spray to a distance of about three meters. The odor of this spray is overpowering and is very hard to remove from clothing. If any of the secretion gets into an

animal's eyes, temporary or permanent blindness may result.

Most other animals give skunks a wide berth, and if unmolested the skunks quietly go their way. Apparently they realize that they have little to fear from other animals of the wild. In fact, they are apt to adopt an attitude that sometimes has disastrous results. For example, they show little disposition to get out of the way of approaching cars on highways at night, and many of them are run over and killed.

The bombardier beetle is a poison-gas specialist among insects. Within its body it secretes a foul-smelling liquid, which turns into vapor as it is discharged from two glands near the anus. There is a popping sound as the gas attack is launched against a pursuer. The beetle generally makes good its escape from its startled foe.

Barbs. The Canada porcupine is well prepared in its normal habitat to defend itself against almost any enemy. This animal, a rodent, lives in Canada and the northern part of the United States. It feeds largely upon bark, buds, and other plant materials. It is armed with a large number of sharply barbed quills, which are especially well-developed on the animal's back, sides, and tail. The quills are loosely inserted in the skin and normally point backward, but can be raised in moments of excitement.

Contrary to popular belief, a porcupine does not throw its quills. However, they are so loosely inserted that they come out readily and fasten themselves upon an attacker. A dog that attempts to harass a porcupine will soon have an unwelcome collection of quills in its face and neck. The

Horns provide many species of animals with a weapon and means of defense. Here a male impala with large lyre-shaped horns.

Visage/Jacana

presence of barbs makes the quills difficult to pull out. In fact, they are likely to be still more firmly embedded when a victim tries to get rid of them.

The defense mechanism of squids and octopuses is also most effective. These animals are soft-bodied, sea-dwelling mollusks, related to the clams and snails. Each squid or octopus develops an internal structure, known as an ink sac, which is filled with a black fluid. When an enemy appears, the mollusk squirts its ink into the surrounding water. Then it beats a hasty retreat while the attacker is trying to locate its prey again.

Camouflage and mimicry. Many of the defense mechanisms encountered in nature are effective because they make animals inconspicuous and therefore less likely to attract the attention of their enemies. The walking-leaf insect of Asia, for instance, shows a remarkable resemblance to a green leaf. The upper surfaces of its green-colored wings bear a brown pattern that looks very much like the veins of a large leaf. Its legs are flattened out and veined so that they look like smaller leaves.

The dead-leaf butterfly of Indonesia does, as its name indicates, look remarkably like a dead leaf. When this butterfly is at rest on a twig, its wings are folded together and only the lower surfaces can be seen. These are dull brown in color and bear a black, veined pattern. The upper wing sur-

faces are brightly colored by comparison with the lower surfaces. In flight, therefore, the dead-leaf butterfly looks like a different insect.

Certain defenseless insects closely resemble insects of other species that are feared by their neighbors. This is called *mimicry*. Bees, as you know, are four-winged insects that have an effective sting. They are hairy and show contrasting colors. Bee flies have only two wings and they have no sting. But since they too are hairy and are colored after the fashion of bees, they may easily be mistaken for these formidable insects. It is possible that potential enemies may avoid bee flies for this reason.

Some animals feign death in order to try to escape from their foes. Sometimes the performance is striking but not particularly convincing. One often sees a Virginia opossum lying inert, with eyes closed, as a dog approaches. The hognose snake of North America also feigns death on occasion. When confronted by an enemy, it will turn over on its back and lie without movement. If the snake is turned over on its belly, the chances are that it will assume its former position on its back, as if this were the only possible position for a dead snake to maintain.

Poisons and spines. Plants, too, possess adaptations that enable them to ward off enemies. Some secrete poisons, which are usually concentrated in the leaves, fruits, or flowers. The leaves of the wild cherry contain prussic acid and will kill the animals feeding upon them in a matter of minutes. The locoweed absorbs the element selenium from the soil. The selenium-containing leaves of this plant will poison animals that graze on it.

Certain plants are not poisonous, but are unappetizing and more or less indigestible because they have absorbed so much silica from the soil. This is true, for example, of horsetails and the leaves of certain evergreen trees. Other plants, like the dog fennel, have a disagreeable odor that discourages would-be browsers or are disagreeable to the taste.

Various plants are protected by sharp

and pointed weapons, such as spines, thorns, and prickles. These structures are so distributed as to protect vital parts—leaves, flowers, and fruits. The spines of cactus plants are a familiar example. Pine needles will cut the mouths and the internal organs of animals feeding upon them. Not only are the leaves of the South African aloe stiff, narrow, and pointed, but they are also protected by small spines. The tiny, hooked bristles along the edges of the cut-grass can inflict nasty wounds. The barbs of the stinging nettle not only pierce an attacking animal but also inject poison into it.

ADAPTATIONS FOR PRESERVING OFFSPRING

A great many adaptations have to do with preservation of offspring.

Survival of plant seeds. There are numerous instances of such adaptations among the members of the plant kingdom. A maple tree, for example, may bear thousands of seeds in spring or early summer. If all of these seeds fell directly to the ground beneath the parent tree, the resulting offspring would be far too crowded for successful growth. Maple seeds, however, are enclosed in winged fruits, called *samaras,* each generally containing one seed.

When samaras break off from the tree, they are carried for some distance by the wind before they come to rest. The seeds of the common dandelion have similar adaptations. Each seed bears an umbrellalike process. This is caught and lifted by air currents so that the seed may be carried far from the mother plant.

The nut of the coconut palm is a great traveler, too, but it voyages by sea. This fruit is protected by a thick but light husk. When the fruit falls into the water from a palm on a tropical beach, it may be carried for great distances. It may be washed up on a distant isle and sprout there.

Even animals serve to transport various plant seeds. Thus the fruits in which the seeds of the burdock are enclosed are covered with prickly envelopes known as *burrs* and provided with innumerable hooks. When an animal brushes against the burrs, they stick to its hair. As a result, the seeds get widely distributed.

Number of offspring. Some carnivores have fewer young for several years after there has been a drop in the numbers of their favorite prey. This kind of reproductive adaptation has been scientifically established for lynxes and snowshoe hares, for example.

J. R. Simon/Bruce Coleman

Some animals adapt to severe climate conditions and to scanty food supplies during the winter by hibernating. A state of dormancy, it allows the animal to save energy. At right, a marmot hibernating.

Protection of animal young. There are many outstanding examples of special adaptations for the care and protection of the young among animals. The eggs of the little marine creatures known as sea horses are protected by a special brood pouch in which they are hatched. It is the male and not the female that develops this pouch, on the lower surface of the body near the tail. The female inserts her eggs into her mate's pouch and there they remain until the young have emerged.

The male of the midwife toad also bears the responsibility for the safety of its mate's eggs. The female lays two strings of eggs, which the male wraps around its hind legs. It takes refuge in out-of-the-way places while the eggs are developing, coming out now and then to moisten the eggs in a pool of water. When the time for hatching is near, the male again takes to the water, and well-developed tadpoles emerge from the eggs. The tadpoles complete their development in ponds before becoming full-fledged adults.

During the egg-laying period, the male Surinam toad grips the female about the lower part of the body. Male and female roll over repeatedly as the eggs are laid. The eggs pop out a few at a time while the male and female are upside down. The eggs fall forward and stick to the swollen skin of the female's back. The male fertilizes the eggs as the pair right themselves and swim to the bottom. In the week that follows the laying, the eggs sink into the back of the female until they are flush with the surface of her skin. The eggs develop in these individual pockets of skin or brood pouches. After three or four months in the skin pockets, small but fully formed toads pop out from the pouches in response to pressure exerted by the mother toad.

A well-known adaptation for the protection of the young is found among the kangaroos and wallabies of Australia and the opossums of North and South America. This is the marsupial pouch, a pocket on the surface of the female's abdomen. The name "marsupial" refers to the order to which these animals belong. The marsupial pouch serves as a place of refuge for the

J. Six

Holton/PR

Animals have many adaptations to preserve their young. Top: a male midwife toad carries its mate's eggs on its back. Lower: a kangaroo carries its young in a pouch.

Gerster/Rapho

Many species of animals migrate—to find suitable climatic conditions, adequate food, and a favorable place for their young to develop. Above, a herd of sea lions has just arrived on a California island after breeding in the North Pacific.

young. These vary in number from one or two, in the case of kangaroos, to the fourteen or more of a Virginia opossum. The young are small, immature, and quite helpless at birth. But they are housed safely in the pouch, where they can get at the milk glands without leaving their shelter.

A kangaroo carries her young around in her pouch until it has grown to considerable size. When pursued by an enemy, a kangaroo will sometimes pull a young one from the pouch and throw it to the ground. It is not known whether she does so in order to lighten her own load so she moves faster, or to provide for the escape of the young kangaroo in case the mother is overtaken by her pursuer.

Some animals safeguard their young by remarkable adaptations of behavior. The female Siamese hornbill makes her nest in a cavity in a tree. After she has begun to sit on her eggs, the male hornbill seals up the entrance to the hole almost entirely with mud. This dries and hardens, leaving a slit-like opening, through which the male feeds his mate while she is nesting. The opening is not large enough to permit the entrance of animal foes that climb about among the trees in warm countries.

Migrating for a nesting site. The *runs,* or migrations, of the king salmon of North America are associated with the finding of a suitable place in which the eggs may hatch and the young may develop. A king salmon spends most of its life in the sea and may grow to a weight of over 40 kilograms. Each year adult salmon four to seven years of age gather in bays along the western coast of North America in preparation for a run up a freshwater stream to the quiet pools at its source. The fish do not stop to feed or rest for any length of time on the way. They always press on upstream, and this despite the fact that some of their journeys are hundreds of kilometers in length. Their courses are made difficult by the presence of numerous rapids and falls. They pass all these barriers, sometimes

Joe Munroe/PR

Salmon are well-known for their annual runs upstream to breed. Above, salmon travelling up a rapid Alaska stream.

leaping out of the water.

Eventually the salmon reach the headwaters of the stream, often in a sadly battered and worn condition. The females deposit their eggs on the bottoms of the small streams or shallow pools of the headwaters. The male salmon discharge milt, or reproductive fluid, over the eggs. Thereupon, the adult salmon usually die.

After hatching from the eggs, the young remain for a time in the fresh water where the eggs were laid. Here there are not likely to be many large enemies and the young salmon have a good chance to survive. As they become larger, they move downstream and finally enter the sea. Most of them are a year old and ten centimeters in length before they see the ocean.

Other fishes that have become famous for their spawning migrations are the freshwater eels, found in freshwater ponds, lakes, and streams of Europe and eastern North America. These bony fishes have long, slender snakelike bodies.

Until recently, the spawning habits of eels were an unsolved mystery. People knew only that fully grown eels went out to sea, never to return, and that young eels came in from the sea. The 4th century B.C. Greek philosopher Aristotle thought that these fishes did not lay eggs but that their young arose in some unexplainable way from the sea itself. It was not until modern times that the facts were revealed.

Eels live in their freshwater homes until they become mature. In some cases, they attain a length of one and one-half to two meters and a weight of several kilograms. In the autumn in which they become mature, the fishes begin a migration to the sea. When they reach salt water, they continue onward for hundreds of kilometers to the Sargasso Sea, a calm stretch in the North Atlantic Ocean. Here the females deposit their eggs in the spring of the year. It is probable that male and female adults die soon afterward.

When the eel eggs hatch, a vast number of almost transparent, threadlike young rise to the surface of the sea. Some of them turn eastward toward Europe, and others begin the shorter journey to North America. The young usually are three years old before they appear on the European coast, but the American eels make the trip from the breeding grounds to North America in two years. They now go up streams to remain until they become adults.

Why should adult eels make this long spawning migration which, as far as we know, ends in their death? According to one theory, at some time in the distant past the coasts of Europe and North America were closer together than they are today. The eels, therefore, had a relatively short trip to their ancestral breeding ground. In the course of time, the two continents became farther removed from each other, but the eels still obeyed the instinctive urge to return to the place of their origin.

We could extend almost indefinitely this account of the special adaptations of animals and plants. We have already provided enough examples, however, to show the importance of these adaptations.

ECOLOGY

by Beth Schultz

Prehistoric man was not a scientist. But in order to survive he had to understand some relationships among plants and animals and the environment. He had to know where his favorite meat animal lived, where it went to drink water, to graze, or to rest. Knowing the habits of the animal made it easier for him to stalk and kill it.

Later, man began to raise crops. He domesticated the animals he needed for food and other products. Thus he no longer had to spend most of his time hunting. But he had to be able to recognize good soils, to predict the changing seasons, and to select plants and animals that would thrive under domestication. Gradually man became a practical ecologist—thousands of years before he developed a formal, orderly study of this science.

Naturalists of the past five or six hundred years made major contributions to the science of ecology. Men such as Carolus Linnaeus, Charles Darwin, and Alfred Wallace collected specimens and made notes of what they saw as they journeyed around the world. Slowly, knowledge accumulated, and scientists began to understand the relationships among organisms and their surrounding environments.

The word *ecology* was coined in the late 1860s. It is derived from two ancient Greek words meaning "study of the home." The science of ecology is the study of the relationships among plants, animals, and their home, or environment. Ecologists use information from many branches of science: biology, chemistry, the earth sciences, and so on. They must deal with all this information because the life of every plant and animal depends upon the complex interaction of its internal environment—its

Plants, animals, and their surroundings exist in a complex give-and-take system. A typical food chain (right) illustrates one part of an ecosystem. Cabbage (top) is a green plant, which is able to manufacture its own food. It is eaten by a rabbit, which is, in turn, preyed on by a fox (middle right). When the fox eats, saprophytes (lower right) in its intestine are also nourished.

Bringe, Jacana

Choussy, Jacana

Lanceau, Jacana

C. Nuridsany

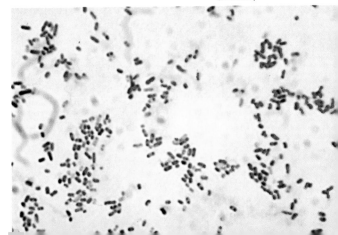

Species within an ecosystem interact in various ways. In parasitism, one species lives on or in another, benefiting at the expense of the host. The broom rape (upper left) is a parasite living on the roots of vegetables. In symbiosis, or mutualism, two species live together, both benefited by the association. A lichen (upper right) consists of an alga and a fungus living together. In yet another type of interaction—commensalism—one species is benefited by the association while the other is unaffected. The clown loach (opposite page) benefits from its life with a sea anemone.

physiology—and its external environment—its habitat.

LIFE AND ENVIRONMENT: A DYNAMIC SYSTEM

The whole earth is an *ecosystem,* a system of give-and-take among plants, animals, and their surroundings. As in any system, whatever happens to one part of an ecosystem affects its other parts. Materials are cycled from soil, water, and air through the plants and animals and then back to the soil, water, and air.

The energy that operates the ecosystem originates in the sun. This solar energy is trapped by green plants in the food they manufacture during the process of photosynthesis. The energy is needed to hold atoms together in the compounds we call food. As the food is used by the plants, by animals that eat the plants, and by animals that eat other animals, energy is released and used. By the time the carbon and other elements found in food are cycled through the plants and animals and back into the soil, water, and air, all of the energy has been dissipated.

Thus the sun's energy is not the only essential to life. Green plants are also essential to our planet's ecosystem. They make the food and trap the energy that operates the entire system.

Gross/Jacana

FOOD IN THE ECOSYSTEM

The paths of food materials through the ecosystem are many and complex. A *food chain* describes how one morsel of food might "travel" through the system. A typical food chain begins with a one-celled green alga that manufactures food. The alga is eaten by a tadpole. The tadpole in turn is eaten by a fish, which later may be devoured by a water snake. A hawk may then swallow the snake. Indirectly, then, the hawk has eaten the green alga.

There are many paths by which food can be cycled through the ecosystem. A *food web* shows the many possible paths. For example, green algae may be eaten by water insects or various fish, as well as by tadpoles. The insects, fish, and tadpoles may be ingested by bigger fish, large frogs, snakes, ducks, raccoons, and other creatures. These in turn may be eaten by still other animals.

How much food and energy pass along a food chain? How many kilograms of grass are needed to produce one kilogram of grasshoppers? How many kilograms of grasshoppers are consumed to form one kilogram of robins? Ecologists use the term *food pyramid* to describe the quantities of materials that pass from one kind of eater to another.

The word "pyramid" is used because the amount of energy associated with the food material becomes smaller at every link, or upward step. More than one kilogram of grass is needed to create one kilogram of grasshoppers. More than one kilogram of grasshoppers goes into making one kilogram of robin flesh. The net losses in weight are due to the fact that part of the material is converted into energy to power the transformations of grass into grasshoppers and grasshoppers into robins.

LINKS IN THE CHAIN

When discussing food relationships in the ecosystem, scientists find it easiest to use words that indicate the roles animals play in the food chain. A *herbivore* is an animal that eats plants. *Carnivores* eat animals—both herbivores and other kinds. *Omnivores* eat a variety of both plant and animal foods. *Saprophytes* and *decomposers* eat the leftovers, plant and animal, of what other animals have largely devoured. *Parasites* live on and in the bodies of living plants and animals.

There is a similar situation among plants. Green plants make their own food through photosynthesis. They are the *producers* of food for all life, including themselves. Other plants, such as bacteria and fungi, are not green and cannot carry on photosynthesis. They therefore depend on green plants and, in many cases, animals for their food. Bacteria and fungi may live as parasites on plants and animals. Other bacteria and fungi thrive on the waste products or dead bodies of organisms. They decompose the wastes and dead bodies and return the materials to the soil, water, and air.

PARTS OF THE ECOSYSTEM

As you can see, an ecosystem is very complex. To study such a system, we must examine its parts and learn how these parts relate to the whole. First, we study an individual organism. Second, we examine a *population*. This comprises all the individuals of a species found in a given area. Then we study all the populations in a given area. These make up a *biotic community*.

If the area is very large and has certain general environmental characteristics, it is called a *biome*. A biome contains many communities. There are forest biomes, grassland biomes, desert biomes, and so on. The characteristics that distinguish one biome from others include its temperature, rainfall, type of soil, and amount of available light.

The term ecosystem is sometimes used as a synonym for biome or for biotic community. Or it may be used to refer to all living things and their environment. All these terms include descriptions of the cycling of materials, the food webs, and the flow of energy. But when ecologists talk about biomes or communities, they describe the habitats and the kind of organisms living in them, as well as the feeding habits of the community's animals. When ecologists investigate ecosystems, they concentrate on the materials and energy that flow through the system. This is a more schematic, abstract manner of describing the relationships between organisms and their environment.

The part of the earth in which life is found is called the *biosphere*. This consists of the outer shell of the earth: the upper part of the crust, bodies of water, and the atmosphere. Birds fly through the air; worms burrow into the ground; microorganisms are found in clouds kilometers above the earth and in waters kilometers underneath the ocean's surface. Not only is the biosphere the part of the earth where life is found, it is also the part where solar energy is used to bring about photosynthesis and other chemical and physical changes—changes that are primarily caused by the organisms in the earth's ecosystem.

In the rest of this article we will consider some of the individuals, populations, communities, and biomes that make up our ecosystem.

CARBON CYCLE IN NATURE

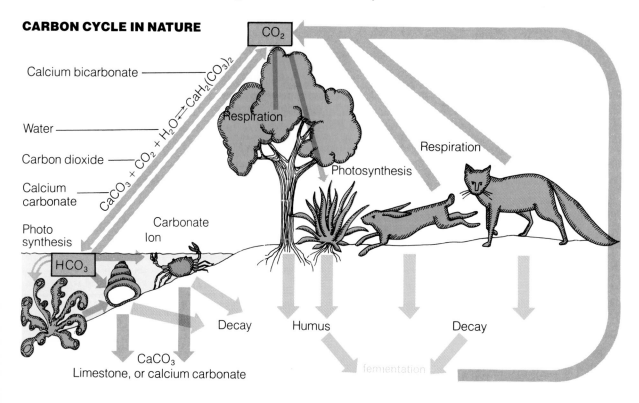

Calcium bicarbonate

Water

Carbon dioxide

Calcium carbonate

$CaCO_3 + CO_2 + H_2O \rightleftarrows CaH_2(CO_3)_2$

Respiration

Photosynthesis

Respiration

Photo synthesis

Carbonate Ion

CO_2

HCO_3

Decay

Humus

Decay

$CaCO_3$
Limestone, or calcium carbonate

fermentation

ECOLOGY OF INDIVIDUALS

A plant or animal lives where it does because that place provides it with at least the minimum requirements for life. These requirements include sufficient space, water, food, and air, and a tolerable range of temperatures. This living place is the organism's *habitat*—its external environment. The most favorable habitat is the one best suited to the organism's structural and physiological needs—its internal environment.

Ecologists try to analyze the organism's internal and external environments to learn how they interact. Most frogs, for example, live in wet places. Their thin skin does not protect them from drying out. Therefore, only frogs that stay in wet or damp places survive.

The frog's method of reproduction also limits it to the vicinity of water. Fertilization of the eggs is external. The male frog embraces the female while she emits several hundred eggs in a jellylike mass. The male then emits swimming sperm that penetrate the mass and enter the eggs. Since the sperm must swim to the eggs, fertilization would be impossible on dry land. Because the fertilized eggs have no protection against water loss, the embryos can only develop if they are in water. After hatching, the immature frogs, or tadpoles, are gill-breathers. They must stay in water until they become adult frogs with lungs.

Do frogs ever stray into a dry field? Do they ever attempt to breed on land? If they do, neither they nor their eggs survive. The dry environment is hostile to frogs. It is said to select against frogs that enter it. In comparison, a wet environment selects in favor of frogs. This is an example of natural selection—survival of those organisms that conform best to environmental conditions in terms of their own internal environments, or physiologies. Only those individuals that remain in the favorable habitat reproduce.

An animal's reactions to its environment, its *behavior,* help it survive. Lizards, which lack any effective means of regulating their body heat, go into their burrows

NITROGEN CYCLE IN NATURE

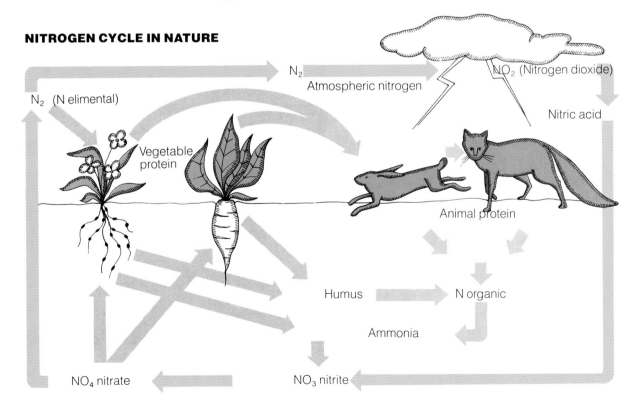

N₂ Atmospheric nitrogen
N₂ (N elimental)
NO₂ (Nitrogen dioxide)
Nitric acid
Vegetable protein
Animal protein
Humus
N organic
Ammonia
NO₄ nitrate
NO₃ nitrite

C. Nuridsany

Blood-sucking ticks sometimes infest some species of vertebrates. Here the tick that parasitizes hedgehogs.

when the outside temperature becomes too high. Earthworms will stay in a damp place and will move away from a dry place. Termites build tunnels that enable them to travel across open places without risking the drying effects of the open air.

Plants also respond to environmental conditions. A bean seedling will grow toward a source of light. When the growing tip of a morning-glory vine touches a stake or taut cord, the stem begins to twine around it. If a seed or spore lands in a spot where few or none of the conditions are suitable for its existence, it will not sprout.

Plants and animals react negatively to conditions that may cause death and positively to those favoring life. If an environment changes drastically, the plants usually die. Many animals, on the other hand, can move to more-favorable habitats. Now, however, man is changing so much of the environment that many kinds of animals can no longer find suitable places to live or breed. Some of these are thus threatened with extinction.

ADAPTATION

The range of environmental conditions within which a given plant or animal can stay alive is called the organism's *range of*

tolerance. Every plant and animal has a built-in (inherited) range of tolerance for each environmental factor: temperature, light, water, and so on. Some individuals and species have remarkably wide ranges of tolerance and can survive extreme changes in conditions. Other organisms have very narrow ranges. A shift, especially a sudden shift, in one or more environmental conditions to a point beyond the tolerance range of a population may cause the death of all or most of the individuals of that population.

A good example of man's deliberate efforts to keep certain organisms from growing is his method of preserving food. Man alters the environment of food to extend some factor beyond the tolerance range of bacteria and molds that spoil food. Canning and freezing extend the temperature ranges to points beyond the tolerance of these microorganisms. Drying removes the water necessary for life. Sealing of food prevents its recontamination. Bacteria and molds present in the foods are thus killed or made inactive; those in the environment cannot develop on the food.

Knowing an organism's range of tolerance helps ecologists understand why that species lives only in a certain habitat, or environment.

The environment is a mixture of many components: heat, light, the gases in the atmosphere, and all the substances in the soil and water. The amounts and kinds of energy and materials are different in different places on earth. Probably no two places are exactly alike. In some places, where one or two of these components are in short supply, only a few kinds of plants and animals can tolerate the environment. In Arctic regions, for example, temperatures are low and determine what can live in these regions. We say that temperature is the *limiting factor*. In temperate and tropical deserts, water is the limiting factor; only plants and animals that can tolerate dryness survive.

Think about the vast number of different environments on the earth and realize that there are few places devoid of life. Consider the variety of life-forms necessary for inhabiting such diverse environments. Think, too, about matching the ranges of tolerance of the organisms with the limiting factors of the various environments. You can begin to understand why our planet supports so many different kinds of plants and animals.

ECOLOGY OF POPULATIONS

Fifty pigeons, 30 house sparrows, 10 gray squirrels and 4 robins live in a city park. The park therefore has a population of pigeons, a population of sparrows, a population of gray squirrels, and a population of robins. There is also a population of maple trees, one of grass, and another of dandelions. The list could go on, including all the insects, spiders, worms, mites, soil bacteria, and so on. How many populations live in the park? All the organisms of a species living in a given area constitute a *population*. Thus the total number of populations equals the number of different species of plants and animals there.

How large can a population grow? What determines how many squirrels, robins, or maple trees can live in the park? Habitat is probably the most important determinant of the size of a wild population. Food supply, predators, space, disease, and environmental factors such as water, soil, and climate are all part of the habitat.

Ecologists use the term *carrying capacity* when they discuss the number of animals a habitat can support. Carrying capacity depends not only on food supply, but also on the number of places to live. For example, an old abandoned field that is growing back to shrubs and trees will support as many rabbits as there are brushy places for them in which to hide and raise their young. The surplus rabbits are liable to be killed by hunters or their natural predators.

Even when the physical environment is favorable and when food is abundant, most animal populations do not keep on growing without limit. There seems to be a maximum population density for each species. If left undisturbed a population will become stable and will not show large increases or decreases in numbers.

ECOLOGY AND SOCIAL BEHAVIOR

One of the most important factors that regulate animal population size is their behavior. Most species of birds and mammals, as well as some lower animals, exhibit some kind of social behavior. That is, they react to the presence of members of their own species. Scientists who specialize in the study of the behavior of wild animals are called *ethologists*.

Let's consider an interesting example of animal behavior. In spring, during the breeding season, birds often become fiercely territorial. Each bird establishes a living space for himself which he defends vigorously. A male robin, for example, arrives in a park in early spring. Soon he begins to sing from two or three perches — always the same two or three perches. As other male robins arrive, the first male's singing keeps them out of the territory he is defending. If his persistent singing does not keep them out, he flies toward the trespassers, threatens them, and sometimes fights them off.

When a female robin arrives, she selects a mate and his territory. Thus the number of robins living in the park will depend upon the number of territories it can support. The size of a territory varies with the species as well as with the habitat. Usually a pair of birds defends an area large enough to provide food for a brood of young.

In most cases a male animal defends his territory against members of his own species but is tolerant of some other species. A pair of chipping sparrows may nest in a bush next to a robin's tree, but there are never two robin nests in the same small area.

Are there male robins "left over" after all available territories in an area are occupied by other robins? Yes, there are. These robins are called bachelors. They live outside the boundaries of the territories of mated pairs. If a mated male robin dies, his female partner can soon find another mate from among the bachelors.

Many kinds of fish are territorial, too. Certain sunfish, for example, build pebble nests on the bottom of a pond. After the eggs are laid, the mated pair stays near the

THE MAJOR BIOMES OF THE WORLD

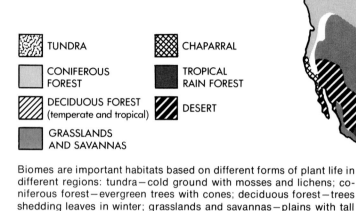

TUNDRA

CONIFEROUS FOREST

DECIDUOUS FOREST (temperate and tropical)

GRASSLANDS AND SAVANNAS

CHAPARRAL

TROPICAL RAIN FOREST

DESERT

Biomes are important habitats based on different forms of plant life in different regions: tundra—cold ground with mosses and lichens; coniferous forest—evergreen trees with cones; deciduous forest—trees shedding leaves in winter; grasslands and savannas—plains with tall grasses; chaparral—dry areas with shrubby growth; tropical rain forest—warm moist jungles; desert—arid to semiarid areas. Blank areas represent regions with many kinds of plants, or no plants, as in the icy interior of Greenland.

nest, fanning it with their fins and tails. This not only keeps silt from smothering the eggs but also prevents other fish from eating the eggs.

HIERARCHIES

Mammals that live in herds or other kinds of groupings develop ranks, or *hierarchies,* among themselves. Reindeer, elk, whales, seals, and domestic cattle are examples of animals with hierarchies.

The individuals, especially the males, of a hierarchy engage in contests for rank. Usually these contests consist only of threats, or shows of force. Rarely is any animal seriously injured or killed in these contests. However, the resulting *dominance hierarchy,* in which the animals become ranked according to their aggressiveness, seriously affects the health and even the survival of the herd.

When food is in short supply, the dominant animals eat while those of low rank must wait and take the leftovers, if there are any. When there is not enough food for all, the weaker animals starve, but the strong survive in good condition. During the breeding season in some species, the males compete for mates. The strongest males protect harems of many females. The weaker and younger males do not mate at all.

Dominance hierarchies therefore have two major effects on populations. When the habitat is poor and food is scarce, the animals do not share it equally. The strong survive and the weak die. In this way the population is reduced to a number that the habitat can support. The second effect is that only the strongest animals reproduce. Thus the young have a better chance of inheriting genetic characteristics that will help them survive in the habitat.

Certain species respond in unique ways to crowding, or high population density. The lemmings of Norway provide an example. When their numbers exceed the

carrying capacity of their habitat, mass emigrations begin. When the lemmings reach and enter the sea, thousands, perhaps millions, of these small rodents perish.

GAME MANAGEMENT

All too often man interferes in nature, wittingly or unwittingly altering the sizes of animal populations. Because man often enjoys hunting as a sport, he has tried to manage and conserve the habitats and populations of game animals—deer, pheasant, grouse, rabbits, and many others. While controlled hunting of animals has often been beneficial for the animals, sometimes it has also had unforeseen consequences.

The management of deer in the northeastern part of the United States is a good example of mistakes made before man understood more fully the habits of animals. At one time, white-tailed deer were abundant in this area. They were extensively hunted to provide meat for individual families and for the commercial market. By 1900, however, hunting, especially the commercial kind, had reduced deer herds to levels approaching extinction.

In man's efforts to save the remaining deer, predators (wolves and wildcats) were slaughtered. Game laws limited the number of deer that a person could kill in a season and protected female deer and their young.

These conservation laws were made in an effort to increase the size of the deer herds. But the habitat was changing. Forests were lumbered and land was cleared for farms. This activity improved the habitat for deer, which thrive on buds and the tender bark of bushes and new, small trees. With a plentiful food supply and few predators, deer populations increased markedly. Today, however, the young trees have grown tall; deer can no longer reach and browse on the soft bark and buds.

The carrying capacity of the deer range has decreased. But the deer-protection laws are still in force. Consequently, thousands of deer starve to death in late winter.

By permission of the Elementary Science Study of Education Development Center, Inc.

Children studying an aquarium. What are the producers? The consumers?

One of the major purposes of game-management research is to study all the factors involved in ecology of game animals. This should ensure that, in the future, man's activities do not lead to situations such as the mass starvation of a deer herd.

ECOLOGY OF COMMUNITIES

A biotic *community* consists of many populations living in one habitat. Some of these plant and animal populations depend on each other. Others are there only because they tolerate the same environment.

Communities are named after some dominant environmental or biological feature. Thus we speak of a cattail marsh, a cypress marsh, a beech and maple forest, a tropical rain forest, an Arctic tundra, and a desert.

Food relationships in a community are the easiest relationships to observe and understand. All plants and animals use food. Even the green plants that manufacture all the world's food consume some of it as they grow and use energy. If we observe a pond community, we will notice that snails, mayfly nymphs, small crustaceans, and some kinds of fish feed on the water plants. Other aquatic insects, crustaceans, and fish prey on the herbivores and on each

other. The crayfish, for example, will eat any animal it can catch. Scavengers, such as the water sow bug, eat the dead organic material.

Each plant and animal plays a definite role in food chains and webs in the pond. We call this role, or function, an *ecological niche*. The water sow bug, for example, fills the ecological niche of scavenging. Water plants fill the niche of food production.

A healthy pond community may have populations of several hundred different types of plants and animals, each filling a particular niche. There are microscopic organisms; floating plants and rooted plants; animals in the mud and debris on the pond's bottom; animals swimming in the open water; animals clinging to the roots of floating plants or burrowing into the stems; and animals crawling or skimming on the surface.

HEALTH AND STABILITY

Freshwater biologists have learned to determine the health of water environments by the kinds of plants and animals they find. If they find huge populations of only a few varieties of life, ecologists suspect pollution. For example, tubifex worms, catfish, and carp can tolerate water with less available oxygen than can dragonflies, mayfly nymphs, sunfish, and bass.

Lakes and rivers that are heavily polluted with organic material, such as the outflow from sewers, sewage-treatment plants, and paper mills, are liable to contain less dissolved oxygen than does nonpolluted water. These waters will support only a few species of tolerant organisms. Sometimes the pollution is so great that not even the most tolerant species can survive. Then the lake or river is said to be dead.

Healthy communities are basically stable. They support many kinds of green plants, many kinds of herbivores, and many kinds of carnivores. Thus there are many paths for the cycling of materials and the flow of energy. If one species of plant disappears, there are other green plants still manufacturing food. If one species of herbivore dies out, there are other herbivores still eating plants. Likewise if a species of carnivore is removed, there are others remaining to continue the cycling of materials and energy through the community.

The community that has only a few kinds of living things is less stable. If one or two species die out, there may not be alternative paths for the flow of materials and energy. If the environment changes suddenly, the whole community may die.

Life, however, is not static, even in the most stable communities. Natural changes occur continuously. Individual organisms are born; they grow, change, and eventually die. Communities also develop and change. Members of a community, by their very activities, tend to change the environment. After a period of time they often make the habitat unfit for themselves. The organisms then die out or migrate elsewhere. But they have made the environment fit for other kinds of plants and animals. A different kind of community develops in place of the old one. This kind of gradual, but continuous, change is called *community succession,* or, in a broader sense, *ecological succession.*

ECOLOGICAL SUCCESSION

A good example of succession may be observed by watching an abandoned farm over a period of many years. During the first few seasons after abandonment, the fields produce only grasses and perennials such as goldenrod and Queen Anne's lace. In another few years briers begin to grow and it is difficult for a person to walk through the fields. Rabbits and other small animals that were chased out by the plow now return. Soon seedlings of so-called "pioneer" trees appear: aspen, thorn apple, and wild cherry.

As these trees grow, the briers and some of the grasses and perennials die out. They cannot survive in the shade of the trees. But seedlings of long-lived trees start to grow in the shade of the first trees. These new trees include oak, sugar maple, hickory, and beech.

After many years, the long-lived trees are taller than the aspen, thorn apple, and wild cherry. The latter trees cannot compete, and gradually they die out. Now a mature forest is forming on the abandoned farm. Seedlings of the beech and maple and other forest trees are able to survive in the shade of their parents. The forest grows taller and thicker and eventually reaches a condition of stability known as an *ecological climax.*

A mature forest of the kind described above is called a *climax community.* Although it will change over the years, the changes will be slow and difficult to see during a person's lifetime.

A forest is not the only kind of climax community. In some geographic areas and climatic zones of the world, a climax community may be a grassland with populations of prairie dogs and coyotes. It may be an Arctic tundra with a bountiful supply of mosses for reindeer to feed on. Or it may be a desert supporting sparse numbers of cacti, lizards, snakes, and rodents. Large, well-defined, climatically determined communities of these kinds are often called *biomes.* Ecological succession also takes place in bodies of inland water. Lakes, for example, gradually fill in as they become older. Aging lakes are called *eutrophic.*

PEOPLE IN THE ECOSYSTEM

Man is a part of nature. Although ecologists and naturalists have known this for many years, the majority of people have only recently become aware of the importance, and implications, of this fact.

Suddenly we notice that the earth is becoming crowded with people. With every generation—that is, every 25 or 30 years—the time it takes to double the world's human population is shortened. This population doubled between 1850 and 1950; by 1990 or 2000 it will double again.

This is a problem. More people require more food and more manufactured products. But the earth is no bigger, nor does it contain any more materials than it did millions of years ago.

More and more people are beginning to realize that wastes must not be dumped into the air, water, and soil unless they are known to be nonpoisonous. In short, man is becoming aware of his dependency on other living things and the he can survive only as long as he ensures their survival.

DNA AND RNA

by Stephen N. Kreitzman

Blueprints, patterns, recipes—what do they have in common? Each contains information needed to make something: a house, a sundress, a batch of chocolate-chip cookies.

Your body is also made according to a blueprint or recipe. This is true for all living things: trees, insects, starfish—even the tiny bacteria and viruses. This blueprint, the "code of life," is carried in large molecules called DNA (deoxyribonucleic acid) and RNA (ribonucleic acid).

Every single living cell contains this code of life. The code contains all the infor-

A computer-generated drawing of a DNA molecule shows its double-helix design, which allows genes to accurately duplicate themselves.

R. Langridge/Rainbow

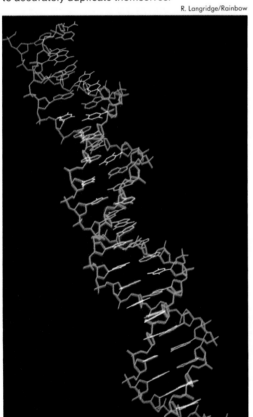

mation needed to make that cell into a nerve cell, muscle cell, or skin cell. Moreover the code contains the information that determines whether the cell will be a mouse cell, dog cell, or human cell.

The code is the thing that makes you different not only from the lower forms of life, but also from your friends and even from your brothers and sisters. It makes you unique, a one-of-a-kind organism.

THE HEREDITARY MECHANISM

Just a little more than one hundred years ago it was realized that many characteristics could be transferred to children from their parents. Although probably this had been noted in a general way since primitive times, it took the observations of an Austrian monk who was working with pea plants to make the point clear. Gregor Mendel was breeding pea plants of different strains. Some of these plants produced smooth seeds; other had wrinkled seeds. Some plants were tall, while others were short. Some produced purple flowers, others white flowers.

Mendel realized that, in some unknown way, the information about each of these characteristics was carried to the next generation of pea plants. And he observed that each characteristic could be transmitted to the next generation independently of the other characteristics.

Since Mendel's time a great deal has been learned about the manner in which characteristics are transferred from generation to generation. Every single characteristic of a pea plant—and every other kind of living thing—was found to be represented in the nucleus of each of its cells, in structures called *chromosomes*. Furthermore, it was found that each characteristic is represented by a specific part of a chromosome—a *gene*.

It is the genes that contain the information that is transmitted from generation to generation. It is the genes that determine the characteristics of an organism. And genes, scientists discovered, are composed of DNA. RNA molecules, it was learned, decipher the code contained in the DNA so that the many parts of a cell can be properly built.

NUCLEOTIDES: LINKS IN A CHAIN MOLECULE

DNA and RNA are made of the same kinds of building blocks. In fact, they are both long strings of these building blocks. When a molecule is composed of a long string of the same kind of smaller unit, it is called a *polymer*.

If you look closely at a chain, you will see that it is, like a polymer, made of a series of links. When you put all the links together you have a chain. The chain can be used to do things that the individual links cannot do—pull a car out of mud, for example.

In an ordinary chain, all the links are exactly alike. In DNA and RNA chains, however, the links have different shapes. Four kinds of links make up a DNA chain. Four slightly different kinds of links make up the RNA chain.

The links in DNA and RNA chains are called *nucleotides*. Let's look more closely at these nucleotides. Each has three parts: a phosphate unit, which is exactly the same for every link; a sugar unit, which explains the names ribonucleic acid and deoxyribonucleic acid, since RNA contains the sugar *ribose,* and DNA the sugar *deoxyribose;* and the unit that has the important shape, the base.

Since there are usually only four different bases that are found in DNA and RNA, there are four different deoxyribonucleotides and four ribonucleotides. The important thing is that there are only four different shapes. All of the meaning of the code is related to the shapes of the bases in the DNA and RNA chains.

PIECES IN A JIGSAW PUZZLE

The four bases found in a DNA chain are adenine, guanine, cytosine, and thymine. In a DNA link (nucleotide) a base is connected to the sugar deoxyribose, which,

in turn, is connected to the phosphate. One nucleotide is fastened to the next nucleotide in the chain by the phosphate unit (see Figure 1). When the whole DNA chain is linked together, it twists into a coil, like a spring. This coiled structure is called a *helix.*

About twenty years ago an Austrian-American biochemist named Erwin Chargaff discovered a very puzzling fact about DNA. When he measured the amount of each base in DNA, he found that there was always the same amount of thymine as there was adenine. Equally interesting was the fact that the amount of guanine was always the same as the amount of cytosine.

The puzzle was solved when scientists found that thymine is always attached, or bonded, to adenine. Similarly, cytosine is always bonded to guanine. Molecular biologists James D. Watson and Francis H. C. Crick showed that DNA does not exist as a single chain but rather as two chains wound around each other, with the bases on one chain attached to the bases on the other. DNA, in other words, was discovered to be in the form of a double helix.

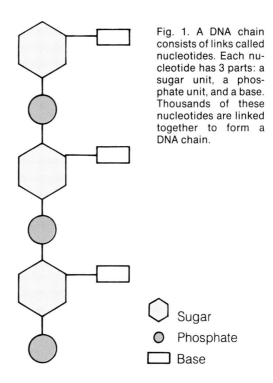

Fig. 1. A DNA chain consists of links called nucleotides. Each nucleotide has 3 parts: a sugar unit, a phosphate unit, and a base. Thousands of these nucleotides are linked together to form a DNA chain.

◯ Sugar

⬤ Phosphate

▭ Base

We can say that one of the DNA chains is *complementary* to the other. You are familiar with many things that are complementary to each other. For example, two pieces of a jigsaw puzzle that fit together have shapes that are complementary. In DNA the shape of adenine is complementary to that of thymine. Every adenine in one of the DNA chains is attached to thymine in the other chain. Similarly, cytosine is complementary to guanine. Cytosine will never attach to adenine or thymine—or to another cytosine. Its shape is such that it can only be attached to its complement, guanine (see Figure 2).

The order in which the nucleotides are linked determines the message contained in the DNA. If a new cell is to be exactly like its parent cell, it is necessary to copy the sequence of bases in exactly the same order as the original DNA. Any changes in the order would change the coded message.

Fig. 2. If a DNA molecule is unwound, a small part can be represented as shown below. The two DNA chains are held together by bonds between the bases. What is the relationship between adenine and thymine? Cytosine and guanine?

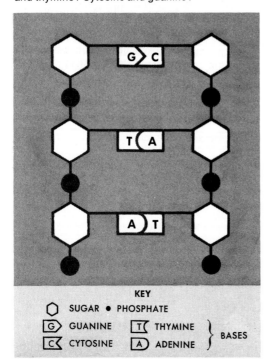

KEY

⬡ SUGAR ● PHOSPHATE

[G⊂ GUANINE   [T⊂ THYMINE ⎫
[C⊂ CYTOSINE [A⊃ ADENINE ⎬ BASES
                           ⎭

Because the DNA molecule is a double helix and the shapes of the bases are so well matched, it is easy to make an exact copy of the DNA. To understand how this works, let's make our own model. Using heavy paper or cardboard, cut out the four base shapes shown in Figure 2. Make at least ten copies of each shape and label each piece A, T, C, or G. These pieces will represent the four nucleotides in DNA.

Now make a single chain of nucleotides using the following sequence: adenine - cytosine - adenine - thymine - guanine - guanine-thymine. Abbreviated, this reads A-C-A-T-G-G-T. Can you make a complementary chain for this sequence? What is the sequence of nucleotides on the complementary chain? The two chains should fit together just like the pieces in any jigsaw puzzle.

During cell division the two chains of the DNA molecule pull apart, or "unzip." The nucleotides on each chain, no longer attached to their complements on the other chain, are free to become attached to complementary nucleotides that are present in the surrounding medium.

You can simulate this process with your model. Pull the two chains apart. Each chain can form its complement from the extra shapes you cut out (see Figure 3). In this manner you get two DNA molecules. How do they compare with each other?

You're right—both molecules are exactly the same. During subsequent division of the cell, each new cell receives one of the DNA molecules. If a mistake occurs and the coded message is changed, whatever information is contained in the molecule is also changed. This is called a *mutation*. The mutation may be harmful to the cell, perhaps causing it to die. Or it may not be important at all, having no effect on the functioning of the cell. In rare cases the mutation offers the cell an advantage over cells that have not mutated.

Good, bad, or unimportant, however, the change will be transmitted to future cell generations. If, for example, a mutation occurs in one of your skin cells, it will be passed on to all future skin cells that develop from that one cell. If the mutation

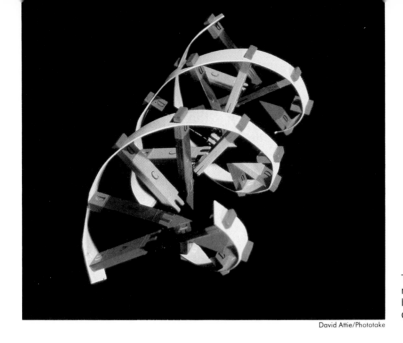

This simple model of a DNA molecule shows the specific links between bases in the coiled structure.

David Attie/Phototake

occurs in a human sperm cell, and this sperm fertilizes an egg cell, the cells of the child that is born from this union will contain the mutation. Such mutations in reproductive cells play a major role in the evolution of all living things.

The DNA molecules in a cell are obviously very precious and must be protected. Thus they are located in the nucleus of the cell where they are least likely to be damaged. But the information contained in these molecules must be gotten to the rest of the cell, where food is broken down, proteins are made, and so on. This job of information transfer is done by RNA.

WORK OF RNA

Like DNA, RNA is a chain of nucleotides. But the structure of RNA is a little different from that of DNA. The RNA molecule is much smaller than the DNA molecule and usually is a single helix. As the name of RNA implies, the sugar units are ribose rather than deoxyribose. And RNA does not usually contain thymine. It contains plenty of adenine, guanine, and cytosine, but no thymine. Instead, RNA has a base called *uracil*. The part of uracil that could attach to adenine has the same shape as thymine, however, so there is really very little difference between thymine and uracil. Although RNA does not usually form a double helix, it is still important that the bases can pair up. We will see why shortly.

RNA performs a number of functions in a cell. Most of these jobs are related in some way to the reading and deciphering of the code contained in the DNA molecules. There are several different kinds of RNA molecules, each with its own function and a chemical structure that differs slightly from the other kinds. In the following paragraphs we will briefly look at two important types of RNA: messenger RNA and transfer RNA.

Messenger RNA. Almost everything that happens in a cell is controlled by *enzymes*. Enzymes are protein substances. Like DNA and RNA, proteins are polymers. But they are made of different kinds of links—links called *amino acids*.

A cell can function properly only when the proper enzyme proteins are present in the cell. A major part of the coded message in DNA tells which proteins are to be made at any particular time. But the DNA is much too important and precious to be used over and over for the processes of building proteins. Therefore the instructions for these processes are *transcribed,* or copied, into a working molecule of RNA. It is this RNA which builds the proteins.

How are the instructions copied? First, the part of a DNA molecule that contains the instructions unzips. Thus the nucleotides at this point are no longer attached to their complements. What follows is basically the same process that occurs when DNA replicates itself. For every

DNA AND RNA   **499**

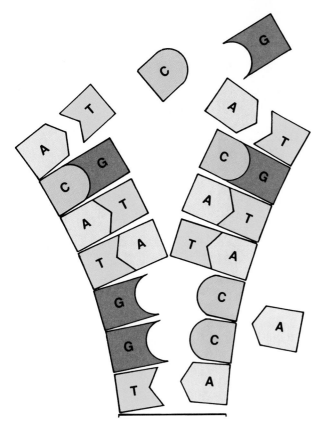

Fig. 3. To reproduce itself, a DNA molecule "unzips." Nucleotides from the surrounding medium attach to their complements on the two chains. Each letter represents a nucleotide. For example, "A" represents the nucleotide containing adenine.

The RNA made in this manner is called *messenger RNA*. After its creation, the RNA molecule moves from the nucleus to the cell area where proteins are made.

Each messenger RNA molecule contains the code needed to make one specific protein. And since a protein consists of amino acids, the job of messenger RNA is to specify the sequence in which the amino acids are linked together.

Transfer RNA. Sometimes known as *soluble RNA,* this relatively short chain of ribonucleotides plays an extremely important part in the reading of the genetic message contained in the messenger RNA. It does this by helping to translate the coded information from the language of nucleic acids into the language of proteins. The language of nucleic acids is the *order* of the nucleotides in the DNA and RNA. The language of proteins is the *order* of amino acids in the protein chains.

Proteins, remember, are polymers consisting of repeating links called amino acids. About twenty different amino acids are usually found in proteins. It is the order of amino acids in the protein chains that determines the twisted, folded shape that is special for each protein. In turn, the shape of the protein determines what job the protein can do in the cell.

If amino acids could match up directly with the nucleotides of the messenger RNA, there would be no need for transfer RNA. However, amino acids cannot attach to the messenger RNA unless they have a special adapter molecule which can do the attaching. Transfer RNA is this adapter molecule. Each amino acid has its own specific transfer RNA-adapter that is able to attach to the messenger RNA.

Only about 80 nucleotides make up a transfer RNA molecule. Because it is so short, it has been the most thoroughly studied of the RNAs. Although no one really knows why, it contains some unusual nucleotides in addition to the four major ones. Scientists have been able to determine the exact nucleotide order for several different transfer RNAs, and some interesting facts are now known about the chemistry of these molecules. All the transfer

adenine on the unzipped part of the DNA chain, a uracil can be matched on an RNA chain. For every thymine on the DNA, an adenine can be put into place on the RNA chain. And so on, until a complete chain of ribonucleotides is put together in the exact complementary sequence to the DNA chain. You can use your model to simulate this. But you will need some pieces to represent uracil. These pieces should be the same shape as thymine. Label each with a U.

You might wonder what makes the ribonucleotides join together to form the RNA polymer. This is done by an enzyme in the nucleus with the informative but long name, *DNA dependent RNA polymerase.* This enzyme helps copy the DNA sequence of nucleotides and then zippers the ribonucleotides up into a complete RNA molecule.

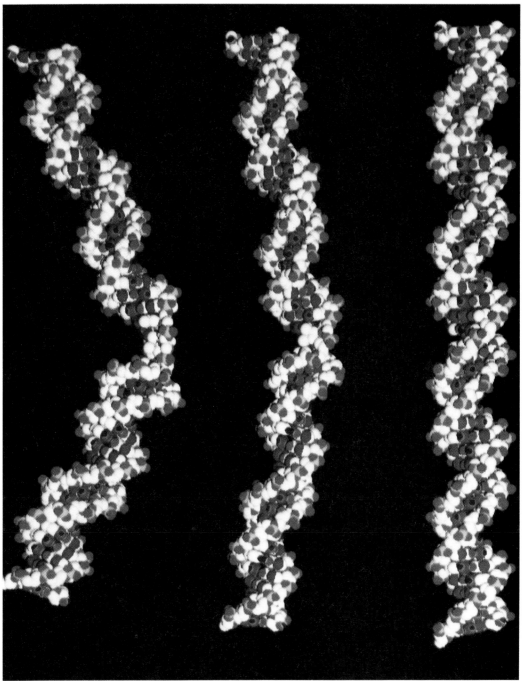

University of California, Lawrence Berkeley Laboratory

Scientists at Lawrence Berkeley Laboratory at the University of California, Berkeley, used a computer to generate these three-dimensional models of DNA. The healthy DNA at far right is straight; when damaged by ultraviolet radiation (left and center), DNA kinks and unwinds near the site of damage.

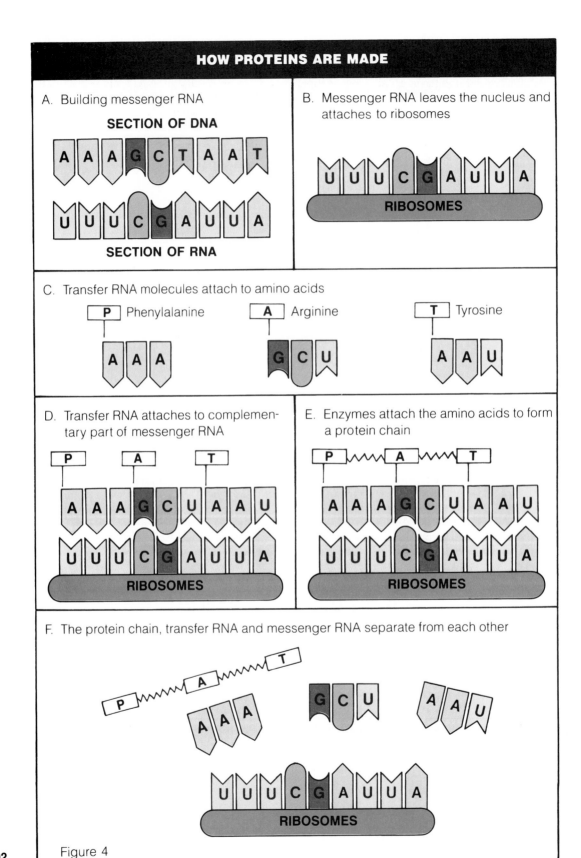

HOW PROTEINS ARE MADE

A. Building messenger RNA

SECTION OF DNA

A A A G C T A A T

U U U C G A U U A

SECTION OF RNA

B. Messenger RNA leaves the nucleus and attaches to ribosomes

U U U C G A U U A

RIBOSOMES

C. Transfer RNA molecules attach to amino acids

P Phenylalanine   A Arginine   T Tyrosine

A A A   G C U   A A U

D. Transfer RNA attaches to complementary part of messenger RNA

P   A   T

A A A G C U A A U

U U U C G A U U A

RIBOSOMES

E. Enzymes attach the amino acids to form a protein chain

P   A   T

A A A G C U A A U

U U U C G A U U A

RIBOSOMES

F. The protein chain, transfer RNA and messenger RNA separate from each other

P   A   T

A A A   G C U   A A U

U U U C G A U U A

RIBOSOMES

Figure 4

RNAs have the same first nucleotide in their chains: the nucleotide containing guanine as a base. Also, the last three nucleotides are always the same; two containing cytosine and, at the very end of the chain, one with adenine. It is this adenine-containing nucleotide that attaches to the amino acid. Somewhere in the middle of the transfer RNA molecule are those nucleotides that determine which amino acid will be transported by the particular transfer RNA molecule and other nucleotides that can interpret the coded genetic message of the messenger RNA.

TRANSLATING THE CODE INTO PROTEIN

The coded message in genes is now understood. Every three nucleotides in an RNA molecule code for a specific amino acid to be placed in a protein chain.

Suppose the sequence in a particular section of DNA is adenine-adenine-adenine-guanine-cytosine-thymine. Abbreviated, this reads A-A-A-G-C-T. When the code is copied into ribonucleotides, the sequence in the messenger RNA will consist of the complements: uracil-uracil-uracil-cytosine-guanine-adenine, or U-U-U-C-G-A.

This messenger RNA moves out of the nucleus and attaches to particles in the cell called *ribosomes*. They act as "workbenches" where messenger RNA and transfer RNA make proteins.

In the nearby protoplasm is the amino acid phenylalanine attached to a transfer RNA with the nucleotide sequence A-A-A at a particular place in its structure. This sequence has the exact complementary shape of the U-U-U of the messenger RNA; the two attach. Another transfer RNA with the sequence G-C-U at the critical place in its molecule will bind to the C-G-A of the messenger RNA. This transfer RNA has the amino acid arginine attached to it.

Enzymes in the cell attach the arginine to the phenylalanine. Thus two amino acids have been put into place as directed by the genetic code. Many, many more amino acids have to be added in this way in order to build a protein. Each is added until the end of the message is signaled—and the protein is complete. (See Figure 4.)

INBORN ERRORS OF METABOLISM

As the story of DNA and RNA is unfolded, it is becoming clear that small changes in nucleic acids can have serious consequences. These changes can result in unusual diseases that are called inborn errors of metabolism. Some hereditary traits are actually metabolic diseases that occur because a particular chemical reaction does not take place. A necessary enzyme is not produced because the gene responsible for it is either missing or altered.

SPLIT, MOVABLE, AND SPLICED GENES

The section of DNA that codes for one protein is called a gene. The structure of a gene can be determined by sequencing, that is, by identifying the order of the bases making up the DNA of the gene. The first genes to be sequenced were globin, immunoglobulin, and ovalbumin.

Genes are known to be split into functional (*exons*) and nonfunctional sections (*introns*). The introns seem to make sure that the coding sequences are kept intact during such procedures as genetic crossing over. The introns are "spliced out" when the primary RNA transcript is processed into messenger RNA. The processed molecule then moves to the cell cytoplasm, where it serves as the template for protein synthesis.

Another major feature of DNA is transposable elements—sections of DNA that can move about the DNA molecule. These elements, also called movable genes or jumping genes, were first discovered as controlling elements in corn in the 1950's by Barbara McClintock and were found in bacteria in the late 1960's. These elements usually are found in closely spaced pairs. When they move, they take the genes lying between them. The entire unit is called a transposon. Scientists believe that these movable elements are one way of creating genetic diversity in an organism.

Scientists have also learned to manipulate DNA sequence by inserting DNA from one organism into that of another—gene splicing—to create a recombinant DNA molecule. See the following article, "Recombinant DNA."

RECOMBINANT DNA

by Jenny Tesar

Recombinant DNA is one of the few revolutionary scientific achievements of the 20th century. Also known as "gene splicing" and "gene cloning," it involves combining genes in new and unusual ways.

Recombinant DNA is rapidly becoming the basis of a growing industry that is creating important medical, industrial, and agricultural products. Using gene-splicing techniques, researchers have:

• Created bacteria that can make human insulin and other substances valuable to our health.

• Constructed yeast that can manufacture vaccine for hepatitis B.

• Created a fusion gene that when injected into fertilized mouse eggs has resulted in offspring twice normal size.

As the technology is refined and improved, scientists expect to be able to cure inherited diseases such as dwarfism, diabetes, and sickle-cell anemia. They also expect to develop plants that can take nitrogen directly from the air, rather than from expensive fertilizers, and they hope to use splicing techniques to turn off cancer-causing oncogenes.

At the same time there is great concern that these techniques might be used to create new disease-causing organisms or to manipulate human genes to create a new superrace.

BITS AND PIECES OF THE GENETIC CODE

Every living cell in every organism contains molecules of the chemical DNA (deoxyribonucleic acid). DNA is the genetic code, telling the cell how to grow, what to do, when to reproduce.

In bacterial cells, in addition to the main DNA molecules, there are additional smaller loops of DNA. These loops are called "plasmids," and they contain the genes for antibiotic resistance.

Plasmids were discovered in 1952. In the succeeding years, two other important types of cell chemicals were discovered. One of these was restriction enzymes.

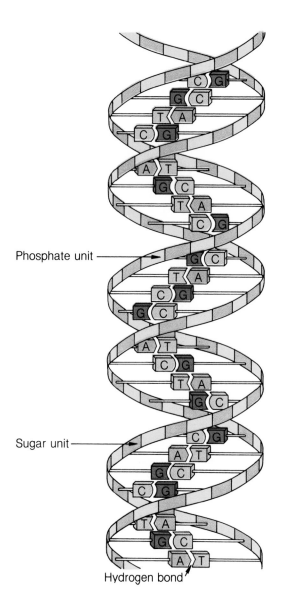

Phosphate unit

Sugar unit

Hydrogen bond

DNA molecules have the form of a helical ladder. Molecules represented by A and G are purines, and those represented by T and C are pyrimidines.

These are chemicals that cut apart DNA molecules. The other was an enzyme called ligase. This acts as a glue to hold together bits of DNA molecules.

CUTTING AND SPLICING GENES

In the early 1970's, Stanley Cohen and his co-workers at Stanford University discovered a comparatively easy way to transfer genes from one kind of organism to another. They worked with *Escherichia coli* (*E. coli*, for short), a harmless bacterium commonly found in the human intestine. As shown in the drawing, they separated the plasmids from the rest of the *E. coli* cell material. Then they placed the plasmids in a solution with a restriction enzyme, which broke the plasmid loops at specific places.

At the same time, the scientists used other restriction enzymes to separate certain genes from the long DNA molecules of an African clawed toad. These genes were mixed with the open plasmid loops from *E. coli*. Then, by using the enzyme ligase, the toad genes and the *E. coli* plasmids were "glued" together.

Thus a completely new type of structure—part bacterial, part toad—was formed. Such a structure has since been named a "plasmid chimera." (The chimera was a mythical animal that was composed of parts from several different organisms.)

The final step was to place the plasmid chimera in a solution containing normal *E. coli* cells. When the solution was warmed, the plasmid chimeras were able to enter the *E. coli* cells and become part of the cells' genetic structure. The toad genes now were part of the bacteria. And when the bacteria reproduced by dividing in half, each of the offspring inherited the toad genes.

The scientists used the same method to enter yeast genes in *E. coli*. And since then, genetic materials from a variety of plants, animals, bacteria, and viruses have been transplanted into bacterial cells.

Bacteria reproduce rapidly. In a 24-hour period, one *E. coli* cell can make a few billion copies of itself. If that cell contains a foreign gene, the gene will be reproduced at the same rate. Hence, there is the possibility of using bacteria as factories.

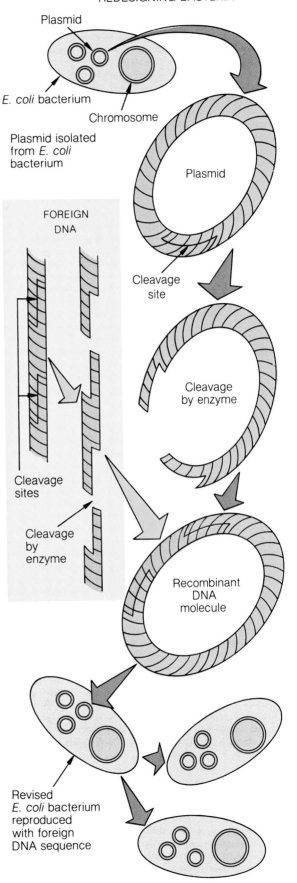

Plasmid

E. coli bacterium

Chromosome

Plasmid isolated from *E. coli* bacterium

Plasmid

Cleavage site

FOREIGN DNA

Cleavage by enzyme

Cleavage sites

Cleavage by enzyme

Recombinant DNA molecule

Revised *E. coli* bacterium reproduced with foreign DNA sequence

POTENTIAL MEDICAL PRODUCTS

Gene-splicing techniques promise major medical benefits. The first pharmaceutical product approved for use in humans is human insulin made by the bacterium *E. coli*.

Insulin is a hormone that controls the amount of sugar in the blood. Diabetics—people whose bodies do not produce insulin—must be given insulin from another source. At the present time, insulin is produced from cows and pigs. However, there are chemical differences between these insulins and human insulin, and some people's bodies reject the foreign substance. The insulin produced by bacteria that have been given the human insulin gene is chemically identical to the insulin produced by human beings. In 1982 human insulin grown in bacteria became the first product of gene-splicing to be approved by the U.S. Food and Drug Administration for marketing for use in human beings.

Another product made by recombinant means is growth hormone, which is produced by the pituitary gland and controls growth. People with insufficient amounts of this hormone do not grow tall. They are dwarfs. Children suffering from dwarfism have been given growth hormone removed from the pituitary glands of people who have died. To treat one child in this way for one year requires pituitary glands from 50 cadavers. Unfortunately, growth hormone obtained like this can be contaminated by viruses, as was discovered in 1985. Recombinant human growth hormone will provide an uncontaminated source of this much-needed hormone.

Other products produced by recombinant techniques are interferon, which has anti-virus properties; interleukin-2; and diagnostic monoclonal antibodies. The first vaccine to be manufactured by genetic engineering is that for hepatitis B. The vaccine originally was purified in a long and costly procedure from the blood of chronic carriers of this disease. Both time and cost have been cut by inserting the gene for the vaccine product into the yeast chromosome, making the yeast cells factories for this product. Yeasts have some advantages over bacteria in manufacturing recombinant products—they have no toxins, and they secrete the product rather than breaking apart to release it. Other safe vaccines are being developed through the use of only the virus' protein coat, which can be synthesized by recombinant yeasts or bacteria.

INDUSTRY AND AGRICULTURE

Gene-splicing also promises to have a major impact on industry and agriculture. In 1981 the development by gene-splicing of a safe and effective vaccine against one form of hoof-and-mouth disease promised a solution to a worldwide problem.

By splicing desirable genes from one plant species to another, it may be possible to develop improved plant breeds at a much faster pace than the traditional cross-breeding methods. Experimental strains of several engineered plant species with improved photosynthetic and nitrogen-fixing abilities have been developed.

TRANSGENIC MICE

In 1982, researchers produced spectacular results with a recombinant DNA technique. They fused a rat growth hormone gene to a regulatory switch in a mouse gene. Then, they injected this fusion gene into a fertilized mouse egg and implanted the egg in a foster mother. The growth hormone gene was incorporated into the DNA of the developing offspring. The result was transgenic mice—offspring two times their normal size and dubbed ''supermice.'' The supermice passed the new growth hormone gene along to their offspring. Scientists hope to perfect this technique to treat cancer, sickle-cell anemia, thalassemia, and enzyme deficiencies in humans.

In 1985, the same technique was used to introduce foreign genes into larger mammals, including rabbits, pigs, and sheep. This method can be used to produce livestock that are larger, more disease resistant, and with improved reproductive capacity.

GENE THERAPY

The development of recombinant DNA techniques has made it possible to

Ted Spiegel/Black Star

In a gene-splicing experiment aimed at producing hardy plant breeds, a technician extracts bean genes from a test tube to place in another plant.

think about correcting defective genes that interfere with production of essential enzymes. These techniques may also be used to correct proteins that cause Lesch-Nyhan disease, immunodeficiency diseases, and blood disorders in humans.

The best targets for such an approach are genes that are always switched on for enzyme production. An example is the gene for the enzyme adenosine deaminase. A deficiency of the enzyme causes a severe immunodeficiency disease. The normal gene for this enzyme is inserted into cells from the patient's bone marrow, and these cells are reimplanted. The normal gene then starts expressing itself, causing production of the needed enzyme, hopefully in amounts sufficient to correct the condition.

Because of the vast ethical concerns implicit in this research—particularly, whether scientists have the right to change the genetic makeup of the human species—the Recombinant DNA Advisory Committee of the National Institutes of Health (NIH) has set up certain guidelines. Scientists proposing to test gene therapy in humans must have their planned research approved by local review committees and by the federal government before embarking on their work.

RESTRICTIONS ON RESEARCH

In the 1970s many scientists expressed concern that there might be some danger in recombinant DNA research. For example, altered organisms might be accidently released from the lab and infect people with an incurable disease. A 1975 meeting of scientists recommended regulatory guidelines in 1976. Certain experiments can be conducted only if specific safety precautions are observed.

Biological controls are designed to minimize survival of an organism outside the laboratory. They limit the chances of a bacterial strain being transmitted from the laboratory host to another host (a person, for example). To do this, special strains of *E. coli* are used that are unable to replicate their DNA except in a carefully controlled laboratory environment.

Physical controls are meant to confine organisms to the laboratory through the use of special equipment and laboratory procedures. Specially designed labs provide a secondary means of protection.

By 1979, the scientific community decided that the initial fears were exaggerated and that restrictions were impeding research. That year, the NIH relaxed guidelines, allowing work on viral DNAs. By 1982, the guidelines were relaxed further, and compliance was made voluntary.

NIH continues to ban certain potentially hazardous experiments. However, in 1987, plant pathologists at the University of California, Berkeley, applying the advances of recombinant DNA technology to agriculture, became the first to release man-made microbes into the environment. The scientists were testing a mutant of a common parasite, *Pseudomonas syringae*, which produces a protein that serves as a site for the formation of ice crystals. By removing the protein-making gene from the microbe's DNA, the scientists created a mutant form of the parasite that resists ice formation. Leaves coated with the man-made microbe have withstood temperatures as low as 23°F. Potatoes coated with the mutant parasite are part of a test to determine how low the temperature must drop before frost endangers the crop.

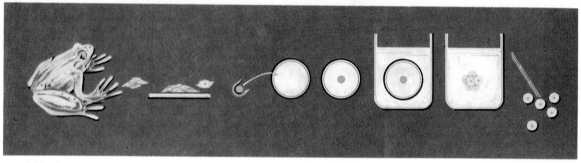

2. Cells from the tissue are cultured.

4. The egg develops into an embryo.

1. A bit of skin tissue is taken from a frog.

3. The nucleus from one of the cells is transplanted into a recipient egg (the nucleus of which has been removed).

5. Embryo cells are separated.

CLONING

by Damon R. Olszowy

CLONING is suddenly in the news, thanks to revolutionary techniques developed by genetic engineers and other new breeds of biologists. The newspapers are full of stories describing how scientists can produce a clone, or an identical copy, of an organism from just a single cell. Biologists now have the ability to clone some plants and small animals. Can people be far behind?

All of this seems frightening to many, not quite right to some, just plain startling to others. Perhaps the basic question is: how can cloning experiments contribute to future human welfare?

WHAT IS A CLONE?

A clone is an organism or a group of organisms created from a single parent. The process of cloning is really a form of asexual reproduction. You know that sexual reproduction involves the union of sex cells: the sperm from the male parent and the egg from the female parent. The nucleus of each sex cell—called a pronucleus—contains only one set of chromosomes with their genes and not the two sets that are found in the nuclei of all other cells, called body cells, and which give each species its characteristic chromosome number. The union of sperm and egg in fertilization produces two full sets of chromosomes. It is the first step in the creation of a new and unique individual with traits and characteristics inherited from both parents.

Cloning is asexual. There is only one parent. And the offspring has the hereditary traits of that single parent.

The word "clone" comes from the ancient Greek root, *klon,* meaning a twig or slip. Taking a twig or cutting from a plant and growing it into another plant is actually cloning the plant.

Today, however, the word "cloning" is used in a slightly different way. It has come to mean the production of an organism from just a single cell taken from the body of a plant or animal. This single cell, being a body cell and not a sex cell, contains two sets of chromosomes—one set from its mother and one set from its father. It thus has all the genetic information necessary to produce a complete individual if it is stimulated to grow.

CLONING PLANTS

Cloning is easier to do in plants than in animals because plant cells are simple in structure. Most young plant cells also have the ability to divide again and again. Individual plant cells are cultured in a special growing medium that contains the proper nutrients. The cells can divide, grow, and develop into new plants with leaves, roots, and flowers. Indeed, scientists have been able to do this in the laboratory with plants such as carrots, pine trees, and African violets.

This technique of cloning plants enables scientists to select a desirable plant and produce as many identical copies of it as they wish from each cell that is cultured.

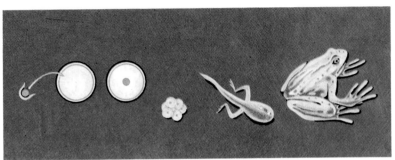

Illustrations by W. Hortens

Cloning a frog is relatively easy because frog eggs develop by themselves in water.

6. *The nucleus from each embryo cell is transplanted into a separate recipient egg.*

7. *The eggs develop into clones of the original frog. (Only one such example is shown above.)*

For example, in 1985 Subhash Minocha, chairman of the botany department at the University of New Hampshire, succeeded in producing cloned copies of the Venus's-flytrap. In America, where the insect-eating plant is heavily collected for use as a houseplant, it is becoming extremely rare in the wild. By cloning endangered species, Minocha hopes to increase their numbers.

A CLONAL FROG

The technique used to clone an animal is more difficult. First the nucleus of an egg cell is removed. Then a nucleus from a body cell of a donor of the same species (either male or female) is placed into the egg cell. The egg cell now has a nucleus that has a complete chromosome complement and not half as its own nucleus had. This technique is called "nuclear transfer."

In 1952 U.S. scientists Robert Briggs and Thomas J. King tried the technique with frogs. They destroyed the nucleus of an unfertilized frog egg cell with radiation. Then they transplanted the nucleus from an ordinary body cell of a young tadpole (the juvenile stage in a frog's life) into the egg cell. The egg with transplanted nucleus survived and developed into a normal frog. The result of this experiment was a tadpole that was a genetic twin of the tadpole that donated the nucleus.

Ten years later British biologist John B. Gurdon tried using a nucleus from a skin cell of an adult frog. The results were unsuccessful: many abnormal and deformed frogs resulted. This happened because the egg cell divided more rapidly than the body cell from which the nucleus was taken. In other words, there was a discrepancy between the division rate of the egg cell and the division rate of the donor nucleus.

The key to success in cloning experiments is to transplant the right kind of nucleus into an egg cell. In this case, "right" means that both divide at the same rate. Gurdon refined his experiment by first transplanting a nucleus from a skin cell into an enucleated egg cell and then growing the embryo until it was a very young tadpole. Then he separated the body cells of the young tadpole and transplanted the nuclei from these tadpole body cells into other enucleated frog egg cells. The new clones grew from these eggs. This time many more normal frogs developed into adults.

This procedure was successful because the nuclei from the younger cells—in this case, from the tadpoles—divided at the same rate as the egg cells. This procedure is called "serial transfer" because it happens in a series of steps involving two or more nuclear transplants.

CLONING MAMMALS

The cloning of mammals is even more difficult. The eggs of mammals are smaller and more fragile than the eggs of frogs. It requires microsurgery to transplant a nucleus without damaging the egg. In 1975 J. Dereck Bromhall of Oxford University, in England, tried to clone rabbits, but the results were unsatisfactory. Since then, however, new techniques and instruments have been developed so that now a nucleus from a donor cell can be transplanted successfully into an egg cell of a mouse, a rabbit, or even a human.

Another major difference that makes cloning mammals more difficult than clon-

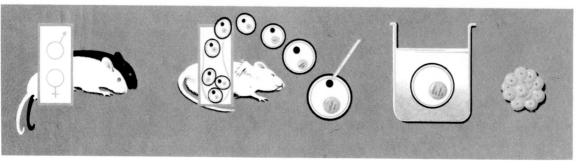

2. The female is killed and the fertilized eggs are extracted.

4. The egg develops into an embryo.

1. A male (black) and female are mated.

3. The sperm nucleus is removed from the egg prior to fusion.

5. The chromosomes in the remaining nucleus (from the female) are doubled by chemical action.

ing frogs is that their embryos grow differently. A frog is an amphibian, and its eggs can develop by themselves in water. In contrast, the embryo of a mammal must be protected inside the uterus of its mother to complete its growth and development.

Today scientists can remove eggs from a female mammal, and replace the egg's nucleus with the nucleus from a body cell of a donor. They then implant the egg back into the uterus of the same female or into another female, which then acts as a substitute mother. In the uterus the embryo can complete its growth and development. Substitute mothers have, in fact, given birth to mice, cows, and even to a baboon. The young produced in this way are not even related to their substitute mothers, since they do not contain any of the substitute mother's genetic material.

Cloning need not require a living organism to duplicate. In the mid-1980s, scientist Allan Wilson of the University of California cloned a pair of gene fragments from a salt-preserved pelt of a South African quagga, a zebra-like animal that has been extinct for hundreds of years. The exact duplication of a small part of an extinct animal is an important development in evolutionary studies.

MICE MADE TO ORDER

U.S. biologists Clement L. Markert and Peter Hoppe of Jackson Laboratory in Bar Harbor, Maine, have worked on the cloning of mice. First they removed an egg from a female mouse just after it had been penetrated by a sperm cell. Then, using the techniques of microsurgery, they removed either the egg pronucleus or the sperm pronucleus before the two pronuclei united in the essential, final part of fertilization.

The egg cell then had one pronucleus. A pronucleus has only one set of chromosomes—or half the normal chromosome number. They stimulated the chromosomes in the pronucleus still in the egg cell to duplicate themselves. They then had an egg cell with the full chromosome complement, but all of it from one parent.

The egg continued to divide and develop into an embryo. At an early stage of its development, the embryo was implanted into a female mouse, which acted as a substitute mother. More than one egg can be implanted and, in fact, several were implanted. Seven offspring were born.

All the baby mice were female. Five of them had all their genes from a female pronucleus, and two had all their genes from a male pronucleus. Males cannot be produced in this experiment because each pronucleus carries only one type of sex chromosome. The production of males in mammals requires the presence of two different types of sex chromosomes—an X and a Y sex chromosome. (A female pronucleus contains only the X sex chromosome; the male pronucleus contains either the X or the Y sex chromosome, but not both. When either pronucleus is stimulated to duplicate its chromosomes, only the pronucleus containing an X chromosome can double and subsequently divide. Thus only females can be produced.)

You may notice that this procedure is different from the method used to clone a frog. It is true that all the genes in the mouse embryo come from a single adult mouse. However, the genes in its embryo consist of two copies of only *half* of the donor's genetic material rather than a complete set of genes representing a single copy of every gene present in the donor. The off-

In this method of mouse cloning normal mating occurs, and then the sperm nucleus is removed. The female chromosomes double, and the developing embryo is implanted into another female.

6. The embryo is implanted in a substitute mother.

7. The substitute mother gives birth to a mouse whose genes are derived solely from the original mother.

spring could be considered semi-clones since they have only one half of all the possible genes that the donor has. Because of this condition, semi-clonal mice are not identical to the donor.

By repeating the experiment and starting with a semi-clone, a completely clonal mouse could be produced. An egg is taken from a semi-clonal mouse, and the chromosomes in its pronucleus are stimulated to duplicate themselves. This produces a complete nucleus with two chromosome sets. Then the embryo is implanted into a substitute mother. The resulting offspring will have exactly the same genes as its semi-clonal mother. It will be a true clone.

CLONING A HUMAN

Scientists believe that much of the technological knowledge already exists to clone a human. The creation of a human by cloning would, however, be a very complex and risky laboratory procedure.

First, a healthy egg would be removed from a woman. Instead of fertilizing the egg with a sperm cell, scientists would destroy the nucleus in the egg and replace it with a nucleus from a donor cell. This step is the biggest challenge.

The nucleus of the egg cell and the nucleus of the donor cell must first be separated from their surrounding cytoplasm. Then the donor's nucleus must be transplanted into the empty cytoplasm of the egg cell. Although this has been done successfully in frogs and mice, the chances for success with human eggs are less because human eggs are much smaller and more delicate. However, cells can now be united by using a method called cell fusion.

In this case, the nucleus is removed from the donor cell with radiation. Then certain chemicals or viruses are used to help make the nucleus fuse with the empty egg cell. But even with cell fusion, some of the cytoplasm from the egg cell can get mixed up. Some scientists believe that the disruption of the cytoplasm can alter the production of clone. (The disruption of egg cytoplasm is easier to avoid in the cloning of other species because non-human eggs are larger, less delicate, and easier to handle.)

For the purposes of this discussion, however, let's assume that the nucleus from the donor was transplanted into the human egg cell without disrupting the egg cytoplasm or damaging the egg. The egg would then be cultured in the laboratory until the embryo reached the blastocyst stage of development in which it is a hollow ball of cells. It would then be implanted into the uterus of a woman—a substitute mother—so that it could complete its development.

The baby born would not be genetically related to its "mother" since it would not have any of her hereditary characteristics. The baby would have only one parent —the donor that provided the nucleus. The baby would be, in fact, a genetic copy of the donor—its identical twin, a generation or more removed.

BUT . . .

Although the cloning of a human seems possible, there are still technological problems involved. More important, there are serious ethical questions involved.

One danger is the possibility of a mutation in the nucleus of the donor. The chance of a mutation occurring depends to a large extent on the kind of cell that is used as the nucleus-donating cell.

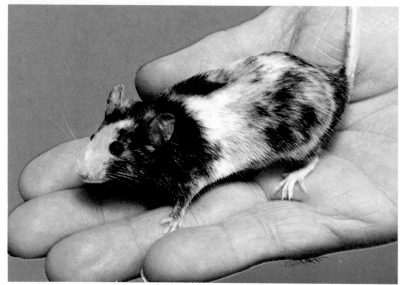

This tricolored mouse was formed by joining three eight-cell mouse embryos from mice of three different colors. By 25 hours the 24 cells had joined together, forming a single embryo, which was transferred to the uterus of a foster mother.

Clement Markert

What kind of cell then should be used to clone a human male? Some scientists believe that the best choice is a spermatogonium, or sperm-producing cell. These cells are found in the testes. They possess all the donor's genes to clone a complete individual.

The appropriate cell to use to clone a female has not yet been identified.

NOT AN EXACT COPY

How similar would a human clone be to its donor? It surely would not resemble the adult donor from the start. The human infant would have to go through the normal stages of growth and development. It might resemble the donor during certain comparable stages of life—for example, the clone at five years old might resemble the donor when he or she was five years old.

In terms of genetics, the clone would be as similar to the donor as an identical twin would be. Mutations could, however, cause differences between the donor and the clone even though they originally had the exact same genetic makeup.

There could also be other differences in the clone. No one, for example, really knows the effects of the egg's cytoplasm on the developing embryo. There are also the largely unknown effects of the substitute mother's hormones and general health and nutrition during pregnancy.

Even more probably, differences between the clone and its human donor would result from the two having different histo-

ries, environments, life-styles, and diseases throughout life. Certainly the notion that a human clone would be an exact duplicate of the donor is very questionable.

THE FUTURE

We are back to our original question—namely, how will cloning contribute to human welfare? The techniques of cloning can improve our methods of producing desirable plants and animals. The day may come when we will be able to select prize animals and plants with superior genetic traits and produce exact copies of them in large quantities in less time than it would take under natural conditions. This could have an important effect on the world's production of food.

Cloning techniques could also be used in basic biological and medical research. Clonal mice and rats, for example, could be used to study genetic diseases, cancer, and other disorders.

But what about human cloning? At the present time, most researchers are not concerned with human cloning. There are serious ethical questions to be considered as well as practical doubts about the consequences of human cloning. For example: who is to decide which hereditary traits are desirable and to be continued via cloning and which traits are not? Is musical talent more valuable than scientific talent? Is blond hair preferable to red hair? These are the unanswered and perhaps unanswerable questions.

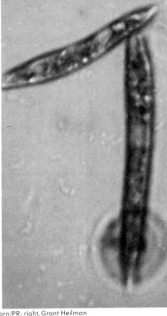

left, Ian Cleghorn/PR; right, Grant Heilman

No one doubts that the lions seen fighting above are animals. But what is on the right? A euglena, a one-celled organism considered by many—but not all—scientists to be a plant. The enormous diversity of life on earth has long presented a challenge to all attempts to classify it.

CLASSIFICATION OF LIFE

An intelligent being from another planet might well be astonished at the wonderful variety of animals and plants on the earth. He would see living things of all sizes and shapes and colors: animals that look like plants and plants that look like animals; plants that tower a hundred or more meters in the air and others so small that thousands would fit on the head of a pin; land animals encased in armor plate and sea animals that resemble floating umbrellas. Our visitor from outer space would wonder how anyone could possibly classify all these creatures. Yet it has been done.

EARLY ATTEMPTS

The great Greek philosopher Aristotle made the first important attempt to classify animals and plants systematically. He divided all living things into three general classes. He put plants in the first class. He held that they had a vegetative soul that governed feeding and reproductive activities. The second class consisted of ani-

mals, who had not only a vegetative soul but also an animal soul, regulating movement. Human beings made up the third class. They had a vegetative soul, an animal soul, and a rational soul, which directed their intellectual activities, or reason.

Aristotle's writings on plants have been lost, but his works on animals have come down to us. He classified over five hundred species, of which most were sea-dwelling forms, abounding in the warm Mediterranean waters. He divided animals into two main groups. The first consisted of animals with red blood. It included the four highest classes: mammals, birds, reptiles, and fish. The second group was made up of animals without blood—that is, with colorless blood—and included squids, shellfish, crayfish, and insects.

Aristotle's disciple, Theophrastus, wrote on plants, using his own observations and maintaining a quite remarkable standard of accuracy. He made no outstanding contribution to classification, however. Neither did the Roman Pliny the Elder (A.D. 23–79), whose *Natural History* contained a vast amount of information, much of it based on hearsay.

For hundreds of years (well into the Renaissance period), little progress was made in the study of animals and plants. Even in the first part of the Renaissance, the naturalists believed that Aristotle, Theophrastus, and Pliny had described most, if not all, of the world's animals and plants. They spent much of their time trying to fit all plants, including those growing in central Europe, into Theophrastus' descriptions of the vegetation of the warm Mediterranean regions.

The geographical discoveries of the sixteenth and seventeenth centuries proved that new species of plants and animals did exist, and scholars included these new types in their books. The chief divisions of plants in those days were trees, shrubs, half-shrubs, and herbs. When listing an individual plant, a botanist would give its name in various languages, the statements of the ancient authors about it, the writer's judgment of it, and, finally, its medicinal powers.

WORK OF LINNAEUS

This haphazard method proved completely unsatisfactory. It was not, however, until the time of the 18th century Swedish scholar Karl von Linné (better known as Linnaeus) that the first effective classification system was set up.

Linnaeus divided all plants into classes, based on the number of stamens, or male elements, in each flower. Each class was further subdivided into orders, according to the number of pistils, or female parts, in the flower. He subdivided the orders into genera (singular, genus) on the basis of certain minor differences. The genera, in turn, were divided into species. Thus *Euphorbia apios,* one of the European species of the plant known as the spurge, was classified by Linnaeus as follows:

Class— Dodecandria (12 stamens).
Order— Trigynia (3 pistils). In this order there were three different genera.

Genus— *Euphorbia.* The generic name (genus name) of the spurges.
Species— *apios.* A spurge found in Crete.

Linnaeus divided all animals into six great classes: (1) quadrupeds (four-legged animals—mammals), (2) birds, (3) reptiles (in which he included frogs and salamanders), (4) fish, (5) insects, and (6) worms. Such very different forms as the clam and the earthworm fell in the same class (that of the worms) in his classification.

BINOMIAL NOMENCLATURE

Linnaeus' great contribution to classification was his use of a genus and species name for each living thing—a method known as binomial nomenclature (two-term naming). He used Latin or Latinized forms because Latin was a common language among scholars of different lands at that time.

The classification method of Linnaeus marked a great advance. However, it became apparent in time that there were serious defects in the system. Biologists tried to find a more natural method of classification than the one employed by Linnaeus.

In 1898, the International Congress of Zoology organized an International Commission on Zoological Nomenclature to establish rules for the naming of animals.

Animals are generally given a genus name and a species name (the binomial nomenclature introduced by Linnaeus). Sometimes the subgenus or subspecies or both are also given; this is called polynomial (many-term) nomenclature. From time to time, the Commission issues new rules and makes decisions on disputed names. The most recent edition of the International Code of Zoological Nomenclature was adopted in 1958 and extensively revised for publication in 1981. Similarly, rules of botanical nomenclature have been established by the International Botanical Congresses. The most recent edition of the botanical code was adopted in 1975.

MODERN SYSTEM

In any modern system of classification, the arrangement of various plants and animals into categories must reflect their evolutionary relationships. Closeness or distance of relationships is established by comparisons of structural or behavioral characters in living forms, coupled with the evidence of the fossil record. Fossils permit the establishment of a geological time frame within which ancient plants and animals lived and evolved into modern representatives, or died out without descendants. During the 1970s and 1980s, the techniques of molecular biology were added to clarify evolutionary relationships in classification. Comparisons of tissue proteins and nucleic acid sequences in the DNA and RNA chains of different species are now used to indicate the time that has elapsed since such species diverged from each other.

In the modern classification system, each living thing belongs to a species, genus, family, order, class, phylum (or division, in plants), and kingdom. These groups are sometimes divided into subgroups, as mentioned above.

To show how the different groups of the modern classification system are related, let us follow a representative form—the dog—from species to kingdom.

Species. As you know, there are many breeds of dogs—dachshunds, poodles, boxers, greyhounds, and so on. Yet, different as these breeds may appear to be, they have much in common in structure and habits, and they interbreed. Hence we lump them together in the same species.

Genus. We note that there is a certain resemblance among dogs, jackals, coyotes, and wolves, all of which belong to different species. We put them in the group called the genus, and we name it the genus *Canis* (dog genus). Members of this group resemble one another more closely than they would resemble the cats, which belong to a different genus. The genus and species are always given in Latin or a Latinized form, after the fashion originated by Linnaeus. The genus name begins with a capital letter, the species name with a small letter; both names are italicized. The scientific name for dog, for example, is *Canis familiaris*. A coyote is *Canis latrans;* a timber wolf,

Carolus Linnaeus

Canis nubilus. If the genus and species of an animal (or plant) are given several times in the course, say, of an article, after the first time the generic name may be abbreviated by using only the first letter. Thus, for *Canis familiaris* we might write *C. familiaris.* The generic name is never omitted.

Family. The animals belonging to the genus *Canis* are grouped with certain other animals in the family Canidae. The name of the family is based on the name of a genus belonging to it; of course, the family name Canidae is derived from *Canis.* The catlike animals, such as lions, tigers, and domestic cats, make up the family of the Felidae (named after the cat genus *Felis*).

Order. Families, in turn, are combined in orders. The dog, cat, and weasel families belong to the order Carnivora ("flesh-eaters," in Latin).

Subclass. The Carnivora order and a number of other orders, including the Rodentia (gnawing animals), Chiroptera (bats), and Primates (apes, monkeys, and man), form part of the subclass called the Eutheria. The unborn babies of the Eutheria are attached to the womb of the mother by a structure called the placenta.

A TERRIER IN THE CLASSIFICATION SYSTEM

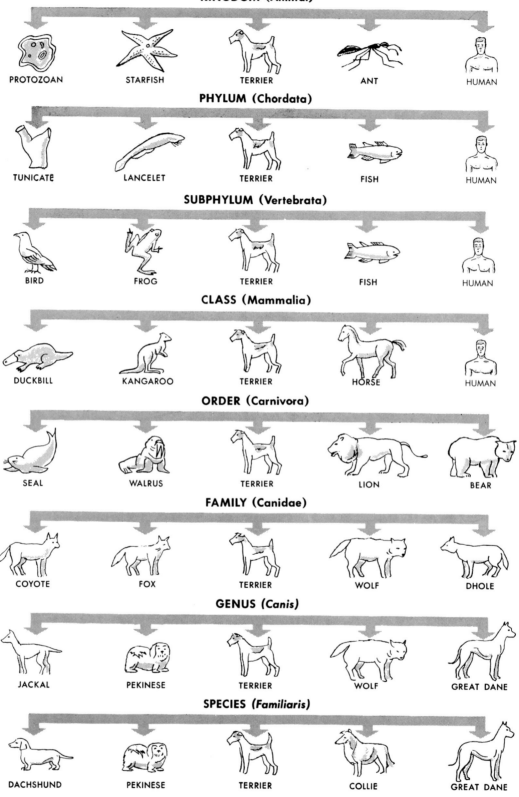

KINGDOM (Animal)

PROTOZOAN — STARFISH — TERRIER — ANT — HUMAN

PHYLUM (Chordata)

TUNICATE — LANCELET — TERRIER — FISH — HUMAN

SUBPHYLUM (Vertebrata)

BIRD — FROG — TERRIER — FISH — HUMAN

CLASS (Mammalia)

DUCKBILL — KANGAROO — TERRIER — HORSE — HUMAN

ORDER (Carnivora)

SEAL — WALRUS — TERRIER — LION — BEAR

FAMILY (Canidae)

COYOTE — FOX — TERRIER — WOLF — DHOLE

GENUS (Canis)

JACKAL — PEKINESE — TERRIER — WOLF — GREAT DANE

SPECIES (Familiaris)

DACHSHUND — PEKINESE — TERRIER — COLLIE — GREAT DANE

Class. The next classification group is called the class. The animals that compose the Carnivora, Rodentia, and several other orders possess glands that produce milk for the young. The orders consisting of animals with milk glands and certain other features make up the class of the Mammalia, also known as the mammals. The dog is a mammal.

Subphylum. The dog and all the other mammals belong to the subphylum Vertebrata. The animals who make up the Vertebrata have a well-developed internal bone skeleton. Fishes, amphibians, reptiles, and birds also belong to this subphylum.

Phylum. Next, we have the phylum (from the Greek *phylon*, meaning "tribe"). A phylum is composed of classes having certain common characteristics. Dogs and snakes do not resemble one another. Yet there is a relationship between them, because both animals have a spinal column, which develops from a rodlike support, the notochord. All the animals that have notochords, either as embryos or adults, are grouped together in the phylum Chordata. Some have no spinal column; they have a notochord or once had a notochord.

Kingdom. Finally, there are the *kingdoms*. In recent classification systems, there are five kingdoms. Kingdom Monera contains the bacteria (including blue-green algae). Kingdom Protista includes single-celled organisms having both plantlike and animal-like characteristics. Kingdom Fungi consists of the fungi. Kingdom Plantae is made up of the plants and Kingdom Animalia, the animals. Some authorities employ somewhat different categories, but all modern classification systems are similar.

Let us see how a typical plant, such as a white oak, would be classified.

Kingdom— Plantae. The white oak is clearly a plant.
Division— Magnoliophyta. The angiosperms, or flowering plants, with fruits and protected seeds. The white oak is a flowering plant.
Class— Magnoliopsida. The dicotyledons, or angiosperms with two cotyledons, or seed leaves. The white oak is a dicotyledon, or dicot.
Subclass— Hamamelidae. Dicots with characteristic structural and chemical features.
Order— Fagales. The beechlike plants.
Family— Fagaceae. The beeches, chestnuts, and oaks.
Genus— *Quercus.* The oaks. Their seed vessels bear single nuts.
Species— *alba* (white). *Quercus alba* has light-colored bark and other features that set it apart from the rest of the oak species.

DECIDING ON NAMES

As in the days of Linnaeus, the scientific names of living things are given in Latin, or in a Latinized form of Greek. In certain cases the scientific name suggests an outstanding feature of the group or species. Arthropoda, for example, means "jointed feet"; this name is applied to a phylum made up of spiders, insects, and other animals with jointed legs.

However, scientific names of living things do not always indicate structures. Sometimes they are the names of persons, particularly scientists who discovered new species. Animals may also be given the names of mythical characters. For example, the arachnids, the group to which the spiders belong, are named after Arachne, a figure in Greek mythology. Sometimes a scientific name indicates a locality; thus *mexicanus* means Mexican, and *virginiensis* means Virginian. Or else the name refers to a color, as in the case of *niger*, which means black.

Well over a million species of living plants and animals (mostly animal species) have already been described by biologists, and additions to the list are being made every year. It would take many hundreds of pages to give a complete list of all these species, and no scientist, however learned, knows more than a small fraction of them well.

selected readings

PHYSICAL SCIENCES

GENERAL WORKS

Cohen, I. Bernard. *Revolution in Science.* Cambridge, Mass.: Harvard University Press, 1985; 704 pp., illus.—A carefully documented and straightforward history of physical science from the 17th century to the present day.

Diagram Group. *Comparisons of Distance, Size, Area, Volume, Mass, Weight, Density, Energy, Temperature, Time, Speed and Number Throughout the Universe.* New York: St. Martin's Press, 1980; 240 pp., illus.—A novel presentation, easily comprehensible.

Filson, Brent. *Famous Experiments and How To Repeat Them.* New York: Julian Messner, 1986; 80 pp., illus.—Easy-to-read book about experimentation and the application of the scientific method to solving problems.

Flatow, Irving. *Rainbows, Curve Balls and Other Wonders of the Natural World Explained.* New York: Morrow, 1988; 240 pp., illus.—An enthusiastic and entertaining look at science organized in terms of everyday experience.

Hazen, Robert M., and James Trefil. *Science Matters: Achieving Scientific Literacy.* Garden City, N.Y.: Doubleday, 1991; 294 pp., illus.—Why we need to be scientifically literate, with a potpourri of down-to-earth explanations of various scientific principles.

PHYSICS

Adair, Robert Kemp. *The Physics of Baseball.* New York: Harper & Row, 1990; 110 pp., illus.—A fascinating look at the physical principles behind the sport.

Asimov, Isaac. *How Did We Find Out About the Speed of Light?* New York: Walker, 1986; 64 pp., illus.—A concise and easy-to-read history of studying the speed of light.

Bernstein, Jeremy. *Quantum Profiles.* Princeton, N.J.: Princeton University Press, 1991; 166 pp.—Profiles of three scientists who struggled with the strangeness of quantum theory; technical but clearly written.

Bohren, Craig F. *Clouds in a Glass of Beer.* New York: John Wiley, 1987; 192 pp., illus.—Simple experiments in atmospheric physics for high school students.

Cole, K. C. *Sympathetic Vibrations: Reflections on Physics as a Way of Life.* New York: Morrow, 1985; 342 pp.—Enthusiastic essays about science and scientists and the relationship of physics to everyday life.

Coveney, Peter, and Roger Highfield. *The Arrow of Time: A Voyage Through Science to Solve Time's Greatest Mystery.* New York: Fawcett Columbine, 1991; repr. 1992; 384 pp.—A clear explanation of time's linear nature and its relationship to living things; for advanced students.

Epstein, Lewis C., and Paul G. Hewitt. *Thinking Physics.* San Francisco: Insight Press, 1980; 262 pp., illus.—A lighthearted collection of perceptual puzzles; good fun for science students who enjoy mental games.

Goodwin, Peter. *Physics Can Be Fun.* Portland, Maine: J. Weston Walch, 1986; 190 pp., illus.—A sourcebook of practical problems in physics; for junior high on up.

Hunter, Mark. *Fantastic Journeys: Five Great Quests of Modern Science.* New York: Walker, 1980; 224 pp., illus.—Makes accessible to the high school student the newest ideas in the "hot" fields of geophysics, particle physics, and astrophysics.

Laithwaite, Eric. *Force: The Power Behind Movement.* New York: Franklin Watts, 1986; 32 pp., illus.—Fascinating explanation of the principles of gravity, inertia, and friction, and the way different forces combine in everyday activities; for grades 5–8.

Lerner, Rita G., and George L. Trigg, eds. *Encyclopedia of Physics.* New York: Addison-Wesley, 1980; 1,157 pp., illus.—Informative work for both the layperson and the physicist.

Rifkin, Jeremy, with Ted Howard. *Entropy: A New World View.* New York: Viking Press, 1980; 324 pp.—An explanation for the layperson of one of the mysterious but basic topics in physics.

Segre, Emilio. *From Falling Bodies to Radio Waves.* New York: W. H. Freeman, 1984; 298 pp.—A very readable history of physics, with an emphasis on scientific personalities.

Swenson, Lloyd S., Jr. *Genesis of Relativity: Einstein in Context.* New York: Burt Franklin, 1980; 266 pp., illus.—Historical background and emergence of the idea of relativity.

Trefil, James S. *From Atoms to Quarks: An Introduction to the Strange World of Particle Physics.* New York: Scribners, 1980; 225 pp., illus.—Clarity is brought to the murky subject of particles smaller than neutrons and electrons.

Weber, Robert L. *Pioneers of Science: Nobel Prize Winners in Physics.* Inst. of Physics (dist. by Hayden), 1980; 280 pp., illus.—Identifies the work for which each award was given and its place in modern physics; for general readers.

White, Laurence B., Jr., and Ray Broekel. *Optical Illusions.* New York: Franklin Watts, 1986; 96 pp., illus.—Enjoyable and informative look at tricks of visual perception.

Wood, Alexander. *The Physics of Music.* New York: Halsted Press, 7th ed., rev. by T. M. Bowsher, 1976; repr. 1981; 273 pp., illus.—A discussion of resonance, pitch, reproduction of sound, acoustics, and other music-related topics; for older readers.

Zee, A. *An Old Man's Toy: Gravity at Work and Play in Einstein's Universe.* New York: Macmillan, 1989; 272 pp., illus.—An authoritative examination of nature's weakest and most taken-for-granted force; for advanced readers.

Zubrowski, Bernie. *Mirrors: Finding Out About the Properties of Light.* New York: Morrow, 1992; 96 pp., illus.—Boston Children's Museum activity book; includes explorations and games.

ATOMIC PHYSICS

Badash, Lawrence. *Radioactivity in America: Growth and Decay of a Science.* Baltimore: Johns Hopkins University Press, 1979; 320 pp., illus.—Describes the development of radioactivity in the 20th century; informative and provocative in its portrayal of the American scientific community.

Broad, William J., et al. *Claiming the Heavens.* New York: Times Books, 1988; 304 pp., illus.—An evenhanded guide to the Star Wars debate; for older readers.

Feldbaum, Carl B., and Ronald J. Bee. *Looking the Tiger in the Eye.* New York: Harper & Row, 1988; 320 pp., illus.—An admirably clear history of the nuclear era.

McGowen, Tom. *Radioactivity: From the Curies to the Atomic Age.* New York: Franklin Watts, 1986; 72 pp., illus.—A concise introduction to turn-of-the-century scientific breakthroughs that led to the atomic age.

Panati, Charles, and Michael Hudson. *The Silent Intruder: Surviving the Radiation Age.* New York: Houghton Mifflin, 1981; 210 pp.—Concise, understandable discussion of the health effects of all kinds of radiation enables the reader to make knowledgeable decisions regarding exposure.

Rhodes, Richard. *The Making of the Atomic Bomb.* New York: Simon & Schuster, 1987; repr. 1988; 928 pp.—Pulitzer Prize-winning account of the origins and development of nuclear weapons; for advanced readers.

von Baeyer, Hans Christian. *Taming the Atom.* New York: Random House, 1992; 223 pp., illus.—A lucid and entertaining account of atoms and the scientists who study them.

CHEMISTRY

Caglioti, Luciano. *The Two Faces of Chemistry.* Cambridge, Mass.: MIT Press, 1983; 218 pp., illus.—Discussions of the positive and negative aspects of a broad range of modern chemical compounds and by-products, including food additives, drugs, pesticides, and fertilizers.

Diagram Group. *Physical Sciences on File.* New York: Facts on File, 1989; 300 pp., illus.—A good basic introduction to chemistry for high school students.

Hill, John W. *Chemistry for Changing Times.* Minneapolis: Burgess, 5th ed., 1988; 456 pp., illus.—Introductory treatment of essentials of chemistry, as a background to exploring the issues of chemistry in the modern world.

Mebane, Robert C., and Thomas R. Rybolt. *Adventures with Atoms and Molecules: Chemistry Experiments for Young People, Book III.* Hillside, N.J.: Enslow, 1991; 96 pp., illus.—A fine collection of hands-on experiments illustrating the fundamentals of atoms and molecules by using common household items.

Rudolph, Joachim. *Chemistry for the Modern Mind.* New York: Macmillan, 1975; 346 pp.—Thorough coverage of basic chemical theories, with clear explanations of such topics as semiconductor diodes and methods of reading chromatograms.

Stark, Norman. *The Formula Book.* Kansas City, Kans.: Sheed & Ward, 1975; 209 pp., illus.—Practical guide to the chemical makeup of numerous common materials used around the home, telling how to make them inexpensively.

Whiteside, Thomas. *The Pendulum and the Toxic Cloud: The Course of Dioxin Contamination.* New Haven: Yale University Press, 1979; 206 pp., illus.—A sobering discussion of the environmental and human-exposure problems resulting from the use of one toxic herbicide.

Woodburn, John H. *Taking Things Apart and Putting Things Together.* Washington, D.C.: American Chemical Society, 1976; 122 pp., illus.—The basic story of what chemistry is about; for the general reader.

LIFE SCIENCES

GENERAL WORKS

Corrick, James A. *Recent Revolutions in Biology.* New York: Franklin Watts, 1987; 128 pp., illus.—Well-organized and absorbing account of recent innovations in the field.

Frisch, Otto von. *Animal Camouflage.* New York: Franklin Watts, 1973; 128 pp., illus.—Very good pictures and clearly written text; for junior high on up.

Goodfield, June. *An Imagined World: A Story of Scientific Discovery.* New York: Harper & Row, 1981; 272 pp., illus.—This detailed record of five years in the life of a biomedical researcher captures the essence of "doing science."

Gould, Stephen J. *Bully for Brontosaurus: Reflections in Natural History.* New York: Norton, 1991; 540 pp., illus.—Provocative and delightful essays on natural history.

———. *The Flamingo's Smile: Reflections in Natural History.* New York: Norton, 1985; 480 pp., illus.—Clearly written essays on a variety of scientific topics.

Jancovy, John, Jr. *On Becoming a Biologist.* New York: Harper & Row, 1986; 160 pp.—A perceptive, candid, and readable look at the field.

Jenkins, Marie M. *Embryos and How They Develop.* New York: Holiday House, 1975; 194 pp., illus.—How life develops, from one-celled organisms to mammals; includes human reproduction; for junior high on up.

Judson, Horace Freeland. *The Eighth Day of Creation: Makers of the Revolution in Biology.* New York: Simon & Schuster, 1979; 686 pp.—Insights into the personalities involved in the rise of molecular biology in the mid-20th century.

Medawar, Peter B., and J. S. Medawar. *Aristotle to Zoo: A Philosophical Dictionary of Biology.* Cambridge, Mass.: Harvard University Press, 1983; 320 pp.—A thoughtful condensation of the universe of modern biology.

Reiger, George. *Wanderer on My Native Shore.* New York: Lyons & Burford, 1983; repr. 1992; 286 pp.—A brilliant guide to the wild and not-so-wild life of the Eastern seaboard.

Silverstein, Alvin, and Virginia Silverstein. *Nature's Living Lights: Fireflies and Other Bioluminescent Creatures.* Boston: Little, Brown, 1988; 45 pp., illus.—An exploration of the chemical reaction that causes some insects, fish, plants, and fungi to light up in the dark, and the uses of bioluminescence; for younger readers.

Weissmann, Gerald. *The Woods Hole Cantata: Essays on Science and Society.* New York: Dodd, Mead, 1985; 224 pp.—A lively collection of essays on medicine and science and their socioeconomic context; for advanced readers.

GENETICS AND EVOLUTION

Arnold, Caroline. *Genetics: From Mendel to Gene Splicing.* New York: Franklin Watts, 1986; 72 pp., illus.—A clear and concise introduction to the history of genetics.

Berry, R. J., and A. Hallam, eds. *The Encyclopedia of Animal Evolution.* New York: Facts on File, 1987; 152 pp., illus.—Takes the reader on a trip through time.

Cole, Joanna. *The Human Body: How We Evolved.* New York: Morrow, 1987; 64 pp., illus.—Excellent treatment of a difficult and controversial topic; for grades 4–8.

Eldredge, Niles. *The Miner's Canary: Unraveling the Mysteries of Extinction.* New York: Prentice Hall, 1991; 246 pp.—For advanced students; the curator of invertebrates at the American Museum of Natural History poses the idea that evolution occurs in fits and starts rather than continuously.

————. *Fossils: The Evolution and Extinction of Species.* New York: Harry N. Abrams, 1992; 240 pp., illus.—An exciting look at earth's earliest history via text and dramatic color photos.

Gallant, Roy A. *Before the Sun Dies: The Story of Evolution.* New York: Macmillan, 1989; 228 pp., illus.—Traces biological evolution from the origin of life through human evolution.

Harris, Marvin. *Our Kind: Who We Are, Where We Came From, Where We Are Going.* New York: Harper & Row, 1990; 448 pp.—Traces the human passage from australopithecine to suburbanite in a series of catchy essays.

Johanson, Donald, and Maitland Edey. *Lucy: The Beginnings of Mankind.* New York: Simon & Schuster, 1981; 409 pp., illus.—The fascinating discovery of the 3-million-year-old *Australopithecus afarensis* skeleton, "Lucy."

Merriman, Nick. *Early Humans.* New York: Knopf, 1989; 64 pp., illus.—A magnificent resource for grade 5 on up.

Pringle, Laurence. *Living Treasure: Saving Earth's Threatened Biodiversity.* New York: Morrow, 1991; 64 pp., illus.—An examination of variety and diversity within the biosphere, including natural selection, genetic variation, and the evolution of species; for younger readers.

Sattler, Helen Roney. *Hominids: A Look Back at Our Ancestors.* New York: Lothrop, Lee & Shepard Books, 1988; 128 pp., illus.—An exciting chronological examination of the ancestry of the human family; for younger readers.

Silverstein, Alvin, and Virginia B. Silverstein. *The Genetics Explosion.* New York: Macmillan, 1985; 160 pp., illus.—A clearly written look at advances in the field.

Stein, Sara. *The Evolution Book.* New York: Workman, 1986; 320 pp., illus.—An abundance of everyday examples contributes to reader interest and understanding.

Whitfield, Phillip. *Macmillan Children's Guide to Dinosaurs and Other Prehistoric Animals.* New York: Macmillan, 1992; 96 pp., illus.—Discusses dinosaurs in the major geological eras.

Wilson, Edward O. *The Diversity of Life.* Boston: Harvard University Press, 1992; 424 pp.—A leading biologist warns of the human threat to life-forms on earth; for advanced readers.

Wingerson, Lois. *Mapping Our Genes: The Genome Project and the Future of Medicine.* New York: Dutton, 1990; 338 pp., illus.—A look at the study of genetically inherited diseases; for the nonspecialist.

ECOLOGY

Cherry, Lynne. *A River Ran Wild.* San Diego: Gulliver Books, 1992; 32 pp., illus.—The historical development and subsequent pollution of the Nashua River; for young readers.

Cone, Molly. *Come Back, Salmon.* San Francisco: Sierra Club, 1992; 48 pp., illus.—Describes how a group of elementary school students worked to save a creek and prepare it for the salmon being "grown" at their school.

DiSilvestro, Roger L. *Audubon Perspectives: Fight for Survival.* New York: John Wiley, 1990; illus.—Richly illustrated plea for the preservation of wildlife and the environment.

Downer, Ann. *Spring Pool: A Guide to the Ecology of Temporary Ponds.* New York: Franklin Watts, 1992; 56 pp., illus.—An introduction to these seasonal, endangered wetlands.

Ehrlich, Paul R. *The Machinery of Nature.* New York: Simon & Schuster, 1986; 320 pp., illus.—An enjoyable introduction to ecology.

Friedman, Judi. *Operation Siberian Crane.* New York: Dillon Press, 1992; 96 pp., illus.—The story behind the international effort to save the endangered Siberian crane.

Harris, Tom. *Death in the Marsh.* Washington, D.C.: Island Press, 1992; 245 pp.—Describes the chemical threat to wildlife and the human food supply posed by selenium concentrations in the soil and water of the western United States; for advanced readers.

McLaughlin, Molly. *Earthworms, Dirt, and Rotten Leaves: An Exploration in Ecology.* New York: Atheneum, 1986; 96 pp., illus.—Explores the structure and functions of the earthworm and includes suggestions for designing original experiments in basic ecology; for grades 4–8.

McPhee, John. *The Control of Nature.* New York: Farrar, Straus & Giroux, 1989; repr. 1991; 288 pp., illus.—A report on human efforts to defy nature; for advanced students.

Maynard, Thane. *Saving Endangered Mammals: A Field Guide to Some of the Earth's Rarest Mammals.* New York: Franklin Watts, 1992; 56 pp., illus.—A two-page spread on each of more than 20 endangered mammals, including what is being done and what needs to be done to save them; includes a list of conservation organizations.

Moore, Peter D., ed. *The Encyclopedia of Animal Ecology.* New York: Facts on File, 1987; 152 pp., illus.—Shows principles of ecology at work in specific environments.

Pringle, Laurence. *Saving Our Wildlife.* Hillside, N.J.: Enslow, 1990; 64 pp., illus.—A look at efforts to protect wildlife habitats, emphasizing the relationship between animals and their environment; for grades 5–8.

Segalof, Nat, and Paul Erickson. *A Reef Comes to Life: Creating an Undersea Exhibit.* New York: Franklin Watts, 1992; 40 pp., illus.—Desribes the design and construction of an exact replica of a Caribbean reef at the New England Aquarium.

Wrightson, Patricia. *Moon-Dark.* New York: Margaret K. McElderry Books, 1989; 169 pp., illus.—Well-written account of a disturbance of the ecological balance; for ages 9–12.